FUNCTIONAL ANALYTIC
METHODS FOR
PARTIAL DIFFERENTIAL EQUATIONS

PURE AND APPLIED MATHEMATICS

A Program of Monographs, Textbooks, and Lecture Notes

MONOGRAPHS AND TEXTBOOKS IN
PURE AND APPLIED MATHEMATICS

110. *G. S. Ladde et al.,* Oscillation Theory of Differential Equations with Deviating Arguments (1987)
111. *L. Dudkin et al.,* Iterative Aggregation Theory (1987)
112. *T. Okubo,* Differential Geometry (1987)
113. *D. L. Stancl and M. L. Stancl,* Real Analysis with Point-Set Topology (1987)
114. *T. C. Gard,* Introduction to Stochastic Differential Equations (1988)
115. *S. S. Abhyankar,* Enumerative Combinatorics of Young Tableaux (1988)
116. *H. Strade and R. Farnsteiner,* Modular Lie Algebras and Their Representations (1988)
117. *J. A. Huckaba,* Commutative Rings with Zero Divisors (1988)
118. *W. D. Wallis,* Combinatorial Designs (1988)
119. *W. Wiesław,* Topological Fields (1988)
120. *G. Karpilovsky,* Field Theory (1988)
121. *S. Caenepeel and F. Van Oystaeyen,* Brauer Groups and the Cohomology of Graded Rings (1989)
122. *W. Kozlowski,* Modular Function Spaces (1988)
123. *E. Lowen-Colebunders,* Function Classes of Cauchy Continuous Maps (1989)
124. *M. Pavel,* Fundamentals of Pattern Recognition (1989)
125. *V. Lakshmikantham et al.,* Stability Analysis of Nonlinear Systems (1989)
126. *R. Sivaramakrishnan,* The Classical Theory of Arithmetic Functions (1989)
127. *N. A. Watson,* Parabolic Equations on an Infinite Strip (1989)
128. *K. J. Hastings,* Introduction to the Mathematics of Operations Research (1989)
129. *B. Fine,* Algebraic Theory of the Bianchi Groups (1989)
130. *D. N. Dikranjan et al.,* Topological Groups (1989)
131. *J. C. Morgan II,* Point Set Theory (1990)
132. *P. Biler and A. Witkowski,* Problems in Mathematical Analysis (1990)
133. *H. J. Sussmann,* Nonlinear Controllability and Optimal Control (1990)
134. *J.-P. Florens et al.,* Elements of Bayesian Statistics (1990)
135. *N. Shell,* Topological Fields and Near Valuations (1990)
136. *B. F. Doolin and C. F. Martin,* Introduction to Differential Geometry for Engineers (1990)
137. *S. S. Holland, Jr.,* Applied Analysis by the Hilbert Space Method (1990)
138. *J. Okniński,* Semigroup Algebras (1990)
139. *K. Zhu,* Operator Theory in Function Spaces (1990)
140. *G. B. Price,* An Introduction to Multicomplex Spaces and Functions (1991)
141. *R. B. Darst,* Introduction to Linear Programming (1991)
142. *P. L. Sachdev,* Nonlinear Ordinary Differential Equations and Their Applications (1991)
143. *T. Husain,* Orthogonal Schauder Bases (1991)
144. *J. Foran,* Fundamentals of Real Analysis (1991)
145. *W. C. Brown,* Matrices and Vector Spaces (1991)
146. *M. M. Rao and Z. D. Ren,* Theory of Orlicz Spaces (1991)
147. *J. S. Golan and T. Head,* Modules and the Structures of Rings (1991)
148. *C. Small,* Arithmetic of Finite Fields (1991)
149. *K. Yang,* Complex Algebraic Geometry (1991)
150. *D. G. Hoffman et al.,* Coding Theory (1991)
151. *M. O. González,* Classical Complex Analysis (1992)
152. *M. O. González,* Complex Analysis (1992)
153. *L. W. Baggett,* Functional Analysis (1992)
154. *M. Sniedovich,* Dynamic Programming (1992)
155. *R. P. Agarwal,* Difference Equations and Inequalities (1992)
156. *C. Brezinski,* Biorthogonality and Its Applications to Numerical Analysis (1992)
157. *C. Swartz,* An Introduction to Functional Analysis (1992)
158. *S. B. Nadler, Jr.,* Continuum Theory (1992)
159. *M. A. Al-Gwaiz,* Theory of Distributions (1992)
160. *E. Perry,* Geometry: Axiomatic Developments with Problem Solving (1992)
161. *E. Castillo and M. R. Ruiz-Cobo,* Functional Equations and Modelling in Science and Engineering (1992)
162. *A. J. Jerri,* Integral and Discrete Transforms with Applications and Error Analysis (1992)
163. *A. Charlier et al.,* Tensors and the Clifford Algebra (1992)
164. *P. Biler and T. Nadzieja,* Problems and Examples in Differential Equations (1992)
165. *E. Hansen,* Global Optimization Using Interval Analysis (1992)

Additional Volumes in Preparation

FUNCTIONAL ANALYTIC METHODS FOR PARTIAL DIFFERENTIAL EQUATIONS

Hiroki Tanabe

Otemon Gakuin University
Osaka, Japan

CRC Press
Taylor & Francis Group
Boca Raton London New York

CRC Press is an imprint of the
Taylor & Francis Group, an **informa** business

CRC Press
Taylor & Francis Group
6000 Broken Sound Parkway NW, Suite 300
Boca Raton, FL 33487-2742

First issued in paperback 2019

© 1997 by Taylor & Francis Group, LLC
CRC Press is an imprint of Taylor & Francis Group, an Informa business

ISBN-13: 978-0-8247-9774-4 (hbk)
ISBN-13: 978-0-367-40122-1 (pbk)

Library of Congress Cataloging–in–Publication Data

Tanabe, Hiroki.
    Functional analytic methods for partial differential equations /
Hiroki Tanabe.
        p. cm. — (Monographs and textbooks in pure and applied
mathematics : 204)
    Includes bibliographical references (p. –) and index.
    ISBN 0–8247–9774–4 (alk. paper)
    1. Functional analysis. 2. Differential equations, Partial.
I. Title. II. Series.
QA321.T36 1996
515'.353—dc20

                                                                96–31585
                                                                   CIP

Visit the Taylor & Francis Web site at
http://www.taylorandfrancis.com

and the CRC Press Web site at
http://www.crcpress.com

Preface

This book is devoted to the functional analytic method for partial differential equations, and includes both classical theory and up-to-date results. The first main object is to present a self-contained proof of Agmon-Douglis-Nirenberg's L^p estimates for elliptic boundary value problems. For that purpose it is required to discuss the theory of singular integrals of A. P. Calderón and A. Zygmund. Although this theory is classical, it still maintains its importance, and its extension is now being studied vigorously. The Hilbert transform, which is the case of the singular integral of a single independent variable, is also described completely. The Hilbert transform plays an important role in the study of maximal regularity results. It is one of the objects of this book to describe such concepts, which generally are not found in other books on our subject. The second half of the book is devoted to evolution equations in Banach spaces. Recent results on equations of parabolic type and of hyperbolic type, and retarded functional differential equations together with control theory related to them, are described. This book is concerned only with linear equations; however, some notes on nonlinear equations as well as degenerate equations, which are also out of our scope, are given in the bibliographical remarks.

Chapter 1 includes preliminary materials which will be used in subsequent chapters.

Chapter 2 is devoted to the most elementary part of the theory of singular integrals of Calderón and Zygmund and the basic theory of the Hilbert transform.

Chapter 3 is devoted to the theory of function spaces, such as interpolation inequalities, the Gagliardo-Nirenberg inequality and Sobolev's imbedding theorems.

Chapter 4 include Agmon-Douglis-Nirenberg's L^p estimates for solutions of elliptic boundary value problems with a slightly simplified proof of the original one.

In Chapter 5 adjoint boundary value problems are described following M. Schechter, and their solvability is discussed following S. Agmon. The estimates of the kernels of the resolvents of elliptic operators and the semi-groups generated by them are established by a functional analytic method, and the result is applied to the problem in L^1 space.

In Chapter 6 parabolic evolution equations are described following P. Acquistapace and B. Terreni. Some remarks on A. Yagi's results are also given.

Chapter 7 is devoted to the study of hyperbolic evolution equations originated by T. Kato and developed by himself, K. Kobayashi, A. Yagi, N. Okazawa, and A. Unai.

Chapter 8 is devoted to retarded functional differential equations in Hilbert spaces. First the solvability is described following G. Di Blasio, K. Kunisch and E. Sinestrari. Next, the results on control problems of C. Bernier, M. C. Delfour, and A. Manitius for equations in finite dimensional spaces and of S. Nakagiri for equations in reflexive Banach spaces are extended to equations in Hilbert spaces with delay terms that are as unbounded as the main term.

Prerequisites are Lebesgue integration and the elementary concepts of Banach and Hilbert spaces.

I express my profound gratitude to Professor Angelo Favini of the University of Bologna and Professor Atsushi Yagi of Osaka University who urged me to write this book and constantly encouraged me during its preparation. I also thank Professor Yasushi Shizuta of Nara Women's University, who motivated me to greatly improve the contents of Chapter 3, Professor Noboru Okazawa of Science University of Tokyo, who kindly offered me his unpublished paper, and Professor Shin-ichi Nakagiri of Kobe University, who read the manuscript of Chapter 8 carefully and gave me kind advice.

Thanks are also due to Professor Yoshihiko Yamamoto of Osaka University for his cordial instruction on how to operate the computer to prepare the manuscript by LaTeX.

The contents of this book considerably overlap with my book *Functional Analysis, II*, published in Japanese by Jikkyo Shuppan Publishing Company, Tokyo, in 1981. I express my deep gratitude to Jikkyo Shuppan for permitting me to reproduce the relevant parts of the book.

I also would like to express gratitude to the staff of Marcel Dekker, Inc., in particular to Ms Maria Allegra for their assistance.

Hiroki Tanabe

Contents

FUNCTIONAL ANALYTIC
METHODS FOR
PARTIAL DIFFERENTIAL EQUATIONS

Chapter 1

Preliminaries

1.1 Preliminaries on Measure and Integration

In this chapter we collect some preliminary results from the theory of measure and integration.

Let E be a Lebesgue measurable subset of the ν-dimensional Euclidean space R^ν. The Lebesgue measure of E is denoted by $|E|$. The supremum of $|E|/|Q|$ over all cubes Q containing E is called the *parameter of regularity* of E and is denoted by $r(E)$.

A sequence $\{E_n\}$ of measurable sets is called a *regular sequence* if there exists a positive number α such that $r(E_n) \geq \alpha$ for all $n = 1, 2, \ldots$.

For a nonempty open subset Ω of R^ν the notation $L^1_{loc}(\Omega)$ stands for the set of all functions which are Lebesgue integrable on each compact subset of Ω.

Theorem 1.1 *For a function $f \in L^1_{loc}(R^\nu)$ there exists a null set N such that for any $x \notin N$ and for any regular sequence $\{e_n\}$ tending to x*

$$\lim_{n \to \infty} \frac{1}{|e_n|} \int_{e_n} f(y) dy = f(x).$$

Especially for any $x \notin N$

$$\lim_{\rho \to 0} \frac{1}{V_\nu \rho^\nu} \int_{|x-y|<\rho} f(y) dy = f(x),$$

where V_ν is the volume of the unit ball of R^ν.

More strongly the following theorem holds.

Theorem 1.2 *For a function $f \in L^1_{loc}(R^\nu)$*

$$\lim_{\rho \to 0} \frac{1}{V_\nu \rho^\nu} \int_{|x-y| < \rho} |f(y) - f(x)| dy = 0 \qquad (1.1)$$

at almost every point x in R^ν.

Definition 1.1 A point at which (1.1) holds is called a *Lebesgue point* of f.

Lemma 1.1 (Young's inequality) *If $a \geq 0, b \geq 0, 1 < p < \infty, q = p/(p-1)$*

$$ab \leq a^p/p + b^q/q.$$

The equality holds if and only if $a^p = b^q$.

Proof. The assertion is obtained by comparing the sum of the area of the figure surrounded by the three curves $x = a, y = 0, y = x^{p-1}$ and that of the figure surrounded by $x = 0, y = b, x = y^{q-1}$ with ab.

Lemma 1.2 (Hausdorff-Young's inequality) *For $f \in L^1(R^\nu), g \in L^p(R^\nu)$, $1 \leq p \leq \infty$, we have*

$$\|f * g\|_p \leq \|f\|_1 \|g\|_p,$$

where $f * g$ is the convolution of f and g:

$$(f * g)(x) = \int_{R^\nu} f(x - y)g(y)dy,$$

and $\| \quad \|_p$ is the norm of $L^p(R^\nu)$.

Proof. Integrating both sides of

$$|(f * g)(x)|^p = \left| \int_{R^n} f(x - y)g(y)dy \right|^p$$

$$\leq \left(\int_{R^n} |f(x - y)|^{1/p'} |f(x - y)|^{1/p} |g(y)| dy \right)^p$$

$$\leq \left(\int_{R^n} |f(x - y)| dy \right)^{p/p'} \int_{R^n} |f(x - y)| |g(y)|^p dy$$

$$= \|f\|_1^{p-1} \int_{R^n} |f(x - y)| |g(y)|^p dy$$

over R^n we readily obtain the desired result.

1.2 Preliminaries from Functional Analysis

Let T be a linear operator with domain in a Banach space X and range in another Banach space Y. The domain and the range of T are denoted by $D(T)$ and $R(T)$ respectively. For a linear closed operator T from a Banach space X into itself the spectrum and the resolvent set of T are denoted by $\sigma(T)$ and $\rho(T)$ respectively, and the point spectrum of T by $\sigma_p(T)$. The adjoint space of X is denoted by X^*. The adjoint operator of a densely defined linear operator T is denoted by T^*.

Theorem 1.3 *Let T be a densely defined closed linear operator from a Banach space X to another Banach space Y.*
(i) A necessary and sufficient condition in order that $R(T) = Y$ is that T^* has a continuous inverse.
(ii) A necessary and sufficient condition in order that $R(T^*) = X^*$ is that T has a continuous inverse.

For the proof see K. Yosida [166: Corollary 1 of VII. 5].

The following lemma is known as Dini's theorem.

Lemma 1.3 *If $f(t)$ is a strongly continuous function defined in the interval $[a, b]$ taking values in a Banach space X and has a strongly continuous right derivative $D^+ f(t)$ in (a, b), then f is strongly continously differentiable in (a, b) and $df(t)/dt = D^+ f(t)$.*

We denote the set of all bounded linear operators from a Banach space X into another Banach space Y by $\mathcal{L}(X, Y)$. We make this space a Banach spaces with the operator norm unless otherwise stated. We write $\mathcal{L}(X)$ instead of $\mathcal{L}(X, X)$. We denote by

$$x = \text{w-}\lim_{n \to \infty} x_n$$

that a sequence $\{x_n\}$ of a Banach space converges weakly to x.

1.3 Semigroups

Let X be a Banach space with norm $\| \cdot \|$.

Definition 1.2 A family of bounded linear operators $\{T(t); t > 0\}$ is called a *semigroup* in X if the following conditions are satisfied:
(i) $T(t + s) = T(t)T(s)$ for $t > 0, s > 0$.
(ii) $T(t)$ is strongly continuous in $t > 0$.
If moreover the following condition is satisfied, it is called a C_0 *-semigroup*.
(iii) $\lim_{t \to 0} T(t) = I$ in the strong operator topology.

Definition 1.3 Let $\{T(t); t \geq 0\}$ be a C_0-semigroup in X. The operator A defined by

$$Ax = \lim_{t \to 0} \frac{T(t)x - x}{t} \tag{1.2}$$

for $x \in D(A) = \{x \in X;$ the limit (1.2) exists$\}$ is called the *infinitesimal generator* of the C_0-semigroup $\{T(t); t \geq 0\}$.

Theorem 1.4 *Let $\{T(t); t \geq 0\}$ ba a C_0-semigroup. Then there exists a constant $M \geq 1$ and a real number β such that*

$$\|T(t)\| \leq Me^{\beta t}, \quad t \geq 0. \tag{1.3}$$

The infinitesimal generator A of $\{T(t); t \geq 0\}$ is a densely defined closed linear operator satisfying the following conditions:
(i) $\rho(A) \supset \{\lambda; \mathrm{Re}\lambda > \beta\}$;
(ii) $\|(\lambda - A)^{-n}\| \leq M(\mathrm{Re}\lambda - \beta)^{-n}, \quad \mathrm{Re}\lambda > \beta, \quad n = 1, 2, \dots.$
Conversely if A is a densely defined closed linear operator satisfying (i) *and* (ii), *then there exists a C_0-semigroup $\{T(t); t \geq 0\}$ with infinitesimal generator A. The semigroup $\{T(t); t \geq 0\}$ is uniquely determined by A and satisfies* (1.3).

   Moreover we have

$$(\lambda - A)^{-1} = \int_0^\infty e^{-\lambda t} T(t) dt \tag{1.4}$$

for $\mathrm{Re}\lambda > \beta$.

The set of all infinitesimal generators which generate C_0-semigroups in X satisfying (1.3) is denoted by $G(X, M, \beta)$. The semigroup with infinitesimal generator A is denoted by e^{tA} or $\exp(tA)$.

Theorem 1.5 *Let $\{T(t); t \geq 0\}$ be a C_0-semigroup in a Banach space X with infinitesimal generator A. Then*

$$T(t) = \lim_{n \to \infty} \left(1 - \frac{t}{n}A\right)^{-n} \quad \textit{Hille's representation} \tag{1.5}$$

$$T(t) = \lim_{n \to \infty} \exp(tA_n) \quad \textit{Yosida's representation,} \tag{1.6}$$

uniformly in any bounded subinterval of $[0, \infty)$ in the strong operator topology of X, where $A_n = A(1 - n^{-1}A)^{-1}$.

Theorem 1.6 *Let A be a closed linear operator with not necessarily dense domain. Suppose that there exist an angle $\theta_0 \in (\pi/2, \pi]$ and positive constants C_1, C_2 such that*

(i) $\rho(A) \supset \Sigma = \{\lambda; -\theta_0 < \arg \lambda < \theta_0, |\lambda| > C_1\}$,
(ii) $\|(\lambda - A)^{-1}\| \leq C_2/|\lambda|, \quad \lambda \in \Sigma$.
Let Γ be a smooth path running from $\infty e^{-i\theta}$ to $\infty e^{i\theta}$ in Σ, where $\pi/2 < \theta < \theta_0$. Set

$$T(t) = \frac{1}{2\pi i} \int_\Gamma e^{\lambda t} (\lambda - A)^{-1} d\lambda, \quad t > 0. \tag{1.7}$$

Then $T(t)$ is a semigroup. $T(t)$ is extended as an analytic function of t in the sector $\{t; |\arg t| < \theta_0 - \pi/2\}$ and the inequality

$$\left\|\frac{d}{dt}T(t)\right\| = \|AT(t)\| \leq \frac{Ce^{\omega t}}{t}, \quad t > 0 \tag{1.8}$$

holds for some constants C and ω. If A is densely defined, then $\{T(t)\}$ is a C_0-semigroup, and A is the infinitesimal generator of $\{T(t)\}$.

Definition 1.4 The semigroup $\{T(t)\}$ of the above theorem is called an *analytic semigroup.*

Remark 1.1 Let A be a closed linear operator in a Banach space X. Suppose that $(\beta, \infty) \subset \rho(A)$ and $\|(\lambda - \beta)(\lambda - A)^{-1}\|$ is bounded for $\lambda \in (\beta, \infty)$. Furthermore if X is reflexive, then $D(A)$ is dense as is shown below. Let x be an arbitrary element of X. Then $\lambda(\lambda - A)^{-1}x$ is bounded as $\lambda \to \infty$. Therefore, there exists a subsequence $\lambda_n \to \infty$ such that $\lambda_n(\lambda_n - A)^{-1}x$ converges weakly to some element $y \in X$. Since

$$A(\lambda_n - A)^{-1}x = \lambda_n(\lambda_n - A)^{-1}x - x \to y - x$$

weakly, $(\lambda_n - A)^{-1}x \to 0$ and A is closed, it follows that $y - x = 0$. Since $\lambda_n(\lambda_n - A)^{-1}x \in D(A)$, we conclude $x = y \in \overline{D(A)}$. Therefore an analytic semigroup in a reflexive Banach space is a C_0-semigroup.

The semigroup (1.7) is also denoted by e^{tA} or $\exp(tA)$. More generally we have for $n = 0, 1, 2, \ldots$

$$\frac{d^n}{dt^n}T(t) = \frac{1}{2\pi i} \int_\Gamma \lambda^n e^{\lambda t}(\lambda - A)^{-1} d\lambda, \tag{1.9}$$

$$\left\|\frac{d^n}{dt^n}T(t)\right\| = \|A^n e^{tA}\| \leq \frac{C_n e^{\omega t}}{t^n}. \tag{1.10}$$

The inequality (1.10) is proved as follows. If we choose a sufficiently large positive number k if necessary, the conditions (i),(ii) of Theorem 1.6 hold for $A_1 = A - k$ with $C_1 = 0$, and $T(t) = e^{tk}T_1(t)$, where $T_1(t)$ is the semigroup generated by A_1. It suffices to show that (1.10) holds for $T_1(t)$ with $\omega = 0$:

$$\left\|\frac{d^n}{dt^n}T_1(t)\right\| = \|A_1^n e^{tA_1}\| \leq \frac{C_n}{t^n}. \tag{1.11}$$

The case $n = 0$ is established by choosing Γ as

$$\{\lambda; \lambda = re^{-i\theta}, r \geq 1/t\} \cup \{\lambda; |\lambda| = 1/t, |\arg \lambda| \leq \theta\} \cup \{\lambda; \lambda = re^{i\theta}, r \geq 1/t\},$$

where $\pi/2 < \theta < \theta_0$. The case $n \geq 1$ is proved by choosing

$$\Gamma = \{\lambda; \lambda = re^{-i\theta}, -\infty < r \leq 0\} \cup \{\lambda; \lambda = re^{i\theta}, 0 \leq r < \infty\}.$$

Theorem 1.7 *Let e^{tA} be an analytic semigroup. Then $e^{tA}x \to x$ as $t \to 0$ if and only if $x \in \overline{D(A)}$.*

Proof. Since e^{tA} is uniformly bounded for t bounded, it suffices to show that for $x \in D(A)$ $e^{tA}x \to x$ as $t \to 0$. If $x \in D(A)$

$$\begin{aligned}
e^{tA}x &= \frac{1}{2\pi i} \int_\Gamma e^{\lambda t}(\lambda - A)^{-1}A^{-1}Ax d\lambda \\
&= \frac{1}{2\pi i} \int_\Gamma \frac{e^{\lambda t}}{\lambda} \{(\lambda - A)^{-1} + A^{-1}\} Ax d\lambda \\
&= \frac{1}{2\pi i} \int_\Gamma \frac{e^{\lambda t}}{\lambda}(\lambda - A)^{-1}Ax d\lambda + \frac{1}{2\pi i} \int_\Gamma \frac{e^{\lambda t}}{\lambda}x d\lambda \\
&= \frac{1}{2\pi i} \int_\Gamma \frac{e^{\lambda t}}{\lambda}(\lambda - A)^{-1}Ax d\lambda + x.
\end{aligned}$$

Hence choosing Γ as in the proof of (1.11) for $n = 0$ we get

$$\|e^{tA}x - x\| \leq Ct\|Ax\|$$

if t is sufficiently small. Hence the assertion follows.

Lemma 1.4 *If A generates a C_0-semigroup or an analytic semigroup in X, and $0 \in \rho(A)$, then for any $x \in X$ and $t > 0$ we have*

$$\int_0^t e^{sA}x ds = (e^{tA} - I)A^{-1}x.$$

Proof. For $0 < \epsilon < t$ we have

$$\int_\epsilon^t e^{sA}x ds = \int_\epsilon^t \frac{d}{ds} e^{sA}A^{-1}x ds = (e^{tA} - e^{\epsilon A})A^{-1}x.$$

If e^{tA} is a C_0-semigroup, then the right hand side goes to $(e^{tA} - I)A^{-1}x$ as $t \to 0$. If e^{tA} is an analytic semigroup, the same holds in view of Theorem 1.7.

Lemma 1.5 *Suppose that A generates an analytic semigroup.*

(i) *For $x \in D(A)$* $\lim\limits_{t \to 0} \dfrac{e^{tA}x - x}{t} = Ax$ *if and only if $Ax \in \overline{D(A)}$.*

(ii) *For $x \in \overline{D(A)}$ we have $\lim\limits_{t \to 0} tAe^{tA}x = 0$.*

Proof. (i) By the proof of Lemma 1.4 we have

$$D(A) \ni \frac{e^{tA}x - x}{t} = \frac{1}{t} \int_0^t e^{sA} Ax \, ds.$$

Hence the assertion follows from Theorem 1.7.

(ii) If $x \in D(A)$, then $tAe^{tA}x = te^{tA}Ax \to 0$ as $t \to 0$. The same fact remains valid for $x \in \overline{D(A)}$, since in view of (1.8) $\|tAe^{tA}\|$ is bounded for t bounded.

For the details for semigroups we refer to E. Hille and R. S. Phillips [74].

1.4 Interpolation Spaces

Let X and Y be a couple of Banach spaces with norm $\| \cdot \|_X$ and $\| \cdot \|_Y$ respectively. We assume that both X and Y are continuously imbedded in some locally convex linear topological space $\mathcal{E}$. Such a pair is called an *interpolation pair*.

For $1 \leq p \leq \infty$ $L_*^p(X)$ is the set of all functions with values in X which are defined and strongly measurable in $(0, \infty)$ such that

$$\|u\|_{L_*^p(X)} = \begin{cases} \left(\displaystyle\int_0^\infty \|u(t)\|_X^p \, \frac{dt}{t} \right)^{1/p} & 1 \leq p < \infty \\ \operatorname{ess\,sup}_{t \in (0,\infty)} \|u(t)\|_X & p = \infty \end{cases} \tag{1.12}$$

are finite. As is easily seen $L_*^p(X)$ is a Banach space with norm (1.12).

For $0 < \theta < 1$, $1 \leq p \leq \infty$ let $V(\theta, p; X, Y)$ be the set of all functions u defined in $(0, \infty)$ and taking values in $\mathcal{E}$ such that

$$t^\theta u \in L_*^p(X) \quad \text{and} \quad t^\theta u' \in L_*^p(Y),$$

where u' is the derivative in the sense of distributions. $V(\theta, p; X, Y)$ is a Banach space with norm

$$\|u\|_{V(\theta,p;X,Y)} = \left(\|t^\theta u\|_{L_*^p(X)}^p + \|t^\theta u'\|_{L_*^p(Y)}^p \right)^{1/p}.$$

It is easily seen that if $u \in V(\theta, p; X, Y)$ then the trace $u(0)$ of u at $t = 0$ is well defined as an element of $X + Y$. We denote the set of all these traces by $(X, Y)_{\theta,p}$:

$$(X, Y)_{\theta,p} = \{u(0); u \in V(\theta, p; X, Y)\}.$$

$(X, Y)_{\theta,p}$ is a Banach space with norm

$$\|a\|_{(X,Y)_{\theta,p}} = \inf\{\|u\|_{V(\theta,p;X,Y)}; u(0) = a, u \in V(\theta, p; X, Y)\}.$$

The space $(X, Y)_{\theta,p}, 0 < \theta < 1, \leq p \leq \infty$ is called a *real interpolation space* between X and Y. For the details of interpolation spaces we refer to P. L. Butzer and H. Berens [25], J. L. Lions and E. Magenes [99], J. L. Lions and J. Peetre [100] and H. Triebel [154].

Let A be a densely defined closed linear operator in a Banach space X. A is called a *positive operator* if $(-\infty, 0] \subset \rho(A)$ and there exists a positive constant C such that for each $t > 0$

$$\|(t + A)^{-1}\| \leq C/(t + 1).$$

We make $D(A)$ a Banach space with the graph norm of A:

$$\|u\|_{D(A)} = \|Au\|_X + \|u\|_X.$$

The proof of the following theorems are found in [25: Chapter 3] and [154: 1.14].

Theorem 1.8 *Let A be a positive operator in a Banach space X and x an element of X. Then $x \in (D(A), X)_{\theta,p}$ if and only if the functions $t \mapsto t^{1-\theta} A(A+t)^{-1}x$ belongs to $L_*^p(X)$. The norm of $(D(A), X)_{\theta,p}$ is equivalent to*

$$\begin{cases} \left(\int_0^\infty \|t^{1-\theta} A(A + t)^{-1}x\|_X^p \frac{dt}{t}\right)^{1/p} + \|x\|_X & 1 \leq p < \infty \\ \text{ess sup}_{t\in(0,\infty)} \|t^{1-\theta} A(A + t)^{-1}x\|_X + \|x\|_X & p = \infty \end{cases}.$$

If A generates a C_0-semigroup in X, then for some real number k, $k - A$ is a positive operator. For the sake of simplicity of notations we assume $k = 0$.

Theorem 1.9 *Let A generate a C_0-semigroup in X. Then an element x of X belongs to $(D(A), X)_{\theta,p}$ if and only if the function $t \mapsto t^{\theta-1}(e^{tA}x - x)$ belongs to $L_*^p(X)$. The norm of $(D(A), X)_{\theta,p}$ is equivalent to*

$$\begin{cases} \left(\int_0^\infty \|t^{\theta-1}(e^{tA}x - x)\|_X^p \frac{dt}{t}\right)^{1/p} + \|x\|_X & 1 \leq p < \infty \\ \text{ess sup}_{t\in(0,\infty)} \|t^{\theta-1}(e^{tA}x - x)\|_X + \|x\|_X & p = \infty \end{cases}.$$

Suppose that A generates an analytic C_0-semigroup in X. Then there exist positive constants C_1, C_2 and an angle $\theta_0 \in (\pi/2, \pi]$ such that

$$\rho(A) \supset \Sigma = \{\lambda; -\theta_0 < \arg \lambda < \theta_0, |\lambda| > C_1\},$$
$$\|(\lambda - A)^{-1}\| \leq C_2/|\lambda|, \quad \lambda \in \Sigma.$$

For the sake of simplicity we assume that $C_1 = 0$ and $-A$ is a positive operator.

Theorem 1.10 *If A generates an analytic C_0-semigroup in X, then an element x of X belongs to $(D(A), X)_{\theta,p}$ if and only if the function $t \mapsto t^\theta A e^{tA} x$ belongs to $L^p_*(X)$. The norm of $(D(A), X)_{\theta,p}$ is equivalent to*

$$\begin{cases} \left(\int_0^\infty \|t^\theta A e^{tA} x\|_X^p \frac{dt}{t} \right)^{1/p} + \|x\|_X & 1 \le p < \infty \\ \text{ess sup}_{t \in (0,\infty)} \|t^\theta A e^{tA} x\|_X + \|x\|_X & p = \infty \end{cases}.$$

Furthermore if $x \in (D(A), X)_{\theta,p}$, then for any $\epsilon \in (0, \theta_0)$

$$\begin{cases} \displaystyle\sup_{|\phi| \le \theta_0 - \epsilon} \int_0^\infty \|t^{1-\theta} A(A - te^{i\phi})^{-1} x\|_X^p \frac{dt}{t} < \infty & 1 \le p < \infty \\ \displaystyle\sup_{|\phi| \le \theta_0 - \epsilon} \text{ess sup}_{t \in (0,\infty)} \|t^{1-\theta} A(A - te^{i\phi})^{-1} x\|_X < \infty & p = \infty \end{cases}.$$

We denote the norm of $\mathcal{L}(X, Y)$ by $\| \cdot \|_{\mathcal{L}(X,Y)}$.

Theorem 1.11 *Let (X_1, Y_1) and (X_2, Y_2) be two interpolation pairs. If $T \in \mathcal{L}(X_1, X_2) \cap \mathcal{L}(Y_1, Y_2)$, then*

$$T \in \mathcal{L}((X_1, Y_1)_{\theta,p}, (X_2, Y_2)_{\theta,p}), \quad 0 < \theta < 1, \quad 1 \le p \le \infty,$$

and

$$\|T\|_{\mathcal{L}((X_1,Y_1)_{\theta,p},(X_2,Y_2)_{\theta,p})} \le \|T\|_{\mathcal{L}(X_1,X_2)}^{1-\theta} \|T\|_{\mathcal{L}(Y_1,Y_2)}^\theta. \tag{1.13}$$

Remark 1.2 In the interpolation theory we often have an inequality of the type (1.13) with the right hand side multiplied by some positive constant c_θ depending on θ. In the present case we have (1.13) with $c_\theta = 1$ which can be shown if we use

$$\|a\|_{(X,Y)_{\theta,p}} = (1 - \theta)^{\theta - 1} \theta^{-\theta} \inf_{u(0)=0} \|t^\theta u\|_{L^p_*(X)}^{1-\theta} \|t^\theta u'\|_{L^p_*(Y)}^\theta.$$

This equality follows from the fact that if $u(0) = a$, $u \in V(\theta, p; X, Y)$, then $u_\lambda(0) = a$, $u_\lambda \in V(\theta, p; X, Y)$, $\lambda > 0$, where $u_\lambda(t) = u(\lambda t)$, and Lemma 1.1.

If $1 < p < \infty$, then $0 < 1/p < 1$ and

$$V(1/p, p; X, Y) = \{u; u \in L^p(0, \infty; X), u' \in L^p(0, \infty; Y)\}, \tag{1.14}$$

where $L^p(0, \infty; X)$ is the set of all strongly measurable functions in $(0, \infty)$ with values in X such that

$$\|u\|_{L^p(0,\infty;X)} = \left(\int_0^\infty \|u(t)\|_X^p dt \right)^{1/p} < \infty. \tag{1.15}$$

As is easily seen $L^p(0,\infty;X)$ is a Banach space with norm (1.15). The norm of $V(1/p,p;X,Y)$ is

$$\|u\|_{V(1/p,p;X,Y)} = \left(\|u\|^p_{L^p(0,\infty;X)} + \|u'\|^p_{L^p(0,\infty;Y)} \right)^{1/p}.$$

If $u \in V(1/p,p;X,Y)$, then for any $t \geq 0$ $u(t) \in (X,Y)_{1/p,p}$, and

$$\|u(t)\|_{(X,Y)_{1/p,p}} \leq \|u(t+\cdot)\|_{V(1/p,p;X,Y)} \leq \|u\|_{V(1/p,p;X,Y)}. \qquad (1.16)$$

Furthermore

$$\|u(t) - u(s)\|^p_{(X,Y)_{1/p,p}} \leq \|u(t+\cdot) - u(s+\cdot)\|^p_{V(1/p,p;X,Y)}$$

$$= \int_0^\infty \left(\|u(t+\tau) - u(s+\tau)\|^p_X + \|u'(t+\tau) - u'(s+\tau)\|^p_Y \right) d\tau \to 0$$

as $t \to s$. Hence we have

$$V(1/p,p;X,Y) \subset C([0,\infty)); (X,Y)_{1/p,p}), \qquad (1.17)$$

where $C([0,\infty)); (X,Y)_{1/p,p})$ is the set of all continuous functions in $[0,\infty)$ with values in $(X,Y)_{1/p,p}$.

We denote by $W^{1,p}(0,\infty;Y)$ the set of all functions u such that $u, u' \in L^p(0,\infty;Y)$. $W^{1,p}(0,\infty;Y)$ is a Banach space with norm

$$\|u\|_{W^{1,p}(0,\infty;Y)} = \left(\|u\|^p_{L^p(0,\infty;Y)} + \|u'\|^p_{L^p(0,\infty;Y)} \right)^{1/p}.$$

If X and Y are Hilbert spaces, then $L^2(0,\infty;X)$ and $W^{1,2}(0,\infty;Y)$ are Hilbert spaces.

If X is a dense subspace of Y and the imbedding $X \subset Y$ is continuous, then in view of (1.14),(1.17) we have

$$L^p(0,\infty;X) \cap W^{1,p}(0,\infty;Y) \subset C([0,\infty)); (X,Y)_{1/p,p}) \qquad (1.18)$$

and

$$\sup_{t\in[0,\infty)} \|u(t)\|_{(X,Y)_{1/p,p}} \leq \left(\|u\|^p_{L^p(0,\infty;X)} + \|u\|^p_{W^{1,p}(0,\infty;Y)} \right)^{1/p}$$

$$\equiv \|u\|_{L^p(0,\infty;X)\cap W^{1,p}(0,\infty;Y)} \qquad (1.19)$$

for each $p \in (1,\infty)$.

Let A be a positive operator. Then $\rho(A)$ contains a closed sector $\Sigma = \{\lambda; \theta_0 \leq \arg\lambda \leq 2\pi-\theta_0\}\cup\{0\}$, where $0 < \theta_0 < \pi$, and $\|(\lambda - A)^{-1}\| \leq \dfrac{C}{1+|\lambda|}$

in Σ, where C is a positive constant. Let $\rho > 0$. The fractional power A^ρ of A is defined as the inverse of the operator

$$A^{-\rho} = \frac{1}{2\pi i} \int_\Gamma \lambda^{-\rho} (\lambda - A)^{-1} d\lambda,$$

where Γ is the path running from $\infty e^{-i\theta_0}$ to $\infty e^{i\theta_0}$ along the boundary of Σ.

Lemma 1.6 *If $0 < \rho < 1 - \theta$, then*

$$(D(A), X)_{\theta, p} \subset D(A^\rho)$$

for any $1 \leq p \leq \infty$.

Proof. For $\lambda \in \Sigma$

$$\|A(\lambda - A)^{-1}\|_{\mathcal{L}(X)} \leq C, \quad \|A(\lambda - A)^{-1}\|_{\mathcal{L}(D(A), X)} \leq \frac{C}{1 + |\lambda|}.$$

Hence in view of Theorem 1.11

$$\|A(\lambda - A)^{-1}\|_{\mathcal{L}((D(A), X)_{\theta, p}, X)} \leq \frac{C}{(1 + |\lambda|)^{1 - \theta}}. \tag{1.20}$$

The conclusion follows from (1.20) and

$$A^\rho x = \frac{1}{2\pi i} \int_\Gamma \lambda^{\rho - 1} A(\lambda - A)^{-1} x \, d\lambda.$$

The following lemma is known as the moment inequality.

Lemma 1.7 *Let A be a positive operator. Then for $0 \leq \alpha < \beta < \gamma$ there exists a constant $C_{\alpha, \beta, \gamma}$ such that for $x \in D(A^\gamma)$*

$$\|A^\beta x\| \leq C_{\alpha, \beta, \gamma} \|A^\gamma x\|^{(\beta - \alpha)/(\gamma - \alpha)} \|A^\alpha x\|^{(\gamma - \beta)/(\gamma - \alpha)}.$$

In particular for $0 \leq \alpha \leq 1, \lambda \in \Sigma$ we have

$$\|A^\alpha (\lambda - A)^{-1}\| \leq \frac{C_\alpha}{(1 + |\lambda|)^{1 - \alpha}}. \tag{1.21}$$

For the details on fractional powers see H. Triebel [154].

Chapter 2

Singular Integrals

2.1 Singular Integrals of A. P. Cålderón and A. Zygmund

In this chapter the most fundamental part of the theory of the singular integrals of A. P. Calderón and A. Zygmund

$$\text{v.p.} \int_{R^n} K(x-y)f(y)dy = \lim_{\epsilon \to 0} \int_{|x-y|>\epsilon} K(x-y)f(y)dy$$

is described. The results are due to A. P. Calderón and A. Zygmund [27]. Throughout this chapter it is assumed that the integral kernel $K(x)$ satisfies the following assumption.

Assumption 1 $K(x)$ is a complex valued measurable function defined in R^n, homogeneous of degree $-n$:

$$K(\lambda x) = \lambda^{-n} K(x) \quad \text{for } \lambda > 0, \quad x \in R^n, \tag{2.1}$$

absolutely integrable on the unit sphere $\Sigma = \{x \in R^n; |x| = 1\}$ and satisfies

$$\int_{\Sigma} K(x)d\sigma = 0, \tag{2.2}$$

where $d\sigma$ is the surface element of Σ.

The Fourier transform is denoted by $\mathcal{F}$:

$$(\mathcal{F}f)(\xi) = (2\pi)^{-n/2} \int_{R^n} e^{-ix\xi} f(x)dx. \tag{2.3}$$

The Lebesgue measure of a subset A of R^n is denoted by $|A|$. The norm of $L^p(R^n)$ is denoted by $\| \ \|_p$ in this chapter.

Lemma 2.1 *For $f \in L^p(R^n), 1 < p < \infty, \epsilon > 0$ the integral*

$$\tilde{f}_\epsilon(x) = \int_{|x-y|>\epsilon} K(x-y)f(y)dy \tag{2.4}$$

exists almost everywhere in R^n.

Proof. By a suitable change of independent variables

$$\int_{|x-y|>\epsilon} K(x-y)f(y)dy = \int_{|y|>\epsilon} K(y)f(x-y)dy$$

$$= \int_\Sigma \int_\epsilon^\infty K(t\sigma)f(x-t\sigma)t^{n-1}dtd\sigma = \int_\Sigma K(\sigma)\int_\epsilon^\infty f(x-t\sigma)\frac{dt}{t}d\sigma.$$

Hence, it suffices to show that for any bounded measurable subset S of R^n

$$\int_S \int_\Sigma |K(\sigma)|\int_\epsilon^\infty |f(x-t\sigma)|\frac{dt}{t}d\sigma\, dx \tag{2.5}$$

is finite. With the aid of Hölder's inequality

$$\int_S \int_\epsilon^\infty |f(x-t\sigma)|\frac{dt}{t}dx \le C_{\epsilon,p}\int_S \left(\int_{-\infty}^\infty |f(x-t\sigma)|^p dt\right)^{1/p}dx$$

$$\le C_{\epsilon,p}|S|^{1-1/p}\left(\int_S \int_{-\infty}^\infty |f(x-t\sigma)|^p dtdx\right)^{1/p}, \tag{2.6}$$

where $C_{\epsilon,p} = (p-1)^{(p-1)/p}\epsilon^{-1/p}$. For a fixed $\sigma \in \Sigma$ write $x = z - s\sigma, s \in R^1, (z,\sigma) = z_1\sigma_1 + \cdots + z_n\sigma_n = 0$. Choose $S' \subset R^{n-1}$ and a, b so that

$$S \subset \{x = z - s\sigma; z \in S', a < s < b\}.$$

Then the last side of (2.6) does not exceed

$$C_{\epsilon,p}|S|^{1-1/p}\left(\int_a^b \int_{S'} \int_{-\infty}^\infty |f(z-(t+s)\sigma)|^p dtdzds\right)^{1/p}$$

$$\le C_{\epsilon,p}|S|^{1-1/p}\left[(b-a)\int_{R^n} |f(x)|^p dx\right]^{1/p} = C_{\epsilon,p}|S|^{1-1/p}(b-a)^{1/p}\|f\|_p.$$

The numbers a, b can be chosen independently of σ, and hence (2.5) does not exceed

$$C_{\epsilon,p}|S|^{1-1/p}(b-a)^{1/p}\|f\|_p \int_\Sigma |K(\sigma)|d\sigma.$$

Remark 2.1 In the above proof (2.2) was not used.

2.2 J. Marcinkiewicz's Interpolation Theorem

Let $(\Omega, \mathcal{B}, \phi), (\hat{\Omega}, \hat{\mathcal{B}}, \hat{\phi})$ be two measure spaces. The norms of $L^p(\Omega), L^p(\hat{\Omega})$ are both denoted by $\| \ \|_p$.

Definition 2.1 Let $1 \le p \le \infty, 1 \le q \le \infty$.
(i) A not necessarily linear mapping T from $L^p(\Omega)$ to $L^q(\hat{\Omega})$ is said to be *of type* (p, q) if there exists a constant K such that $\|Tf\|_q \le K\|f\|_p$ for any $f \in L^p(\Omega)$.
(ii) Let T be a not necessarily linear mapping from $L^p(\Omega)$ to a set of measurable functions defined in $\hat{\Omega}$. In case $q < \infty$ T is said to be *of weak type* (p, q) if there exists a constant K such that

$$\hat{\phi}(\{y \in \hat{\Omega}; |(Tf)(y)| > s\}) \le (K\|f\|_p/s)^q \qquad (2.7)$$

for any $f \in L^p(\Omega)$ and $s > 0$. In case $q = \infty$ a mapping of type (p, ∞) is said to be *of weak type* (p, ∞).

As is easily seen a mapping of type (p, q) is of weak type (p, q).

The totality of functions which are expressed as a sum of a function in $L^p(\Omega)$ and a function in $L^q(\Omega)$ is denoted by $L^p(\Omega) + L^q(\Omega)$. Suppose $1 \le p < q < r < \infty$ and $f \in L^q(\Omega)$. Let f_1, f_2 be functions such that $f_1(x) = 0, f_2(x) = f(x)$ if $|f(x)| \le 1$ and $f_1(x) = f(x), f_2(x) = 0$ if $|f(x)| > 1$. Then $f = f_1 + f_2, |f_1(x)|^p \le |f(x)|^q, |f_2(x)|^r \le |f(x)|^q$. Hence $f_1 \in L^p(\Omega), f_2 \in L^r(\Omega)$. Therefore $L^q(\Omega) \subset L^p(\Omega) + L^r(\Omega)$.

Lemma 2.2 *Let f be a nonnegative function defined in Ω, and $1 \le p < \infty$. Set $E(\tau) = \{x \in \Omega; f(x) > \tau\}$ for $\tau \ge 0$. A necessary and sufficient condition in order that $f \in L^p(\Omega)$ is*

$$\int_0^\infty \tau^{p-1}\phi(E(\tau))d\tau < \infty. \qquad (2.8)$$

In this case

$$\int_\Omega f(x)^p \phi(dx) = p \int_0^\infty \tau^{p-1}\phi(E(\tau))d\tau. \qquad (2.9)$$

Proof. Suppose (2.8) is true. Then it can be shown that $\tau^p\phi(E(\tau)) \to 0$ as $\tau \to 0, \infty$ as follows. Suppose $\tau^p\phi(E(\tau))$ does not tend to 0 as $\tau \to 0$. Then there exists a sequence $\{\tau_j\}$ such that $\tau_j^p\phi(E(\tau_j)) \ge \delta > 0, \tau_{j+1} < \tau_j/2$. Then we have

$$p \int_{\tau_{j+1}}^{\tau_j} \tau^{p-1}\phi(E(\tau))d\tau \ge \phi(E(\tau_j))(\tau_j^p - \tau_{j+1}^p)$$

$$\ge \frac{\delta}{\tau_j^p}(\tau_j^p - \tau_{j+1}^p) = \delta\left(1 - \left(\frac{\tau_{j+1}}{\tau_j}\right)^p\right) \ge \delta\left(1 - \frac{1}{2^p}\right) > 0,$$

which contradicts (2.8). If $\tau^p \phi(E(\tau))$ does not tend to 0 as $\tau \to \infty$, then we can find a sequence $\{\tau_j\}$ such that $\tau_j^p \phi(E(\tau_j)) \geq \delta > 0, 2\tau_j < \tau_{j+1}$. This leads us to a contradiction analogously. Therefore we obtain with the aid of an integration by parts

$$p \int_0^\infty \tau^{p-1} \phi(E(\tau)) d\tau = - \int_0^\infty \tau^p d\phi(E(\tau)) = \int_\Omega f(x)^p \phi(dx).$$

Conversely suppose $f \in L^p(\Omega)$. Since

$$\tau^p \phi(E(\tau)) \leq \int_{E(\tau)} f(x)^p \phi(dx),$$

we see that $\tau^p \phi(E(\tau)) \to 0$ as $\tau \to \infty$. Consequently

$$\int_\Omega f(x)^p \phi(dx) = - \lim_{\epsilon \to 0} \int_\epsilon^\infty \tau^p d\phi(E(\tau))$$
$$= \lim_{\epsilon \to 0} \left[\epsilon^p \phi(E(\epsilon)) + p \int_\epsilon^\infty \tau^{p-1} \phi(E(\tau)) d\tau \right] \geq p \int_0^\infty \tau^{p-1} \phi(E(\tau)) d\tau,$$

from which (2.8) follows.

Theorem 2.1 Let $1 \leq p < q < r$ and T be a linear mapping from $L^p(\Omega) + L^r(\Omega)$ to a set of measurable functions defined in $\hat{\Omega}$. If T is of weak type (p,p) and also of weak type (r,r), then T is of type (q,q).

Proof. For a measurble functions defined in Ω and $\tau > 0$ put

$$(f)_\tau = \phi(\{x \in \Omega; |f(x)| > \tau\}).$$

We use the same notation for a measurable function defined in $\hat{\Omega}$ with $\hat{\phi}$ in place of ϕ. By the assumption there exists a positive constant K such that for any $\tau > 0, f \in L^p(\Omega)$ or $f \in L^r(\Omega)$

$$(Tf)_\tau \leq (K\|f\|_p/\tau)^p \quad \text{or} \quad (Tf)_\tau \leq (K\|f\|_r/\tau)^r. \qquad (2.10)$$

Let f be a function belonging to $L^q(\Omega)$. Put

$$f_1(x) = \begin{cases} f(x) & |f(x)| \leq \tau \\ \tau \operatorname{sign} f(x) & |f(x)| > \tau \end{cases}, \quad f_2(x) = f(x) - f_1(x),$$

Then $Tf = Tf_1 + Tf_2$, and hence

$$\{y; |(Tf)(y)| > \tau\} \subset \{y; |(Tf_1)(y)| > \tau/2\} \cup \{y; |(Tf_2)(y)| > \tau/2\}.$$

Therefore by (2.10)

$$(Tf)_\tau \le (Tf_1)_{\tau/2} + (Tf_2)_{\tau/2} \le (2K\|f_1\|_r/\tau)^r + (2K\|f_2\|_p/\tau)^p. \quad (2.11)$$

Noting $(f_1)_\sigma = (f)_\sigma$ for $\sigma < \tau$, $(f_1)_\sigma = 0$ for $\sigma \ge \tau$, $(f_2)_\sigma = (f)_{\sigma+\tau}$ for any $\sigma \ge 0$, we get in view of Lemma 2.2

$$\|f_1\|_r^r = r \int_0^\infty \sigma^{r-1}(f_1)_\sigma d\sigma = r \int_0^\tau \sigma^{r-1}(f)_\sigma d\sigma, \quad (2.12)$$

$$\|f_2\|_p^p = p \int_0^\infty \sigma^{p-1}(f_2)_\sigma d\sigma$$

$$= p \int_0^\infty \sigma^{p-1}(f)_{\sigma+\tau} d\sigma = p \int_\tau^\infty (\sigma - \tau)^{p-1}(f)_\sigma d\sigma. \quad (2.13)$$

With the aid of (2.11),(2.12),(2.13)

$$(Tf)_\tau \le \left(\frac{2K}{\tau}\right)^r r \int_0^\tau \sigma^{r-1}(f)_\sigma d\sigma + \left(\frac{2K}{\tau}\right)^p p \int_\tau^\infty (\sigma - \tau)^{p-1}(f)_\sigma d\sigma.$$

Again using Lemma 2.2

$$\|Tf\|_q^q = q \int_0^\infty \tau^{q-1}(Tf)_\tau d\tau$$

$$\le q \int_0^\infty \tau^{q-1} \left(\frac{2K}{\tau}\right)^r r \int_0^\tau \sigma^{r-1}(f)_\sigma d\sigma d\tau$$

$$+ q \int_0^\infty \tau^{q-1} \left(\frac{2K}{\tau}\right)^p p \int_\tau^\infty (\sigma - \tau)^{p-1}(f)_\sigma d\sigma d\tau$$

$$= (2K)^r qr \int_0^\infty \tau^{q-r-1} \int_0^\tau \sigma^{r-1}(f)_\sigma d\sigma d\tau$$

$$+ (2K)^p pq \int_0^\infty \tau^{q-p-1} \int_\tau^\infty (\sigma - \tau)^{p-1}(f)_\sigma d\sigma d\tau$$

$$= (2K)^r qr \int_0^\infty \int_\sigma^\infty \tau^{q-r-1} d\tau \sigma^{r-1}(f)_\sigma d\sigma$$

$$+ (2K)^p pq \int_0^\infty \int_0^\sigma \tau^{q-p-1}(\sigma - \tau)^{p-1} d\tau (f)_\sigma d\sigma$$

$$= \frac{(2K)^r qr}{r-q} \int_0^\infty \sigma^{q-1}(f)_\sigma d\sigma + (2K)^p pq B(q-p,p) \int_0^\infty \sigma^{q-1}(f)_\sigma d\sigma$$

$$= \left\{ \frac{(2K)^r r}{r-q} + (2K)^p p B(q-p,p) \right\} \|f\|_q^q.$$

2.3   Case of Bounded Kernels

In this section we assume that the kernel $K(x)$ satisfies the following assumption.

Assumption 2 $K(x)$ satisfies Assumtion 1 and is bounded on the unit sphere Σ.

For a function $f \in L^2(R^n)$ we consider the following integral

$$\lim_{\epsilon \to 0} \tilde{f}_\epsilon(x) = \lim_{\epsilon \to 0} \int_{|x-y|>\epsilon} K(x-y)f(y)dy. \tag{2.14}$$

For $0 < \epsilon < \mu$ put

$$K_{\epsilon,\mu}(x) = \begin{cases} K(x) & \epsilon < |x| < \mu \\ 0 & \text{otherwise} \end{cases}, \quad K_\epsilon(x) = \begin{cases} K(x) & |x| > \epsilon \\ 0 & |x| \le \epsilon \end{cases}. \tag{2.15}$$

Then

$$\tilde{f}_\epsilon(x) = \int_{R^n} K_\epsilon(x-y)f(y)dy = (K_\epsilon * f)(x).$$

We calculate the Fourier transform of $K_{\epsilon,\mu}$. Note that $K_{\epsilon,\mu} \in L^1(R^n)$. Write $|x| = r, |\xi| = \rho, x\xi = r\rho \cos\phi, x = r\sigma$ for $x, \xi \in R^n$. In view of (2.1),(2.2)

$$(2\pi)^{n/2}(\mathcal{F}K_{\epsilon,\mu})(\xi) = \int_{R^n} K_{\epsilon,\mu}(x)e^{-ix\xi}dx$$

$$= \int_{\epsilon<|x|<\mu} K(x)e^{-ix\xi}dx = \int_\epsilon^\mu \int_\Sigma K(\sigma)e^{-ir\rho\cos\phi}d\sigma\frac{dr}{r}$$

$$= \int_{\epsilon\rho}^{\mu\rho} \int_\Sigma K(\sigma)e^{-is\cos\phi}d\sigma\frac{ds}{s} = \int_{\epsilon\rho}^{\mu\rho} \int_\Sigma K(\sigma)(e^{-is\cos\phi} - e^{-s})d\sigma\frac{ds}{s}$$

$$= \int_\Sigma K(\sigma)\int_{\epsilon\rho}^{\mu\rho}(e^{-is\cos\phi} - e^{-s})\frac{ds}{s}d\sigma. \tag{2.16}$$

Write

$$\int_{\epsilon\rho}^{\mu\rho}(e^{-is\cos\phi} - e^{-s})\frac{ds}{s} = I_1 + I_2 i, \tag{2.17}$$

$$I_1 = \int_{\epsilon\rho}^{\mu\rho}(\cos(s\cos\phi) - e^{-s})\frac{ds}{s}, \quad I_2 = -\int_{\epsilon\rho}^{\mu\rho}\sin(s\cos\phi)\frac{ds}{s}.$$

For $0 < a < b$

$$\int_a^b \frac{\cos s}{s}ds = \int_{a+\pi/2}^{b+\pi/2} \frac{\cos(s-\pi/2)}{s-\pi/2}ds = \int_{a+\pi/2}^{b+\pi/2} \frac{\sin s}{s-\pi/2}ds$$

$$= \int_{a+\pi/2}^{b+\pi/2} \frac{\sin s}{s}ds + \int_{a+\pi/2}^{b+\pi/2}\left(\frac{1}{s-\pi/2} - \frac{1}{s}\right)\sin s\,ds. \tag{2.18}$$

Since $\lim\limits_{b\to\infty}\int_0^b \dfrac{\sin x}{x}dx$ exists, we see from (2.18) that $\lim\limits_{b\to\infty}\int_a^b \dfrac{\cos s}{s}ds$ exists. Hence if $\cos\phi\neq 0$,

$$\lim_{\epsilon\to 0,\mu\to\infty}\int_{\epsilon\rho}^{\mu\rho}\left(e^{-is\cos\phi}-e^{-s}\right)\frac{ds}{s}$$

exists. Next we obtain the estimates of (2.17) independent of ϵ,μ. Let A be a constant such that

$$\left|\int_0^a \frac{\sin x}{x}dx\right|\le A \tag{2.19}$$

for any $a>0$. As is easily seen $|I_2|\le 2A$. In view of (2.18),(2.19)

$$\left|\int_a^b \frac{\cos s}{s}ds\right|\le 2A+\int_{a+\pi/2}^{b+\pi/2}\left(\frac{1}{s-\pi/2}-\frac{1}{s}\right)ds$$

$$=2A+\log\frac{b(a+\pi/2)}{a(b+\pi/2)}\le 2A+\log\left(1+\frac{\pi}{2a}\right). \tag{2.20}$$

First assume $\cos\phi>0$. Then

$$\int_0^1 \left|\cos(s\cos\phi)-e^{-s}\right|\frac{ds}{s}$$

$$\le \int_0^1 (1-\cos(s\cos\phi))\frac{ds}{s}+\int_0^1 (1-e^{-s})\frac{ds}{s}$$

$$\le \int_0^1 (1-\cos s)\frac{ds}{s}+\int_0^1 (1-e^{-s})\frac{ds}{s}\equiv B. \tag{2.21}$$

Hence
(i) if $\mu\rho<1$, $|I_1|\le B$,
(ii) if $\epsilon\rho>1$, in view of (2.20)

$$|I_1|=\left|\int_{\epsilon\rho\cos\phi}^{\mu\rho\cos\phi}\frac{\cos s}{s}ds-\int_{\epsilon\rho}^{\mu\rho}e^{-s}\frac{ds}{s}\right|$$

$$\le 2A+\log\left(1+\frac{\pi}{2\epsilon\rho\cos\phi}\right)+\int_1^\infty e^{-s}\frac{ds}{s}$$

$$\le 2A+\log\left(1+\frac{\pi}{2\cos\phi}\right)+\int_1^\infty e^{-s}\frac{ds}{s},$$

(iii) if $\epsilon\rho<1<\mu\rho$, in view of (2.20),(2.21)

$$|I_1|=\left|\left(\int_{\epsilon\rho}^1+\int_1^{\mu\rho}\right)(\cos(s\cos\phi)-e^{-s})\frac{ds}{s}\right|$$

$$\leq B + \left| \int_{\cos\phi}^{\mu\rho\cos\phi} \frac{\cos s}{s} ds \right| + \int_1^\infty e^{-s} \frac{ds}{s}$$

$$\leq B + 2A + \log\left(1 + \frac{\pi}{2\cos\phi}\right) + \int_1^\infty e^{-s} \frac{ds}{s}.$$

The case $\cos\phi < 0$ can be estimated similarly. Summing up we see that there exists a constant C such that

$$\left| \int_{\epsilon\rho}^{\mu\rho} \left(e^{-is\cos\phi} - e^{-s}\right) \frac{ds}{s} \right| \leq C + \log\frac{1}{|\cos\phi|}. \qquad (2.22)$$

Consequently

$$(2\pi)^{2n} |(\mathcal{F}K_{\epsilon\mu})(\xi)| \leq \int_\Sigma \left| K(\sigma) \int_{\epsilon\rho}^{\mu\rho} \left(e^{-is\cos\phi} - e^{-s}\right) \frac{ds}{s} \right| d\sigma$$

$$\leq \sup_{\sigma\in\Sigma} |K(\sigma)| \int_\Sigma \left(C + \log\frac{1}{|\cos\phi|}\right) d\sigma.$$

Thus there exists a constant M such that for any ϵ, μ, ξ

$$|(\mathcal{F}K_{\epsilon\mu})(\xi)| \leq M. \qquad (2.23)$$

As $\mu \to \infty$, $K_{\epsilon\mu}$ converges strongly to K_ϵ in $L^2(R^n)$, and hence $\mathcal{F}K_{\epsilon\mu}$ tends to $\mathcal{F}K_\epsilon$ strongly in $L^2(R^n)$. On the other hand in view of (2.16),(2.22) for any $\xi \neq 0$

$$\lim_{\mu\to\infty} (2\pi)^{n/2} (\mathcal{F}K_{\epsilon\mu})(\xi) = \int_\Sigma K(\sigma) \int_{\epsilon\rho}^\infty \left(e^{-is\cos\phi} - e^{-s}\right) \frac{ds}{s} d\sigma,$$

$$\left| \int_{\epsilon\rho}^\infty \left(e^{-is\cos\phi} - e^{-s}\right) \frac{ds}{s} \right| \leq C + \log\frac{1}{|\cos\phi|}. \qquad (2.24)$$

Therefore

$$(\mathcal{F}K_\epsilon)(\xi) = (2\pi)^{-n/2} \int_\Sigma K(\sigma) \int_{\epsilon\rho}^\infty \left(e^{-is\cos\phi} - e^{-s}\right) \frac{ds}{s} d\sigma. \qquad (2.25)$$

In view of (2.24), (2.25) for any $\xi \neq 0$

$$\lim_{\epsilon\to 0} (\mathcal{F}K_\epsilon)(\xi) = (2\pi)^{-n/2} \int_\Sigma K(\sigma) \int_0^\infty \left(e^{-is\cos\phi} - e^{-s}\right) \frac{ds}{s} d\sigma. \qquad (2.26)$$

The right hand side of (2.26) is defined to be the Fourier transform of $K(x)$, and is denoted by $(\mathcal{F}K)(\xi)$. Obviously $\mathcal{F}K$ is a homogeneous function of order 0, and by virtue of (2.23) we have

$$|(\mathcal{F}K_\epsilon)(\xi)| \leq M, \quad |(\mathcal{F}K)(\xi)| \leq M. \qquad (2.27)$$

If $f \in L^2(R^n)$, then $K_{\epsilon\mu} * f \in L^2(R^n)$ since $K_{\epsilon\mu} \in L^1(R^n)$. As $\mu \to \infty$

$$\mathcal{F}(K_{\epsilon\mu} * f) = (2\pi)^{n/2}\mathcal{F}K_{\epsilon\mu} \cdot \mathcal{F}f \to (2\pi)^{n/2}\mathcal{F}K_\epsilon \cdot \mathcal{F}f$$

a.e. and also in the strong topology of $L^2(R^n)$. Therefore as $\mu \to 0$ $\{K_{\epsilon\mu}*f\}$ is a Cauchy sequence in $L^2(R^n)$. The limit is $K_\epsilon * f$ since $K_{\epsilon\mu} * f$ tends to $K_\epsilon * f$ a.e., and

$$\mathcal{F}(K_\epsilon * f) = \text{l.i.m.}_{\mu \to \infty}\mathcal{F}(K_{\epsilon\mu} * f) = (2\pi)^{n/2}\mathcal{F}K_\epsilon \cdot \mathcal{F}f.$$

Since $\mathcal{F}K_\epsilon \cdot \mathcal{F}f$ is strongly convergent to $\mathcal{F}K \cdot \mathcal{F}f$ in $L^2(R^n)$ as $\epsilon \to 0$, $K_\epsilon * f$ is also strongly convergeent in $L^2(R^n)$. We denote the limit by $K * f$. Then

$$\mathcal{F}(K * f) = \text{l.i.m.}_{\epsilon \to 0}\mathcal{F}(K_\epsilon * f) = (2\pi)^{n/2}\mathcal{F}K \cdot \mathcal{F}f.$$

Consequently

$$\|K * f\|_2 = \|\mathcal{F}(K * f)\|_2 = (2\pi)^{n/2}\|\mathcal{F}K \cdot \mathcal{F}f\|_2$$
$$\leq (2\pi)^{n/2}M\|\mathcal{F}f\|_2 = (2\pi)^{n/2}M\|f\|_2.$$

Summing up we have established the following theorem.

Theorem 2.2 *Let $K(x)$ be an integral kernel satisfying Assumption 2. For $f \in L^2(R^n), \epsilon > 0$ put*

$$\tilde{f}_\epsilon(x) = \int_{|x-y|>\epsilon} K(x - y)f(y)dy.$$

Then there exists a constant C independent of ϵ and f such that $\|\tilde{f}_\epsilon\|_2 \leq C\|f\|_2$. $\{\tilde{f}_\epsilon\}$ is strongly convergent in $L^2(R^n)$ as $\epsilon \to 0$. The mapping $f \mapsto K * f \equiv \lim_{\epsilon \to 0} \tilde{f}_\epsilon$ is a bounded linear operator from $L^2(R^n)$ to itself. The Fourier transform of $K * f$ is expressed as

$$\mathcal{F}(K * f) = (2\pi)^{n/2}\mathcal{F}K \cdot \mathcal{F}f, \tag{2.28}$$

where $\mathcal{F}K$ is an essentially bounded measurable function defined by the right hand side of (2.26). $\mathcal{F}K$ is a homogeneous function of order 0.

2.4 Case of Continuous Kernels

In this section we assume

Assumption 3 In addition to Assumption 2 $K(x)$ satisfies the following conditions. There exists a function $\Omega(x)$ homogeneous of order 0 such that

$$K(x) = \Omega\left(\frac{x}{|x|}\right)\frac{1}{|x|^n}, \quad x \in R^n \setminus \{0\}, \tag{2.29}$$

$$|\Omega(x) - \Omega(y)| \le \omega(|x - y|) \quad \text{for} \quad x, y \in \Sigma, \tag{2.30}$$

where $\omega(t)$ is a nonnegative continuous increasing function defined in $[0, \infty)$ satisfying

$$\int_0^1 \frac{\omega(t)}{t}dt < \infty, \quad \frac{\omega(t)}{t} \ge c_0 > 0. \tag{2.31}$$

The following lemma is known as the Calderón-Zygmund decomposition ([27]).

Lemma 2.3 Suppose $f \ge 0, f \in L^p(R^n), 1 \le p < \infty$. Then for any positive number s there exists a sequence of non-overlapping cubes $\{I_k\}$ such that

$$s \le \frac{1}{|I_k|}\int_{I_k} f(x)dx < 2^n s. \tag{2.32}$$

Put $D_s = \cup_k I_k$. Then $f(x) \le s$ a.e. outside D_s, and

$$s \le \frac{1}{|D_s|}\int_{D_s} f(x)dx < 2^n s. \tag{2.33}$$

Proof. Let I be a cube. Since

$$\frac{1}{|I|}\int_I f(x)dx \le |I|^{-1/p}\|f\|_p \to 0$$

as $|I| \to \infty$, there exists a positive number ν such that if $|I| \ge \nu$

$$\frac{1}{|I|}\int_I f(x)dx < s.$$

Let $R^n = \cup_{k=1}^\infty I_{0,k}$ be a decomposition of R^n into a sum of cubes of volume ν. Then for each k

$$\frac{1}{|I_{0,k}|}\int_{I_{0,k}} f(x)dx < s.$$

Denote by $\{I_{1,k}\}$ the sequence of cubes obtained by dividing each $I_{0,k}$ into 2^n equal parts. We classify $\{I_{1,k}\}$ as follows:

$$\{I_{1,k}\} = \{I'_{1,k}\} \cup \{I''_{1,k}\}, \quad \frac{1}{|I'_{1,k}|}\int_{I'_{1,k}} f(x)dx \ge s, \quad \frac{1}{|I''_{1,k}|}\int_{I''_{1,k}} f(x)dx < s.$$

Denote by $\{I_{2,k}\}$ the sequence of cubes obtained by dividing each $I''_{1,k}$ into 2^n equal parts, and we classify it as follows:

$$\{I_{2,k}\} = \{I'_{2,k}\} \cup \{I''_{2,k}\}, \quad \frac{1}{|I'_{2,k}|} \int_{I'_{2,k}} f(x)dx \geq s, \quad \frac{1}{|I''_{2,k}|} \int_{I''_{2,k}} f(x)dx < s.$$

Repeating this process we get a sequence of cubes $\{I_{m,k}\}$ such that

$$\{I_{m,k}\} = \{I'_{m,k}\} \cup \{I''_{m,k}\}, \quad \frac{1}{|I'_{m,k}|} \int_{I'_{m,k}} f(x)dx \geq s, \quad \frac{1}{|I''_{m,k}|} \int_{I''_{m,k}} f(x)dx < s$$

for each m. Then

$$R^n = (\cup I'_{1,k}) \cup (\cup I'_{2,k}) \cup \cdots \cup (\cup I'_{m,k}) \cup (\cup I''_{m,k}).$$

If $m \geq 1$, each $I'_{m,k}$ is obtained by dividing some $I''_{m-1,l}$, where $I''_{0,l} = I_{0,l}$. Hence $|I'_{m,k}| = 2^{-n}|I''_{m-1,l}|$. Therefore

$$\frac{1}{|I'_{m,k}|} \int_{I'_{m,k}} f(x)dx \leq \frac{2^n}{|I''_{m-1,l}|} \int_{I''_{m-1,l}} f(x)dx < 2^n s.$$

Ler $\{I_k\}$ be the totality of $I'_{m,k}, m = 1, 2, \ldots, k = 1, 2, \ldots$. Then (2.32) holds for any k. Put $D_s = \cup_k I_k$. Since

$$|I_k| \leq \frac{1}{s} \int_{I_k} f(x)dx \leq \frac{1}{s}|I_k|^{1-1/p} \left(\int_{I_k} f(x)^p dx \right)^{1/p},$$

we get

$$|I_k| \leq s^{-p} \int_{I_k} f(x)^p dx.$$

Hence

$$|D_s| = \sum |I_k| \leq s^{-p} \int_{D_s} f(x)^p dx < \infty.$$

It is easy to show that (2.33) holds. Let x_0 be a Lebesgue point of f which does not belong to D_s. Then for any $m \geq 1$ there exists an integer $k(m)$ such that $x_0 \in I''_{m,k(m)}$. Since $\{I''_{m,k(m)}\}$ is a sequence of regular closed sets tending to x_0, we get by Theorem 1.1

$$s > \frac{1}{|I''_{m,k(m)}|} \int_{I''_{m,k(m)}} f(x)dx \to f(x_0).$$

Thus we conclude $f(x_0) \leq s$.

Lemma 2.4 *Let $K(x)$ be an integral kernel satisfying Assumption 3. Suppose $0 \le f \in L^p(R^n), 1 \le p \le 2, \epsilon > 0$. Let K_ϵ be the function defined by (2.15), and let $\tilde{f}_\epsilon = K_\epsilon * f$ be the function defined by (2.4). Let s be an arbitrary positive number. If we put $E_s = \{x; |\tilde{f}_\epsilon(x)| > s\}$, then there exist positive constants C_1, C_2 independent of f, ϵ, s such that*

$$|E_s| \le \frac{C_1}{s^2} \int_{R^n} [f(x)]_s^2 dx + C_2 |D_s|, \qquad (2.34)$$

where D_s is the set in Lemma 2.3, and $[f(x)]_s$ is the function defined by

$$[f(x)]_s = \begin{cases} f(x) & f(x) \le s \\ s & f(x) > s. \end{cases}$$

Proof. We denote various constants independent of f, ϵ, s by C. Let $\{I_k\}$ be the sequence of cubes of Lemma 2.3, and g, h be the functions defined by

$$h(x) = \begin{cases} \dfrac{1}{|I_k|} \displaystyle\int_{I_k} f(y) dy & x \in I_k \\ f(x) & x \notin D_s \end{cases}, \quad g(x) = f(x) - h(x).$$

Then $s \le h(x) \le 2^n s$ in D_s, and $h(x) = f(x) = [f(x)]_s$ in $R^n \setminus D_s$. For each k

$$\int_{I_k} g(y) dy = 0, \qquad (2.35)$$

and $g(x) = 0$ in $R^n \setminus D_s$. If we put $\tilde{h}_\epsilon = K_\epsilon * h, \tilde{g}_\epsilon = K_\epsilon * g$,

$$E_1 = \{x; |\tilde{h}_\epsilon(x)| \ge \frac{s}{2}\}, \quad E_2 = \{x; |\tilde{g}_\epsilon(x)| \ge \frac{s}{2}\},$$

then

$$E_s \subset E_1 \cup E_2. \qquad (2.36)$$

With the aid of the proof of Theorem 2.2

$$\int_{R^n} |\tilde{h}_\epsilon(x)|^2 dx = (2\pi)^n \int_{R^n} |(\mathcal{F}K_\epsilon)(\xi)(\mathcal{F}h)(\xi)|^2 d\xi$$

$$\le (2\pi)^n M^2 \int_{R^n} |(\mathcal{F}h)(\xi)|^2 d\xi = (2\pi)^n M^2 \int_{R^n} h(x)^2 dx.$$

Hence

$$\frac{s^2}{4} |E_1| \le (2\pi)^n M^2 \int_{R^n} h(x)^2 dx. \qquad (2.37)$$

On the other hand

$$\int_{R^n} h(x)^2 dx = \int_{D_s} h(x)^2 dx + \int_{R^n \setminus D_s} h(x)^2 dx \le 2^{2n} s^2 |D_s| + \int_{R^n} [f(x)]_s^2 dx.$$

Combining this with (2.37) we get

$$|E_1| \le 2^{2n+2}(2\pi)^n M^2 |D_s| + \frac{4}{s^2}(2\pi)^n M^2 \int_{R^n} [f(x)]_s^2 dx. \qquad (2.38)$$

Letting S_k be the ball centered at the center of I_k and of radius the diameter of I_k we put $\tilde{D}_s = \cup_k S_k$, $S_k' = R^n \setminus S_k$, $\tilde{D}_s' = R^n \setminus \tilde{D}_s$. Then

$$|\tilde{D}_s| \le C|D_s|. \qquad (2.39)$$

Noting

$$\tilde{g}_\epsilon(x) = \int_{D_s} K_\epsilon(x - y)g(y)dy = \sum_k \int_{I_k} K_\epsilon(x - y)g(y)dy$$

we have

$$\int_{\tilde{D}_s'} |\tilde{g}_\epsilon(x)| dx \le \sum_k \int_{\tilde{D}_s'} \left| \int_{I_k} K_\epsilon(x - y)g(y)dy \right| dx$$

$$\le \sum_k \int_{S_k'} \left| \int_{I_k} K_\epsilon(x - y)g(y)dy \right| dx. \qquad (2.40)$$

Let x be fixed in S_k'. Denote the center of I_k by y_k. In case where $\{y; |y-x| \le \epsilon\} \cap I_k$ is empty we get in view of (2.35)

$$\int_{I_k} K_\epsilon(x - y)g(y)dy = \int_{I_k} K(x - y)g(y)dy$$

$$= \int_{I_k} (K(x - y) - K(x - y_k))g(y)dy. \qquad (2.41)$$

We estimate each term of the rightmost side of

$$|K(x - y) - K(x - y_k)|$$

$$= \left| \frac{1}{|x - y|^n} \Omega\left(\frac{x - y}{|x - y|}\right) - \frac{1}{|x - y_k|^n} \Omega\left(\frac{x - y_k}{|x - y_k|}\right) \right|$$

$$\le \left| \frac{1}{|x - y|^n} - \frac{1}{|x - y_k|^n} \right| \left| \Omega\left(\frac{x - y}{|x - y|}\right) \right|$$

$$+ \frac{1}{|x - y_k|^n} \left| \Omega\left(\frac{x - y}{|x - y|}\right) - \Omega\left(\frac{x - y_k}{|x - y_k|}\right) \right|. \qquad (2.42)$$

If $y \in I_k$, we have $|x-y_k|/2 \leq |x-y| \leq 3|x-y_k|/2$ since $|y-y_k| \leq |x-y_k|/2$. Hence

$$
\left| \frac{1}{|x-y|^n} - \frac{1}{|x-y_k|^n} \right|
$$

$$
= \left| \frac{1}{|x-y|} - \frac{1}{|x-y_k|} \right| \left| \sum_{i=0}^{n-1} \frac{1}{|x-y|^{n-1-i}|x-y_k|^i} \right|
$$

$$
\leq \frac{|y-y_k|}{|x-y||x-y_k|} \sum_{i=0}^{n-1} \frac{1}{|x-y|^{n-1-i}|x-y_k|^i} \tag{2.43}
$$

$$
\leq \frac{C|y-y_k|}{|x-y_k|^{n+1}} \leq \frac{C|I_k|^{1/n}}{|x-y_k|^{n+1}}.
$$

Noting

$$
\left| \frac{x-y}{|x-y|} - \frac{x-y_k}{|x-y_k|} \right|
$$

$$
\leq \left| \frac{x-y}{|x-y|} - \frac{x-y}{|x-y_k|} \right| + \left| \frac{x-y}{|x-y_k|} - \frac{x-y_k}{|x-y_k|} \right| \leq \frac{2|y-y_k|}{|x-y_k|} \leq \frac{\sqrt{n}|I_k|^{1/n}}{|x-y_k|},
$$

we have in view of (2.30)

$$
\left| \Omega \left(\frac{x-y}{|x-y|} \right) - \Omega \left(\frac{x-y_k}{|x-y_k|} \right) \right| \leq \omega \left(\frac{\sqrt{n}|I_k|^{1/n}}{|x-y_k|} \right). \tag{2.44}
$$

Combining (2.42),(2.43),(2.44) and using the latter inequality in (2.31) we get

$$
|K(x-y) - K(x-y_k)|
$$

$$
\leq \frac{C|I_k|^{1/n}}{|x-y_k|^{n+1}} + \frac{1}{|x-y_k|^n} \omega \left(\frac{\sqrt{n}|I_k|^{1/n}}{|x-y_k|} \right) \leq \frac{C}{|x-y_k|^n} \omega \left(\frac{\sqrt{n}|I_k|^{1/n}}{|x-y_k|} \right).
$$

It follows from this inequality and (2.41) that

$$
\left| \int_{I_k} K_\epsilon(x-y)g(y)dy \right| \leq \frac{C}{|x-y_k|^n} \omega \left(\frac{\sqrt{n}|I_k|^{1/n}}{|x-y_k|} \right) \int_{I_k} |g(y)|dy. \tag{2.45}
$$

Next consider the case where $\{y; |y-x| \leq \epsilon\} \cap I_k$ is not empty. Note that in this case

$$
I_k \subset \{y; |y-x| \leq 3\epsilon\}. \tag{2.46}
$$

Let γ be the characteristic function of the interval $[0,3]$. If $y \in I_k$, then $\gamma(|x-y|/\epsilon) = 1$ in view of (2.46). Hence, noting that $|K_\epsilon(x-y)| \leq C\epsilon^{-n}$

$$
\left| \int_{I_k} K_\epsilon(x-y)g(y)dy \right| \leq C\epsilon^{-n} \int_{I_k} \gamma \left(\frac{|x-y|}{\epsilon} \right) |g(y)|dy. \tag{2.47}
$$

It follows from (2.45),(2.47) that

$$\left| \int_{I_k} K_\epsilon(x-y)g(y)dy \right| \leq \frac{C}{|x-y_k|^n} \omega \left(\frac{\sqrt{n}|I_k|^{1/n}}{|x-y_k|} \right) \int_{I_k} |g(y)|dy$$

$$+C\epsilon^{-n} \int_{I_k} \gamma \left(\frac{|x-y|}{\epsilon} \right) |g(y)|dy \qquad (2.48)$$

holds for $x \in S'_k$. By virtue of (2.40),(2.48)

$$\int_{\tilde{D}'_s} |\tilde{g}_\epsilon(x)|dx \leq \sum_k \int_{S'_k} \frac{C}{|x-y_k|^n} \omega \left(\frac{\sqrt{n}|I_k|^{1/n}}{|x-y_k|} \right) \int_{I_k} |g(y)|dydx$$

$$+\sum_k \int_{S'_k} C\epsilon^{-n} \int_{I_k} \gamma \left(\frac{|x-y|}{\epsilon} \right) |g(y)|dydx$$

$$=\sum_k \int_{S'_k} \frac{C}{|x-y_k|^n} \omega \left(\frac{\sqrt{n}|I_k|^{1/n}}{|x-y_k|} \right) dx \int_{I_k} |g(y)|dy$$

$$+C\epsilon^{-n} \sum_k \int_{I_k} |g(y)| \int_{S'_k} \gamma \left(\frac{|x-y|}{\epsilon} \right) dxdy. \qquad (2.49)$$

Since

$$\int_{S'_k} \frac{C}{|x-y_k|^n} \omega \left(\frac{\sqrt{n}|I_k|^{1/n}}{|x-y_k|} \right) dx = \int_\Sigma \int_{\sqrt{n}|I_k|^{1/n}}^\infty \frac{C}{r^n} \omega \left(\frac{\sqrt{n}|I_k|^{1/n}}{r} \right) r^{n-1} dr d\sigma$$

$$= \int_\Sigma d\sigma \int_{\sqrt{n}|I_k|^{1/n}}^\infty \frac{C}{r} \omega \left(\frac{\sqrt{n}|I_k|^{1/n}}{r} \right) dr = C \int_\Sigma d\sigma \int_0^1 \frac{\omega(t)}{t} dt,$$

$$\epsilon^{-n} \int_{S'_k} \gamma \left(\frac{|x-y|}{\epsilon} \right) dx \leq \epsilon^{-n} \int_{R^n} \gamma \left(\frac{|x|}{\epsilon} \right) dx = \int_{R^n} \gamma(|x|)dx,$$

we obtain from (2.49) that

$$\int_{\tilde{D}'_s} |\tilde{g}_\epsilon(x)|dx \leq C \sum_k \int_{I_k} |g(y)|dy = C \int_{D_s} |g(y)|dy. \qquad (2.50)$$

On the other hand

$$\int_{D_s} |g(y)|dy \leq \int_{D_s} f(y)dy + \int_{D_s} h(y)dy = 2 \int_{D_s} f(y)dy \leq 2^{n+1}s|D_s|.$$

Combining this with (2.50) we get

$$\int_{\tilde{D}'_s} |\tilde{g}_\epsilon(x)|dx \leq Cs|D_s|.$$

This implies $|\tilde{D}'_s \cap E_2| \leq C|D_s|$. Hence noting (2.39)

$$|E_2| \leq |\tilde{D}'_s \cap E_2| + |\tilde{D}_s \cap E_2| \leq C|D_s|. \tag{2.51}$$

We conclude (2.34) combining (2.36),(2.38),(2.51).

Theorem 2. 3 *Suppose $K(x)$ is an integral kernel satisfying Assumption 3. Let $f \in L^p(R^n), 1 < p < \infty$. For $\epsilon > 0$ let $\tilde{f}_\epsilon(x)$ be the function defined by (2.4). Then there exists a constant C_p independent of ϵ, f such that*

$$\|\tilde{f}_\epsilon\|_p \leq C_p\|f\|_p. \tag{2.52}$$

As $\epsilon \to 0$ $\{\tilde{f}_\epsilon\}$ is strongly convergent in $L^p(R^n)$. If we denote the limit by $K * f$, then the mapping $f \mapsto K * f$ is a bounded linear operator from $L^p(R^n)$ to itself.

Proof. Write $T_\epsilon f = \tilde{f}_\epsilon$. In view of Theorem 2.2 there exists a constant C such that $\|T_\epsilon f\|_2 \leq C\|f\|_2$ for $f \in L^2(R^n), \epsilon > 0$. Therefore T_ϵ is of type (2,2) uniformly with respect to ϵ. If $0 \leq f \in L^1(R^n)$, we have by (2.33),(2.34) and $[f(x)]_s^2 \leq sf(x)$

$$|\{x; |(T_\epsilon f)(x)| > s\}| \leq C\|f\|_1/s.$$

If f is an arbitrary element of $L^1(R^n)$, then noting $f = f_1 - f_2, f_1 = \max\{f,0\}, f_2 = -\min\{f,0\}$ we get

$$|\{x; |(T_\epsilon f)(x)| > s\}|$$
$$\leq |\{x; |(T_\epsilon f_1)(x)| > s/2\}| + |\{x; |(T_\epsilon f_2)(x)| > s/2\}|$$
$$\leq 2C\|f_1\|_1/s + 2C\|f_2\|_1/s = 2C\|f\|_1/s.$$

Therefore T_ϵ is of weak type (1,1) uniformly in ϵ. Hence by virtue of Theorem 2.1 and its proof we have for any $f \in L^p(R^n), 1 < p \leq 2, \epsilon > 0$

$$\|T_\epsilon f\|_p \leq C_p\|f\|_p \tag{2.53}$$

for some constant C_p. Considering the adjoint operator we see that (2.53) holds also for $2 < p < \infty$.

Next, let $f \in L^p(R^n), 1 < p < \infty$. If $g \in C_0^1(R^n)$, i.e. g is a continuously differentiable function having a compact support in R^n, and $\epsilon, \delta > 0$, then we have

$$\|T_\epsilon f - T_\delta f\|_p \leq 2C_p\|f - g\|_p + \|T_\epsilon g - T_\delta g\|_p \tag{2.54}$$

by virtue of (2.53). If $0 < \epsilon < 1$, then by (2.2)

$$(T_\epsilon g)(x) = \int_{|x-y|>1} K(x-y)g(y)dy + \int_{\epsilon<|x-y|<1} K(x-y)(g(y) - g(x))dy. \tag{2.55}$$

Since $K \in L^p(\{x; |x| > 1\}), g \in L^1(R^n)$, the first term on the right of (2.55) belongs to $L^p(R^n)$ in view of Lemma 1.2. The second term on the right of (2.55) is a function with support contained in a fixed compact set, and converges uniformly as $\epsilon \to 0$ since g satisfies a uniform Lipschitz condition. Hence it converges strongly in $L^p(R^n)$. Combining this with (2.54) we conclude that $T_\epsilon f$ is strongly convergent in $L^p(R^n)$ as $\epsilon \to 0$. Hence using (2.53) we obtain

$$\|K * f\|_p \le C_p \|f\|_p.$$

2.5 Hilbert Transform

The mapping $f \mapsto K * f$ in the previous sections essentially reduces to

$$\frac{1}{\pi} \int_{-\infty}^{\infty} \frac{f(t)}{x - t} dt = \frac{1}{\pi} \lim_{\epsilon \to 0} \int_{|x-t|>\epsilon} \frac{f(t)}{x - t} dt \tag{2.56}$$

when $n = 1$. This is also expressed as

$$\frac{1}{\pi} \left(\text{Pf.} \frac{1}{x} \right) * f,$$

where Pf.$1/x$ is the distribution defined by

$$\left(\text{Pf.} \frac{1}{x} \right)(\phi) = \lim_{\epsilon \to +0} \int_{|x|>\epsilon} \frac{\phi(x)}{x} dx.$$

Definition 2. 2 The function defined by (2.56) is called the *Hilbert transform* of f.

In what follows we calculate the Fourier transform of π^{-1}Pf.$1/x$ following the proof of Theorem 2.2. For $0 < \epsilon < \mu$ we put

$$h_{\epsilon,\mu}(x) = \begin{cases} 1/(\pi x) & \epsilon < |x| < \mu \\ 0 & \text{otherwise} \end{cases}, \quad h_\epsilon(x) = \begin{cases} 1/(\pi x) & |x| > \epsilon \\ 0 & |x| \le \epsilon \end{cases}.$$

Then

$$(\mathcal{F}h_{\epsilon\mu})(\xi) = \frac{1}{\sqrt{2\pi}} \int_{\epsilon < |x| < \mu} \frac{e^{-ix\xi}}{\pi x} dx = -\sqrt{\frac{2}{\pi}} \frac{i}{\pi} \int_\epsilon^\mu \frac{\sin(x\xi)}{x} dx$$

$$= \begin{cases} -\sqrt{\dfrac{2}{\pi}} \dfrac{i}{\pi} \displaystyle\int_{\epsilon\xi}^{\mu\xi} \dfrac{\sin x}{x} dx & \xi > 0 \\[4mm] \sqrt{\dfrac{2}{\pi}} \dfrac{i}{\pi} \displaystyle\int_{-\epsilon\xi}^{-\mu\xi} \dfrac{\sin x}{x} dx & \xi < 0 \end{cases}.$$

Letting $\mu \to \infty$ we get

$$(\mathcal{F}h_\epsilon)(\xi) = -\sqrt{\frac{2}{\pi}}\frac{i}{\pi}\int_{\epsilon|\xi|}^\infty \frac{\sin x}{x}dx\text{sign}\xi. \tag{2.57}$$

Put $h = \pi^{-1}\text{Pf}.1/x$. Letting $\epsilon \to 0$ in (2.57) yields

$$(\mathcal{F}h)(\xi) = -(2\pi)^{-1/2}i\text{sign}\xi. \tag{2.58}$$

In view of (2.57) we have

$$|(\mathcal{F}h_\epsilon)(\xi)| \leq (2/\pi)^{3/2}A,$$

where A is the constant in (2.19). Hence for $f \in L^2(R)$

$$\|h_\epsilon * f\|_2 \leq 4A\pi^{-1}\|f\|_2.$$

According to Theorem 2.2

$$g(x) = \frac{1}{\pi}\lim_{\epsilon \to 0}\int_{|x-t|>\epsilon}\frac{f(t)}{x-t}dt \tag{2.59}$$

exists in the strong topology of $L^2(R)$, and in view of (2.28),(2.58)

$$(\mathcal{F}g)(\xi) = -i(\mathcal{F}f)(\xi)\text{sign}\xi. \tag{2.60}$$

Consequently

$$\|g\|_2 = \|f\|_2, \tag{2.61}$$

$$(\mathcal{F}f)(\xi) = i(\mathcal{F}g)(\xi)\text{sign}\xi. \tag{2.62}$$

By virtue of (2.62)

$$f(x) = -\frac{1}{\pi}\lim_{\epsilon \to 0}\int_{|x-t|>\epsilon}\frac{g(t)}{x-t}dt \tag{2.63}$$

holds in the strong topology of $L^2(R)$.

Suppose $1 < p < \infty$. The Hilbert transform is a bounded linear operator from $L^p(R)$ to itself in view of Theorem 2.3. Let $f \in L^p(R)$ and g be its Hilbert transform. If $f_n \in L^p(R) \cap L^2(R)$, $f_n \to f$ in $L^p(R)$, g_n is the Hilbert transform of f_n, then $g_n \in L^p(R) \cap L^2(R)$, $g_n \to g$ in $L^p(R)$. Since

$$-f_n(x) = \frac{1}{\pi}\lim_{\epsilon \to 0}\int_{|x-t|>\epsilon}\frac{g_n(t)}{x-t}dt$$

holds in the strong topology of $L^2(R)$ and the right hand side is the Hilbert transform of g_n, the above limit exists also in the strong topology of $L^p(R)$.

Hence letting $n \to \infty$ we see that (2.63) holds also in the strong topology of $L^p(R)$.

Suppose $f \in L^p(R), \hat{f} \in L^{p'}(R), p' = p/(p-1)$, and $g, \hat{g}$ are the Hilbert transforms of $f, \hat{f}$ respectively. Let $f_n \in L^p(R) \cap L^2(R), \hat{f}_n \in L^{p'}(R) \cap L^2(R), f_n \to f$ in $L^p(R), \hat{f}_n \to \hat{f}$ in $L^{p'}(R)$, and $g_n, \hat{g}_n$ be the Hilbert transforms of $f_n, \hat{f}_n$ respectively. Then, $g_n \in L^p(R) \cap L^2(R), \hat{g}_n \in L^{p'}(R) \cap L^2(R), g_n \to g$ in $L^p(R), \hat{g}_n \to \hat{g}$ in $L^{p'}(R)$. Since the Hilbert transform is a unitary operator in $L^2(R)$ in view of (2.61) and (2.63), we have

$$\int_{-\infty}^{\infty} f_n(x)\hat{f}_n(x)dx = \int_{-\infty}^{\infty} g_n(x)\hat{g}_n(x)dx.$$

Letting $n \to \infty$ we get

$$\int_{-\infty}^{\infty} f(x)\hat{f}(x)dx = \int_{-\infty}^{\infty} g(x)\hat{g}(x)dx. \tag{2.64}$$

Summing up we have proved the following theorem.

Theorem 2. 4 *For $1 < p < \infty$ the Hilbert transform is a bounded linear operator from $L^p(R)$ to itself. If g is the Hilbert transform of $f \in L^p(R)$, then the inverse transform is given by (2.63), and the limits in the right hand sides of (2.59) and (2.63) exist in the strong topology of $L^p(R)$. If $f \in L^p(R), \hat{f} \in L^{p'}(R), p' = p/(p-1)$ and $g, \hat{g}$ are the Hilbert transforms of $f, \hat{f}$ respectively, then the equality (2.64) holds. In particular the Hilbert transform is a unitary operator in $L^2(R)$, and (2.60) holds if g is the Hilbert transform of f.*

2.6   Equimeasurable Functions

Let $f(x)$ be a nonnegative measurable function defined in R^n. For each $\tau \geq 0$ put

$$m(\tau) = |\{x \in R^n; f(x) > \tau\}|. \tag{2.65}$$

In this section we consider only functions such that $m(\tau) < \infty$ for any $\tau > 0$ and $\lim_{\tau \to \infty} m(\tau) = 0$. It is easy to see that functions f such that $0 \leq f \in L^p(R^n), 1 \leq p < \infty$, satisfy this condition.

Definition 2. 3 Let f be a nonnegative measurable function defined in R^n. A nonnegative measurable function f^* defined in $[0, \infty)$ is called an *equimeasurable function* of f if

$$m^*(\tau) \equiv |\{t; f^*(t) > \tau\}| = m(\tau) \tag{2.66}$$

holds for any $\tau \geq 0$.

Lemma 2.5 *The function $m(\tau)$ defined by (2.65) is monotone decreasing and right continuous.*

Proof. It is obvious that $m(\tau)$ is monotone decreasing. The right continuity follows from

$$\{x; f(x) > \tau\} = \cup_{k=1}^{\infty}\{x; f(x) > \tau_k\} \quad \text{if} \quad \tau_k \downarrow \tau.$$

Put

$$t_0 = m(0), \quad \tau_0 = \sup\{\tau; m(\tau) > 0\}. \tag{2.67}$$

Clearly $0 < t_0 \leq \infty, 0 < \tau_0 \leq \infty$.

Lemma 2.6 *The function f^* defined by*

$$f^*(t) = \inf\{\tau; t \geq m(\tau)\}, \quad f^*(0) = \tau_0$$

is the unique monotone decreasing, right continuous equimeasurable function of f.

Proof. First we show that $\tau \geq f^*(t)$ and $t \geq m(\tau)$ are equivalent. It is obvious that $t \geq m(\tau)$ implies $\tau \geq f^*(t)$. Conversely, if $\tau > f^*(t)$ then $t \geq m(\tau)$ by the definition of f^*. Suppose $\tau = f^*(t)$. If $\tau_k \downarrow \tau$, then $t \geq m(\tau_k)$. Since $m(\tau)$ is right continuous, we get $t \geq m(\tau)$. Thus, we see that $\tau < f^*(t)$ and $t < m(\tau)$ are equivalent. Hence for any $\tau > 0$

$$\{t; f^*(t) > \tau\} = [0, m(\tau)).$$

Consequently if $m^*(\tau)$ is the function defined by (2.66), then we have $m^*(\tau) = m(\tau)$. Clearly $f^*(t)$ is monotone decreasing. For each $\tau > 0$ we have $t_0 \geq m(\tau)$, or $\tau \geq f^*(t_0)$, and hence $f^*(t_0) = 0$. Therefore, if $t_0 < \infty$, then $f^*(t) = 0$ for any $t > t_0$. If $t < t_0$, then $t < m(\tau)$ for some $\tau > 0$. Since $\tau < f^*(t)$ then, we have $f^*(t) > 0$. Hence, $f^*(t) > 0$ for $0 < t < t_0$. Suppose $0 < t < t_0$ and $t_k \downarrow t$. If $0 < \tau < f^*(t)$, then $t < m(\tau)$. Hence $t_k < m(\tau)$ if k is sufficiently large, which implies $\tau < f^*(t_k) \leq f^*(t)$. This shows $f^*(t) = \lim_{k\to\infty} f^*(t_k)$. Consequently f^* is right continuous. Finally suppose g^* is another monotone decreasing right continuous equimeasurable function of f. Let t_1 be an arbitrary positive number. Put $\tau = f^*(t_1), t_2 = \min\{t; f^*(t) = \tau\} = \inf\{t; f^*(t) = \tau\}$. Since $\{t; f^*(t) > \tau\} = [0, t_2)$, we have

$$|\{t; g^*(t) > \tau\}| = m(\tau) = |\{t; f^*(t) > \tau\}| = t_2.$$

Since g^* is monotone decreasing and right continuous, we have $\{t; g^*(t) > \tau\} = [0, t_2)$. Hence

$$g^*(t_1) \leq g^*(t_2) \leq \tau = f^*(t_1).$$

Similarly we can show $f^*(t_1) \leq g^*(t_1)$. Thus $f^*(t) \equiv g^*(t)$.

Lemma 2. 7 *If f^* is an equimeasurable function of f, then for $1 \leq p < \infty$*

$$\int_{R^n} f(x)^p dx = \int_0^\infty f^*(t)^p dt. \tag{2.68}$$

Proof. In view of (2.66)

$$\int_{R^n} f(x)^p dx = -\int_0^\infty \tau^p dm(\tau) = -\int_0^\infty \tau^p dm^*(\tau) = \int_0^\infty f^*(t)^p dt.$$

Lemma 2. 8 *If A is a measurable subset of R^n of finite measure, then*

$$\int_A f(x) dx \leq \int_0^{|A|} f^*(t) dt. \tag{2.69}$$

Proof. Let f_1 be the function such that $f_1(x) = f(x)$ for $x \in A$ and $f_1(x) = 0$ for $x \notin A$. Since $f_1(x) \leq f(x)$ we have

$$m_1(\tau) \equiv |\{x; f_1(x) > \tau\}| \leq |\{x; f(x) > \tau\}| = m(\tau).$$

Therefore if f_1^* is the monotone decreasing right continuous equimeasurable function of f_1, then

$$f_1^*(t) = \inf\{\tau; t \geq m_1(\tau)\} \leq f^*(t).$$

Since $\{x; f_1(x) > \tau\} \subset \{x; f_1(x) > 0\} \subset A$ for any $\tau > 0$, we have $m_1(\tau) \leq |A|$. Hence $f_1^*(t) = 0$ for $t \geq |A|$. With the aid of Lemma 2.7

$$\int_A f(x) dx = \int_{R^n} f_1(x) dx = \int_0^\infty f_1^*(t) dt = \int_0^{|A|} f_1^*(t) dt \leq \int_0^{|A|} f^*(t) dt.$$

In what follows in this section we suppose that $0 \leq f \in L^p(R^n), 1 \leq p < \infty$, and f^* is the monotone decreasing, right continuous equimeasurable function of f. For $t > 0$ we put

$$\beta_f(t) = \frac{1}{t} \int_0^t f^*(s) ds. \tag{2.70}$$

The right hand side of (2.70) is finite since $f^* \in L^p(0, \infty)$. Since

$$\frac{d}{dt} \beta_f(t) = \frac{1}{t^2} \int_0^t (f^*(t) - f^*(s)) ds \leq 0,$$

if $\beta_f'(t_1) = 0$ for some $t_1 > 0$, then $f^*(t) = f^*(t_1)$ in $(0, t_1)$, and hence $\beta_f(t) = f^*(t_1)$ there. Therefore we see that either $\beta_f(t)$ is strictly decreasing in $(0, \infty)$ or there exists a positive number t_1 such that $\beta_f(t)$ is constant in $(0, t_1]$ and strictly decreasing in $[t_1, \infty)$. In the former case we put $t_1 = 0$. Put $\beta_f(0) = s_1 \equiv \lim_{t \to 0} \beta_f(t)$. As is easily seen $0 < s_1 \leq \infty$ and $s_1 = \beta_f(t_1)$. It is easy to see that $\beta_f(t) \to 0$ as $t \to \infty$.

We denote the inverse function of $s = \beta_f(t)$ by $t = \beta^f(s)$. If $t_1 = 0, s_1 = \infty$, then $\beta^f(s)$ is uniquely determined in $0 < s < \infty$. If $t_1 = 0, s_1 < \infty$, we put $\beta^f(s) = 0$ for $s > s_1$. If $t_1 > 0$, we put $\beta^f(s_1) = t_1$ and $\beta^f(s) = 0$ for $s > s_1$.

Lemma 2.9 *If $f \in L^p(R^n), 1 < p < \infty$, we have*

$$\left(\int_0^\infty \beta_f(t)^p dt \right)^{1/p} \leq \frac{p}{p-1} \left(\int_{R^n} f(x)^p dx \right)^{1/p}. \qquad (2.71)$$

Proof. By virtue of Lemma 2.7 $f^* \in L^p(0, \infty)$. For $0 < a < b$ we get by integrating by parts

$$\int_a^b \beta_f(t)^p dt = \int_a^b t^{-p} \left(\int_0^t f^*(s) ds \right)^p dt \leq \frac{a^{1-p}}{p-1} \left(\int_0^a f^*(s) ds \right)^p$$

$$+ \frac{p}{p-1} \int_a^b t^{1-p} f^*(t) \left(\int_0^t f^*(s) ds \right)^{p-1} dt. \qquad (2.72)$$

The first term of the right hand side of (2.72) tends to 0 as $a \to 0$ since

$$\frac{a^{1-p}}{p-1} \left(\int_0^a f^*(s) ds \right)^p \leq \frac{1}{p-1} \int_0^a f^*(s)^p ds$$

by Hölder's inequality. Hence letting $a \to 0, b \to \infty$ in (2.72) we obtain

$$\int_0^\infty \beta_f(t)^p dt \leq \frac{p}{p-1} \int_0^\infty f^*(t) \beta_f(t)^{p-1} dt$$

$$\leq \frac{p}{p-1} \left(\int_0^\infty f^*(t)^p dt \right)^{1/p} \left(\int_0^\infty \beta_f(t)^p dt \right)^{1-1/p},$$

from which (2.71) readily follows.

Lemma 2.10 *Suppose $0 \leq f \in L^p(R), 1 < p < \infty$. For $\epsilon > 0$ put*

$$F_\epsilon(x) = \frac{1}{\epsilon} \int_0^\epsilon f(x+y) dy, \quad G(x) = \sup_{\epsilon > 0} F_\epsilon(x). \qquad (2.73)$$

Then

$$\int_{-\infty}^\infty G(x)^p dx \leq \left(\frac{p}{p-1} \right)^p \int_{-\infty}^\infty f(x)^p dx. \qquad (2.74)$$

Proof. First note that $G(x)$ is measurable since the supremum of the right hand side of the second equality of (2.73) may be taken only over rational ϵ's. Put $H(\tau) = \{x; G(x) > \tau\}$ for $\tau > 0$. Since

$$G(x) = \sup_{y > x} \frac{1}{y - x} \int_x^y f(t)dt,$$

$x \in H(\tau)$ if and only if

$$\int_0^y f(t)dt - y\tau > \int_0^x f(t)dt - x\tau$$

for some $y > x$. Hence, if we set $F(x) = \int_0^x f(t)dt - x\tau$, we have

$$H(\tau) = \{x; F(y) > F(x) \quad \text{for some} \quad y > x\}. \tag{2.75}$$

Since $H(\tau)$ is an open set, it is the sum of disjoint open intervals: $H(\tau) = \cup(a_k, b_k)$. With the aid of Hölder's inequality we see

$$\lim_{x \to \pm\infty} F(x) = \mp\infty. \tag{2.76}$$

We show $-\infty < a_k < b_k < \infty$ for any k. Suppose $a_k = -\infty$ for some k. Then $(-\infty, b_k) \subset H(\tau)$. By (2.76) we have $F(c) > F(b_k)$ for some $c < b_k$. Let c_1 be such that $F(c_1) = \max_{c \le x \le b_k} F(x), c_1 \in [c, b_k]$. Then $c \le c_1 < b_k$ since $F(c_1) \ge F(c) > F(b_k)$. Hence $c_1 \in (-\infty, b_k) \subset H(\tau)$. If $c_1 < x \le b_k$, then $F(x) \le F(c_1)$. If $x > b_k$, then $F(x) \le F(b_k) < F(c_1)$ since $b_k \notin H(\tau)$. Hence $F(x) \le F(c_1)$ for any $x > c_1$. This contradicts $c_1 \in H(\tau)$. Next suppose $b_k = \infty$. Then $(a_k, \infty) \subset H(\tau)$. Let $c > a_k$. In view of (2.76) $F(c_1) = \max_{x \ge c} F(x)$ for some $c_1 \ge c$. Then $F(x) \le F(c_1)$ for $x > c_1$. This is a contradiction since $c_1 \in (a_k, \infty) \subset H(\tau)$. Thus we have proved $-\infty < a_k < b_k < \infty$.

Next we show $F(a_k) = F(b_k)$ for any k. Since $a_k \notin H(\tau)$ and $b_k > a_k$, we have $F(b_k) \le F(a_k)$. Suppose $F(b_k) < F(a_k)$ for some k. Then $F(b_k) < F(c)$ for some $c \in (a_k, b_k)$. Let c_1 be such that $F(c_1) = \max_{c \le x \le b_k} F(x), c \le c_1 \le b_k$. Since $F(c_1) \ge F(c) > F(b_k)$, we have $c \le c_1 < b_k$. Hence $c_1 \in (a_k, b_k) \subset H(\tau)$. If $c_1 < x \le b_k, F(x) \le F(c_1)$. If $x > b_k, F(x) \le F(b_k) < F(c_1)$ since $b_k \notin H(\tau)$. Hence $F(x) \le F(c_1)$ for any $x > c_1$. This contradicts $c_1 \in H(\tau)$. Thus we conclude $F(a_k) = F(b_k)$.

Consequently

$$(b_k - a_k)\tau = \int_{a_k}^{b_k} f(t)dt \le (b_k - a_k)^{1-1/p} \left(\int_{a_k}^{b_k} f(t)^p dt \right)^{1/p},$$

and hence

$$(b_k - a_k)^{1/p}\tau \leq \left(\int_{a_k}^{b_k} f(t)^p dt\right)^{1/p}.$$

Therefore

$$|H(\tau)|\tau^p = \sum (b_k - a_k)\tau^p \leq \sum \int_{a_k}^{b_k} f(t)^p dt < \infty.$$

In view of Lemma 2.8

$$|H(\tau)|\tau = \sum (b_k - a_k)\tau = \sum \int_{a_k}^{b_k} f(t)dt = \int_{H(\tau)} f(t)dt \leq \int_0^{|H(\tau)|} f^*(t)dt.$$

Consequently

$$\tau \leq \frac{1}{|H(\tau)|} \int_0^{|H(\tau)|} f^*(t)dt = \beta_f(|H(\tau)|),$$

which implies $|H(\tau)| \leq \beta^f(\tau)$. From this and Lemma 2.2

$$\int_{-\infty}^{\infty} G(x)^p dx = p \int_0^{\infty} \tau^{p-1}|H(\tau)|d\tau$$

$$\leq p \int_0^{\infty} \tau^{p-1}\beta^f(\tau)d\tau = \int_0^{s_1} \beta^f(\tau)d\tau^p. \qquad (2.77)$$

As was shown in the proof of Lemma 2.9

$$\lim_{t \to 0} t\beta_f(t)^p = \lim_{t \to 0} t^{1-p} \left(\int_0^t f^*(s)ds\right)^p = 0.$$

Hence making the change of the independent variable $t = \beta^f(\tau)$ in the last integral of (2.77)

$$\int_{-\infty}^{\infty} G(x)^p dx \leq -\int_{t_1}^{\infty} td\beta_f(t)^p = -[t\beta_f(t)]_{t_1}^{\infty} + \int_{t_1}^{\infty} \beta_f(t)^p dt$$

$$\leq t_1\beta_f(t_1)^p + \int_{t_1}^{\infty} \beta_f(t)^p dt$$

$$= \int_0^{t_1} \beta_f(t)^p dt + \int_{t_1}^{\infty} \beta_f(t)^p dt = \int_0^{\infty} \beta_f(t)^p dt.$$

With the aid of this inequality and Lemma 2.9 we obtain (2.74).

Lemma 2. 11 *Under the assumptions of Lemma 2.10 set*

$$\bar{f}(x) = \sup_{\epsilon>0} \frac{1}{2\epsilon} \int_{-\epsilon}^{\epsilon} f(x+y)dy.$$

Then

$$\int_{-\infty}^{\infty} \bar{f}(x)^p dx \le 2 \left(\frac{p}{p-1}\right)^p \int_{-\infty}^{\infty} f(x)^p dx. \tag{2.78}$$

Proof. Set $\bar{E}(\tau) = \{x; \bar{f}(x) > \tau\}$ for $\tau > 0$, and

$$F_\epsilon^-(x) = \frac{1}{\epsilon} \int_0^\epsilon f(x-y)dy, \quad G^-(x) = \sup_{\epsilon>0} F_\epsilon^-(x).$$

Let $H(\tau)$ be the set in the proof of Lemma 2.10 and $H^-(\tau)$ $= \{x; G^-(x) > \tau\}$. Then by the proof of Lemma 2.10 we get $|H^-(\tau)| \le \beta^f(\tau)$. Since $\bar{f}(x) \le (G(x) + G^-(x))/2$ we have $\bar{E}(\tau) \subset H(\tau) \cap H^-(\tau)$. Hence

$$|\bar{E}(\tau)| \le |H(\tau)| + |H^-(\tau)| \le 2\beta^f(\tau).$$

Following the proof of the previous lemma we conclude (2.78).

2.7 Hilbert Transform (Continued)

Let ϕ be an even function in $C^1(R)$ such that $\phi(x) = 1$ for $x \ge 3/4, \phi(x) = 0$ for $0 \le x \le 1/4$ and $0 \le \phi(x) \le 1$ in R. We denote by $\psi(x)$ the Hilbert transform of $\phi(x)/x$. Since

$$\frac{d}{dx} \int_{|x-t|>\epsilon} \frac{\phi(t)}{t} \frac{dt}{x-t} = \frac{d}{dx} \int_{|t|>\epsilon} \frac{\phi(x-t)}{x-t} \frac{dt}{t}$$

$$= \int_{|t|>\epsilon} \frac{\partial}{\partial x} \frac{\phi(x-t)}{x-t} \frac{dt}{t} = \int_{|x-t|>\epsilon} \frac{d}{dt} \frac{\phi(t)}{t} \frac{dt}{x-t},$$

we see that

$$\frac{d}{dx}\psi(x) = \lim_{\epsilon\to0} \frac{1}{\pi} \int_{|x-t|>\epsilon} \frac{d}{dt} \frac{\phi(t)}{t} \frac{dt}{x-t}$$

exists and is continuous. It is also easily seen that ψ is an even function. Next we show that

$$\psi(x) = O(x^{-2}), \quad \psi'(x) = O(|x|^{-3}) \tag{2.79}$$

as $x \to \pm\infty$. Suppose $x > 1$. Then,

$$\pi\psi(x) = \lim_{N\to\infty,\epsilon\to0} \left(\int_{-N}^{-1} + \int_{-1}^{0} + \int_{0}^{1} + \int_{1}^{x-\epsilon} + \int_{x+\epsilon}^{N} \right) \frac{\phi(t)}{t} \frac{dt}{x-t},$$

$$\int_{-N}^{-1} \frac{\phi(t)}{t} \frac{dt}{x-t} = \int_{-N}^{-1} \frac{dt}{t(x-t)} = -\frac{1}{x} \left[\log \frac{N}{x+N} + \log(x+1) \right],$$

$$\int_{-1}^{0} \frac{\phi(t)}{t} \frac{dt}{x-t} = -\int_{0}^{1} \frac{\phi(t)}{t} \frac{dt}{x+t},$$

$$\int_{1}^{x-\epsilon} \frac{\phi(t)}{t} \frac{dt}{x-t} = \int_{1}^{x-\epsilon} \frac{dt}{t(x-t)} = \frac{1}{x} \left[\log \frac{x-\epsilon}{\epsilon} + \log(x-1) \right],$$

$$\int_{x+\epsilon}^{N} \frac{\phi(t)}{t} \frac{dt}{x-t} = \int_{x+\epsilon}^{N} \frac{dt}{t(x-t)} = \frac{1}{x} \left(\log \frac{N}{N-x} - \log \frac{x+\epsilon}{\epsilon} \right).$$

Hence we have

$$\pi\psi(x) = \frac{1}{x} \log \frac{x-1}{x+1} + \int_{0}^{1} \frac{2\phi(t)}{(x-t)(x+t)} dt$$

for $x > 1$, from which (2.79) follows.

Theorem 2. 5 *Suppose* $f \in L^p(R), 1 < p < \infty$. *Set*

$$\tilde{f}_\epsilon(x) = \frac{1}{\pi} \int_{|x-t|>\epsilon} \frac{f(t)}{x-t} dt$$

for $\epsilon > 0$. *Then, there exists a constant* C_p *depending only on* p *such that*

$$\int_{-\infty}^{\infty} \sup_{\epsilon>0} |\tilde{f}_\epsilon(x)|^p dx \le C_p^p \int_{-\infty}^{\infty} |f(x)|^p dx. \tag{2.80}$$

As $\epsilon \to 0, \{\tilde{f}_\epsilon\}$ *converges both almost everywhere and in the strong topology of* $L^p(R)$.

Proof. Let ϕ, ψ be functions as above. Then

$$\tilde{f}_\epsilon(x) = \frac{1}{\pi} \int_{|x-t|>\epsilon} \phi\left(\frac{x-t}{\epsilon}\right) \frac{f(t)}{x-t} dt$$

$$= \frac{1}{\pi} \int_{-\infty}^{\infty} \phi\left(\frac{x-t}{\epsilon}\right) \frac{f(t)}{x-t} dt - \frac{1}{\pi} \int_{|x-t|<\epsilon} \phi\left(\frac{x-t}{\epsilon}\right) \frac{f(t)}{x-t} dt$$

$$\equiv \tilde{f}_{1,\epsilon}(x) + \tilde{f}_{2,\epsilon}(x). \tag{2.81}$$

The Hilbert transform of $\phi\left(\dfrac{x-t}{\epsilon}\right)\dfrac{1}{x-t}$ considered as a function of t for

a fixed x is $-\dfrac{1}{\epsilon}\psi\left(\dfrac{x-t}{\epsilon}\right)$. Hence, if we denote the Hilbert transform of f

by g, then in view of Theorem 2.4

$$\tilde{f}_{1,\epsilon}(x) = -\frac{1}{\pi\epsilon}\int_{-\infty}^{\infty}\psi\left(\frac{x-t}{\epsilon}\right)g(t)dt$$

$$= -\frac{1}{\pi\epsilon}\int_0^{\infty}\psi\left(\frac{t}{\epsilon}\right)(g(x-t)+g(x+t))dt. \tag{2.82}$$

Set

$$I(x;t) = \int_{-t}^{t}g(x+y)dy, \qquad \bar{g}(x) = \sup_{\epsilon>0}\frac{1}{2\epsilon}\int_{-\epsilon}^{\epsilon}|g(x+y)|dy.$$

According to Lemma 2.11 $\bar{g}\in L^p(R)$. Let x be a point such that $\bar{g}(x)<\infty$.
Then

$$|I(t;x)| \le \int_{-t}^{t}|g(x+y)|dy \le 2t\bar{g}(x), \tag{2.83}$$

$$dI(t;x) = (g(x-t)+g(x+t))dt. \tag{2.84}$$

Hence, in view of (2.82),(2.84)

$$\tilde{f}_{1,\epsilon}(x) = -\frac{1}{\pi\epsilon}\int_0^{\infty}\psi\left(\frac{t}{\epsilon}\right)dI(x;t).$$

Integrating by parts and noting (2.79)

$$\tilde{f}_{1,\epsilon}(x) = \frac{1}{\pi\epsilon^2}\int_0^{\infty}\psi'\left(\frac{t}{\epsilon}\right)I(x;t)dt. \tag{2.85}$$

By virtue of (2.83),(2.85)

$$|\tilde{f}_{1,\epsilon}(x)| \le \frac{1}{\pi\epsilon^2}\int_0^{\infty}\left|\psi'\left(\frac{t}{\epsilon}\right)\right|2t\bar{g}(x)dt = \frac{2}{\pi}\bar{g}(x)\int_0^{\infty}t|\psi'(t)|dt. \tag{2.86}$$

On the other hand

$$|\tilde{f}_{2,\epsilon}(x)| \le \frac{1}{\pi}\int_{\epsilon/4<|x-t|<\epsilon}\phi\left(\frac{x-t}{\epsilon}\right)\left|\frac{f(t)}{x-t}\right|dt \le \frac{4}{\pi\epsilon}\int_{|x-t|<\epsilon}|f(t)|dt.$$

Hence if we put

$$\bar{f}(x) = \sup_{\epsilon>0}\frac{1}{2\epsilon}\int_{-\epsilon}^{\epsilon}|f(x+y)|dy,$$

we have

$$|\tilde{f}_{2,\epsilon}(x)| \leq \frac{8}{\pi}\bar{f}(x). \tag{2.87}$$

From (2.81),(2.86),(2.87) it follows that

$$\sup_{\epsilon>0}|\tilde{f}_{\epsilon}(x)| \leq \frac{2}{\pi}\bar{g}(x)\int_0^\infty t|\psi'(t)|dt + \frac{8}{\pi}\bar{f}(x).$$

Applying Theorem 2.4 and Lemma 2.11 to the right hand side of the above inequality we conclude (2.80).

The strong convergence of $\{\tilde{f}_{\epsilon}\}$ in $L^p(R)$ was already shown. Let f be an arbitrary element of $L^p(R)$ and $g \in C_0^1(R)$. Set $h = f - g$. If we define $\tilde{g}_{\epsilon}, \tilde{h}_{\epsilon}$ as we defined $\tilde{f}_{\epsilon}$, then by the argument at the end of the proof of Theorem 2.3 we see that $\tilde{g}_{\epsilon}$ converges uniformly. Consequently

$$\limsup_{\epsilon\to0}\tilde{f}_{\epsilon}(x) - \liminf_{\epsilon\to0}\tilde{f}_{\epsilon}(x)$$
$$= \limsup_{\epsilon\to0}\tilde{h}_{\epsilon}(x) - \liminf_{\epsilon\to0}\tilde{h}_{\epsilon}(x) \leq 2\sup_{\epsilon>0}|\tilde{h}_{\epsilon}(x)| \tag{2.88}$$

By virtue of (2.80),(2.88) we get

$$\int_{-\infty}^\infty |\limsup_{\epsilon\to0}\tilde{f}_{\epsilon}(x) - \liminf_{\epsilon\to0}\tilde{f}_{\epsilon}(x)|^p dx \leq 2C_p\int_{-\infty}^\infty |h(x)|^p dx.$$

The right hand side of this inequality can be made arbitrarily small by choosing g appropriately. Therefore the integrand of the left hand side vanishes almost everywhere.

2.8   Case of Odd Kernels

In this section we make the following hypothesis.

Assumption 4 $K(x)$ is an odd function satisfying Assumption 1.

If $K(x)$ is an odd function, (2.2) clearly holds.

Theorem 2. 6 *Suppose $K(x)$ is a kernel satisfying Assumption 4. For $f \in L^p(R^n), 1 < p < \infty, \epsilon > 0$ set*

$$\tilde{f}_{\epsilon}(x) = \int_{|x-y|>\epsilon} K(x-y)f(y)dy.$$

Then

$$\int_{R^n}\sup_{\epsilon>0}|\tilde{f}_{\epsilon}(x)|^p dx \leq \left(\frac{\pi}{2}C_p\right)^p\left(\int_\Sigma |K(\sigma)|d\sigma\right)^p\int_{R^n}|f(x)|^p dx, \tag{2.89}$$

where C_p is the constant in Theorem 2.5. *The singular integral*

$$K * f = \lim_{\epsilon \to 0} \tilde{f}_\epsilon \tag{2.90}$$

exists both almost everywhere and in the strong topology of $L^p(R^n)$. The mapping $f \mapsto K * f$ is a bounded linear transformation from $L^p(R^n)$ to itself.

Proof. As was shown in the proof of Lemma 2.1 we have

$$\tilde{f}_\epsilon(x) = \int_\Sigma K(\sigma) \int_\epsilon^\infty f(x - t\sigma) \frac{dt}{t} d\sigma.$$

Since $K(x)$ is an odd function we get

$$\tilde{f}_\epsilon(x) = \frac{1}{2} \int_\Sigma K(\sigma) \int_{|t|>\epsilon} f(x - t\sigma) \frac{dt}{t} d\sigma.$$

With the aid of Hölder's inequality

$$|\tilde{f}_\epsilon(x)| \leq \frac{1}{2} \int_\Sigma |K(\sigma)|^{1/p'} |K(\sigma)|^{1/p} \left| \int_{|t|>\epsilon} f(x - t\sigma) \frac{dt}{t} \right| d\sigma$$

$$\leq \frac{1}{2} \left(\int_\Sigma |K(\sigma)| d\sigma \right)^{1/p'} \left(\int_\Sigma |K(\sigma)| \left| \int_{|t|>\epsilon} f(x - t\sigma) \frac{dt}{t} \right|^p d\sigma \right)^{1/p}.$$

Hence

$$\sup_{\epsilon>0} |\tilde{f}_\epsilon(x)|^p \leq \frac{1}{2^p} \left(\int_\Sigma |K(\sigma)| d\sigma \right)^{p-1} \int_\Sigma |K(\sigma)| \sup_{\epsilon>0} \left| \int_{|t|>\epsilon} f(x - t\sigma) \frac{dt}{t} \right|^p d\sigma.$$

Integrating both sides over R^n

$$\int_{R^n} \sup_{\epsilon>0} |\tilde{f}_\epsilon(x)|^p dx \leq \frac{1}{2^p} \left(\int_\Sigma |K(\sigma)| d\sigma \right)^{p-1}$$

$$\times \int_\Sigma |K(\sigma)| \int_{R^n} \sup_{\epsilon>0} \left| \int_{|t|>\epsilon} f(x - t\sigma) \frac{dt}{t} \right|^p dx d\sigma. \tag{2.91}$$

Making the change of the variables $x = y + s\sigma, s \in R, y_1\sigma_1 + \cdots + y_n\sigma_n = 0$ for a fixed $\sigma \in \Sigma$ and applying Theorem 2.5

$$\int_{R^n} \sup_{\epsilon>0} \left| \int_{|t|>\epsilon} f(x - t\sigma) \frac{dt}{t} \right|^p dx$$

$$= \int_{R^{n-1}} \int_{-\infty}^{\infty} \sup_{\epsilon > 0} \left| \int_{|t| > \epsilon} f(y + (s-t)\sigma) \frac{dt}{t} \right|^p ds dy$$

$$\leq (\pi C_p)^p \int_{R^{n-1}} \int_{-\infty}^{\infty} |f(y + s\sigma)|^p ds dy = (\pi C_p)^p \int_{R^n} |f(x)|^p dx.$$

Combining this with (2.91)

$$\int_{R^n} \sup_{\epsilon > 0} |\tilde{f}_\epsilon(x)|^p dx \leq \left(\frac{\pi}{2} C_p \right)^p \left(\int_\Sigma |K(\sigma)| d\sigma \right)^p \int_{R^n} |f(x)|^p dx.$$

Thus (2.89) has been shown. Let f be a function in $C_0^1(R^n)$. If $f(x) = 0$ for $|x| > M$, then $\tilde{f}_\epsilon(x)$ is independent of ϵ for $|x| \geq M + 1, 0 < \epsilon < 1$. If $|x| < M + 1$

$$\tilde{f}_\epsilon(x) = \int_{\epsilon < |x-y| < 2M+1} K(x - y) f(y) dy$$

$$= \int_{\epsilon < |x-y| < 2M+1} K(x - y)(f(y) - f(x)) dy.$$

If we set $L = \max |\nabla f(x)|$, then

$$\int_{|x-y| < \epsilon} |K(x - y)(f(y) - f(x))| dy$$

$$\leq L \int_{|x-y| < \epsilon} |x - y||K(x - y)| dy = \epsilon L \int_\Sigma |K(\sigma)| d\sigma.$$

Hence, $\tilde{f}_\epsilon$ converges uniformly to some function $\tilde{f}$ as $\epsilon \to 0$. By virtue of (2.89)

$$\int_{R^n} |\tilde{f}(x)|^p dx \leq \int_{R^n} \sup_{\epsilon > 0} |\tilde{f}_\epsilon(x)|^p dx < \infty.$$

Consequently $\tilde{f} \in L^p(R^n)$ and

$$\|\tilde{f}_\epsilon - \tilde{f}\|_p^p = \int_{|x| < M+1} |\tilde{f}_\epsilon(x) - \tilde{f}(x)|^p dx \to 0.$$

In case f is an arbitrary element of $L^p(R^n)$, let $g \in C_0^1(R^n)$ and $h = f - g$. As in the last part of the proof of Theorem 2.5 we can show that $\tilde{f}_\epsilon$ converges almost everywhere to some function which we denote by $K * f$. The first term of the right hand side of

$$\|\tilde{f}_\epsilon - \tilde{f}_\delta\|_p \leq \|\tilde{g}_\epsilon - \tilde{g}_\delta\|_p + \|\tilde{h}_\epsilon - \tilde{h}_\delta\|_p$$

tends to 0 as $\epsilon, \delta \to 0$. Noting in addition that

$$\|\tilde{h}_\epsilon - \tilde{h}_\delta\|_p \leq 2 \left(\int_{R^n} \sup_{\epsilon>0} |\tilde{h}_\epsilon(x)|^p dx \right)^{1/p} \leq \pi C_p \int_\Sigma |K(\sigma)| d\sigma \|h\|_p,$$

we see that $\{\tilde{f}_\epsilon\}$ converges in $L^p(R^n)$ as $\epsilon \to 0$. By virtue of (2.89) we have

$$\|K * f\|_p \leq \frac{\pi}{2} C_p \int_\Sigma |K(\sigma)| d\sigma \|f\|_p.$$

2.9 Riesz Kernels

The *Riesz kernels* are functions defined by

$$R_j(x) = -\frac{\Gamma((n+1)/2)}{\pi^{(n+1)/2}} \frac{x_j}{|x|^{n+1}} \tag{2.92}$$

for $j = 1, 2. \ldots, n$. Clearly the Riesz kernels satisfy Assumptions 3 and 4.

We calculate the Fourier transforms of the Riesz kernels. Let $0 < \epsilon < 1 < \mu$. With the aid of an integration by parts

$$(1-n) \int_{\epsilon<|x|<1} e^{-ix\xi} \frac{x_j}{|x|^{n+1}} dx = \int_{\epsilon<|x|<1} e^{-ix\xi} \frac{\partial}{\partial x_j} |x|^{1-n} dx$$

$$= \int_{|x|=1} e^{-ix\xi} \frac{x_j}{|x|^n} dS - \int_{|x|=\epsilon} e^{-ix\xi} \frac{x_j}{|x|^n} dS + \int_{\epsilon<|x|<1} i\xi_j e^{-ix\xi} |x|^{1-n} dx$$

$$= \int_{|x|=1} x_j e^{-ix\xi} dS - \epsilon^{-n} \int_{|x|=\epsilon} x_j e^{-ix\xi} dS + i\xi_j \int_{\epsilon<|x|<1} e^{-ix\xi} |x|^{1-n} dx. \tag{2.93}$$

As is easily seen

$$\lim_{\epsilon \to 0} \epsilon^{-n} \int_{|x|=\epsilon} x_j e^{-ix\xi} dS = \lim_{\epsilon \to 0} \int_\Sigma \sigma_j e^{-i\epsilon\sigma\xi} d\sigma = \int_\Sigma \sigma_j d\sigma = 0. \tag{2.94}$$

With the aid of an orthogonal transformation which maps ξ to $(|\xi|, 0, \ldots, 0)$

$$\int_{|x|=1} x_j e^{-ix\xi} dS = i \frac{\partial}{\partial \xi_j} \int_{|x|=1} e^{-ix\xi} dS = i \frac{\partial}{\partial \xi_j} \int_{|x|=1} \cos(x\xi) dS$$

$$= i \frac{\partial}{\partial \xi_j} \int_{|x|=1} \cos(|\xi|x_1) dS = -i \frac{\xi_j}{|\xi|} \int_{|x|=1} x_1 \sin(|\xi|x_1) dS, \tag{2.95}$$

$$\lim_{\epsilon \to 0} \int_{\epsilon<|x|<1} e^{-ix\xi} |x|^{1-n} dx = \int_{|x|<1} e^{-ix\xi} |x|^{1-n} dx$$

$$= \int_{|x|<1} |x|^{1-n} \cos(x\xi) dx = \int_{|x|<1} |x|^{1-n} \cos(|\xi|x_1) dx. \tag{2.96}$$

From (2.93),(2.94),(2.95),(2.96) it follows that

$$\lim_{\epsilon \to 0} \int_{\epsilon < |x| < 1} e^{-ix\xi} \frac{x_j}{|x|^{n+1}} dx = \frac{1}{1-n} \left(-i\frac{\xi_j}{|\xi|} \int_{|x|=1} x_1 \sin(|\xi||x_1|) dS \right.$$

$$\left. +i\xi_j \int_{|x|<1} |x|^{1-n} \cos(|\xi||x_1|) dx \right). \tag{2.97}$$

Hence, if we set

$$c_1 = \frac{i}{1-n} \left(-\int_{|x|=1} x_1 \sin x_1 \, dS + \int_{|x|<1} |x|^{1-n} \cos x_1 \, dx \right),$$

the equality

$$\lim_{\epsilon \to 0} \int_{\epsilon < |x| < 1} e^{-ix\xi} \frac{x_j}{|x|^{n+1}} dx = c_1 \xi_j \tag{2.98}$$

holds for $|\xi| = 1$. Next

$$\int_{1 < |x| < \mu} e^{-ix\xi} \frac{x_j}{|x|^{n+1}} dx = i\frac{\partial}{\partial \xi_j} \int_{1 < |x| < \mu} \frac{e^{-ix\xi}}{|x|^{n+1}} dx. \tag{2.99}$$

Evidently

$$\lim_{\mu \to \infty} \int_{1 < |x| < \mu} \frac{e^{-ix\xi}}{|x|^{n+1}} dx = \int_{|x|>1} \frac{e^{-ix\xi}}{|x|^{n+1}} dx \tag{2.100}$$

uniformly in R^n. If $\xi_j \neq 0$

$$\int_{1 < |x| < \mu} e^{-ix\xi} \frac{x_j}{|x|^{n+1}} dx = \frac{i}{\xi_j} \int_{1 < |x| < \mu} \frac{x_j}{|x|^{n+1}} \frac{\partial}{\partial x_j} e^{-ix\xi} dx$$

$$= \frac{i}{\xi_j} \left(\int_{|x|=\mu} e^{-ix\xi} \frac{x_j^2}{|x|^{n+2}} dS \right.$$

$$- \int_{|x|=1} e^{-ix\xi} \frac{x_j^2}{|x|^{n+2}} dS - \int_{1 < |x| < \mu} e^{-ix\xi} \frac{\partial}{\partial x_j} \frac{x_j}{|x|^{n+1}} dx \right)$$

$$= \frac{i}{\xi_j} \left(\frac{1}{\mu} \int_{|x|=1} x_j^2 e^{-i\mu x\xi} dS \right.$$

$$- \int_{|x|=1} x_j^2 e^{-ix\xi} dS - \int_{1 < |x| < \mu} e^{-ix\xi} \frac{\partial}{\partial x_j} \frac{x_j}{|x|^{n+1}} dx \right).$$

Hence

$$\lim_{\mu \to \infty} \int_{1 < |x| < \mu} e^{-ix\xi} \frac{x_j}{|x|^{n+1}} dx$$

$$= \frac{i}{\xi_j} \left(-\int_{|x|=1} x_j^2 e^{-ix\xi} dS - \int_{|x|>1} e^{-ix\xi} \frac{\partial}{\partial x_j} \frac{x_j}{|x|^{n+1}} dx \right) \quad (2.101)$$

uniformly on any compact subset of $\{\xi \in R^n; \xi_j \neq 0\}$. By virtue of (2.99),(2.100),(2.101)

$$\int_{|x|>1} \frac{e^{-ix\xi}}{|x|^{n+1}} dx \in C^1(\{\xi \in R^n; \xi_1 \cdots \xi_n \neq 0\}), \quad (2.102)$$

$$\lim_{\mu \to \infty} \int_{1<|x|<\mu} e^{-ix\xi} \frac{x_j}{|x|^{n+1}} dx = i\frac{\partial}{\partial \xi_j} \int_{|x|>1} \frac{e^{-ix\xi}}{|x|^{n+1}} dx \quad (2.103)$$

for $j = 1, \ldots, n$. Set

$$f(|\xi|) = \int_{|x|>1} \frac{e^{-ix\xi}}{|x|^{n+1}} dx = \int_{|x|>1} \frac{\cos(x\xi)}{|x|^{n+1}} dx = \int_{|x|>1} \frac{\cos(|\xi|x_1)}{|x|^{n+1}} dx.$$

Then $f(|\xi|)$ is of class C^1 in $\{\xi; \xi_1 \cdots \xi_n \neq 0\}$. Hence $f \in C^1((0, \infty))$, and $f(|\xi|)$ is continuously differentiable in $R^n \setminus \{0\}$. In view of (2.103)

$$\lim_{\mu \to \infty} \int_{1<|x|<\mu} e^{-ix\xi} \frac{x_j}{|x|^{n+1}} dx = i\frac{\partial}{\partial \xi_j} f(|\xi|) = i\frac{\xi_j}{|\xi|} f'(|\xi|).$$

Hence if we set $c_2 = if'(1)$ we see that

$$\lim_{\mu \to \infty} \int_{1<|x|<\mu} e^{-ix\xi} \frac{x_j}{|x|^{n+1}} dx = c_2\xi_j \quad (2.104)$$

holds when $|\xi| = 1$. From (2.98) and (2.104) it follows that

$$\lim_{\epsilon \to 0, \mu \to \infty} \int_{\epsilon<|x|<\mu} e^{-ix\xi} \frac{x_j}{|x|^{n+1}} dx = c_3\xi_j \quad (2.105)$$

holds when $|\xi| = 1$ where $c_3 = c_1 + c_2$. Since the left hand side of (2.105) is a homogeneous function of ξ of degree 0 we have

$$\lim_{\epsilon \to 0, \mu \to \infty} \int_{\epsilon<|x|<\mu} e^{-ix\xi} \frac{x_j}{|x|^{n+1}} dx = c_3\frac{\xi_j}{|\xi|} \quad (2.106)$$

when $\xi \neq 0$. In order to obtain the value of c_3 we let $j = 1, \xi = (1, 0, \ldots, 0)$. Then

$$\lim_{\epsilon \to 0, \mu \to \infty} \int_{\epsilon<|x|<\mu} e^{-ix_1} \frac{x_1}{|x|^{n+1}} dx = c_3, \quad (2.107)$$

$$\int_{\epsilon<|x|<\mu} e^{-ix_1} \frac{x_1}{|x|^{n+1}} dx$$

$$= -i \int_{\epsilon<|x|<\mu} \frac{x_1 \sin x_1}{|x|^{n+1}} dx = -i \int_\Sigma \sigma_1 \int_\epsilon^\mu \frac{\sin(r\sigma_1)}{r} dr d\sigma. \quad (2.108)$$

Since

$$\lim_{\epsilon \to 0, \mu \to \infty} \int_\epsilon^\mu \frac{\sin r\sigma_1}{r} dr = \frac{\pi}{2}\text{sign}\sigma_1, \quad \left| \int_\epsilon^\mu \frac{\sin(r\sigma_1)}{r} dr \right| \le 2A,$$

we get from (2.108)

$$\lim_{\epsilon \to 0, \mu \to \infty} \int_{\epsilon < |x| < \mu} e^{-ix_1} \frac{x_1}{|x|^{n+1}} dx = -\frac{\pi}{2} i \int_\Sigma \sigma_1 \text{sign}\sigma_1 d\sigma = -\pi i \int_{\sigma_1 > 0} \sigma_1 d\sigma.$$

If we denote the area of the unit sphere of R^n by Ω_n,

$$\int_{\sigma_1 > 0} \sigma_1 d\sigma = \Omega_n \int_0^1 t(1-t^2)^{(n-3)/2} dt$$

$$= \frac{1}{2}\Omega_n \int_0^1 (1-s)^{(n-3)/2} ds = \frac{\Omega_n}{n-1}.$$

Hence

$$c_3 = -\pi i \frac{\Omega_n}{n-1} = -\frac{\pi i}{n-1} \frac{2\pi^{(n-1)/2}}{\Gamma((n-1)/2)} = -\frac{\pi^{(n+1)/2} i}{\Gamma((n+1)/2)}.$$

Thus the Fourier transform of the Riesz kernels are

$$(\mathcal{F}R_j)(\xi) = (2\pi)^{-n/2} i \frac{\xi_j}{|\xi|}. \tag{2.109}$$

Let $f \in L^2(R^n)$ and

$$g_j(x) = -\lim_{\epsilon \to 0} \int_{|x-y| > \epsilon} R_j(x-y)f(y)dy = -(R_j * f)(x).$$

Then in view of (2.109)

$$(\mathcal{F}\sum_{j=1}^n R_j * g_j)(\xi) = (2\pi)^{n/2} \sum_{j=1}^n (\mathcal{F}R_j)(\xi)(\mathcal{F}g_j)(\xi)$$

$$= -(2\pi)^n \sum_{j=1}^n (\mathcal{F}R_j)(\xi)^2 (\mathcal{F}f)(\xi) = (\mathcal{F}f)(\xi).$$

Hence

$$-\sum_{j=1}^n R_j * (R_j * f) = f. \tag{2.110}$$

By virtue of Theorem 2.3 we see that (2.110) also holds for $f \in L^p(R^n)$, $1 < p < \infty$. Thus we have established the following theorem.

Theorem 2.7 *The Fourier transforms of the Riesz kernels are expressed as (2.109). The equality (2.110) holds for $f \in L^p(R^n)$, $1 < p < \infty$.*

Riesz kernels are used to transform even kernels to odd kernels so that we are able to establish the boundedness of operators with general kernels with the aid of Theorem 2.6.

2.10   Case of Even Kernels

In this section we make the following assumption.

Assumption 5 *$K(x)$ is an even function satisfying Assumption 1 and*

$$\int_\Omega |K(\sigma)| \log^+ |K(\sigma)| d\sigma < \infty. \tag{2.111}$$

Here

$$\log^+ x = \begin{cases} \log x & x > 1 \\ 0 & 0 < x \le 1 \end{cases}. \tag{2.112}$$

Lemma 2.12 *For $a > 0, b > 0$*

$$ab \le a \log^+ a + e^{b-1}. \tag{2.113}$$

Proof. Let $\phi(x), \psi(y)$ be functions defined by

$$\phi(x) = \begin{cases} \log x + 1 & x > 1 \\ 0 & 0 \le x \le 1 \end{cases}, \quad \psi(y) = \begin{cases} e^{y-1} & y > 1 \\ 1 & 0 \le y \le 1 \end{cases}.$$

Since ψ is the inverse function of ϕ in the region $x > 1, y > 1$, we get following the proof of Lemma 1.1 that

$$ab \le \int_0^a \phi(x) dx + \int_0^b \psi(y) dy.$$

The inequality (2.113) follows from this inequality and

$$\int_0^a \phi(x) dx = a \log^+ a, \quad \int_0^b \psi(y) dy \le e^{b-1}.$$

Lemma 2.13 *Let $K(x)$ be a kernel satisfying Assumption 3. Suppose f is a measurable function with compact support in R^n satisfying*

$$\int_{R^n} |f(x)| \log^+ |f(x)| dx < \infty. \tag{2.114}$$

If S is a measurable set of R^n with $|\text{supp} f| < |S| < \infty$, where $\text{supp} f$ stands for the support of f, then there exist constants C_3, C_4 such that for $\epsilon > 0$

$$\int_S |\tilde{f}_\epsilon(x)| dx \leq C_3 \int_{R^n} |f(x)| \log^+ |f(x)| dx + C_4 |S|, \qquad (2.115)$$

where $\tilde{f}_\epsilon$ is the function defined by (2.4).

Proof. Since

$$\int_{R^n} |f(x)| dx = \int_{|f|>e} |f(x)| dx + \int_{|f|\leq e} |f(x)| dx$$

$$\leq \int_{R^n} |f(x)| \log^+ |f(x)| dx + e |\text{supp} f|, \qquad (2.116)$$

we see that $f \in L^1(R^n)$. We begin with the case $f \geq 0$. Set

$$E_s = \{x; |\tilde{f}_\epsilon(x)| > s\}, \quad E'_s = E_s \cap S.$$

Let β_f, β^f be the functions defined in section 6 and $s_0 = \beta_f(|S|)$. Then by Lemma 2.2

$$\int_S |\tilde{f}_\epsilon(x)| dx = \int_0^\infty |E'_s| ds \leq |S| s_0 + \int_{s_0}^\infty |E_s| ds. \qquad (2.117)$$

In view of Lemma 2.7

$$|S| s_0 = |S| \beta_f(|S|) = \int_0^{|S|} f^*(s) ds \leq \int_{R^n} f(x) dx. \qquad (2.118)$$

From Lemmas 2.2 and 2.8 we get $|D_s| \leq \beta^f(s)$. In view of this and Lemma 2.4

$$\int_{s_0}^\infty |E_s| ds \leq C_1 \int_0^\infty \frac{1}{s^2} \int_{R^n} [f(x)]_s^2 dx ds + C_2 \int_{s_0}^\infty \beta^f(s) ds, \qquad (2.119)$$

$$\int_0^\infty \frac{1}{s^2} \int_{R^n} [f(x)]_s^2 dx ds = \int_{R^n} \int_0^\infty \frac{1}{s^2} [f(x)]_s^2 ds dx$$

$$= \int_{R^n} \left[\int_0^{f(x)} ds + \int_{f(x)}^\infty \frac{1}{s^2} f(x)^2 ds \right] dx = 2 \int_{R^n} f(x) dx. \qquad (2.120)$$

Let $s_1 = \beta_f(t_1)$ be the number defined in section 6. If we make the change of the variable $t = \beta^f(s)$, then noting $s = \beta_f(t), \beta^f(s_0) = |S|, \beta^f(s_1) = t_1 \leq |\text{supp} f| < |S|$ we get

$$\int_{s_0}^\infty \beta^f(s) ds = \int_{s_0}^{s_1} \beta^f(s) ds = \int_{|S|}^{t_1} t d\beta_f(t)$$

$$= [t\beta_f(t)]_{|S|}^{t_1} + \int_{t_1}^{|S|} \beta_f(t)dt = \left[\int_0^t f^*(s)ds\right]_{|S|}^{t_1} + \int_{t_1}^{|S|} \beta_f(t)dt$$

$$= \int_0^{t_1} f^*(s)ds - \int_0^{|S|} f^*(s)ds + \int_{t_1}^{|S|} \beta_f(t)dt.$$

Since $f^*(s) = \beta_f(s) = \beta_f(t_1)$ for $0 < s \le t_1$, the last side of the above equalities

$$= \int_0^{|S|} \beta_f(t)dt - \int_0^{|S|} f^*(s)ds \le \int_0^{|S|} \beta_f(t)dt$$

$$= \int_0^{|S|} \frac{1}{t} \int_0^t f^*(s)ds dt = \int_0^{|S|} f^*(s) \int_s^{|S|} \frac{dt}{t} ds$$

$$= \int_0^{|S|} f^*(t) \log \frac{|S|}{t} dt = 2 \int_0^{|S|} f^*(t) \log \left(\frac{|S|}{t}\right)^{1/2} dt.$$

With the aid of Lemma 2.12 we can show that this does not exceed

$$2 \int_0^{|S|} \left[f^*(t) \log^+ f^*(t) + \exp\left(\log \left(\frac{|S|}{t}\right)^{1/2} - 1 \right) \right] dt$$

$$\le 2 \int_0^{\infty} f^*(t) \log^+ f^*(t)dt + \frac{2}{e} \int_0^{|S|} \left(\frac{|S|}{t}\right)^{1/2} dt$$

$$= 2 \int_{R^n} f(x) \log^+ f(x)dx + \frac{4}{e}|S|.$$

From this inequality and (2.116),(2.117),(2.118),(2.119),(2.120) we obtain

$$\int_S |\tilde{f}_\epsilon(x)|dx$$

$$\le (1 + 2C_1 + 2C_2) \int_{R^n} f(x) \log^+ f(x)dx + \left((1 + 2C_1)e + \frac{4C_2}{e} \right) |S|,$$

In the general case dividing f into its positive and negative parts we can show that (2.115) holds with

$$C_3 = 1 + 2C_1 + 2C_2, \quad C_4 = 2(1 + 2C_1)e + 8C_2/e.$$

Lemma 2.14 *For $\alpha > 0, \beta > 0$*

$$\log^+(\alpha\beta) \le \log^+ \alpha + \log^+ \beta, \tag{2.121}$$

$$(\alpha + \beta) \log^+ \frac{\alpha + \beta}{2} \le \alpha \log^+ \alpha + \beta \log^+ \beta. \tag{2.122}$$

Proof. If $\alpha\beta \leq 1$, (2.121) clearly holds. If $\alpha\beta > 1$,

$$\log^+(\alpha\beta) = \log(\alpha\beta) = \log\alpha + \log\beta \leq \log^+\alpha + \log^+\beta.$$

The inequality (2.122) follows from the fact that $x\log^+ x$ is a convex function.

Lemma 2.15 *Suppose that $K(x)$ is a kernel satisfying Assumption 3. If f is a function satisfying the hypotheses of Lemma 2.13, then $\tilde{f}_\epsilon$ converges in $L^1_{loc}(R^n)$ as $\epsilon \to 0$.*

Proof. Let S be a bounded measurable set with $|\mathrm{supp} f| < |S|$. For $k > 0$ set

$$[f(x)]_k = \begin{cases} f(x) & |f(x)| \leq k \\ k\,\mathrm{sign} f(x) & |f(x)| > k \end{cases}.$$

For $0 < \delta < 2$ choose k so that

$$\int_{R^n} |f - [f]_k| \log^+ |f - [f]_k| dx < \frac{\delta}{4}, \tag{2.123}$$

$$\int_{R^n} |f - [f]_k| dx < \frac{\delta}{4}\frac{1}{\log^+(2/\delta)}. \tag{2.124}$$

With the aid of Lemma 2.14 and (2.123),(2.124)

$$\int_{R^n} \left|\frac{f - [f]_k}{\delta/2}\right| \log^+ \left|\frac{f - [f]_k}{\delta/2}\right| dx \leq \int_{R^n} \left|\frac{f - [f]_k}{\delta/2}\right| \log^+ |f - [f]_k| dx$$

$$+ \int_{R^n} \left|\frac{f - [f]_k}{\delta/2}\right| \log^+ \frac{2}{\delta} dx < 1. \tag{2.125}$$

Let g be a function in $C_0^1(R^n)$ such that

$$|\mathrm{supp} g| < |S|, \quad \int_{R^n} |g - [f]_k|^2 dx < \left(\frac{\delta}{2}\right)^2.$$

Noting $\alpha\log^+\alpha \leq \alpha^2$

$$\int_{R^n} \left|\frac{g - [f]_k}{\delta/2}\right| \log^+ \left|\frac{g - [f]_k}{\delta/2}\right| dx \leq \int_{R^n} \left|\frac{g - [f]_k}{\delta/2}\right|^2 dx < 1. \tag{2.126}$$

Applying (2.122) to $\alpha = |f - [f]_k|/(\delta/2)$, $\beta = |[f]_k - g|/(\delta/2)$

$$\frac{|f - g|}{\delta/2} \log^+ \frac{|f - g|}{\delta} \leq (\alpha + \beta) \log^+ \frac{\alpha + \beta}{2} \leq \alpha \log^+\alpha + \beta \log^+\beta$$

$$\leq \frac{|f - [f]_k|}{\delta/2} \log^+ \frac{|f - [f]_k|}{\delta/2} + \frac{|[f]_k - g|}{\delta/2} \log^+ \frac{|[f]_k - g|}{\delta/2}. \tag{2.127}$$

Set $h = f - g$. From (2.125),(2.126),(2.127)

$$\int_{R^n} \frac{|h|}{\delta/2} \log^+ \frac{|h|}{\delta} dx < 2. \qquad (2.128)$$

As was shown in the proof of Theorem 2.3 $\tilde{g}_\epsilon(x)$ converges uniformly as $\epsilon \to 0$. Since $|\text{supp} h| < |S|$ we get from Lemma 2.13 and (2.128)

$$\int_S \left|\frac{\tilde{h}_\epsilon}{\delta}\right| dx \le C_3 \int_{R^n} \left|\frac{h}{\delta}\right| \log^+ \left|\frac{h}{\delta}\right| dx + C_4|S| \le C_3 + C_4|S|.$$

Therefore

$$\int_S |\tilde{f}_\epsilon - \tilde{f}_{\epsilon'}| dx \le \int_S |\tilde{g}_\epsilon - \tilde{g}_{\epsilon'}| dx + \int_S |\tilde{h}_\epsilon| dx + \int_S |\tilde{h}_{\epsilon'}| dx$$

$$\le \int_S |\tilde{g}_\epsilon - \tilde{g}_{\epsilon'}| dx + 2\delta(C_3 + C_4|S|).$$

Thus we conclude

$$\lim_{\epsilon,\epsilon' \to 0} \int_S |\tilde{f}_\epsilon - \tilde{f}_{\epsilon'}| dx = 0.$$

In what follows we suppose that $K(x)$ is a kernel satisfying Assumption 5. Let $\phi(t)$ be a continuously differentiable function satisfying

$$\phi(t) = \begin{cases} 0 & 0 \le t \le 1/4 \\ 1 & t > 3/4 \end{cases}.$$

Let $R(x) = (R_1(x), \ldots, R_n(x))$ be the Riesz kernels. Set

$$K_1(x) = \lim_{\epsilon \to 0} \lim_{\delta \to 0} \int_{|x-y|>\epsilon, |y|>\delta} R(x - y)K(y)dy, \qquad (2.129)$$

$$K_2(x) = \lim_{\epsilon \to 0} \int_{|x-y|>\epsilon} R(x - y)K(y)\phi(|y|)dy. \qquad (2.130)$$

Lemma 2.16 *The right hand side of (2.129) converges in $L^1_{loc}(R^n \setminus \{0\})$ and that of (2.130) in $L^1_{loc}(R^n)$. K_1 and K_2 are both odd functions, and K_1 is homogeneous of degree $-n$. If $K \in L^q(\Sigma)$ for some $1 < q < \infty$, then $K_1 \in L^q(\Sigma)$ and for some constant C_q*

$$\int_\Sigma |K_1(\sigma)|^q d\sigma \le C_q \int_\Sigma |K(\sigma)|^q d\sigma. \qquad (2.131)$$

Proof. Suppose $1/2 \le |x| \le 1, 0 < \epsilon < 1/4$. Then

$$\int_{|x-y|>\epsilon} R(x-y)K(y)dy = \lim_{\delta \to 0} \int_{|x-y|>\epsilon, |y|>\delta} R(x-y)K(y)dy$$

$$= \lim_{\delta \to 0} \int_{\delta<|y|<1/4} R(x-y)K(y)dy + \int_{|x-y|>\epsilon, 1/4<|y|<2} R(x-y)K(y)dy$$

$$+ \int_{|y|>2} R(x-y)K(y)dy = I_1 + I_2 + I_3. \tag{2.132}$$

Since for $|x| \ge 1/2, |y| < 1/4$

$$|R(x-y) - R(x)| \le C|y||x|^{-n-1}, \tag{2.133}$$

we have

$$|I_1| = \left| \int_{|y|<1/4} (R(x-y) - R(x))K(y)dy \right|$$

$$\le \frac{C}{|x|^{n+1}} \int_{|y|<1/4} |y||K(y)|dy = \frac{C}{|x|^{n+1}} \int_{\Sigma} |K(\sigma)|d\sigma. \tag{2.134}$$

Noting $|x-y| \ge |y|/2$ if $1/2 \le |x| \le 1, |y| > 2$, we get

$$|I_3| \le \int_{|y|>2} \frac{C}{|x-y|^n} |K(y)|dy \le C \int_{|y|>2} \frac{|K(y)|}{|y|^n} dy. \tag{2.135}$$

If we set

$$f(x) = \begin{cases} K(x) & 1/4 < |x| < 2 \\ 0 & \text{otherwise} \end{cases},$$

then

$$\int_{R^n} |f(x)| \log^+ |f(x)|dx = \int_{1/4<|x|<2} |K(x)| \log^+ |K(x)|dx$$

$$= \int_{\Sigma} \int_{1/4}^{2} |K(\sigma)| \log^+ \left(t^{-n}|K(\sigma)| \right) \frac{dt}{t} d\sigma$$

$$\le \int_{\Sigma} \int_{1/4}^{2} |K(\sigma)| \left(\log^+ |K(\sigma)| + \log^+ t^{-n} \right) \frac{dt}{t} d\sigma$$

$$= \int_{\Sigma} |K(\sigma)| \log^+ |K(\sigma)|d\sigma \int_{1/4}^{2} \frac{dt}{t} + \int_{\Sigma} |K(\sigma)|d\sigma \int_{1/4}^{2} \log^+ t^{-n} \frac{dt}{t}$$

$$\le C \int_{\Sigma} |K(\sigma)| \log^+ |K(\sigma)|d\sigma + C \int_{\Sigma} |K(\sigma)|d\sigma < \infty.$$

Hence by Lemma 2.15 I_2 converges in $L^1_{loc}(R^n)$ as $\epsilon \to 0$. Consequently we see that (2.132) converges in $L^1(\{x; 1/2 \le |x| \le 1\})$. Next suppose $1 \le |x| \le 2$. Then $1/2 \le |x/2| \le 1$ and

$$\int_{|x-y|>\epsilon} R(x-y)K(y)dy = \frac{1}{2^n} \int_{|x/2-y|>\epsilon/2} R\left(\frac{x}{2} - y\right) K(y)dy.$$

Hence we see that (2.132) converges in $L^1(\{x; 1 \le |x| \le 2\})$. Repeating this process we conclude that (2.132) converges in $L^1_{loc}(R^n \setminus \{0\})$. It is easy to show that K_1 is homogeneous of degree $-n$.

Suppose $K \in L^q(\Sigma), 1 < q < \infty$. Then in view of (2.134),(2.135)

$$\int_{1/2<|x|<1} |I_1|^q dx, \quad \int_{1/2<|x|<1} |I_3|^q dx \le C \int_\Sigma |K(\sigma)|^q d\sigma. \qquad (2.136)$$

By virtue of Theorem 2.3 or 2.6

$$\int_{R^n} |I_2|^q dx \le C_q \int_{1/4<|x|<2} |K(x)|^q dx = C_q \int_\Sigma |K(\sigma)|^q d\sigma. \qquad (2.137)$$

From (2.136),(2.137) it follows that

$$\int_{1/2<|x|<1} |K_1(x)|^q dx \le C_q \int_\Sigma |K(\sigma)|^q d\sigma.$$

From this inequality and the homogeneity of K_1 the inequality (2.131) follows.

Finally we investigate the convergence of (2.130). Let N be an arbitrary positive number. If $|x| < N, 0 < \epsilon < 1$, then

$$\int_{|x-y|>\epsilon} R(x-y)K(y)\phi(|y|)dy$$

$$= \int_{|x-y|>\epsilon,|y|<N+1} R(x-y)K(y)\phi(|y|)dy + \int_{|y|>N+1} R(x-y)K(y)dy$$

$$= I_4 + I_5.$$

Since the function defined by

$$f(x) = \begin{cases} K(x)\phi(|x|) & |x| < N+1 \\ 0 & |x| > N+1 \end{cases}$$

satisfies the assumption of Lemma 2.15, I_4 converges in $L^1_{loc}(R^n)$ as $\epsilon \to 0$. We have

$$|I_5| \le \int_{|y|>N+1} \frac{C}{|x-y|^n} |K(y)|dy$$

$$\le C \int_{|y|>N+1} \frac{(N+1)^n}{|y|^n} |K(y)|dy \le C_N \int_\Sigma |K(\sigma)|d\sigma.$$

Thus we conclude that (2.130) converges in $L^1_{loc}(R^n)$.

Lemma 2.17 *There exists a constant C such that for $|x| \geq 1$*

$$|K_1(x) - K_2(x)| \leq C|x|^{-n-1} \int_\Sigma |K(\sigma)| d\sigma. \qquad (2.138)$$

There exists a function $G(x)$ which is homogeneous of degree 0 and integrable on Σ such that $|K_2(x)| \leq G(x)$ for $|x| \leq 1$. If $K \in L^q(\Sigma)$ for some $1 < q < \infty$, then

$$\int_\Sigma G(\sigma)^q d\sigma \leq C_q \int_\Sigma |K(\sigma)|^q d\sigma \qquad (2.139)$$

for some constant C_q, and (2.130) holds in the strong topology of $L^q(R^n)$.

Proof. In view of Lemma 2.16 $K_1 \in L^1_{loc}(R^n \setminus \{0\})$, $K_2 \in L^1_{loc}(R^n)$, and in $L^1_{loc}(R^n \setminus \{0\})$

$$K_1(x) - K_2(x) = \lim_{\epsilon \to 0} \int_{|x-y|>\epsilon} R(x-y)K(y)(1 - \phi(|y|))dy. \qquad (2.140)$$

Suppose $|x| > 1$. Then if $|y| < 3/4$

$$|R(x-y) - R(x)| \leq C|y||x|^{-n-1}.$$

Hence

$$|K_1(x) - K_2(x)| = \left| \int_{R^n} (R(x-y) - R(x))K(y)(1 - \phi(|y|))dy \right|$$

$$\leq C|x|^{-n-1} \int_{|y|<3/4} |y||K(y)|dy \leq C|x|^{-n-1} \int_\Sigma |K(\sigma)| d\sigma.$$

Thus (2.138) is established.
   If $|x| \leq 1/8$,

$$|K_2(x)| = \left| \int_{|y|>1/4} R(x-y)K(y)\phi(|y|)dy \right|$$

$$\leq C \int_{|y|>1/4} |x-y|^{-n}|K(y)|dy$$

$$\leq C \int_{|y|>1/4} |y|^{-n}|K(y)|dy \leq C \int_\Sigma |K(\sigma)| d\sigma. \qquad (2.141)$$

If $1/8 \leq |x| \leq 1$, letting χ be the characteristic function of the interval $[0,1]$

$$\int_{|x-y|>\epsilon} R(x-y)K(y)\phi(|y|)dy = \phi(|x|) \int_{|x-y|>\epsilon} R(x-y)K(y)dy$$

$$+ \int_{|x-y|>\epsilon} R(x-y)K(y)(\phi(|y|) - \phi(|x|))dy$$

$$= \phi(|x|) \int_{|x-y|>\epsilon} R(x-y)K(y)dy$$

$$+ \int_{|x-y|>\epsilon} (R(x-y) - \chi(|y|)R(x))K(y)(\phi(|y|) - \phi(|x|))dy$$

$$+R(x) \int_{|x-y|>\epsilon} \chi(|y|)K(y)(\phi(|y|) - \phi(|x|))dy. \qquad (2.142)$$

If $0 < \epsilon < 1/16$

$$\int_{|x-y|>\epsilon} \chi(|y|)K(y)(\phi(|y|) - \phi(|x|))dy$$

$$= \int_{|y|<1/16} K(y)(\phi(|y|) - \phi(|x|))dy$$

$$+ \int_{|x-y|>\epsilon, 1/16<|y|<1} K(y)(\phi(|y|) - \phi(|x|))dy$$

$$= \int_{|x-y|>\epsilon, 1/16<|y|<1} K(y)(\phi(|y|) - \phi(|x|))dy$$

$$\to \int_{1/16<|y|<1} K(y)(\phi(|y|) - \phi(|x|))dy = 0$$

as $\epsilon \to 0$. Hence in view of (2.142) if $1/8 \leq |x| \leq 1$

$$K_2(x) = \phi(|x|)K_1(x) + \int_{R^n} (R(x-y) - \chi(|y|)R(x))K(y)(\phi(|y|) - \phi(|x|))dy. \qquad (2.143)$$

If $1/8 \leq |x| \leq 1, |y| \leq 1$, then $|x-y| \leq 9|x|$ and hence

$$|R(x-y) - \chi(|y|)R(x)| = |R(x-y) - R(x)| \leq C|y||x-y|^{-n-1},$$
$$|R(x-y) - \chi(|y|)R(x)| \leq |R(x-y)| + |R(x)| \leq C|x-y|^{-n}.$$

Therefore

$$|R(x-y) - \chi(|y|)R(x)|$$
$$\leq C(|y||x-y|^{-n-1})^{1/2}|x-y|^{-n/2} = C|y|^{1/2}|x-y|^{-n-1/2} \qquad (2.144)$$

if $1/8 \leq |x| \leq 1, |y| \leq 1$. If $1/8 \leq |x| \leq 1, |y| > 1$, then $|x-y| \leq 2|y|$, and hence

$$|R(x-y) - \chi(|y|)R(x)| = |R(x-y)|$$
$$= C|x-y|^{-n} \leq C|y|^{1/2}|x-y|^{-n-1/2}. \qquad (2.145)$$

Combining (2.144) and (2.145) we get that if $1/8 \le |x| \le 1$

$$|R(x-y) - \chi(|y|)R(x)| \le C|y|^{1/2}|x-y|^{-n-1/2}.$$

Therefore in view of (2.143)

$$|K_2(x)| \le |K_1(x)| + C \int_{R^n} |y|^{1/2}|K(y)||x-y|^{-n+1/2}dy$$

$$\le 8^n|x|^n|K_1(x)| + 8^{n-1}C|x|^{n-1} \int_{R^n} |y|^{1/2}|K(y)||x-y|^{-n+1/2}dy$$

$$(2.146)$$

holds for $1/8 \le |x| \le 1$. Now we show that

$$\int_{1/2<|x|<3/2} \int_{R^n} |y|^{1/2}|K(y)||x-y|^{-n+1/2}dydx < \infty. \qquad (2.147)$$

From this result it follows that the last integral of (2.146) is finite for almost every x. We have

$$\int_{1/2<|x|<3/2} \int_{1/4<|y|<3} |y|^{1/2}|K(y)||x-y|^{-n+1/2}dydx$$

$$\le \int_{1/4<|y|<3} |y|^{1/2}|K(y)| \int_{|x-y|<9/2} |x-y|^{-n+1/2}dxdy$$

$$\le C \int_{1/4<|y|<3} |y|^{1/2}|K(y)|dy < \infty,$$

$$\int_{1/2<|x|<3/2} \int_{|y|<1/4} |y|^{1/2}|K(y)||x-y|^{-n+1/2}dydx$$

$$\le 4^{n-1/2} \int_{1/2<|x|<3/2} dx \int_{|y|<1/4} |y|^{1/2}|K(y)|dy < \infty,$$

$$\int_{1/2<|x|<3/2} \int_{|y|>3} |y|^{1/2}|K(y)||x-y|^{-n+1/2}dydx$$

$$\le 2^{n-1/2} \int_{1/2<|x|<3/2} dx \int_{|y|>3} |y|^{1-n}|K(y)|dy < \infty.$$

Combining these three inequalities we obtain (2.147). If we set

$$G(x) = C_0 \left(\int_\Sigma |K(\sigma)|d\sigma + |x|^n|K_1(x)| \right.$$

$$\left. + |x|^{n-1} \int_{R^n} |y|^{1/2}|K(y)||x-y|^{-n+1/2}dy \right) \qquad (2.148)$$

with some suitable constant C_0, then $G(x)$ is homogeneous of degree 0 and by virtue of (2.141),(2.146) $|K_2(x)| \leq G(x)$ holds for $|x| \leq 1$. Furthermore by (2.147) we have

$$\int_{1/2<|x|<3/2} G(x)dx < \infty, \tag{2.149}$$

from which $G \in L^1(\Sigma)$ follows.

Finally suppose $K \in L^q(\Sigma)$ for $1 < q < \infty$. In order to show (2.139) it suffices to investigate the last term in the bracket of (2.148). Integrating both sides of

$$\left(\int_{|y|<1/4} |y|^{1/2}|K(y)||x-y|^{-n+1/2}dy\right)^q$$

$$\leq \left(\int_{|y|<1/4} |y|^{1/2}|K(y)|dy\right)^{q-1} \int_{|y|<1/4} |y|^{1/2}|K(y)||x-y|^{(-n+1/2)q}dy$$

over $1/2 < |x| < 3/2$ we get

$$\int_{1/2<|x|<3/2} \left(\int_{|y|<1/4} |y|^{1/2}|K(y)||x-y|^{-n+1/2}dy\right)^q dx$$

$$\leq \left(\int_{|y|<1/4} |y|^{1/2}|K(y)|dy\right)^q \int_{1/4<|x|<7/4} |x|^{(-n+1/2)q}dx$$

$$\leq C\left(\int_\Sigma |K(\sigma)|d\sigma\right)^q \leq C\int_\Sigma |K(\sigma)|^q d\sigma. \tag{2.150}$$

Since for $1/2 < |x| < 3/2$

$$\left(\int_{1/4<|y|<3} |y|^{1/2}|K(y)||x-y|^{-n+1/2}dy\right)^q$$

$$\leq \int_{1/4<|y|<3} \left(|y|^{1/2}|K(y)|\right)^q |x-y|^{-n+1/2}dy$$

$$\times \left(\int_{1/4<|y|<3} |x-y|^{-n+1/2}dy\right)^{q-1}$$

$$\leq \int_{1/4<|y|<3} \left(|y|^{1/2}|K(y)|\right)^q |x-y|^{-n+1/2}dy \left(\int_{|y|<9/2} |y|^{-n+1/2}dy\right)^{q-1}$$

we have

$$\int_{1/2<|x|<3/2} \left(\int_{1/4<|y|<3} |y|^{1/2}|K(y)||x-y|^{-n+1/2}dy\right)^q dx$$

$$\leq \int_{1/4<|y|<3} \left(|y|^{1/2}|K(y)|\right)^q dy \left(\int_{|y|<9/2}|y|^{-n+1/2}dy\right)^q$$

$$\leq C \int_\Sigma |K(\sigma)|^q d\sigma. \tag{2.151}$$

Analogously

$$\int_{1/2<|x|<3/2} \left(\int_{|y|>3}|y|^{1/2}|K(y)||x-y|^{-n+1/2}dy\right)^q dx$$

$$\leq C \int_{1/2<|x|<3/2} \left(\int_{|y|>3}|y|^{1-n}|K(y)|dy\right)^q dx$$

$$\leq C \left(\int_\Sigma |K(\sigma)|d\sigma\right)^q \leq C \int_\Sigma |K(\sigma)|^q d\sigma. \tag{2.152}$$

Combining (2.150),(2.151),(2.152) we conclude (2.139).

Lemma 2.18 *For $0 \leq f \in L^p(R), 1 < p < \infty$,*

$$\int_{-\infty}^\infty \left(\sup_{\epsilon>0} \epsilon^{-n} \int_0^\epsilon f(t+s)s^{n-1}ds\right)^p dt \leq \left(\frac{p}{p-1}\right)^p \int_{-\infty}^\infty f(t)^p dt, \tag{2.153}$$

$$\int_{-\infty}^\infty \left(\sup_{\epsilon>0} \epsilon \int_\epsilon^\infty f(t+s)s^{-2}ds\right)^p dt \leq \left(\frac{2p}{p-1}\right)^p \int_{-\infty}^\infty f(t)^p dt. \tag{2.154}$$

Proof. For $\epsilon > 0$ define the functions F_ϵ and G by (2.73). Then we have

$$\epsilon^{-n} \int_0^\epsilon f(t+s)s^{n-1}ds \leq \epsilon^{-1} \int_0^\epsilon f(t+s)ds = F_\epsilon(t) \leq G(t). \tag{2.155}$$

With the aid of Lemma 2.10 and (2.155) we obtain (2.153). Since $\epsilon F_\epsilon(t) \leq \epsilon^{1-1/p}\|f\|_p$, we have

$$\epsilon \int_\epsilon^\infty f(t+s)s^{-2}ds = \epsilon \int_\epsilon^\infty \frac{d}{ds}\left(sF_s(t)\right)s^{-2}ds$$

$$= \epsilon \left[s^{-1}F_s(t)\right]_\epsilon^\infty + 2\epsilon \int_\epsilon^\infty s^{-2}F_s(t)ds$$

$$= -F_\epsilon(t) + 2\epsilon \int_\epsilon^\infty s^{-2}F_s(t)ds \leq 2\epsilon G(t) \int_\epsilon^\infty s^{-2}ds = 2G(t). \tag{2.156}$$

The inequality (2.154) follows from Lemma 2.10 and (2.156).

Theorem 2.8 *Suppose $K(x)$ is a kernel satisfying Assumption 5. For $f \in L^p(R^n), 1 < p < \infty$, set*

$$\tilde{f}_\epsilon(x) = \int_{|x-y|>\epsilon} K(x-y)f(y)dy.$$

Then there exists a constant C_p such that

$$\int_{R^n} \sup_{\epsilon>0} |\tilde{f}_\epsilon(x)|^p dx \le C_p \int_{R^n} |f(x)|^p dx. \qquad (2.157)$$

The singular integral $K * f = \lim_{\epsilon \to 0} \tilde{f}_\epsilon$ exists both almost everywhere and in the strong topology of $L^p(R^n)$. The mapping $f \mapsto K * f$ is a bounded linear operator from $L^p(R^n)$ to itself.

Proof. Let $K_1(x), K_2(x)$ be the functions defined by (2.129),(2.130) respectively. Let $g = (g_1, \ldots, g_n)$ be the set of functions defined by

$$g(x) = -(R * f)(x) = -\lim_{\epsilon \to 0} \int_{|x-y|>\epsilon} R(x - y) f(y) dy,$$

where $R(x) = (R_1(x), \ldots, R_n(x))$ are the Riesz kernels. Then in view of Theorems 2.6 and 2.7

$$f(x) = (R * g)(x) = \lim_{\epsilon \to 0} \sum_{j=1}^{n} \int_{|x-y|>\epsilon} R_j(x - y) g_j(y) dy$$

almost everywhere and in the strong topology of $L^p(R^n)$. Now we show

$$\int_{R^n} K(x - y)\phi\left(\frac{|x - y|}{\epsilon}\right) f(y) dy = \epsilon^{-n} \int_{R^n} K_2\left(\frac{x - y}{\epsilon}\right) g(y) dy, \qquad (2.158)$$

where ϕ is the function in the definition of $K_2(x)$. First suppose $g \in C_0^1(R^n)^n$ and supp$g \subset \{x; |x| \le N\}$. In $\{y; |y| \le N + 1\}$

$$\int_{|y-z|>\delta} R(y - z) g(z) dz = \int_{|y-z|>\delta} R(y - z)(g(z) - g(y)) dz$$

converges to $f(y)$ uniformly as $\delta \to 0$. If $|y| > N + 1, 0 < \delta < 1$, we have

$$\left| \int_{|y-z|>\delta} R(y - z) g(z) dz \right| \le \int_{|z|<N} \frac{C|g(z)|}{|y - z|^n} dz \le \frac{C}{(|y| + 1)^n} \|g\|_1,$$

$$\int_{R^n} |K(x - y)|\phi\left(\frac{|x - y|}{\epsilon}\right) \frac{dy}{(|y| + 1)^n}$$

$$= \int_{R^n} |K(y)|\phi\left(\frac{|y|}{\epsilon}\right) \frac{dy}{(|x - y| + 1)^n}$$

$$\le \int_{\epsilon/4<|y|<2|x|} |K(y)| dy + 2^n \int_{|y|>2|x|} \frac{|K(y)|}{|y|^n} dy < \infty. \qquad (2.159)$$

Hence

$$\lim_{\delta \to 0} \int_{R^n} K(x-y)\phi\left(\frac{|x-y|}{\epsilon}\right) \int_{|y-z|>\delta} R(y-z)g(z)dzdy$$

$$= \int_{R^n} K(x-y)\phi\left(\frac{|x-y|}{\epsilon}\right) f(y)dy. \qquad (2.160)$$

If $|y-z| > \delta$, we have for some constant C_δ depending on δ

$$\left| K(x-y)\phi\left(\frac{|x-y|}{\epsilon}\right) R(y-z)g(z) \right|$$

$$\leq C_\delta |K(x-y)|\phi\left(\frac{|x-y|}{\epsilon}\right) \frac{|g(z)|}{(|y-z|+1)^n}. \qquad (2.161)$$

Since for $|z| \leq N$

$$(N+1)(|y-z|+1) \geq |y-z|+N+1 \geq |y|-|z|+N+1 \geq |y|+1,$$

the right hand side of (2.161) does not exceed

$$C_\delta(N+1)^n|K(x-y)|\phi\left(\frac{|x-y|}{\epsilon}\right) \frac{|g(z)|}{(|y|+1)^n}.$$

By virtue of (2.159) this is integrable in $(y,z) \in R^n \times R^n$. Therefore in view of Lemma 2.16 the left hand side of (2.160) is equal to

$$\lim_{\delta \to 0} \iint_{|y-z|>\delta} K(x-y)\phi\left(\frac{|x-y|}{\epsilon}\right) R(y-z)dyg(z)dz$$

$$= \lim_{\delta \to 0} \iint_{|x-\epsilon y-z|>\delta} K(\epsilon y)\phi(|y|)R(x-\epsilon y-z)\epsilon^n dyg(z)dz$$

$$= \epsilon^{-n} \lim_{\delta \to 0} \iint_{|(x-z)/\epsilon-y|>\delta/\epsilon} K(y)\phi(|y|)R\left(\frac{x-z}{\epsilon}-y\right) dyg(z)dz$$

$$= \epsilon^{-n} \int_{R^n} K_2\left(\frac{x-z}{\epsilon}\right) g(z)dz.$$

Thus we see that (2.158) holds if $g \in C_0^1(R^n)^n$. Furthermore from the proof it is clear that (2.158) holds if we define $f = R * g$ for a given $g \in C_0^1(R^n)^n$. In the general case we choose a sequence $\{g^k\} \subset C_0^1(R^n)^n$ so that

$$\|g^k - g\|_p \to 0, \qquad \sum_{k=1}^{\infty} \|g^{k+1} - g^k\|_p < \infty$$

and put $f_k = R * g^k$. Then by virtue of Theorem 2.3 or 2.6

$$\|f_k - f\|_p \to 0, \qquad \sum_{k=1}^{\infty} \|f_{k+1} - f_k\|_p < \infty.$$

If we set

$$\bar{g}(x) = |g^1(x)| + \sum_{k=1}^{\infty} |g^{k+1}(x) - g^k(x)|,$$

$$\bar{f}(x) = |f_1(x)| + \sum_{k=1}^{\infty} |f_{k+1}(x) - f_k(x)|,$$

then $\bar{g}, \bar{f} \in L^p(R^n), |g^k(x)| \le \bar{g}(x), |f_k(x)| \le \bar{f}(x), g^k(x) \to g(x), f_k(x) \to f(x)$ almost everywhere. In view of Lemma 2.1

$$\int_{R^n} |K(x-y)| \phi \left(\frac{|x-y|}{\epsilon} \right) \bar{f}(y) dy \le \int_{|x-y|>\epsilon/4} |K(x-y)| \bar{f}(y) dy < \infty$$

almost everywhere. With the aid of (2.138)

$$\int_{R^n} \left| K_2 \left(\frac{x-y}{\epsilon} \right) \right| \bar{g}(y) dy \le \int_{|x-y|<\epsilon} \left| K_2 \left(\frac{x-y}{\epsilon} \right) \right| \bar{g}(y) dy$$

$$+ \int_{|x-y|>\epsilon} \left| K_1 \left(\frac{x-y}{\epsilon} \right) \right| \bar{g}(y) dy + C\epsilon^{n+1} \int_{|x-y|>\epsilon} |x-y|^{-n-1} \bar{g}(y) dy.$$

The first term of the right hand side of this inequality is finite for almost every x by virtue of Lemma 1.2 and

$$\int_{|x|<\epsilon} \left| K_2 \left(\frac{x}{\epsilon} \right) \right| dx = \epsilon^n \int_{|x|<1} |K_2(x)| dx \le \epsilon^n \int_{|x|<1} G(x) dx < \infty.$$

The second and third terms are also finite almost everywhere by Lemma 2.1 and 1.2 respectively. Hence applying (2.158) to f_k, g^k and letting $k \to \infty$ we see that (2.158) holds in the general case. In view of (2.158)

$$\tilde{f}_\epsilon(x) = \int_{|x-y|>\epsilon} K(x-y)f(y) dy = \int_{|x-y|>\epsilon} K(x-y) \phi \left(\frac{|x-y|}{\epsilon} \right) f(y) dy$$

$$= \int_{R^n} K(x-y) \phi \left(\frac{|x-y|}{\epsilon} \right) f(y) dy$$

$$- \int_{|x-y|<\epsilon} K(x-y) \phi \left(\frac{|x-y|}{\epsilon} \right) f(y) dy$$

$$= \epsilon^{-n} \int_{R^n} K_2 \left(\frac{x-y}{\epsilon} \right) g(y) dy - \int_{|x-y|<\epsilon} K(x-y) \phi \left(\frac{|x-y|}{\epsilon} \right) f(y) dy$$

$$= \epsilon^{-n} \int_{|x-y|>\epsilon} K_1 \left(\frac{x-y}{\epsilon} \right) g(y) dy$$

$$+ \epsilon^{-n} \int_{|x-y|>\epsilon} \left(K_2 \left(\frac{x-y}{\epsilon} \right) - K_1 \left(\frac{x-y}{\epsilon} \right) \right) g(y) dy$$

$$+ \epsilon^{-n} \int_{|x-y|<\epsilon} K_2 \left(\frac{x-y}{\epsilon} \right) g(y) dy$$

$$- \int_{|x-y|<\epsilon} K(x-y) \phi \left(\frac{|x-y|}{\epsilon} \right) f(y) dy = \sum_{i=1}^{4} I_i.$$

Since K_1 is homogeneous of degree $-n$

$$I_1 = \int_{|x-y|>\epsilon} K_1(x-y) g(y) dy.$$

By virtue of Theorem 2.6

$$\int_{R^n} \sup_{\epsilon>0} |I_1|^p dx \leq C_p \left(\int_{\Sigma} |K_1(\sigma)| d\sigma \right)^p \int_{R^n} |g(x)|^p dx. \qquad (2.162)$$

With the aid of (2.138)

$$|I_2| \leq C\epsilon \int_{|x-y|>\epsilon} |x-y|^{-n-1} |g(y)| dy \int_{\Sigma} |K(\sigma)| d\sigma.$$

Denoting the area of the unit sphere of R^n by Ω_n

$$\epsilon \int_{|x-y|>\epsilon} |x-y|^{-n-1} |g(y)| dy = \epsilon \int_{\Sigma} \int_{\epsilon}^{\infty} s^{-2} |g(x-s\sigma)| ds d\sigma$$

$$\leq \Omega_n^{1-1/p} \left[\int_{\Sigma} \left(\epsilon \int_{\epsilon}^{\infty} s^{-2} |g(x-s\sigma)| ds \right)^p d\sigma \right]^{1/p}.$$

Hence

$$\int_{R^n} \left(\sup_{\epsilon>0} \epsilon \int_{|x-y|>\epsilon} |x-y|^{-n-1} |g(y)| dy \right)^p dx$$

$$\leq \Omega_n^{p-1} \int_{\Sigma} \int_{R^n} \left(\sup_{\epsilon>0} \epsilon \int_{\epsilon}^{\infty} s^{-2} |g(x-s\sigma)| ds \right)^p dx d\sigma.$$

Making the change of the variables $x = y - t\sigma$, $\sum_{j=1}^{n} y_j \sigma_j = 0$, and applying Lemma 2.18

$$\int_{R^n} \left(\sup_{\epsilon > 0} \epsilon \int_{\epsilon}^{\infty} s^{-2} |g(x - s\sigma)| ds \right)^p dx$$

$$= \int_{R^{n-1}} \int_{-\infty}^{\infty} \left(\sup_{\epsilon > 0} \epsilon \int_{\epsilon}^{\infty} s^{-2} |g(y - (t+s)\sigma)| ds \right)^p dt dy$$

$$\leq \left(\frac{2p}{p-1} \right)^p \int_{R^{n-1}} \int_{-\infty}^{\infty} |g(y - t\sigma)|^p dt dy = \left(\frac{2p}{p-1} \right)^p \int_{R^n} |g(x)|^p dx.$$

Hence we obtain

$$\int_{R^n} \left(\sup_{\epsilon > 0} \epsilon \int_{|x-y|>\epsilon} |x - y|^{-n-1} |g(y)| dy \right)^p dx \leq \left(\frac{2p \Omega_n}{p-1} \right)^p \int_{R^n} |g(x)|^p dx.$$

Therefore

$$\int_{R^n} \sup_{\epsilon > 0} |I_2|^p dx \leq \left(\frac{2p}{p-1} \Omega_n C \int_{\Sigma} |K(\sigma)| d\sigma \right)^p \int_{R^n} |g(x)|^p dx. \qquad (2.163)$$

Next

$$|I_3| \leq \epsilon^{-n} \int_{|x-y|<\epsilon} G(x - y) |g(y)| dy$$

$$= \int_{\Sigma} G(\sigma) \epsilon^{-n} \int_0^{\epsilon} |g(x - s\sigma)| s^{n-1} ds d\sigma$$

$$\leq \left(\int_{\Sigma} G(\sigma) d\sigma \right)^{1-1/p} \left[\int_{\Sigma} G(\sigma) \left(\epsilon^{-n} \int_0^{\epsilon} |g(x - s\sigma)| s^{n-1} ds \right)^p d\sigma \right]^{1/p}.$$

By virtue of Lemma 2.18

$$\int_{R^n} \left(\sup_{\epsilon > 0} \epsilon^{-n} \int_0^{\epsilon} |g(x - s\sigma)| s^{n-1} ds \right)^p dx$$

$$= \int_{R^{n-1}} \int_{-\infty}^{\infty} \left(\sup_{\epsilon > 0} \epsilon^{-n} \int_0^{\epsilon} |g(y - (t+s)\sigma)| s^{n-1} ds \right)^p dt dy$$

$$\leq \int_{R^{n-1}} \left(\frac{p}{p-1} \right)^p \int_{-\infty}^{\infty} |g(y - t\sigma)|^p dt dy = \left(\frac{p}{p-1} \right)^p \int_{R^n} |g(x)|^p dx.$$

Hence

$$\int_{R^n} \sup_{\epsilon > 0} |I_3|^p dx \leq \left(\frac{p}{p-1} \right)^p \left(\int_{\Sigma} G(\sigma) d\sigma \right)^p \int_{R^n} |g(x)|^p dx. \qquad (2.164)$$

Finally

$$|I_4| = \left| \int_{\epsilon/4 < |x-y| < \epsilon} |x-y|^{-n} K\left(\frac{x-y}{|x-y|}\right) \phi\left(\frac{|x-y|}{\epsilon}\right) f(y) dy \right|$$

$$\leq \left(\frac{4}{\epsilon}\right)^n \int_{|x-y| < \epsilon} \left| K\left(\frac{x-y}{|x-y|}\right) \right| |f(y)| dy$$

$$= \left(\frac{4}{\epsilon}\right)^n \int_{\Sigma} |K(\sigma)| \int_0^{\epsilon} |f(x-s\sigma)| s^{n-1} ds d\sigma \leq 4^n \left(\int_{\Sigma} |K(\sigma)| d\sigma\right)^{1-1/p}$$

$$\times \left[\int_{\Sigma} |K(\sigma)| \left(\epsilon^{-n} \int_0^{\epsilon} |f(x-s\sigma)| s^{n-1} ds\right)^p d\sigma \right]^{1/p}.$$

Arguing as in the proof of (2.164) we obtain

$$\int_{R^n} \sup_{\epsilon > 0} |I_4|^p dx \leq \left(\frac{4^n p}{p-1}\right)^p \left(\int_{\Sigma} |K(\sigma)| d\sigma\right)^p \int_{R^n} |f(x)|^p dx. \qquad (2.165)$$

From (2.162),(2.163),(2.164),(2.165) we conclude (2.157). The proof of the remaining part is anologous to that of Theorem 2.5.

2.11   General Case

The following is the main theorem of this chapter.

Theorem 2. 9 *Suppose that $K(x)$ is a measurable, homogeneous function of degree $-n$ defined in R^n satisfying*

$$\int_{\Sigma} |K(x)| \log^+ |K(x)| d\sigma < \infty, \qquad \int_{\Sigma} K(x) d\sigma = 0, \qquad (2.166)$$

where Σ is the unit sphere of R^n and $d\sigma$ is the areal element of Σ. For $f \in L^p(R^n), 1 < p < \infty, \epsilon > 0$ set

$$\tilde{f}_\epsilon(x) = \int_{|x-y| > \epsilon} K(x-y) f(y) dy.$$

Then there exists a constant C_p which does not depend on f such that

$$\int_{R^n} \sup_{\epsilon > 0} |\tilde{f}_\epsilon(x)|^p dx \leq C_p \int_{R^n} |f(x)|^p dx.$$

The singular integral $K * f = \lim_{\epsilon \to 0} \tilde{f}_\epsilon$ exists both almost everywhere and in the strong topology of $L^p(R^n)$, and the mapping $f \mapsto K * f$ is a bounded linear operator from $L^p(R^n)$ to itself.

Proof. We have only to apply Theorem 2.6 and 2.8 to the odd part and the even part of K respectively.

2.12 Derivatives of Homogeneous Functions

Theorem 2.10 *Suppose that L is a homogeneous function of degree $1-n$ belonging to $C^1(R^n \setminus \{0\})$. Then for $i = 1, \ldots, n$, $D_i L(x) = (\partial/\partial x_i)L(x)$ is a homogeneous function of degree $-n$ and satisfyies (2.2). For $1 < p < \infty$ there exists a constant C_p such that for $f \in C_0^1(R^n)$*

$$\int_{R^n} \left| D_i \int_{R^n} L(x-y)f(y)dy \right|^p dx \le C_p \int_{R^n} |f(x)|^p dx. \qquad (2.167)$$

Proof. Clearly $D_i L$ is homogeneous of degree $-n$. Let $0 < a < b < \infty$. Since

$$\int_{|x|=a} L(x)x_i dS = a \int_{|y|=1} L(y)y_i dS,$$

we have

$$\int_{a<|x|<b} D_i L(x)dx = \int_{|x|=b} L(x)\frac{x_i}{b}dS - \int_{|x|=a} L(x)\frac{x_i}{a}dS = 0.$$

On the other hand using the polar coordinates $x = r\sigma$

$$\int_{a<|x|<b} D_i L(x)dx = \int_\Sigma \int_a^b D_i L(r\sigma)r^{n-1}dr d\sigma$$

$$= \int_\Sigma D_i L(\sigma)d\sigma \int_a^b \frac{dr}{r} = \int_\Sigma D_i L(\sigma)d\sigma \log\frac{b}{a}.$$

Hence we see that $D_i L$ satisfies (2.2). We have for $f \in C_0^1(R^n)$

$$D_i \int_{|x-y|>\epsilon} L(x-y)f(y)dy = D_i \int_{|y|>\epsilon} L(y)f(x-y)dy$$

$$= \int_{|y|>\epsilon} L(y)D_i f(x-y)dy = -\int_{|y|>\epsilon} L(y)\frac{\partial}{\partial y_i}f(x-y)dy$$

$$= \int_{|y|=\epsilon} L(y)f(x-y)\frac{y_i}{|y|}dS + \int_{|y|>\epsilon} D_i L(y)f(x-y)dy$$

$$= \int_\Sigma L(\sigma)f(x-\epsilon\sigma)\sigma_i d\sigma + \int_{|x-y|>\epsilon} D_i L(x-y)f(y)dy.$$

The first term on the last side converges to $\int_\Sigma L(\sigma)\sigma_i d\sigma f(x)$ as $\epsilon \to 0$. The second term converges to $D_i L * f$. In view of Theorem 2.9 we have $\|D_i L * f\|_p \le C_p \|f\|_p$. Hence

$$D_i \int_{R^n} L(x-y)f(y)dy = \int_\Sigma L(\sigma)\sigma_i d\sigma f(x) + (D_i L * f)(x),$$

and (2.167) holds.

Chapter 3

Sobolev Spaces

3.1 Sobolev Spaces

In this chapter we use the notations $D = (D_1, \ldots, D_n) = (\partial/\partial x_1, \ldots, \partial/\partial x_n)$ and $D^\alpha = D_1^{\alpha_1} \cdots D_n^{\alpha_n}$, $|\alpha| = \alpha_1 + \cdots + \alpha_n$ for a vector $\alpha = (\alpha_1, \ldots, \alpha_n)$ with integral components $\alpha_i \geq 0$. We often write D^m to denote mth order derivatives, i.e. D^m is one of D^α with $|\alpha| = m$. Let Ω be a nonempty open subset of R^n and m be a nonnegative integer. Then $C^m(\Omega)$ denotes the set of all functions whose derivatives of order up to m are all continuous in Ω, and $C_0^m(\Omega)$ the totality of functions belonging to $C^m(\Omega)$ and with compact support in Ω. We denote by $B^m(\Omega)$ the set of all functions which are bounded and continuous in Ω together with their derivatives of order up to m. For $0 < h < 1$ we denote by $B^{m+h}(\Omega)$ the set of all functions belonging to $B^m(\Omega)$ whose mth order derivatives are all uniformly Hölder continuous in Ω with exponent h. Similarly the sets $B^m(\bar{\Omega})$ and $B^{m+h}(\bar{\Omega})$ are defined replacing Ω by $\bar{\Omega}$. For $u \in B^m(\Omega)$ we put

$$|u|_{m,\infty,\Omega} = \max_{|\alpha|=m} \sup_{x \in \Omega} |D^\alpha u(x)|, \tag{3.1}$$

$$\|u\|_{m,\infty,\Omega} = \max_{j=0,\ldots,m} |u|_{j,\infty,\Omega}. \tag{3.2}$$

In particular

$$\|u\|_{0,\infty,\Omega} = |u|_{0,\infty,\Omega} = \sup_{x \in \Omega} |u(x)|. \tag{3.3}$$

If $a = m + h$ with an integer m and $0 < h < 1$ we set

$$|u|_{a,\infty,\Omega} = \max_{|\alpha|=m} \sup_{x,y \in \Omega, x \neq y} \frac{|D^\alpha u(x) - D^\alpha u(y)|}{|x-y|^h} \tag{3.4}$$

$$\|u\|_{a,\infty,\Omega} = \max\{\|u\|_{m,\infty,\Omega}, |u|_{a,\infty,\Omega}\}. \tag{3.5}$$

$B^m(\Omega)$, $B^m(\bar{\Omega})$, $B^a(\Omega)$, $B^a(\bar{\Omega})$ are Banach spaces with norm (3.2) or (3.5).
If $\Omega = R^n$ we write $|\ |_{m,\infty}, \|\ \|_{m,\infty}, |\ |_{a,\infty}, \|\ \|_{a,\infty}$ instead of $|\ |_{m,\infty,R^n}$,
$\|\ \|_{m,\infty,R^n}, |\ |_{a,\infty,R^n}, \|\ \|_{a,\infty,R^n}$ respectively.

3.2 Interpolation Inequalities (1)

In this section we prove interpolation inequalities in case $\Omega = R^n$.

Theorem 3. 1 *For $0 \le a < b < c < \infty$ there exists a constant $\gamma_{a,b,c}$ such
that for $u \in B^c(R^n)$ the following inequality holds:*

$$|u|_{b,\infty} \le \gamma_{a,b,c} |u|_{c,\infty}^{(b-a)/(c-a)} |u|_{a,\infty}^{(c-b)/(c-a)}. \tag{3.6}$$

Proof. For the sake of simplicity we write $|\ |_a$ and γ omitting ∞ and a, b, c
in $|\ |_{a,\infty}$ and $\gamma_{a,b,c}$ respectively.
(i) Case $0 \le a < b < c \le 1$. Since

$$\frac{|u(x) - u(y)|}{|x - y|^b} = \left(\frac{|u(x) - u(y)|}{|x - y|^c}\right)^{(b-a)/(c-a)} \left(\frac{|u(x) - u(y)|}{|x - y|^a}\right)^{(c-b)/(c-a)}$$

we get if $0 < a, c < 1$

$$|u|_b \le |u|_c^{(b-a)/(c-a)} |u|_a^{(c-b)/(c-a)}.$$

For $u \in B^1(R^n)$

$$|u(x) - u(y)| = \left|\int_0^1 \frac{d}{dt} u(y + t(x - y)) dt\right|$$

$$= \left|\int_0^1 \sum_{i=1}^n D_i u(y + t(x - y))(x_i - y_i) dt\right|$$

$$\le \sum_{i=1}^n |D_i u|_0 |x_i - y_i| \le |u|_1 \sqrt{n} |x - y|,$$

which implies

$$|u|_1 \le \sup_{x \ne y} \frac{|u(x) - u(y)|}{|x - y|} \le \sqrt{n} |u|_1. \tag{3.7}$$

From this and $|u(x) - u(y)| \le 2|u|_0$ we see that (3.6) also holds in case $a = 0$
or $c = 1$.
(ii) Case $0 \le a < b = 1 < c \le 2$. If $n = 1$, we have for $\rho > 0$

$$u'(x) = \frac{1}{\rho} \int_x^{x+\rho} (u'(x) - u'(y)) dy + \frac{1}{\rho}(u(x + \rho) - u(x)),$$

$$|u(x+\rho) - u(x)| \leq \begin{cases} |u|_a \rho^a & a > 0 \\ 2|u|_0 & a = 0 \end{cases},$$

$$|u'(x) - u'(y)| \leq |u|_c (y-x)^{c-1}.$$

Hence if $a > 0$

$$|u'(x)| \leq \frac{1}{\rho} \int_x^{x+\rho} |u|_c (y-x)^{c-1} dy + |u|_a \rho^{a-1} = c^{-1} |u|_c \rho^{c-1} + |u|_a \rho^{a-1}.$$

Minimizing the last side with respect to ρ we obtain

$$|u|_1 \leq \gamma |u|_c^{(1-a)/(c-a)} |u|_a^{(c-1)/(c-a)}. \tag{3.8}$$

The case $a = 0$ is similarly proved. If $n > 1$, we obtain the desired result applying (3.8) to each independent variable.

(iii) Case $a = k, b = j, c = m$ are integers. What should be established is

$$|u|_j \leq \gamma |u|_m^{(j-k)/(m-k)} |u|_k^{(m-j)/(m-k)}. \tag{3.9}$$

We have only to consider the case $k = 0$, since (3.9) is implied by

$$|D^k u|_{j-k} \leq \gamma |D^k u|_{m-k}^{(j-k)/(m-k)} |D^k u|_0^{(m-j)/(m-k)}.$$

Hence we shall prove

$$|u|_j \leq C |u|_m^{j/m} |u|_0^{(m-j)/m} \tag{3.10}$$

for $0 < j < m$. First we consider the case $j = m - 1$:

$$|u|_{m-1} \leq C |u|_m^{(m-1)/m} |u|_0^{1/m}. \tag{3.11}$$

If $m = 2$ (3.11) is

$$|u|_1 \leq C |u|_2^{1/2} |u|_0^{1/2}, \tag{3.12}$$

which follows from (ii) with $a = 0$ and $c = 2$. Suppose (3.11) is true for $m - 1$ in place of m. Then

$$|u|_{m-2} \leq C |u|_{m-1}^{(m-2)/(m-1)} |u|_0^{1/(m-1)}. \tag{3.13}$$

Applying (3.12) to $D^{m-2} u$

$$|D^{m-2} u|_1 \leq C |D^{m-2} u|_2^{1/2} |D^{m-2} u|_0^{1/2},$$

which implies

$$|u|_{m-1} \leq C |u|_m^{1/2} |u|_{m-2}^{1/2}. \tag{3.14}$$

(3.11) follows from (3.13) and (3.14). Finally we consider the general case. Suppose (3.10) is true with j replaced by $j+1$. Then

$$|u|_{j+1} \leq C|u|_m^{(j+1)/m}|u|_0^{(m-j-1)/m}. \tag{3.15}$$

In view of (3.11)

$$|u|_j \leq C|u|_{j+1}^{j/(j+1)}|u|_0^{1/(j+1)}. \tag{3.16}$$

The desired inequality follows from (3.15) and (3.16).

(iv) Case $0 \leq a < b = 1 < c = m, m$ an integer. We proceed by induction. The case $m = 2$ is already proved in (ii). Suppose the desired inequality is true for m:

$$|u|_1 \leq C|u|_m^{(1-a)/(m-a)}|u|_a^{(m-1)/(m-a)}. \tag{3.17}$$

By the previous step we have

$$|u|_m \leq C|u|_{m+1}^{(m-1)/m}|u|_1^{1/m}.$$

The desired result for $m+1$ follows from this and (3.17).

(v) Case $0 = a < b = j < c \leq j+1, j$ an integer. The case $j = 1$ is proved in (ii). Suppose $j > 1$. Since $1 < c - j + 1 \leq 2$ we get by (ii)

$$|D^{j-1}u|_1 \leq C|D^{j-1}u|_{c-j+1}^{1/(c-j+1)}|D^{j-1}u|_0^{(c-j)/(c-j+1)},$$

which implies

$$|u|_j \leq C|u|_c^{1/(c-j+1)}|u|_{j-1}^{(c-j)/(c-j+1)}. \tag{3.18}$$

In view of (iii)

$$|u|_{j-1} \leq C|u|_j^{(j-1)/j}|u|_0^{1/j}.$$

The desired result follows from this and (3.18).

(vi) Case $0 \leq a < b = 1 < c$. If $c \leq 2$ or c is an integer, the result is already established in (ii) or (iv) respectively. So we assume that c is not an integer and $c > 2$. Put $[c] = m$. Since

$$0 \leq a < b = 1 < m < c < m+1$$

we have (3.17). Since $0 < m - 1 < c - 1 < m$ we have by (v)

$$|Du|_{m-1} \leq C|Du|_{c-1}^{(m-1)/(c-1)}|Du|_0^{(c-m)/(c-1)}$$

or

$$|u|_m \leq C|u|_c^{(m-1)/(c-1)}|u|_1^{(c-m)/(c-1)}.$$

The desired inequality follows from this and (3.17).

(vii) Case $0 \leq a < b = j < c \leq j+1$, j an integer. If $j - 1 \leq a$, we

have $0 \leq a - j + 1 < 1 < c - j + 1 \leq 2$, and hence applying the result of
(ii) to $D^{j-1}u$ we get the desired inequality. Suppose $a < j - 1$. If we put
$k = [a] + 1$, then $0 < j - k < c - k \leq j - k + 1$. Hence, applying the result
of (v) to $D^k u$ we get

$$|u|_j \leq C|u|_c^{(j-k)/(c-k)}|u|_k^{(c-j)/(c-k)}. \tag{3.19}$$

Noting $0 \leq a - k + 1 < 1 < j - k + 1$ we have by (iv)

$$|u|_k \leq C|u|_j^{(k-a)/(j-a)}|u|_a^{(j-k)/(j-a)}. \tag{3.20}$$

The desired inequality follows from (3.19) and (3.20).
(viii) Case $b = j$ is an integer. If $j - 1 \leq a$, then $0 \leq a - j + 1 < 1 < c - j + 1$.
Hence, we get the desired inequality applying the result of (vi) to $D^{j-1}u$.
The case $c \leq j + 1$ is shown in (vii). In case $a < j - 1, c > j + 1$ let k, m be
integers satisfying $k - 1 \leq a < k, m < c \leq m + 1$. Then $k < j < m$. In view
of (iii) we have

$$|u|_j \leq C|u|_m^{(j-k)/(m-k)}|u|_k^{(m-j)/(m-k)}. \tag{3.21}$$

Noting $0 \leq a - k + 1 < 1 < j - k + 1$ and applying (iv) to $D^{k-1}u$ we get

$$|u|_k \leq C|u|_j^{(k-a)/(j-a)}|u|_a^{(j-k)/(j-a)}. \tag{3.22}$$

Since $0 < m - j < c - j \leq m - j + 1$ we get

$$|u|_m \leq C|u|_c^{(m-j)/(c-j)}|u|_j^{(c-m)/(c-j)} \tag{3.23}$$

applying (v) to $D^j u$. The desired result follows from (3.21),(3.22),(3.23).
(ix) The general case. The case where b is an integer was established in
the previous step. So we assume that b is not an integer. Let $j = [b]$, then
$j < b < j + 1$. If $a < j$ and $c > j + 1$, we have in view of (viii)

$$|u|_j \leq C|u|_c^{(j-a)/(c-a)}|u|_a^{(c-j)/(c-a)}, \tag{3.24}$$

$$|u|_{j+1} \leq C|u|_c^{(j+1-a)/(c-a)}|u|_a^{(c-j-1)/(c-a)}, \tag{3.25}$$

and in view of (i)

$$|u|_b \leq |u|_{j+1}^{b-j}|u|_j^{j+1-b}. \tag{3.26}$$

The desired result follows from (3.24),(3.25),(3.26). If $a \geq j$ and $c > j + 1$,
we have $j \leq a < b < j + 1 < c$. Hence, it follows from (i) and (viii) that

$$|u|_b \leq C|u|_{j+1}^{(b-a)/(j+1-a)}|u|_a^{(j+1-b)/(j+1-a)}, \tag{3.27}$$

$$|u|_{j+1} \leq C|u|_c^{(j+1-b)/(c-b)}|u|_b^{(c-j-1)/(c-b)} \tag{3.28}$$

respectively. The desired result follows from (3.27),(3.28). Analogously, if $a < j$ and $c \leq j + 1$, we obtain the desired result from

$$|u|_b \leq C|u|_c^{(b-j)/(c-j)}|u|_j^{(c-b)/(c-j)}, \quad |u|_j \leq C|u|_b^{(j-a)/(b-a)}|u|_a^{(b-j)/(b-a)},$$

which follow from (i) and (viii). The case $a \geq j, c \leq j + 1$ is reduced to (i).

3.3 Interpolation Inequalities (2)

Let Ω be a nonempty open subset of R^n. Following A. Friedman [65] we define

$$|u|_{p,\Omega} = \begin{cases} \left(\int_\Omega |u|^p \, dx \right)^{1/p} & 0 < p < \infty \\ |u|_{-n/p,\infty,\Omega} & -\infty < p < 0 \\ |u|_{0,\infty,\Omega} & p = \pm\infty \end{cases} \quad (3.29)$$

When $\Omega = R^n$, we write $|\ |_p$ instead of $|\ |_{p,R^n}$.

Lemma 3.1 *There exists a constant C such that for $-\infty < p \leq -n, q > 0$*

$$|u|_{0,\infty} \leq C|u|_p^{p/(p-q)}|u|_q^{q/(q-p)}. \quad (3.30)$$

Proof. Suppose first $p < -n$. Then $0 < -n/p < 1$, and hence

$$|u(x) - u(y)| \leq |u|_{-n/p,\infty}|x - y|^{-n/p} = |u|_p|x - y|^{-n/p}.$$

Therefore

$$|u(x)| \leq |u(x) - u(y)| + |u(y)| \leq |u|_p|x - y|^{-n/p} + |u(y)|. \quad (3.31)$$

When $q \geq 1$, integrating (3.31) over $\{y; |x - y| < \rho\}$ and applying Hölder's inequality to the last term we get

$$|u(x)|V_n\rho^n \leq |u|_p\Omega_n\frac{\rho^{n-n/p}}{n - n/p} + (V_n\rho^n)^{1-1/q}|u|_q, \quad (3.32)$$

where V_n and Ω_n are the volume of the unit ball and the area of the unit shere in R^n. Dividing both side of (3.32) by $V_n\rho^n$ and taking the supremum of the left hand side we get

$$|u|_{0,\infty} \leq |u|_p\frac{\rho^{-n/p}}{1 - 1/p} + (V_n\rho^n)^{-1/q}|u|_q. \quad (3.33)$$

Minimizing the right hand side of (3.33) with respect to ρ we obtain (3.30). When $q < 1$, the result follows analogously starting from

$$|u(x)| \leq |u|_p |x - y|^{-n/p} + |u|_{0,\infty}^{1-q} |u(y)|^q,$$

which is a simple consequence of (3.31). If $p = -n$, we can show (3.30) in the same manner using (3.7).

Lemma 3.2 *There exists a constant C such that for $-\infty < r < p \leq -n, q > 0$*

$$|u|_r \leq C|u|_p^{p(r-q)/r(p-q)} |u|_q^{q(r-p)/r(q-p)}. \tag{3.34}$$

Proof. Noting that $0 < p/r < 1, 0 < -n/r < -n/p \leq 1$ and (3.7)

$$\frac{|u(x) - u(y)|}{|x - y|^{-n/r}} = \left(\frac{|u(x) - u(y)|}{|x - y|^{-n/p}} \right)^{p/r} |u(x) - u(y)|^{(r-p)/r}$$

$$\leq \begin{cases} |u|_p^{p/r}(2|u|_{0,\infty})^{(r-p)/r} & p < -n \\ (\sqrt{n}|u|_p)^{p/r}(2|u|_{0,\infty})^{(r-p)/r} & p = -n \end{cases}. \tag{3.35}$$

We obtain (3.34) combining (3.30) and (3.35).

Lemma 3.3 *For $-\infty < p \leq -n, 0 < q < r$ the inequality (3.34) holds.*

Proof. The result follows from (3.30) and

$$|u|_r = \left(\int |u|^r dx \right)^{1/r} = \left(\int |u|^{r-q} |u|^q dx \right)^{1/r}$$

$$\leq |u|_{0,\infty}^{(r-q)/r} \left(\int |u|^q dx \right)^{1/r} = |u|_{0,\infty}^{(r-q)/r} |u|_q^{q/r}.$$

Theorem 3.2 *For $\lambda < \mu < \nu$, then there exists a constant $\gamma_{\lambda,\mu,\nu}$ such that for $u \in C_0^\infty(R^n)$*

$$|u|_{1/\mu} \leq \gamma_{\lambda,\mu,\nu} |u|_{1/\lambda}^{(\nu-\mu)/(\nu-\lambda)} |u|_{1/\nu}^{(\mu-\lambda)/(\nu-\lambda)}. \tag{3.36}$$

Proof. (i) Case $0 \leq \lambda < \mu < \nu$. If we put $p = \mu(\nu - \lambda)/\lambda(\nu - \mu)$, then $1 < p \leq \infty, p' = p/(p-1) = \mu(\nu - \lambda)/\nu(\mu - \lambda)$. We obtain (3.36) applying Hölder's inequality in

$$|u|_{1/\mu} = \left(\int |u|^{1/\mu} dx \right)^\mu = \left(\int |u|^{(\nu-\mu)/\mu(\nu-\lambda)} |u|^{(\mu-\lambda)/\mu(\nu-\lambda)} dx \right)^\mu.$$

(ii) Case $\lambda < \mu < \nu \leq 0$. What is to be proved is

$$|u|_{-n\mu,\infty} \leq \gamma |u|_{-n\lambda,\infty}^{(\nu-\mu)/(\nu-\lambda)} |u|_{-n\nu,\infty}^{(\mu-\lambda)/(\nu-\lambda)},$$

which is already shown in Theorem 3.1.

(iii) Case $-1/n \leq \lambda < \mu \leq 0 < \nu$. We put $p = 1/\lambda, r = 1/\mu, q = 1/\nu$. Then $-\infty \leq r < p \leq -n, q > 0$. The result follows from Lemma 3.1 or 3.2 according as $\mu = 0$ or $\mu < 0$.

(iv) Case $\lambda < -1/n < \mu \leq 0 < \nu$. From the previous case with $\lambda = -1/n$ it follows that

$$|u|_{1/\mu} \leq C|u|_{-n}^{n(\nu-\mu)/(n\nu+1)} |u|_{1/\nu}^{(n\mu+1)/(n\nu+1)}. \tag{3.37}$$

In view of Theorem 3.1

$$|u|_{-n} = |u|_{1,\infty} \leq C|u|_{-n\lambda,\infty}^{(n\mu+1)/n(\mu-\lambda)} |u|_{-n\mu,\infty}^{(n\lambda+1)/n(\lambda-\mu)}$$
$$= C|u|_{1/\lambda}^{(n\mu+1)/n(\mu-\lambda)} |u|_{1/\mu}^{(n\lambda+1)/n(\lambda-\mu)}. \tag{3.38}$$

Combining (3.37) and (3.38) we obtain (3.36).

(v) Case $\lambda < \mu = -1/n < 0 < \nu$. Since $-n\lambda > 1$, we get by virtue of Theorem 3.1

$$|u|_{-n} = |u|_{1,\infty} \leq C|u|_{-n\lambda,\infty}^{-1/n\lambda} |u|_{0,\infty}^{1+1/n\lambda} = C|u|_{1/\lambda}^{-1/n\lambda} |u|_{0,\infty}^{1+1/n\lambda}.$$

By Lemma 3.1

$$|u|_{0,\infty} \leq C|u|_{-n}^{n\nu/(n\nu+1)} |u|_{1/\nu}^{1/(n\nu+1)}.$$

The desired inequality follows from these two inequalities.

(vi) Case $\lambda < \mu < -1/n < 0 < \nu$. By virtue of (v) with λ replaced by μ

$$|u|_{-n} \leq C|u|_{1/\mu}^{(n\nu+1)/n(\nu-\mu)} |u|_{1/\nu}^{-(n\mu+1)/n(\nu-\mu)}.$$

Noting $1 < -n\mu < -n\lambda$ we get in view of Theorem 3.1

$$|u|_{1/\mu} \leq C|u|_{1/\lambda}^{(n\mu+1)/(n\lambda+1)} |u|_{-n}^{-n(\mu-\lambda)/(n\lambda+1)}.$$

We obtain the desired result from these two inequalities.

(vii) Case $-1/n \leq \lambda < 0 < \mu < \nu$. Since $0 < 1/\nu < 1/\mu, 1/\lambda \leq -n$, we obtain (3.36) with the aid of Lemma 3.3.

(viii) Case $\lambda < -1/n < 0 < \mu < \nu$. We have in view of (v) and (vii)

$$|u|_{-n} \leq C|u|_{1/\lambda}^{(n\mu+1)/n(\mu-\lambda)} |u|_{1/\mu}^{-(n\lambda+1)/n(\mu-\lambda)},$$
$$|u|_{1/\mu} \leq C|u|_{-n}^{n(\nu-\mu)/(n\nu+1)} |u|_{1/\nu}^{(n\mu+1)/(n\nu+1)},$$

respectively. Combining these two inequalities we obtain the desired inequality.

3.4 Interpolation Inequalities (3)

The purpose of this section is to establish the Gagliardo-Nirenberg inequality for R^n. We mainly follow L. Nirenberg [118].

Let Ω be a nonempty open subset of R^n. For $1 \leq p \leq \infty$ and a nonnegative integer m we put

$$|u|_{m,p,\Omega} = \begin{cases} \left(\displaystyle\int_{\Omega} \sum_{|\alpha|=m} |D^\alpha u|^p dx\right)^{1/p} & 1 \leq p < \infty, \\ \displaystyle\max_{|\alpha|=m} \text{ess sup}_{x\in\Omega} |D^\alpha u(x)| & p = \infty, \end{cases} \tag{3.39}$$

$$\|u\|_{m,p,\Omega} = \begin{cases} \left(\displaystyle\sum_{k=0}^{m} |u|_{k,p,\Omega}^p\right)^{1/p} & 1 \leq p < \infty, \\ \displaystyle\max_{k=0,\ldots,m} |u|_{k,\infty,\Omega} & p = \infty, \end{cases} \tag{3.40}$$

if the values of the right hand sides are defined and finite. If $1 \leq p < \infty$, $|u|_{0,p,\Omega}$ coincides with $|u|_{p,\Omega}$ defined by (3.29). For $u \in B^m(\Omega)$, $|u|_{m,\infty,\Omega}$ coincides with the one defined by (3.1). When $\Omega = R^n$, we write $|u|_{m,p}$, $\|u\|_{m,p}$ short for $|u|_{m,p,R^n}$, $\|u\|_{m,p,R^n}$.

The totality of functions whose distribution derivatives of order up to m all belong to $L^p(\Omega)$ is denoted by $W^{m,p}(\Omega)$ which is a Banach space with norm (3.40).

Lemma 3.4 *Let $1 \leq p < \infty, 1 \leq q < \infty, 1/r = (1/q + 1/p)/2$, and $I = [a,b]$ be a finite interval with $|I| = b - a$. Then for any function $u \in C^2(I)$*

$$\left(\int_I |u'|^r dx\right)^{1/r} \leq |I|^{1+1/r-1/p} \left(\int_I |u''|^p dx\right)^{1/p}$$

$$+ 8|I|^{-1-1/r+1/p} \left(\int_I |u|^q dx\right)^{1/q}. \tag{3.41}$$

Proof. Let $u = v + iw$ with real valued functions v and w. Let $\alpha = (b-a)/4, a \leq x_1 \leq a+\alpha, a+3\alpha \leq x_2 \leq b$. Then by virtue of the mean value theorem

$$\frac{v(x_2) - v(x_1)}{x_2 - x_1} = v'(x_{12}), \quad x_1 < x_{12} < x_2,$$

$$\frac{w(x_2) - w(x_1)}{x_2 - x_1} = w'(y_{12}), \quad x_1 < y_{12} < x_2.$$

Hence, for $a \le x \le b$

$$v'(x) = v'(x_{12}) + \int_{x_{12}}^{x} v''(y)dy = \frac{v(x_2) - v(x_1)}{x_2 - x_1} + \int_{x_{12}}^{x} v''(y)dy,$$

$$w'(x) = w'(y_{12}) + \int_{y_{12}}^{x} w''(y)dy = \frac{w(x_2) - w(x_1)}{x_2 - x_1} + \int_{y_{12}}^{x} w''(y)dy.$$

From these two equalities we get

$$|u'(x)| \le \frac{|u(x_2) - u(x_1)|}{x_2 - x_1} + \left| \int_{x_{12}}^{x} v''(y)dy + i \int_{y_{12}}^{x} w''(y)dy \right|. \qquad (3.42)$$

The first term of the right hand side of (3.42) does not exceed $(|u(x_2)| + |u(x_1)|)/2\alpha$ since $x_2 - x_1 \ge 2\alpha$.

Consider the case $x \le x_{12} < y_{12}$. The second term of the right hand side of (3.42) is not greater than

$$\left| \int_{x}^{x_{12}} u''(y)dy + i \int_{x_{12}}^{y_{12}} w''(y)dy \right|$$

$$\le \int_{x}^{x_{12}} |u''(y)|dy + \int_{x_{12}}^{y_{12}} |w''(y)|dy \le \int_{x}^{y_{12}} |u''(y)|dy \le \int_{a}^{b} |u''(y)|dy.$$

Other cases are handled similarly, and we see that the inequality

$$|u'(x)| \le \frac{|u(x_2)| + |u(x_1)|}{2\alpha} + \int_{a}^{b} |u''(y)|dy$$

is true for $a \le x \le b$. Integrating both sides over $a \le x_1 \le a + \alpha$ and $a + 3\alpha \le x_2 \le b$ we get

$$\alpha^2 |u'(x)| \le \frac{1}{2} \int_{a}^{b} |u(y)|dy + \alpha^2 \int_{a}^{b} |u''(y)|dy. \qquad (3.43)$$

With the aid of Hölder's inequality we get from (3.43)

$$\alpha^2 \left(\int_{a}^{b} |u'|^r dx \right)^{1/r} \le \frac{1}{2}(b-a)^{1/r} \int_{a}^{b} |u|dx + \alpha^2 (b-a)^{1/r} \int_{a}^{b} |u''|dx$$

$$\le \frac{1}{2}(b-a)^{1/r+1-1/q} \left(\int_{a}^{b} |u|^q dx \right)^{1/q} + \alpha^2 (b-a)^{1/r+1-1/p} \left(\int_{a}^{b} |u''|^p dx \right)^{1/p}.$$

Dividing by α^2 and noting $1/q = 2/r - 1/p$ we obtain (3.41).

Lemma 3.5 *Let* $1 \leq p \leq \infty, 1 \leq q \leq \infty, 1/r = (1/q + 1/p)/2$. *Then for any function u in $C^2([0, \infty))$ vanishing outside some bounded set*

$$\left(\int_0^\infty |u'|^r dx \right)^{1/r} \leq 4\sqrt{2} \left(\int_0^\infty |u''|^p dx \right)^{1/2p} \left(\int_0^\infty |u|^q dx \right)^{1/2q}. \quad (3.44)$$

Hence, for functions $u \in C_0^\infty(-\infty, \infty)$

$$\left(\int_{-\infty}^\infty |u'|^r dx \right)^{1/r} \leq 4\sqrt{2} \left(\int_{-\infty}^\infty |u''|^p dx \right)^{1/2p} \left(\int_{-\infty}^\infty |u|^q dx \right)^{1/2q}. \quad (3.45)$$

Proof. First consider the case $1 < p < \infty, 1 \leq q < \infty$. Let l be a positive number such that $u(x) = 0$ for $x > l$, and k be a natural number ≥ 2. If the first term of the right hand side of (3.41) with $I = [0, l/k]$ is greater than the second term, we let $I_1 = [0, a_1] = [0, l/k]$. Otherwise, we choose $I_1 = [0, a_1]$ so that $I_1 \supset [0, l/k]$ and both terms of the right hand side of (3.41) with $I = I_1$ are equal. That is possible since $1 + 1/r - 1/p > 0$. Then we have

$$\left(\int_{I_1} |u'|^r dx \right)^{1/r} \leq \begin{cases} 2\left(\dfrac{l}{k}\right)^{1+1/r-1/p} \left(\displaystyle\int_{I_1} |u''|^p dx \right)^{1/p} & a_1 = \dfrac{l}{k} \\[3mm] 4\sqrt{2}\left(\displaystyle\int_{I_1} |u''|^p \right)^{1/2p} \left(\displaystyle\int_{I_1} |u|^q dx \right)^{1/2q} & a_1 > \dfrac{l}{k}. \end{cases}$$

The proof of (3.44) is complete if $a_1 \geq l$. If $a_1 < l$, we define $I_2 = [a_1, a_2], \ldots, I_i = [a_{i-1}, a_i], \ldots$ as follows. Suppose $I_{i-1} = [a_{i-2}, a_{i-1}]$ is defined. If the first term of the right hand side of (3.41) with $I = [a_{i-1}, a_{i-1} + l/k]$ is greater than the second term, we set $I_i = [a_{i-1}, a_i] = [a_{i-1}, a_{i-1} + l/k]$, and otherwise we define $I_i = [a_{i-1}, a_i]$ so that $I_i \supset [a_{i-1}, a_{i-1} + l/k]$ and both terms of the right hand side of (3.41) with $I = I_i$ are equal. After proceeding finite times in this manner, we get $a_{i-1} < l \leq a_i$. Denote this number i by j. Then for $i = 1, \ldots, j$

$$\left(\int_{I_i} |u'|^r dx \right)^{1/r}$$

$$\leq \begin{cases} 2\left(\dfrac{l}{k}\right)^{1+1/r-1/p} \left(\displaystyle\int_{I_i} |u''|^p dx \right)^{1/p} & a_i = a_{i-1} + \dfrac{l}{k} \\[3mm] 4\sqrt{2}\left(\displaystyle\int_{I_i} |u''|^p dx \right)^{1/2p} \left(\displaystyle\int_{I_i} |u|^q dx \right)^{1/2q} & a_i > a_{i-1} + \dfrac{l}{k}. \end{cases}$$

Since $j \leq k, r/p - r < 0$, we have as $k \to \infty$

$$A \equiv \sum_{i=1}^j 2^r \left(\frac{l}{k}\right)^{r+1-r/p} \left(\int_{I_i} |u''|^p dx \right)^{r/p}$$

$$\leq 2^r l^{r+1-r/p} k^{r/p-r} \left(\int_0^\infty |u''|^p dx \right)^{r/p} \to 0.$$

On the other hand, noting $r/2q + r/2p = 1$ and using Hölder's inequality

$$B \equiv \sum_{i=1}^j (4\sqrt{2})^r \left(\int_{I_i} |u''|^p dx \right)^{r/2p} \left(\int_{I_i} |u|^q dx \right)^{r/2q}$$

$$\leq (4\sqrt{2})^r \left(\sum_{i=1}^j \int_{I_i} |u''|^p dx \right)^{r/2p} \left(\sum_{i=1}^j \int_{I_i} |u|^q dx \right)^{r/2q}$$

$$\leq (4\sqrt{2})^r \left(\int_0^\infty |u''|^p dx \right)^{r/2p} \left(\int_0^\infty |u|^q dx \right)^{r/2q}.$$

Combining these two statements and the inequality

$$\int_0^\infty |u'|^r dx = \int_0^l |u'|^r dx = \sum_{i=1}^j \int_{I_i} |u'|^r dx \leq A + B$$

we obtain (3.44). Letting $p \to 1$, we get the result of the case $p = 1$. If $p = \infty$ or $q = \infty$, we obain (3.44) letting $p \to \infty$ or $q \to \infty$.

Lemma 3.6 *If $n < p < \infty$, there exists a constant C depending only on n and p such that for any $u \in C_0^1(R^n)$*

$$|u|_{1-n/p,\infty} \leq C|u|_{1,p}. \tag{3.46}$$

Proof. Let x, y be two distinct points of R^n and $d = |x - y|$. Set

$$G = \{z \in R^n; |z - x| \leq d, |z - y| \leq d\}.$$

Then the volume of G is cd^n with some positive constant c depending only on n. Integrating both sides of

$$|u(x) - u(y)| \leq |u(x) - u(z)| + |u(z) - u(y)|$$

with respect to z in G, we get

$$cd^n |u(x) - u(y)| \leq \int_G |u(x) - u(z)| dz + \int_G |u(z) - u(y)| dz. \tag{3.47}$$

Expressing z by polar coodinates about x:

$$z = x + \rho\sigma, 0 < \rho < \infty, \sigma = (\sigma_1, \ldots, \sigma_n) \in \Sigma = \text{the unit sphere},$$

we get

$$\int_G |u(x) - u(z)| dz \le \int_0^d \int_\Sigma |u(x + \rho\sigma) - u(x)| d\sigma \rho^{n-1} d\rho,$$

$$|u(x + \rho\sigma) - u(x)| = \left| \int_0^\rho \frac{d}{dt} u(x + t\sigma) dt \right|$$

$$= \left| \int_0^\rho \sum_{i=1}^n \sigma_i D_i u(x + t\sigma) dt \right| \le \int_0^\rho |\nabla u(x + t\sigma)| dt.$$

Hence, if $p < \infty$, with the aid of Hölder's inequality

$$\int_G |u(x) - u(z)| dz \le \int_0^d \int_\Sigma \int_0^\rho |\nabla u(x + t\sigma)| dt d\sigma \rho^{n-1} d\rho$$

$$= \int_0^d \int_\Sigma \int_0^\rho |\nabla u(x + t\sigma)| t^{(n-1)/p} t^{(1-n)/p} dt d\sigma \rho^{n-1} d\rho$$

$$\le \int_0^d \left(\int_\Sigma \int_0^\rho |\nabla u(x + t\sigma)|^p t^{n-1} dt d\sigma \right)^{1/p}$$

$$\times \left(\int_\Sigma \int_0^\rho t^{(1-n)/(p-1)} dt d\sigma \right)^{(p-1)/p} \rho^{n-1} d\rho$$

$$= \left(\frac{p-1}{p-n} \Omega_n \right)^{(p-1)/p} \int_0^d \rho^{n-n/p} \left(\int_{|z-x|<\rho} |\nabla u(z)|^p dz \right)^{1/p} d\rho$$

$$\le \left(\frac{p-1}{p-n} \Omega_n \right)^{(p-1)/p} \frac{d^{n+1-n/p}}{n+1-n/p} \left(\int_{R^n} |\nabla u(z)|^p dz \right)^{1/p}.$$

Estimating the second term of (3.47) analogously we obtain (3.46).

Lemma 3.7 If $1 \le p < n$, then for any $u \in C_0^1(R^n)$

$$|u|_{np/(n-p)} \le \frac{p(n-1)}{2(n-p)} \prod_{i=1}^n |D_i u|_p^{1/n}. \tag{3.48}$$

Proof. We begin with the case $p = 1$ and prove

$$|u|_{n/(n-1)} \le \frac{1}{2} \prod_{i=1}^n |D_i u|_1^{1/n}. \tag{3.49}$$

Since

$$u(x) = u(x_1, \ldots, x_n) = \int_{-\infty}^{x_j} D_j u(x_1, \ldots, x_{j-1}, t, x_{j+1}, \ldots, x_n) dt$$

$$= - \int_{x_j}^\infty D_j u(x_1, \ldots, x_{j-1}, t, x_{j+1}, \ldots, x_n) dt,$$

we have

$$2|u(x)| \le \int_{-\infty}^{\infty} |D_j u(x_1, \ldots, x_{j-1}, t, x_{j+1}, \ldots, x_n)| dt.$$

Therefore

$$(2|u(x)|)^{n/(n-1)} = [(2|u(x)|)^n]^{1/(n-1)}$$
$$\le \prod_{j=1}^{n} \left(\int_{-\infty}^{\infty} |D_j u(x_1, \ldots, x_{j-1}, t, x_{j+1}, \ldots, x_n)| dt \right)^{1/(n-1)}.$$

Integrating both sides with respect to x_1 and applying Hölder's inequality

$$\int_{-\infty}^{\infty} (2|u(x)|)^{n/(n-1)} dx_1 \le \left(\int_{-\infty}^{\infty} |D_1 u(t, x_2, \ldots, x_n)| dt \right)^{1/(n-1)}$$
$$\times \int_{-\infty}^{\infty} \prod_{j=2}^{n} \left(\int_{-\infty}^{\infty} |D_j u(x_1, \ldots, x_{j-1}, t, x_{j+1}, \ldots, x_n)| dt \right)^{1/(n-1)} dx_1$$
$$\le \left(\int_{-\infty}^{\infty} |D_1 u(t, x_2, \ldots, x_n)| dt \right)^{1/(n-1)}$$
$$\times \prod_{j=2}^{n} \left(\int_{-\infty}^{\infty} \int_{-\infty}^{\infty} |D_j u(x_1, \ldots, x_{j-1}, t, x_{j+1}, \ldots, x_n)| dt dx_1 \right)^{1/(n-1)}$$
$$= \left(\int_{-\infty}^{\infty} |D_1 u| dx_1 \right)^{1/(n-1)} \prod_{j=2}^{n} \left(\int_{-\infty}^{\infty} \int_{-\infty}^{\infty} |D_j u| dx_1 dx_j \right)^{1/(n-1)}$$
$$= \left(\int_{-\infty}^{\infty} \int_{-\infty}^{\infty} |D_2 u| dx_1 dx_2 \right)^{1/(n-1)} \left(\int_{-\infty}^{\infty} |D_1 u| dx_1 \right)^{1/(n-1)}$$
$$\times \prod_{j=3}^{n} \left(\int_{-\infty}^{\infty} \int_{-\infty}^{\infty} |D_j u| dx_1 dx_j \right)^{1/(n-1)}.$$

Next, integrating with respect to x_2 and applying Hölder's inequality

$$\int_{-\infty}^{\infty} \int_{-\infty}^{\infty} (2|u|)^{n/(n-1)} dx_1 dx_2 \le \prod_{i=1}^{2} \left(\int_{-\infty}^{\infty} \int_{-\infty}^{\infty} |D_i u| dx_1 dx_2 \right)^{1/(n-1)}$$
$$\times \prod_{j=3}^{n} \left(\int_{-\infty}^{\infty} \int_{-\infty}^{\infty} \int_{-\infty}^{\infty} |D_j u| dx_1 dx_2 dx_j \right)^{1/(n-1)}.$$

Repeating this process we obtain (3.49). When $p > 1$, we put $v = |u|^{(n-1)p/(n-p)}$. Since $(n-1)p/(n-p) > 1, v \in C_0^1(R^n)$ and

$$|D_i v| = |D_i(|u|^2)^{(n-1)p/2(n-p)}|$$

$$= \frac{(n-1)p}{n-p}|u|^{(np+p-2n)/(n-p)}|\mathrm{Re}(uD_i\bar{u})|$$

$$\le \frac{(n-1)p}{n-p}|u|^{n(p-1)/(n-p)}|D_iu|.$$

Hence

$$|D_iv|_1 \le \frac{(n-1)p}{n-p}\left(\int_{R^n}|u|^{np/(n-p)}dx\right)^{(p-1)/p}\left(\int_{R^n}|D_iu|^pdx\right)^{1/p}$$

$$= \frac{(n-1)p}{n-p}|u|^{n(p-1)/(n-p)}_{np/(n-p)}|D_iu|_p.$$

Applying (3.49) to v and noting

$$|v|_{n/(n-1)} = |u|^{(n-1)p/(n-p)}_{np/(n-p)}$$

we obtain (3.48).

Now we prove Gagliardo-Nirenberg's inequality ([67],[68],[118]).

Theorem 3. 3 *Let $1 \le p \le \infty, 1 \le q \le \infty$, and let j,m be integers satisfying $0 \le j < m$. If $m-j-n/p$ is not a nonnegative integer, then for any a satisfying $j/m \le a \le 1$ there exists a constant γ depending only on n,m,j,p,q,a such that*

$$|D^ju|_r \le \gamma|u|^a_{m,p}|u|^{1-a}_{0,q} \tag{3.50}$$

for any $u \in C^m_0(R^n)$, where

$$\frac{1}{r} = \frac{j}{n} + a\left(\frac{1}{p} - \frac{m}{n}\right) + \frac{1-a}{q}. \tag{3.51}$$

If $m-j-n/p$ is a nonnegative integer, (3.50) holds for $a = j/m$.

Remark 3. 1 *Since $a \ge j/m$*

$$\frac{1}{r} = \frac{m}{n}\left(\frac{j}{m} - a\right) + \frac{a}{p} + \frac{1-a}{q} \le \frac{a}{p} + \frac{1-a}{q} \le 1.$$

Hence, either $r < 0$ or $r \ge 1$. If $r \ge 1$ (3.50) is

$$|u|_{j,r} \le \gamma|u|^a_{m,p}|u|^{1-a}_{0,q}.$$

If in particular $p = q, a = j/m$, this reduces to

$$|u|_{j,p} \le \gamma|u|^{j/m}_{m,p}|u|^{1-j/m}_{0,p}. \tag{3.52}$$

Remark 3. 2 Further results in case $m - j - n/p$ is a nonnegative integer will be given later.

Proof of Theorem 3.3. (i) Case $a = j/m$. This case (3.51) reduces to

$$\frac{1}{r} = \frac{j}{m}\frac{1}{p} + \left(1 - \frac{j}{m}\right)\frac{1}{q}.$$

Since $r \geq 1$ what should be proved is

$$|u|_{j,r} \leq \gamma |u|_{m,p}^{j/m} |u|_{0,q}^{1-j/m}. \tag{3.53}$$

This is obvious if $j = 0$, and so we assume $0 < j < m$. We begin with the case $j = 1$, and show

$$|u|_{1,r} \leq C|u|_{m,p}^{1/m}|u|_{0,q}^{1-1/m} \tag{3.54}$$

by induction with respect to m. If $m = 2$, this reduces to

$$|u|_{1,r} \leq C|u|_{2,p}^{1/2}|u|_{0,q}^{1/2} \tag{3.55}$$

for $1/r = (1/p + 1/q)/2$. This is due to Lemma 3.5 if $n = 1$. If $n > 1$, writing $x = (x_1, x'), x' = (x_2, \ldots, x_n)$, we apply Lemma 3.5 to $u(\cdot, x')$. If $p < \infty, q < \infty$,

$$\int_{-\infty}^{\infty} |D_1 u(x_1, x')|^r dx_1$$

$$\leq (4\sqrt{2})^r \left(\int_{-\infty}^{\infty} |D_1^2 u(x_1, x')|^p dx_1\right)^{r/2p} \left(\int_{-\infty}^{\infty} |u(x_1, x')|^q dx_1\right)^{r/2q}.$$

With the aid of Hölder's inequality

$$|D_1 u|_{0,r}^r = \int_{R^{n-1}} \int_{-\infty}^{\infty} |D_1 u(x_1, x')|^r dx_1 dx'$$

$$\leq (4\sqrt{2})^r \left(\int_{R^n} |D_1^2 u|^p dx\right)^{r/2p} \left(\int_{R^n} |u|^q dx\right)^{r/2q}.$$

We can estimate $|D_2 u|_{0,r}, \ldots, |D_n u|_{0,r}$ analogously. The inequalities in case of $p = \infty$ or $q = \infty$ are established similarly, and the proof of (3.54) in case of $m = 2$ is complete.

Suppose (3.54) is true with $m - 1$ in place of m. Let

$$\frac{1}{r} = \frac{1}{m}\frac{1}{p} + \left(1 - \frac{1}{m}\right)\frac{1}{q}, \quad \frac{1}{p_1} = \frac{1}{m-1}\frac{1}{p} + \left(1 - \frac{1}{m-1}\right)\frac{1}{r}.$$

In view of the induction hypothesis

$$|Du|_{1,p_1} \le C|Du|_{m-1,p}^{1/(m-1)}|Du|_{0,r}^{(m-2)/(m-1)},$$

or

$$|u|_{2,p_1} \le C|u|_{m,p}^{1/(m-1)}|u|_{1,r}^{(m-2)/(m-1)}. \tag{3.56}$$

Noting $1/r = (1/p_1 + 1/q)/2$ we get in view of (3.55)

$$|u|_{1,r} \le C|u|_{2,p_1}^{1/2}|u|_{0,q}^{1/2}. \tag{3.57}$$

Combining (3.56),(3.57) we obtain (3.54). Thus (3.53) has been proved when $j = 1$.

Suppose (3.53) is known to hold for $0, \ldots, j-1$ instead of j. If we set

$$\frac{1}{r} = \frac{j}{m}\frac{1}{p} + \left(1 - \frac{j}{m}\right)\frac{1}{q}, \quad \frac{1}{q_1} = \frac{1}{m}\frac{1}{p} + \left(1 - \frac{1}{m}\right)\frac{1}{q},$$

then

$$\frac{1}{r} = \frac{j-1}{m-1}\frac{1}{p} + \left(1 - \frac{j-1}{m-1}\right)\frac{1}{q_1}.$$

Therefore, making use of the induction hypothesis with $m - 1$ in place of m

$$|Du|_{j-1,r} \le C|Du|_{m-1,p}^{(j-1)/(m-1)}|Du|_{0,q_1}^{(m-j)/(m-1)},$$

or

$$|u|_{j,r} \le C|u|_{m,p}^{(j-1)/(m-1)}|u|_{1,q_1}^{(m-j)/(m-1)}. \tag{3.58}$$

In view of (3.54)

$$|u|_{1,q_1} \le C|u|_{m,p}^{1/m}|u|_{0,q}^{1-1/m}. \tag{3.59}$$

The inequality (3.53) follows from (3.58),(3.59).

(ii) Case $a = 1$. Since $m - j - n/p$ is not a nonnegative integer, we have $1 \le p < \infty$. What is to be shown is

$$|D^j u|_r \le C|u|_{m,p} \tag{3.60}$$

for $1/r = j/n + 1/p - m/n$. We begin with the case $j = 0$, i.e. with showing

$$|u|_r \le C|u|_{m,p} \tag{3.61}$$

for $1/r = 1/p - m/n$ assuming $m - n/p$ is not a nonnegative integer. If $m = 1, n \neq p < \infty$ since $1 - n/p$ is not a nonnegative integer. If $p > n$, then $r = np/(n - p) < 0$. In view of Lemma 3.6 we get

$$|u|_r = |u|_{-n/r,\infty} = |u|_{1-n/p,\infty} \leq C|u|_{1,p}.$$

If $p < n$, then $r > p$. In view of Lemma 3.7

$$|u|_r = |u|_{0,r} = |u|_{np/(n-p)} \leq \frac{p(n-1)}{2(n-p)} \prod_{i=1}^{n} |D_i u|_{0,p}^{1/n} \leq \frac{p(n-1)}{2(n-p)} |u|_{1,p}.$$

Hence, we see that (3.61) holds if $m = 1$.

Suppose (3.61) has been established up to $m - 1$. Set

$$\frac{1}{r} = \frac{1}{p} - \frac{m}{n}, \quad \frac{1}{r_1} = \frac{1}{p} - \frac{m-1}{n} = \frac{1}{r} + \frac{1}{n}. \tag{3.62}$$

Noting that $m - 1 - n/p$ is not a nonnegative integer we apply the induction hypothesis to Du. Then we get

$$|Du|_{r_1} \leq C|Du|_{m-1,p} \leq C|u|_{m,p}. \tag{3.63}$$

If $r > 0$, then $r > r_1 > p \geq 1$. So, with the aid of (3.61) for the case of $m = 1$ we get

$$|u|_r \leq C|u|_{1,r_1}. \tag{3.64}$$

We obtain (3.61) from (3.63) and (3.64). Next consider the case $r < 0$. If r_1 determined by (3.62) is positive, then $r_1 > n$. In view of Lemma 3.6 we get

$$|u|_r = |u|_{-n/r,\infty} = |u|_{1-n/r_1,\infty} \leq C|u|_{1,r_1}.$$

Combining this with (3.63) we obtain (3.61). If $r_1 < 0$, we set $l = [-n/r], h = -n/r - [-n/r]$. Since $-n/r = m - n/p$ is not a nonnegative integer, $h > 0$. Noting $-n/r_1 = l - 1 + h$ we get

$$|u|_r = |u|_{-n/r,\infty} = \sup_{x \neq y} \frac{|D^l u(x) - D^l u(y)|}{|x - y|^h}$$

$$= \sup_{x \neq y} \frac{|D^{l-1} Du(x) - D^{l-1} Du(y)|}{|x - y|^h} \leq \max_i |D_i u|_{-n/r_1,\infty} \leq \max_i |D_i u|_{r_1}.$$

The inequality (3.61) follows from this and (3.63). The proof of the case $j = 0$ is complete.

Suppose $j > 0$ and (3.60) is known to hold with j replaced by $j - 1$. If $1/r = j/n + 1/p - m/n$, then $1/r = (j-1)/n + 1/p - (m-1)/n$ and $(m-1) - (j-1) - n/p$ is not a nonnegative integer. Therefore, by the induction hypothesis

$$|D^j u|_r = |D^{j-1} Du|_r \leq C|Du|_{m-1,p} \leq C|u|_{m,p}.$$

Thus, the proof of the case $a = 1$ is complete.

(iii) Case $j/m < a < 1$. If we define r_1, r_2, θ by

$$\frac{1}{r_1} = \frac{j}{m}\frac{1}{p} + \left(1 - \frac{1}{m}\right)\frac{1}{q}, \quad \frac{1}{r_2} = \frac{j}{n} + \frac{1}{p} - \frac{m}{n}, \quad \theta = \frac{1-a}{1 - j/m},$$

then by virtue of (i) and (ii)

$$|D^j u|_{r_1} \leq C|u|_{m,p}^{j/m}|u|_{0,q}^{1-j/m}, \quad |D^j u|_{r_2} \leq C|u|_{m,p}. \tag{3.65}$$

Since $0 < \theta < 1, 1/r = \theta/r_1 + (1-\theta)/r_2$, we have in view of Theorem 3.2

$$|D^j u|_r \leq C|D^j u|_{r_1}^{\theta}|D^j u|_{r_2}^{1-\theta}. \tag{3.66}$$

The desired result follows from (3.65),(3.66).

Lemma 3.8 *Let m and j be integers satisfying $0 \leq j < m$. Let ϕ be a function in $C_0^m(R^n)$ such that $\phi \geq 0, \phi(x) = 1$ if $|x| \leq 1/2, \phi(x) = 0$ if $|x| > 1$. Then, for any function u in $C^m(R^n)$ and β with $|\beta| = j$*

$$D^\beta u = \sum_{|\alpha|=m} K_\alpha * D^\alpha u + L * u, \tag{3.67}$$

where

$$K_\alpha(x) = \frac{1}{(m-1)!\Omega_n}\phi(x)D^\beta\left(\frac{x^\alpha}{|x|^n}\right) \tag{3.68}$$

and $L(x)$ is a continuous function with support contained in the set $\{x; 1/2 \leq |x| \leq 1\}$.

Proof. Let $\sigma \in \Sigma$. By induction we can show

$$u(x) = \phi(0)u(x) = (-1)^m \int_0^\infty \frac{t^{m-1}}{(m-1)!}\phi(t\sigma)\left(\frac{\partial}{\partial t}\right)^m u(x - t\sigma)dt$$

$$- \sum_{k=0}^{m-1} \int_0^\infty \left(\frac{\partial}{\partial t}\right)^k \left(\frac{t^k}{k!}\frac{\partial}{\partial t}\phi(t\sigma)\right) \cdot u(x - t\sigma)dt$$

$$= \int_0^\infty \frac{t^{m-1}}{(m-1)!} \phi(t\sigma) \sum_{|\alpha|=m} \sigma^\alpha D^\alpha u(x - t\sigma) dt$$

$$- \sum_{k=0}^{m-1} \int_0^\infty \sum_{j=0}^k \binom{k}{j} \left(\frac{\partial}{\partial t}\right)^{k-j} \frac{t^k}{k!} \cdot \left(\frac{\partial}{\partial t}\right)^{j+1} \phi(t\sigma) \cdot u(x - t\sigma) dt$$

$$= \int_0^\infty \frac{t^{m-1}}{(m-1)!} \phi(t\sigma) \sum_{|\alpha|=m} \sigma^\alpha D^\alpha u(x - t\sigma) dt$$

$$- \sum_{k=0}^{m-1} \sum_{j=0}^k \int_0^\infty \binom{k}{j} \frac{t^j}{j!} \sum_{|\alpha|=j+1} \sigma^\alpha D^\alpha \phi(t\sigma) \cdot u(x - t\sigma) dt.$$

Integrating over Σ

$$\Omega_n u(x) = \int_\Sigma \int_0^\infty \frac{t^{m-1}}{(m-1)!} \phi(t\sigma) \sum_{|\alpha|=m} \sigma^\alpha D^\alpha u(x - t\sigma) dt d\sigma$$

$$- \sum_{k=0}^{m-1} \sum_{j=0}^k \int_\Sigma \int_0^\infty \binom{k}{j} \frac{t^j}{j!} \sum_{|\alpha|=j+1} \sigma^\alpha D^\alpha \phi(t\sigma) \cdot u(x - t\sigma) dt d\sigma$$

$$= \frac{1}{(m-1)!} \sum_{|\alpha|=m} \int_{R^n} |y|^{-n} y^\alpha \phi(y) D^\alpha u(x - y) dy$$

$$- \sum_{k=0}^{m-1} \sum_{j=0}^k \binom{k}{j} \frac{1}{j!} \int_{R^n} |y|^{-n} \sum_{|\alpha|=j+1} y^\alpha D^\alpha \phi(y) \cdot u(x - y) dy.$$

Dividing by Ω_n and letting D^β operate

$$D^\beta u(x) = \frac{1}{(m-1)!\Omega_n} \sum_{|\alpha|=m} \int_{R^n} D^\beta \left(\phi(y) y^\alpha |y|^{-n}\right) \cdot D^\alpha u(x - y) dy$$

$$- \frac{1}{\Omega_n} \int_{R^n} \sum_{k=0}^{m-1} \sum_{j=0}^k \frac{1}{j!} \binom{k}{j} \sum_{|\alpha|=j+1} D^\beta \left(y^\alpha |y|^{-n} D^\alpha \phi(y)\right) \cdot u(x - y) dy$$

$$= \frac{1}{(m-1)!\Omega_n} \sum_{|\alpha|=m} \int_{R^n} \phi(y) D^\beta \left(y^\alpha |y|^{-n}\right) \cdot D^\alpha u(x - y) dy$$

$$+ \frac{1}{(m-1)!\Omega_n} \sum_{|\alpha|=m} \int_{R^n} \sum_{\gamma \neq \beta} \binom{\beta}{\gamma} D^{\beta-\gamma} \phi(y)$$

$$\cdot D^\gamma \left(y^\alpha |y|^{-n}\right) \cdot D^\alpha u(x - y) dy$$

$$- \frac{1}{\Omega_n} \int_{R^n} \sum_{k=0}^{m-1} \sum_{j=0}^k \frac{1}{j!} \binom{k}{j} \sum_{|\alpha|=j+1} D^\beta \left(y^\alpha |y|^{-n} D^\alpha \phi(y)\right) \cdot u(x - y) dy.$$

Hence (3.67) holds with K_α defined by (3.68) and with

$$L(x) = \frac{1}{(m-1)!\Omega_n} \sum_{|\alpha|=m} \sum_{\gamma \neq \beta} \binom{\beta}{\gamma} D^\alpha \left(D^{\beta-\gamma}\phi(x) \cdot D^\gamma \left(x^\alpha |x|^{-n} \right) \right)$$

$$- \frac{1}{\Omega_n} \sum_{k=0}^{m-1} \sum_{j=0}^{k} \frac{1}{j!} \binom{k}{j} \sum_{|\alpha|=j+1} D^\beta \left(x^\alpha |x|^{-n} D^\alpha \phi(x) \right).$$

By the definition (3.68) of K_α there exists a constant C such that

$$|K_\alpha(x)| \leq C\phi(x)|x|^{m-n-j}. \tag{3.69}$$

Next we consider the case $m - j - n/p$ is a nonnegative integer. We begin with the following lemma.

Lemma 3. 9 Let A and B be measurable subsets of R^n, and let G be an integral operator

$$(Gf)(x) = \int_B G(x,y)f(y)dy$$

with kernel $G(x,y)$ which is a measurable function defined in $A \times B$. Let $1 \leq p,q,r \leq \infty$ and $1/r = 1/p + 1/q - 1$. If

$$\left(\int_A |G(x,y)|^q dx \right)^{1/q} \leq K \quad \text{for all } y \in B$$

$$\left(\int_B |G(x,y)|^q dy \right)^{1/q} \leq K \quad \text{for all } x \in A,$$

then

$$\left(\int_A |(Gf)(x)|^r dx \right)^{1/r} \leq K \left(\int_B |f(y)|^p dy \right)^{1/p}.$$

If $q = \infty, r = \infty$ or $p = \infty$, the corresponding integrals are understood as the essential supremum.

Proof. We follow the proof of Lemma 2.6.1 of N. Kerzman [94]. We may assume $G, f \geq 0$. We consider only the case in which p, q, r are all finite, since the proof of other cases is immediate. We denote the L^p norm by $\| \ \|_p$. With the aid of Hölder's inequality

$$(Gf)(x) = \int_B G^{q/r} f^{p/r} G^{(r-q)/r} f^{(r-p)/r} dy$$

$$\leq \left(\int_B G^q f^p dy \right)^{1/r} \left(\int_B G^q dy \right)^{(r-q)/qr} \|f\|_p^{(r-p)/r}$$

$$\leq K^{(r-q)/r} \left(\int_B G(x,y)^q f(y)^p dy \right)^{1/r} \|f\|_p^{(r-p)/r}.$$

Taking the rth power and integrating over A we get

$$\|Gf\|_r^r \leq K^{r-q} \int_B \int_A G(x,y)^q dx f(y)^p dy \|f\|_p^{r-p} \leq K^r \|f\|_p^r.$$

Theorem 3.4 *Let m, j be integers satisfying $0 \leq j < m$, and let $1 \leq p < \infty$, $1 \leq q < \infty$. Suppose $m - j - n/p = 0$, or $p = n/(m-j)$. Then for any a satisfying $j/m \leq a < 1$ there exists a constant γ depending only on m, j, q, a, n such that for any $u \in C_0^m(R^n)$*

$$|u|_{j,r} \leq \gamma |u|_{m,p}^a |u|_{0,q}^{1-a}, \tag{3.70}$$

where

$$\frac{1}{r} = \frac{j}{n} + a \left(\frac{1}{p} - \frac{m}{n} \right) + \frac{1-a}{q} = (1-a) \left(\frac{j}{n} + \frac{1}{q} \right). \tag{3.71}$$

Proof. Put $s = mnq/(m-j)(n+jq)$. Then the set of values of r when a ranges over $[j/m, 1)$ is $[s, \infty)$. So we are going to show (3.70) for $s \leq r < \infty$ with a defined by (3.71). In view of Theorem 3.3 we know that

$$|u|_{j,s} \leq C |u|_{m,p}^{j/m} |u|_{0,q}^{(m-j)/m} \tag{3.72}$$

is true (the case $a = j/m$). Here we note that $s \geq 1$ by Remark 3.1. Since $1/s$ is a convex combination of $1/p$ and $1/q$, we have

$$s \leq \max \{p, q\}. \tag{3.73}$$

Suppose $\max \{p, q\} \leq r < \infty$. We use Lemma 3.8:

$$D^\beta u = \sum_{|\alpha|=m} K_\alpha * D^\alpha u + L * u,$$

where $|\beta| = j$. Let q_1 be a real number defined by $1/r = 1/p + 1/q_1 - 1$. Then, we have

$$1 \leq q_1 < n/(n-m+j). \tag{3.74}$$

By virtue of (3.69) and (3.74)

$$\int_{R^n} |K_\alpha(x)|^{q_1} dx \leq C \int_{|x| \leq 1} |x|^{(m-n-j)q_1} dx < \infty.$$

Therefore, in view of Lemma 3.9

$$|D^\beta u|_{0,r} \le \sum_{|\alpha|=m} |K_\alpha * D^\alpha u|_{0,r} + |L * u|_{0,r} \le C \left(\sum_{|\alpha|=m} |D^\alpha u|_{0,p} + |u|_{0,q} \right).$$

Hence, we get

$$|u|_{j,r} \le C(|u|_{m,p} + |u|_{0,q}). \tag{3.75}$$

Applying this inequality to the function $u_\lambda(x) = u(\lambda x)$, where $\lambda > 0$, and noting

$$|u_\lambda|_{j,r} = \lambda^{j-n/r}|u|_{j,r}, |u_\lambda|_{m,p} = \lambda^{m-n/p}|u|_{m,p}, |u_\lambda|_{0,q} = \lambda^{-n/q}|u|_{0,q} \tag{3.76}$$

we obtain

$$|u|_{j,r} \le C \left(\lambda^{m-n/p-j+n/r}|u|_{m,p} + \lambda^{-n/q-j+n/r}|u|_{0,q} \right).$$

Minimizing the right hand side with respect to λ we conclude (3.70).

Next, suppose $s < r < \max\{p,q\}$. Let t be a number such that $t \ge \max\{p,q\}$. Then, by what was just shown above

$$|u|_{j,t} \le C|u|_{m,p}^b |u|_{0,q}^{1-b}, \tag{3.77}$$

$$\frac{1}{t} = (1-b) \left(\frac{j}{n} + \frac{1}{q} \right). \tag{3.78}$$

Since $s < r < t$ we have with the aid of Hölder's inequality

$$|u|_{j,r} \le |u|_{j,t}^{t(r-s)/r(t-s)} |u|_{j,s}^{s(t-r)/r(t-s)}. \tag{3.79}$$

Combining (3.72),(3.77),(3.79) we get

$$|u|_{j,r} \le C|u|_{m,p}^a |u|_{0,q}^{1-a}, \quad a = b\frac{t(r-s)}{r(t-s)} + \frac{j}{m}\frac{s(t-r)}{r(t-s)}.$$

In view of (3.78)

$$\begin{aligned}
\frac{j}{n} + a\left(\frac{1}{p} - \frac{m}{n}\right) &+ \frac{1-a}{q} = (1-a)\left(\frac{j}{n} + \frac{1}{q}\right) \\
&= \left(\frac{t(r-s)}{r(t-s)} + \frac{s(t-r)}{r(t-s)} - b\frac{t(r-s)}{r(t-s)} - \frac{j}{m}\frac{s(t-r)}{r(t-s)}\right)\left(\frac{j}{n} + \frac{1}{q}\right) \\
&= \frac{t(r-s)}{r(t-s)}(1-b)\left(\frac{j}{n} + \frac{1}{q}\right) + \frac{s(t-r)}{r(t-s)}\left(1 - \frac{j}{m}\right)\left(\frac{j}{n} + \frac{1}{q}\right) \\
&= \frac{r-s}{r(t-s)} + \frac{t-r}{r(t-s)} = \frac{t-s}{r(t-s)} = \frac{1}{r}.
\end{aligned}$$

Next consider the case $m - j - n/p > 0$. In the following theorem $m - j - n/p$ need not be an integer.

Theorem 3. 5 *Let m, j be integers satisfying $0 \le j < m$, and let $1 \le p \le \infty, 1 \le q \le \infty$. Suppose $m - j - n/p > 0$. Then, for any a satisfying*

$$\frac{j}{m} \le a \le \left(\frac{j}{n} + \frac{1}{q} \right) \Big/ \left(\frac{m}{n} - \frac{1}{p} + \frac{1}{q} \right) \tag{3.80}$$

there exists a constant γ depending only on n, m, j, p, q, a such that

$$|u|_{j,r} \le \gamma |u|_{m,p}^a |u|_{0,q}^{1-a} \tag{3.81}$$

for any $u \in C_0^m(R^n)$, where

$$\frac{1}{r} = \frac{j}{n} + a \left(\frac{1}{p} - \frac{m}{n} \right) + \frac{1-a}{q}. \tag{3.82}$$

Remark 3. 3 The rightmost side of (3.80) is less than 1. (3.80) is the range of a for which r defined by (3.82) ranges over $[s, \infty]$, where s is the value of r corresponding to $a = j/m$:

$$\frac{1}{s} = \frac{j}{m}\frac{1}{p} + \left(1 - \frac{j}{m} \right) \frac{1}{q}.$$

Proof of Theorem 3.5. If we prove (3.81) for $r = s$ and $r = \infty$, then the case $r \in (s, \infty)$ can be established following the proof of Theorem 3.4, where s is the number defined in the above remark. The case $r = s$ is proved in Theorem 3.3. The case $r = \infty$ can be established with the aid of (3.67),(3.69) and Lemma 3.9.

Next we consider the case

$$\left(\frac{j}{n} + \frac{1}{q} \right) \Big/ \left(\frac{m}{n} - \frac{1}{p} + \frac{1}{q} \right) < a < 1, \tag{3.83}$$

or

$$\frac{j}{n} + \frac{1}{p} - \frac{m}{n} < \frac{1}{r} < 0. \tag{3.84}$$

Lemma 3. 10 *Let m be a positive integer and $1 \le p \le \infty, 1 \le q \le \infty$. Suppose $m - n/p > 0$. Then for each integer k satisfying $0 \le k < m - n/p$ there exists a constant C such that for $u \in C_0^m(R^n)$*

$$|u|_{k,\infty} \le C \begin{cases} |u|_{m,p}^{(n+kq)p/(mpq+np-nq)} |u|_{0,q}^{(mp-n-kp)q/(mpq+np-nq)} \\ \qquad\qquad\qquad\qquad\qquad\qquad 1 \le p < \infty, 1 \le q < \infty \\[6pt] |u|_{m,\infty}^{(n+kq)/(n+mq)} |u|_{0,q}^{(m-k)q/(n+mq)} \quad p = \infty, 1 \le q < \infty \\[6pt] |u|_{m,p}^{kp/(mp-n)} |u|_{0,\infty}^{(mp-n-kp)/(mp-n)} \quad 1 \le p < \infty, q = \infty \\[6pt] |u|_{m,\infty}^{k/m} |u|_{0,\infty}^{(m-k)/m} \qquad\qquad p = q = \infty \end{cases}.$$

Proof. In view of Lemma 3.8 we have

$$D^\beta u = \sum_{|\alpha|=m} K_\alpha * D^\alpha u + L * u$$

for $|\beta| = k$, and

$$|K_\alpha(x)| \leq C\phi(x)|x|^{m-n-k}.$$

With the aid of Hölder's inequality

$$|D^\beta u(x)| \leq \sum_{|\alpha|=m} |K_\alpha|_{0,p'}|D^\alpha u|_{0,p} + |L|_{0,q'}|u|_{0,q}.$$

If $1 < p \leq \infty$, then

$$(m - n - k)p' > (-n + n/p)p' = -n.$$

Hence $K_\alpha \in L^{p'}(R^n)$. If $p = 1$, then $k \leq m-n-1$, and hence $K_\alpha \in L^\infty(R^n)$. Therefore

$$|D^\beta u(x)| \leq C(|u|_{m,p} + |u|_{0,q}).$$

Using the argument by which we deduced (3.70) from (3.75) we obtain the desired result.

Lemma 3.11 *Let m be a positive integer and $1 \leq p \leq \infty, 1 \leq q \leq \infty$. Suppose that $m - n/p$ is a positive integer. Then there exists a constant C depending only on m, n, p, q such that for $u \in C_0^m(R^n)$ and $|\beta| = m-n/p-1$*

$$|D^\beta u(x) - D^\beta u(y)|$$
$$\leq C|x - y|\left\{\left[1 + \left(\log^+ \frac{2}{|x - y|}\right)^{(p-1)/p}\right]|u|_{m,p} + |u|_{0,q}\right\} \quad (3.85)$$

holds for $x, y \in R^n$, where $\log^+ t = \log t$ if $t > 1$ and 0 if $t \leq 1$.

Proof. According to Lemma 3.8

$$D^\beta u = \sum_{|\alpha|=m} K_\alpha * D^\alpha u + L * u$$

for $|\beta| = m - n/p - 1$, where

$$K_\alpha(x) = \phi(x)P_\alpha(x)|x|^{-k}, \quad (3.86)$$

$P_\alpha(x)$ is a homogenous polynomial of degree $2m - n/p - 1$, $k = n + 2m - 2n/p - 2$, and hence

$$|K_\alpha(x)| \leq C\phi(x)|x|^{n/p-n+1}. \quad (3.87)$$

Consequently for $x, y \in R^n$

$$D^\beta u(x) - D^\beta u(y) = \sum_{|\alpha|=m} \int_{R^n} \left(K_\alpha(x-z) - K_\alpha(y-z) \right) D^\alpha u(z) dz$$

$$+ \int_{R^n} \left(L(x-z) - L(y-z) \right) u(z) dz.$$

Noting that $|y - z| \le 3|x - y|/2$ in the region $|x - z| \le |x - y|/2$, we see if $1 < p \le \infty$ that

$$\left| \int_{|x-z| \le |x-y|/2} \left(K_\alpha(x-z) - K_\alpha(y-z) \right) D^\alpha u(z) dz \right|$$

$$\le \int_{|x-z| \le |x-y|/2} |K_\alpha(x-z) D^\alpha u(z)| dz$$

$$+ \int_{|y-z| \le 3|x-y|/2} |K_\alpha(y-z) D^\alpha u(z)| dz$$

$$\le \left(\int_{|x-z| \le |x-y|/2} |K_\alpha(x-z)|^{p'} dz \right)^{1/p'} |D^\alpha u|_{0,p}$$

$$+ \left(\int_{|y-z| \le 3|x-y|/2} |K_\alpha(y-z)|^{p'} dz \right)^{1/p'} |D^\alpha u|_{0,p}. \qquad (3.88)$$

Noting $(n/p - n + 1)p' = -n + p'$

$$\left(\int_{|x-z| \le |x-y|/2} |K_\alpha(x-z)|^{p'} dz \right)^{1/p'}$$

$$\le C \left(\int_{|z| \le |x-y|/2} |z|^{-n+p'} dz \right)^{1/p'} \le C|x - y|.$$

Estimating the last integral of (3.88) analogously we get

$$\left| \int_{|x-z| \le |x-y|/2} \left(K_\alpha(x-z) - K_\alpha(y-z) \right) D^\alpha u(z) dz \right| \le C|x - y| |u|_{m,p}.$$

$$(3.89)$$

It is easy to see that this inequality is true also when $p = 1$. By the same method as above

$$\left| \int_{|y-z| \le |x-y|/2} \left(K_\alpha(x-z) - K_\alpha(y-z) \right) D^\alpha u(z) dz \right| \le C|x - y| |u|_{m,p}.$$

$$(3.90)$$

Next, we estimate

$$\left| \iint_{|x-z|>|x-y|/2, |y-z|>|x-y|/2} (K_\alpha(x-z) - K_\alpha(y-z)) D^\alpha u(z) dz \right|.$$

In the region

$$|x - z| > |x - y|/2 \quad \text{and} \quad |y - z| > |x - y|/2 \tag{3.91}$$

we have $|y - z|/3 < |x - z| < 3|y - z|$. In what follows for the time being we consider in the region (3.91).

$$K_\alpha(x-z) - K_\alpha(y-z) = (\phi(x-z) - \phi(y-z)) \frac{P_\alpha(x-z)}{|x-z|^k}$$

$$+ \phi(y-z) \left(\frac{P_\alpha(x-z)}{|x-z|^k} - \frac{P_\alpha(y-z)}{|y-z|^k} \right), \tag{3.92}$$

$$|P_\alpha(x-z)|/|x-z|^k \le C|x-z|^{n/p+1-n}. \tag{3.93}$$

Since ∇P_α is homogeneous of degree $2m - n/p - 2 \ge 0$, it is easy to see that

$$|P_\alpha(x-z) - P_\alpha(y-z)| \le C|x-y||y-z|^{2m-n/p-2}.$$

It is also easily seen that

$$\left| |x-z|^{-k} - |y-z|^{-k} \right| \le C|x-y||y-z|^{-k-1}.$$

Hence, we see that

$$\left| P_\alpha(x-z)|x-z|^{-k} - P_\alpha(y-z)|y-z|^{-k} \right| \le C|x-y||y-z|^{n/p-n}. \tag{3.94}$$

From (3.92),(3.93),(3.94) it follows that

$$|K_\alpha(x-z) - K_\alpha(y-z)| \le C|\phi(x-z) - \phi(y-z)||x-z|^{n/p+1-n}$$

$$+ C\phi(y-z)|x-y||y-z|^{n/p-n}. \tag{3.95}$$

Suppose $|x - y| \le 2$. With the aid of (3.95) we see that if $1 < p \le \infty$

$$\left| \iint_{|x-z|>|x-y|/2, |y-z|>|x-y|/2} (K_\alpha(x-z) - K_\alpha(y-z)) D^\alpha u(z) dz \right|$$

$$\le C \int_{|x-z| \le 3} |\phi(x-z) - \phi(y-z)| \, |x-z|^{n/p+1-n} |D^\alpha u(z)| dz$$

$$+ C|x-y| \int_{|y-z|>|x-y|/2} \phi(y-z)|y-z|^{n/p-n} |D^\alpha u(z)| dz$$

$$\leq C|x-y| \int_{|x-z|\leq 3} |x-z|^{n/p+1-n} |D^\alpha u(z)| dz$$

$$+C|x-y| \int_{|x-y|/2<|y-z|<1} |y-z|^{n/p-n} |D^\alpha u(z)| dz$$

$$\leq C|x-y| \left(\int_{|x-z|<3} |x-z|^{p'-n} dz \right)^{1/p'} |u|_{m,p}$$

$$+C|x-y| \left(\int_{|x-y|/2<|y-z|<1} |y-z|^{-n} dz \right)^{1/p'} |u|_{m,p}$$

$$\leq C|x-y||u|_{m,p} + C|x-y| \left(\log \frac{2}{|x-y|} \right)^{1/p'} |u|_{m,p}.$$

It is easy to see that this inequality is true also when $p = 1$. Hence, combining (3.89),(3.90) and this inequality we conclude that if $|x-y| \leq 2$

$$\left| \int_{R^n} \left(K_\alpha(x-z) - K_\alpha(y-z) \right) D^\alpha u(z) dz \right|$$

$$\leq C|x-y||u|_{m,p} + C|x-y| \left(\log \frac{2}{|x-y|} \right)^{(p-1)/p} |u|_{m,p}. \quad (3.96)$$

As is easily seen

$$\left| \int_{R^n} \left(L(x-z) - L(y-z) \right) u(z) dz \right|$$

$$\leq C|x-y| \int_{|x-z|\wedge|y-z|\leq 1} |u(z)| dz \leq C|x-y||u|_{0,q}. \quad (3.97)$$

It follows from (3.96) and (3.97) that if $|x-y| \leq 2$

$$|D^\beta u(x) - D^\beta u(y)|$$

$$\leq C|x-y| \left\{ \left[1 + \left(\log \frac{2}{|x-y|} \right)^{(p-1)/p} \right] |u|_{m,p} + |u|_{0,q} \right\}.$$

Finally consider the case $|x-y| > 2$. In view of Lemma 3.10 or its proof

$$|D^\beta u(x) - D^\beta u(y)| \leq |D^\beta u(x)| + |D^\beta u(y)|$$

$$\leq C \left(|u|_{m,p} + |u|_{0,q} \right) \leq C|x-y|(|u|_{m,p} + |u|_{0,q}).$$

Thus the proof of the lemma is complete.

Theorem 3.6 *Let m, j be integers satisfying $0 \le j < m$, and let $1 \le p \le \infty$, $1 \le q \le \infty$. Suppose that $m - j - n/p$ is a positive integer. Then for any a satisfying*

$$\left(\frac{j}{n} + \frac{1}{q}\right) \Big/ \left(\frac{m}{n} - \frac{1}{p} + \frac{1}{q}\right) < a < 1 \tag{3.98}$$

there exists a constant γ depending only on m, j, n, p, q, a such that for any $u \in C_0^m(R^n)$

$$|D^j u|_r \le \gamma |u|_{m,p}^a |u|_{0,q}^{1-a}, \tag{3.99}$$

where

$$\frac{1}{r} = \frac{j}{n} + a\left(\frac{1}{p} - \frac{m}{n}\right) + \frac{1-a}{q}. \tag{3.100}$$

Proof. In the present case the range of r is (3.84), and

$$|D^j u|_r = |D^j u|_{-n/r,\infty} \le |u|_{j-n/r,\infty}.$$

In view of (3.100)

$$j - n/r = a(m - n/p + n/q) - n/q < m - n/p. \tag{3.101}$$

If $-n/r$ is an integer, then by Lemma 3.10

$$|u|_{j-n/r,\infty} \le C|u|_{m,p}^a |u|_{0,q}^{1-a},$$

where $a = (n + jq - nq/r)p/(mpq + np - nq)$, with a suitable modification when $p = \infty$ or $q = \infty$. It is easily seen that a satisfies (3.98) and (3.100).

Next consider the case where $-n/r$ is not an integer. First suppose that

$$j + [-n/r] + 1 \le m - n/p - 1.$$

Set $k = j + [-n/r]$. Then

$$k < j - n/r < k + 1 \le m - n/p - 1. \tag{3.102}$$

By virtue of Lemma 3.10

$$|u|_{k,\infty} \le C|u|_{m,p}^b |u|_{0,q}^{1-b}, \tag{3.103}$$

$$|u|_{k+1,\infty} \le C|u|_{m,p}^c |u|_{0,q}^{1-c}, \tag{3.104}$$

where

$$b = (n + kq)p/(mpq + np - nq),$$
$$c = (n + kq + q)p/(mpq + np - nq) = b + pq/(mpq + np - nq).$$

By Theorem 3.1 and (3.102)

$$|u|_{j-n/r,\infty} \leq C|u|_{k+1,\infty}^{j-n/r-k}|u|_{k,\infty}^{k+1-j+n/r}. \tag{3.105}$$

Combining (3.103),(3.104),(3.105) we obtain

$$|u|_{j-n/r,\infty} \leq C|u|_{m,p}^{a}|u|_{0,q}^{1-a},$$

where

$$
\begin{aligned}
a &= c(j-n/r-k)+b(k+1-j+n/r) \\
&= [b+pq/(mpq+np-nq)]\,(j-n/r-k)+b(k+1-j+n/r) \\
&= b+pq(j-n/r-k)/(mpq+np-nq) \\
&= [(n+kq)p+pq(j-n/r-k)]\,/(mpq+np-nq) \\
&= (n+jq-nq/r)p/(mpq+np-nq).
\end{aligned}
$$

Finally suppose that

$$j+[-n/r]+1 > m-n/p-1. \tag{3.106}$$

It follows from (3.101) and (3.106) that $j+[-n/r] = m-n/p-1$. Set $h = -n/r-[-n/r]$. Then $-n/r = [-n/r]+h$ and $0 < h < 1$.

$$
\begin{aligned}
|D^j u|_{-n/r,\infty} &= \sup_{x \neq y} |D^{j+[-n/r]}u(x) - D^{j+[-n/r]}u(y)|/|x-y|^h \\
&= \sup_{x \neq y} |D^{m-n/p-1}u(x) - D^{m-n/p-1}u(y)|/|x-y|^h.
\end{aligned}
$$

If $|x-y| \leq 2$, then in view of Lemma 3.11

$$|D^{m-n/p-1}u(x) - D^{m-n/p-1}u(y)|/|x-y|^h$$
$$\leq C|x-y|^{1-h}\left\{\left[1+\left(\log\frac{2}{|x-y|}\right)^{(p-1)/p}\right]|u|_{m,p} + |u|_{0,q}\right\}$$
$$\leq C(|u|_{m,p} + |u|_{0,q}).$$

If $|x-y| \geq 2$, then by Lemma 3.10 or its proof

$$|D^{m-n/p-1}u(x) - D^{m-n/p-1}u(y)|/|x-y|^h$$
$$\leq C(|D^{m-n/p-1}u(x)| + |D^{m-n/p-1}u(y)|) \leq C(|u|_{m,p} + |u|_{0,q}).$$

Hence

$$|D^{m-n/p-1}u(x) - D^{m-n/p-1}u(y)|/|x-y|^h \leq C(|u|_{m,p} + |u|_{0,q})$$

whenever $x \neq y$. The desired result follows if we apply the argument by which we deduced (3.70) from (3.75).

It is easy to verify that the inequalities established so far hold also for functions for which the seminorms in the inequalities are all finite.

Theorem 3.7 *Let m, j be integers satisfying $0 \leq j < m$, and let $1 \leq p < \infty$. Suppose $m - j - n/p$ is not a nonnegative integer. Then, if $j/m \leq a \leq 1$ and $1/r = j/n + 1/p - am/n \geq 0$, we have $W^{m,p}(R^n) \subset W^{j,r}(R^n)$ and*

$$\|u\|_{j,r} \leq \gamma \left(|u|_{m,p}^a |u|_{0,p}^{1-a} + |u|_{0,p} \right). \tag{3.107}$$

If $a = j/m$ in particular

$$\|u\|_{j,p} \leq \gamma \left(|u|_{m,p}^{j/m} |u|_{0,p}^{1-j/m} + |u|_{0,p} \right). \tag{3.108}$$

This holds also when $m - j - n/p$ is a nonnegative integer.

Proof. It follows from the hypothesis that $r \geq p$. Applying Theorem 3.3 with $q = p$,

$$|u|_{j,r} \leq C|u|_{m,p}^a |u|_{0,p}^{1-a}. \tag{3.109}$$

If $a = j/m$ in particular, $r = p$, and hence

$$|u|_{j,p} \leq C|u|_{m,p}^{j/m} |u|_{0,p}^{1-j/m}, \tag{3.110}$$

which holds also when $m - j - n/p$ is a nonnegative integer. For an integer k satisfying $0 \leq k < j$ put $b = (am - j + k)/(m - j + k)$. Then

$$\frac{k}{m-j+k} \leq b \leq a \leq 1, \quad \frac{1}{r} = \frac{k}{n} + \frac{1}{p} - \frac{b(m-j+k)}{n}.$$

Hence, again by Theorem 3.3 with $q = p$

$$|u|_{k,r} \leq C|u|_{m-j+k,p}^b |u|_{0,p}^{1-b}. \tag{3.111}$$

Replacing j by $m - j + k$ in (3.110)

$$|u|_{m-j+k,p} \leq C|u|_{m,p}^{(m-j+k)/m} |u|_{0,p}^{(j-k)/m}.$$

Substituting this into (3.111) and applying Young's inequality

$$\begin{aligned}
|u|_{k,r} &\leq C|u|_{m,p}^{(am-j+k)/m} |u|_{0,p}^{(m-am+j-k)/m} \\
&= C \left(|u|_{m,p}^a |u|_{0,p}^{1-a} \right)^{(am-j+k)/am} |u|_{0,p}^{(j-k)/am} \tag{3.112} \\
&\leq C \left(|u|_{m,p}^a |u|_{0,p}^{1-a} + |u|_{0,p} \right).
\end{aligned}$$

The inequality (3.107) follows from (3.109),(3.112). Analogously (3.108) is established.

Corollary 3. 1 *If m is a positive integer, $1 \leq p < \infty$ and $m - n/p < 0$, then $W^{m,p}(R^n) \subset L^r(R^n)$ for $1/r = 1/p - m/n$.*

Theorem 3. 8 *Let m, j be integers satisfying $0 \leq j < m$ and $1 \leq p < \infty$. Suppose $m - j - n/p = 0$. Then for any r satisfying $p \leq r < \infty$, we have $W^{m,p}(R^n) \subset W^{j,r}(R^n)$ and*

$$\|u\|_{j,r} \leq \gamma \left(|u|^a_{m,p} |u|^{1-a}_{0,p} + |u|_{0,p} \right), \tag{3.113}$$

where a is a number such that $1/r = j/n + 1/p - am/n$, i.e. $j/m \leq a = 1 - n/mr < 1$.

Proof. Applying Theorem 3.4 with $q = p$ we get

$$|u|_{j,r} \leq C|u|^a_{m,p} |u|^{1-a}_{0,p}. \tag{3.114}$$

Let k be an integer such that $0 \leq k < j$. Put $b = m/k + n(1/p - 1/r)/m$. Then,

$$b < a, \quad k/m \leq b < k/m + n/mp, \quad 1/r = k/n + 1/p - bm/n.$$

Hence, in view of Theorem 3.5 and Lemma 1.1 (Young's inequality)

$$|u|_{k,r} \leq C|u|^b_{m,p} |u|^{1-b}_{0,p} \leq C \left(|u|^a_{m,p} |u|^{1-a}_{0,p} \right)^{b/a} |u|^{1-b/a}_{0,p}$$
$$\leq C \left(|u|^a_{m,p} |u|^{1-a}_{0,p} + |u|_{0,p} \right). \tag{3.115}$$

The inequality (3.113) follows from (3.114) and (3.115).

Theorem 3. 9 *Let m, j be integers satisfying $0 \leq j < m$, and let $1 \leq p \leq \infty$. Suppose $m - j - n/p > 0$. Then for any r satisfying $p \leq r \leq \infty$ we have $W^{m,p}(R^n) \subset W^{j,r}(R^n)$ and for any $u \in W^{m,p}(R^n)$*

$$\|u\|_{j,r} \leq \gamma \left(|u|^a_{m,p} |u|^{1-a}_{0,p} + |u|_{0,p} \right),$$

where a is a number satisfying $1/r = j/n + 1/p - am/n$.

Proof. The theorem can be shown with the aid of Theorem 3.5 in the same manner as Theorem 3.7.

Remark 3. 4 *Under the assumption of Theorem 3.9 a satisfies $j/m \leq a \leq j/m + n/mp < 1$.*

Theorem 3. 10 *Let m be a positive integer and $1 \leq p < \infty$. Suppose $m - n/p > 0$.*

(i) *If n/p is not an integer, then $W^{m,p}(R^n) \subset B^{m-n/p}(R^n)$. Put $l = m - [n/p] - 1$. Then, for each integer j satisfying $0 \le j \le l$*

$$\|u\|_{j,\infty} \le \gamma \left(|u|_{m,p}^{(n+jp)/mp} |u|_{0,p}^{(m-n-jp)/mp} + |u|_{0,p} \right). \tag{3.116}$$

For $0 \le h \le [n/p] + 1 - n/p$

$$|D^l u|_{h,\infty} \le \gamma |u|_{m,p}^{(n+lp+hp)/mp} |u|_{0,p}^{(mp-n-lp-hp)/mp}. \tag{3.117}$$

Hence for $0 \le s \le m - n/p$

$$\|u\|_{s,\infty} \le \gamma \left(|u|_{m,p}^{(n+sp)/mp} |u|_{0,p}^{(mp-n-sp)/mp} + |u|_{0,p} \right). \tag{3.118}$$

In particular

$$\|u\|_{m-n/p,\infty} \le \gamma \|u\|_{m,p}. \tag{3.119}$$

(ii) *If n/p is an integer, then for each s satisfying $0 \le s < m - n/p$ we have $W^{m,p}(R^n) \subset B^s(R^n)$. (3.116) holds for each integer j such that $0 \le j \le l = m - n/p - 1$. For $0 \le h < 1$ (3.117) holds with a constant γ depending also on h. For $0 \le s < m - n/p$ (3.118) holds with a constant γ depending also on s.*

Proof. (i) We apply Theorem 3.3 with $j = 0$ and $q = p$. For an integer k satisfying $0 \le k < m - n/p$ we take $a = (n + kp)/mp$. Then $n/mp \le a < 1$ and r defined by (3.51) is $-n/k$. Hence

$$|u|_{k,\infty} = |u|_{-n/k} \le C |u|_{m,p}^{(n+kp)/mp} |u|_{0,p}^{(mp-n-kp)/mp}.$$

Thus the inequality (3.116) is shown with the aid of the argument by which we derived (3.107) from (3.109), (3.112). We next apply Theorem 3.3 with $j = l$ and $p = q$. If we take $a = (n + lp + hp)/mp$, then $l/m < a \le 1$ and $r = -n/k$. Hence we obtain (3.117). (3.118) is a consequence of (3.116) and (3.117).

(ii) We simply apply Theorem 3.5 and 3.6 instead of Theorem 3.3 in the proof of (i).

3.5 Sobolev Spaces in General Domains

In this section we prove some preliminary results concerning Sobolev spaces in general nonempty open sets of R^n.

Theorem 3.11 *If Ω is a nonempty open set of R^n, m is a nonnegative integer and $1 \le p < \infty$, then $W^{m,p}(\Omega) \cap C^\infty(\Omega)$ is dense in $W^{m,p}(\Omega)$.*

Proof. We follow the proof of Theorem 6.3 of A. Friedman [65]. For $a > 0$ set

$$\Omega_a = \{x \in \Omega; \text{dist}(x, \partial\Omega) > a^{-1}, |x| < a\}. \tag{3.120}$$

For each natural number k let G_k, G'_k be the sets defined by

$$G_1 = \Omega_3, G'_1 = \Omega_{8/3}, G_k = \Omega_{k+2} \setminus \bar{\Omega}_k, G'_k = \Omega_{k+5/3} \setminus \bar{\Omega}_{k+1/3} \quad \text{for} \quad k \geq 2.$$

It is easy to show that $\{G'_k\}$ is an open covering of Ω, $\bar{G}'_k \subset G_k$ and any three of $\{G_k\}$ have an empty intersection. In case $G'_1 = \Omega_{8/3}$ is empty we renumber $\{\Omega_k\}$ appropriately. Let η_k be a function in $C_0^\infty(R^n)$ with support contained in G_k such that $\eta_k = 1$ in G'_k and $0 \leq \eta_k \leq 1$ in R^n. Set $\eta = \sum_{k=1}^\infty \eta_k$. For each $x \in \Omega$ at least one of $\{G'_k\}$ contains x and at most two of $\{G_k\}$ contain x. Hence, we have $1 \leq \eta(x) \leq 2$. Therefore if we set $\zeta_k(x) = \eta_k(x)/\eta(x)$, then $\zeta_k \in C_0^\infty(\Omega)$, $\text{supp}\zeta_k \subset G_k$, $\sum_{k=1}^\infty \zeta(x) \equiv 1$, i.e. $\{\zeta_k\}$ is a partition of unity subordinate to $\{G_k\}$. Let u be an arbitrary element of $W^{m,p}(\Omega)$ and $\epsilon > 0$. Set $w_k = \rho_{\epsilon_k} * (\zeta_k u)$, where ρ_ϵ is a mollifier. If ϵ_k is sufficiently small, then the support of w_k is contained in Ω_4 if $k = 1$ and in $\Omega_{k+3} \setminus \bar{\Omega}_{k-1}$ if $k \geq 2$, and

$$\|w_k - \zeta_k u\|_{m,p,\Omega} < 2^{(2-k)/p-2}\epsilon. \tag{3.121}$$

Set $w = \sum_{k=1}^\infty w_k$. Since five of the sets $\Omega_{k+3} \setminus \bar{\Omega}_{k-1}$ have an empty intersection, $w \in C^\infty(\Omega)$ and for $|\alpha| \leq m$

$$|D^\alpha u(x) - D^\alpha w(x)|^p = \left|\sum_{k=1}^\infty D^\alpha(\zeta_k(x)u(x) - w_k(x))\right|^p$$

$$\leq 4^{p-1} \sum_{k=1}^\infty |D^\alpha(\zeta_k(x)u(x) - w_k(x))|^p.$$

With the aid of this and (3.121) we obtain

$$\|u - w\|_{m,p,\Omega}^p \leq 4^{p-1} \sum_{k=1}^\infty \|\zeta_k u - w_k\|_{m,p,\Omega}^p$$

$$< 4^{p-1} \sum_{k=1}^\infty 2^{2-k-2p}\epsilon^p = \sum_{k=1}^\infty 2^{-k}\epsilon^p = \epsilon^p,$$

and the proof is complete.

By $H^{m,\infty}(\Omega)$ we denote the set of all functions whose derivatives of order up to $m - 1$ are all bounded and uniformly Lipschitz continuous in Ω. $H^{m,\infty}(\Omega)$ is a Banach space with norm

$$\|u\|'_{m,\infty,\Omega} = \|u\|_{m-1,\infty,\Omega} + \max_{|\alpha| \leq m-1} \sup_{x,y \in \Omega, x \neq y} \frac{|D^\alpha u(x) - D^\alpha u(y)|}{|x - y|}. \tag{3.122}$$

If $\Omega = R^n$, we simply write $\|u\|'_{m,\infty}$ short for $\|u\|'_{m,\infty,R^n}$.

Lemma 3.12 *For any nonempty open set $\Omega \subset R^n$,*

$$H^{m,\infty}(\Omega) \subset W^{m,\infty}(\Omega).$$

Proof. It suffices to show in case $m = 1$. Let $u \in H^{1,\infty}(\Omega)$ and suppose for all $x, y \in \Omega$

$$|u(x) - u(y)| \le L|x - y|. \tag{3.123}$$

Let e_j be the n-dimensional vector whose jth component is 1 and other components are 0. Let $\phi \in C_0^\infty(\Omega)$. If $|h|$ is sufficiently small, then by (3.123)

$$\left| \int_\Omega u(x)(\phi(x + he_j) - \phi(x))dx \right|$$

$$= \left| \int_\Omega (u(x - he_j) - u(x))\phi(x)dx \right| \le L|h| \int_\Omega |\phi(x)|dx.$$

Therefore

$$|(D_j u)(\phi)| = \left| \int_\Omega u(x)D_j\phi(x)dx \right|$$

$$= \lim_{h \to 0} \left| \int_\Omega u(x)h^{-1}(\phi(x + he_j) - \phi(x))\,dx \right| \le L \int_\Omega |\phi(x)|dx.$$

This means that $D_j u \in L^\infty(\Omega)$ since $C_0^\infty(\Omega)$ is dense in $L^1(\Omega)$.

Lemma 3.13 *Let Ω be a nonempty open set of R^n and $u \in W^{1,\infty}(\Omega)$. Then, if we modify the values of u on some null set, u is Lipschitz continuous in each compact subset of Ω. If $x \in \Omega$ and $|x - y| < \text{dist}(x, \partial\Omega)$*

$$|u(x) - u(y)| \le \left(\sum_{i=1}^n |D_i u|_{0,\infty,\Omega}^2 \right)^{1/2} |x - y|. \tag{3.124}$$

Proof. For $\epsilon > 0$ set

$$\Omega_\epsilon = \{x \in \Omega; \text{dist}(x, \partial\Omega) > \epsilon\}.$$

Let ρ_ϵ be a mollifier and set $u_\epsilon = \rho_\epsilon * u$. Then $u_\epsilon \in C^\infty(\Omega_\epsilon)$. Let $x \in \Omega$ and $|x - y| < \text{dist}(x, \partial\Omega)$. If ϵ is so small that $0 < \epsilon < \text{dist}(x, \partial\Omega)$ and $|x - y| < \text{dist}(x, \partial\Omega) - \epsilon$, then

$$u_\epsilon(y) - u_\epsilon(x) = \int_0^1 \frac{d}{dt} u_\epsilon(x + t(y - x))dt$$

$$= \int_0^1 \sum_{i=1}^n D_i u_\epsilon(x + t(y - x))(y_i - x_i)dt.$$

Therefore noting that $|D_i u_\epsilon|_{0,\infty,\Omega_\epsilon} \leq |D_i u|_{0,\infty,\Omega}$ we get

$$|u_\epsilon(x) - u_\epsilon(y)| \leq \sum_{i=1}^{n} |D_i u_\epsilon|_{0,\infty,\Omega_\epsilon} |x_i - y_i|$$

$$\leq \sum_{i=1}^{n} |D_i u|_{0,\infty,\Omega} |x_i - y_i| \leq \left(\sum_{i=1}^{n} |D_i u|_{0,\infty,\Omega}^2 \right)^{1/2} |x - y|.$$

The desired result is obtained by letting $\epsilon \to 0$.

We get from Lemmas 3.12 and 3.13 the following theorem.

Theorem 3. 12 *For any positive integer m*

$$W^{m,\infty}(R^n) = H^{m,\infty}(R^n).$$

Remark 3. 5 From the proof of Lemma 3.13 it follows that $W^{m,\infty}(\Omega) = H^{m,\infty}(\Omega)$ is true for any convex open set Ω.

3.6 Uniformly Regular Open Sets

Definition 3. 1 Let Ω be a nonempty open subset of R^n. If for any point a of the boundary $\partial\Omega$ there exists a neighborhood O of a and a homeomorphism Φ of class C^m from O to the open unit ball $\{y; |y| < 1\}$ of R^n such that

$$\Phi(a) = 0, \quad \Phi(O \cap \Omega) = \{y \in R^n; |y| < 1, y_n > 0\},$$
$$\Phi(O \cap \partial\Omega) = \{y \in R^n; |y| < 1, y_n = 0\}, \tag{3.125}$$

then Ω is called an *open set of class C^m*.

If Ω is an open set of class C^m with bounded boundary, then there exist finite points $a_1, \ldots, a_N$ on the boundary $\partial\Omega$ and a neighborhood O_i of a_i, and a homeomorphism Φ_i of class C^m from O_i to the open unit ball of $R^n, i = 1, \ldots, N$ satisfying the conditions of Definition 3.1 for $a = a_i, i = 1, \cdots, N$, and

$$\cup_{i=1}^N \Phi^{-1}(\{y; |y| < 1/2\}) \supset \partial\Omega.$$

When $\partial\Omega$ is not bounded, we consider following F. E. Browder [21] open sets with the following properties:

Definition 3. 2 Suppose Ω is a nonempty open subset of R^n whose boundary is not bounded. Then Ω is said to be *uniformly regular of class C^m* if there exist a family of open sets $\{O_i; i = 1, 2, \ldots\}$ and of homeomorphisms

$\{\Phi_i\}$ of O_i onto the unit ball $\{y; |y| < 1\}$ in R^n, and an integer N such that the following conditions are satisfied:

(1) Let $O_i' = \Phi_i^{-1}(\{y \in R^n; |y| < 1/2\})$. Then $\cup_{i=1}^{\infty} O_i'$ contains the N^{-1} neighborhood of $\partial\Omega$.

(2) For each i

$$\Phi_i(O_i \cap \Omega) = \{y; |y| < 1, y_n > 0\}, \tag{3.126}$$

$$\Phi_i(O_i \cap \partial\Omega) = \{y; |y| < 1, y_n = 0\}. \tag{3.127}$$

(3) Any $N + 1$ distinct sets of $\{O_i\}$ have an empty intersection.

(4) The family $\{O_i\}$ is locally finite, i.e. only a finite number of O_i have a nonempty intersection with some neighborhood of each point of R^n.

(5) Let $\Psi_i = \Phi_i^{-1}$ be the inverse mapping of Φ_i. Then for each $i = 1, 2, \ldots$ and $|y| < 1$

$$|\Psi_i(y) - \Psi_i(0)| < M. \tag{3.128}$$

Let $\Phi_{ik}(x), \Psi_{ik}(y)$ be the kth components of $\Phi_i(x), \Psi_i(y)$ respectively. Then

$$|D^\alpha \Phi_{ik}(x)| \le M, \quad |D^\alpha \Psi_{ik}(y)| \le M, \quad |\Phi_{in}(x)| \le M \mathrm{dist}(x, \partial\Omega) \tag{3.129}$$

for $|\alpha| \le m, x \in O_i, |y| < 1, k = 1, \ldots, n$, and $i = 1, 2, \ldots$.

When $m \ge 2$ or also when $m = 1$ if $\partial\Psi_{ik}/\partial y_j$ are equicontinuous, then (4) is implied by (3) and (5) as is shown in the following lemma.

Lemma 3.14 *Let $f(y) = (f_1(y), \ldots, f_n(y))$ be of class C^1 in the unit ball of R^n and suppose the Jacobian Df/Dy is different from 0 at $y = 0$. Then there exists a constant δ depending only on the norm of the inverse of the Jacobian matrix $f_y = \partial f/\partial y$ at $y = 0$ and the moduli of continuity of $\partial f_i/\partial y_j, i, j = 1, \ldots, n$, at $y = 0$ such that*

$$\{f(y); |y| < 1\} \supset \{x; |x - f(0)| < \delta\}.$$

Proof. Denote the norm of the matrix A be $|A|$. Let ϵ be a positive number such that $|f_y(y) - f_y(0)| \le 1/(2|f_y(0)^{-1}|)$ for $|y| \le \epsilon$. Set

$$g(y) = y - f_y(0)^{-1}(f(y) - x),$$

$$f(y) - f(z) = f_y(0)(y - z) + h(y, z)(y - z),$$

$$h(y, z) = \int_0^1 (f_y(z + t(y - z)) - f_y(0)) \, dt.$$

Then $|h(y, z)| \le 1/(2|f_y(0)^{-1}|)$ for $|y| \le \epsilon, |z| \le \epsilon$. Put $\delta = \epsilon/(2|f_y(0)^{-1}|)$. If $|x - f(0)| \le \delta$, then for each y such that $|y| \le \epsilon$

$$|g(y)| = |y - f_y(0)^{-1}(f(y) - f(0)) - f_y(0)^{-1}(f(0) - x)|$$

$$= |-f_y(0)^{-1}h(y, 0)y - f_y(0)^{-1}(f(0) - x)|$$

$$\le |f_y(0)^{-1}||h(y, 0)||y| + |f_y(0)^{-1}|\delta \le \epsilon.$$

Furthermore if $|y| \le \epsilon, |z| \le \epsilon$, then $|g(y) - g(z)| \le |y - z|/2$. Hence, if $|x - f(0)| \le \delta, g$ has a unique fixed point y which clearly satisfies $x = f(y)$.

Applying Lemma 3.14 to Ψ_i we see that each O_i contains a ball B_i of some fixed radius. In view of (3.128) the sets O_i which have a nonempty intersection with some bounded neighborhood of a point are located in some bounded region. If there exist infinite number of such O_i, then a limit point of the centers of the balls B_i contained in these O_i is contained in infinite number of B_i and hence of O_i. This contradicts (3).

Let Ω be an open set uniformly regular of class C^m, and $\{O_i\}, \{\Phi_i\}, N$ be as in Definition 3.2. Put

$$O_0 = \{x \in \Omega; \operatorname{dist}(x, \partial\Omega) > (4N)^{-1}\},$$
$$O_{-1} = \{x \notin \Omega; \operatorname{dist}(x, \partial\Omega) > (4N)^{-1}\}.$$

Then $\{O_i; i = -1, 0, 1, 2, \ldots\}$ is an open covering of R^n. Let λ be a function in $C_0^m(R^n)$ satisfying $\operatorname{supp}\lambda \subset \{y; |y| < 1\}, \lambda(y) = 1$ for $|y| < 1/2, 0 \le \lambda(y) \le 1$ in R^n. If we set

$$\eta_i(x) = \begin{cases} \lambda(\Phi_i(x)) & x \in O_i \\ 0 & x \notin O_i \end{cases}$$

for $i = 1, 2, \ldots$. Then $\eta_i \in C_0^m(R^n), \eta_i(x) = 1$ in $O_i', \operatorname{supp}\eta_i \subset O_i$ and $0 \le \eta_i(x) \le 1$ in R^n. Let η_0 and η_{-1} be functions in $C^m(R^n)$ such that

$$\operatorname{supp}\eta_0 \subset \Omega, \quad \operatorname{supp}\eta_{-1} \subset R^n \setminus \Omega,$$
$$\eta_0(x) = 1 \quad \text{for} \quad x \in \Omega \quad \text{and} \quad \operatorname{dist}(x, \partial\Omega) > 3/(4N),$$
$$\eta_0(x) = 0 \quad \text{for} \quad x \in \Omega \quad \text{and} \quad \operatorname{dist}(x, \partial\Omega) < 1/(2N),$$
$$\eta_{-1}(x) = 1 \quad \text{for} \quad x \notin \Omega \quad \text{and} \quad \operatorname{dist}(x, \partial\Omega) > 3/(4N),$$
$$\eta_{-1}(x) = 0 \quad \text{for} \quad x \notin \Omega \quad \text{and} \quad \operatorname{dist}(x, \partial\Omega) < 1/(2N).$$

Put $\eta(x) = \sum_{i=-1}^{\infty} \eta_i(x)$. Since $\{O_i; i = -1, 0, 1, 2, \ldots\}$ is an open covering of R^n such that at most $N + 2$ of O_i have a nonempty intersection, we have $1 \le \eta(x) \le N + 2$ in R^n and $\eta \in C^m(R^n)$. If we put

$$\zeta_i(x) = \eta_i(x)/\eta(x), \tag{3.130}$$

then $\zeta_i \in C^m(R^n), \operatorname{supp}\zeta_i \subset O_i, \sum_{i=-1}^{\infty} \zeta_i(x) = 1$ and $\{D^\alpha \zeta_i\}$ is uniformly bounded for each $|\alpha| \le m$.

In what follows in this chapter $\{\zeta_i\}_{i=-1}^{\infty}$ will always stand for this partition of unity.

3.7 Sobolev Spaces in Uniformly Regular Open Sets

Let $C^m(\bar{\Omega})$ be the set of all m times continuously differentiable functions in $\bar{\Omega}$, and $C_0^m(\bar{\Omega})$ be the totality of functions in $C^m(\bar{\Omega})$ which vanish outside a bounded set. Put

$$R_+^n = \{x = (x_1, \ldots, x_n) \in R^n; x_n > 0\},$$
$$\bar{R}_+^n = \{x = (x_1, \ldots, x_n) \in R^n; x_n \geq 0\}.$$

For $x = (x_1, \ldots, x_n) \in R^n$ we write $x' = (x_1, \ldots, x_{n-1})$ and $x = (x', x_n)$.

Theorem 3.13 *Let Ω be an open subset of R^n uniformly regular of class C^m. Then $C_0^\infty(\bar{\Omega})$ is dense in $W^{m,p}(\Omega)$ if $1 \leq p < \infty$.*

Proof. According to Theorem 3.11 it suffices to show that each function u in $W^{m,p}(\Omega) \cap C^\infty(\Omega)$ can be approximated by a sequence of functions in $C_0^\infty(\bar{\Omega})$ in the norm of $W^{m,p}(\Omega)$. Let ϕ be a function in $C_0^\infty(R^n)$ such that $\phi(x) = 1$ for $|x| < 1$ and $\phi(x) = 0$ for $|x| > 2$, and put $\phi_R(x) = \phi(x/R)$ for $R > 0$. Since $\phi_R u \to u$ in $W^{m,p}(\Omega)$ as $R \to \infty$, we may assume that u vanishes outside some bounded set. If we put $u_i = \zeta_i u$ for $i = 0, 1, 2, \ldots$, then we have $u = \sum_{i=0}^k u_i$ for some integer k. If we put

$$v_i(y) = \begin{cases} u_i(\Psi_i(y)) & |y| < 1, y_n > 0 \\ 0 & |y| \geq 1, y_n > 0 \end{cases}$$

for $i = 0, 1, \ldots, k$, then $v_i \in W^{m,p}(R_+^n) \cap C^m(R_+^n)$. Let $\psi(t)$ be a function in $C^\infty(-\infty, \infty)$ satisfying $\psi(t) = 1$ for $t \geq 0$ and $\psi(t) = 0$ for $t \leq -1/2$. Put

$$v_{i,\delta}(y) = \begin{cases} \psi(y_n/\delta)v_i(y', y_n + \delta) & y_n > -\delta \\ 0 & y_n \leq -\delta \end{cases}$$

Evidently $v_{i,\delta} \in W^{m,p}(R^n) \cap C^m(R^n)$. Since $v_{i,\delta}(y) = v_i(y', y_n + \delta)$ for $y \in R_+^n$, we have $v_{i,\delta} \to v_i$ in $W^{m,p}(R_+^n)$ as $\delta \to 0$. If δ is sufficiently small, then the support of $v_{i,\delta}$ is contained in the unit ball. Put

$$u_{i,\delta}(x) = \begin{cases} v_{i,\delta}(\Phi_i(x)) & x \in O_i \\ 0 & x \in R^n \setminus O_i \end{cases}.$$

Then $u_{i,\delta} \in W^{m,p}(R^n) \cap C^m(R^n)$, $\text{supp}\, u_{i,\delta} \subset O_i$ and $u_{i,\delta} \to u_i$ in $W^{m,p}(\Omega)$ as $\delta \to 0$. Hence if we put $u_\delta = u_0 + \sum_{i=1}^k u_{i,\delta}$ after extending u_0 as 0 outside Ω, $u_\delta \in W^{m,p}(R^n) \cap C^m(R^n)$ and $u_\delta \to u$ in $W^{m,p}(\Omega)$ as $\delta \to 0$. We obtain a desired sequence by appropriately numbering the restriction of

$\rho_\epsilon * u_\delta$ to Ω where ρ_ϵ is a mollifier.

Next we state a method of extending a function defined in $\bar{R}^n_+$ to R^n following L. Nirenberg [118].

Lemma 3.15 *For a nonnegative integer m there exist real numbers $\lambda_1, \ldots, \lambda_{m+1}$ such that*

$$\sum_{k=1}^{m+1} (-k)^j \lambda_k = 1 \tag{3.131}$$

for $j = 0, 1, \ldots, m$. For a function u defined in $\bar{R}^n_+$ put

$$\tilde{u}(x_1, \ldots, x_{n-1}, x_n) = \begin{cases} u(x_1, \ldots, x_{n-1}, x_n) & x_n \geq 0 \\ \displaystyle\sum_{k=1}^{m+1} \lambda_k u(x_1, \ldots, x_{n-1}, -kx_n) & x_n < 0 \end{cases}. \tag{3.132}$$

Then
(i) $\tilde{u} \in C^m(R^n)$ if $u \in C^m(\bar{R}^n_+)$,
(ii)$\tilde{u} \in C^m_0(R^n)$ if $u \in C^m_0(\bar{R}^m_+)$.
There exists a constant c_m depending only on m such that for any $1 \leq p \leq \infty$ and $0 \leq j \leq m$

$$|\tilde{u}|_{j,p} \leq c_m |u|_{j,p,R^n_+}. \tag{3.133}$$

Proof. The existence of $\lambda_1, \ldots, \lambda_{m+1}$ follows from the nonvanishing of the determinant of Vandermonde. The remaining part of the lemma is easily shown.

Theorem 3.14 *Let m be a natural number. If Ω is an open set uniformly regular of class C^1, then $W^{m,\infty}(\Omega) = H^{m,\infty}(\Omega)$.*

Proof. We may confine ourselves to the case $m = 1$. In view of Lemma 3.12 it suffices to show that $W^{1,\infty}(\Omega) \subset H^{1,\infty}(\Omega)$. Let u be an arbitrary element of $W^{1,\infty}(\Omega)$. Then by Lemma 3.13 u is Lipschitz continuous on each compact subset of Ω if we modify the values of u on a null set, and (3.124) holds for $|x - y| < \text{dist}(x, \partial\Omega)$. Let x and y be two points of Ω belonging to some $O_i, i \geq 1$. If we set

$$x(t) = \Psi_i(\Phi_i(x) + t(\Phi_i(y) - \Phi_i(x))),$$

then $\gamma = \{x(t); 0 \leq t \leq 1\}$ is a path of class C^1 connecting x and y in O_i. In view of (3.129) there exists a constant C independent of i such that

$$|\gamma| = \text{the length of } \gamma \leq C|x - y|. \tag{3.134}$$

Taking points $x_0 = x, x_1, \cdots, x_{k-1}, x_k = y$ on γ so that $|x_{i+1} - x_i| <$ dist(γ, Ω) we get in view of Lemma 3.13 that

$$|u(x_{i+1}) - u(x_i)| \leq \left(\sum_{i=0}^{n} |D^i u|_{0,\infty,\Omega}^2 \right)^{1/2} |x_{i+1} - x_i| \equiv A|x_{i+1} - x_i|.$$

Hence,

$$|u(y) - u(x)| \leq \sum_{i=0}^{k-1} |u(x_{i+1}) - u(x_i)|$$

$$\leq A \sum_{i=0}^{k-1} |x_{i+1} - x_i| \leq A|\gamma| \leq AC|x - y|. \tag{3.135}$$

Suppose that u is not uniformly Lipschitz continuous. Then there exist sequences $\{x_\nu\}, \{y_\nu\}$ of points of Ω so that $x_\nu \neq y_\nu$ and

$$\lim_{\nu \to \infty} \frac{|u(x_\nu) - u(y_\nu)|}{|x_\nu - y_\nu|} = \infty. \tag{3.136}$$

If there exists a positive number δ such that $|x_\nu - y_\nu| \geq \delta$ for any ν, then

$$\frac{|u(x_\nu) - u(y_\nu)|}{|x_\nu - y_\nu|} \leq \frac{2|u|_{0,\infty,\Omega}}{\delta},$$

which contradicts (3.136). Hence replacing by a subsequence if necessary we may assume that $|x_\nu - y_\nu| \to 0$. If dist$(x_\nu, \partial\Omega) \geq \delta > 0$ for any ν, then $|x_\nu - y_\nu| < \delta$ if ν is large. In view of Lemma 3.13 we have

$$|u(x_\nu) - u(y_\nu)|/|x_\nu - y_\nu| \leq A,$$

which contradicts (3.136). Hence we may assume dist$(x_\nu, \partial\Omega) < N^{-1}$. By the condition (1) of the uniform regularity of class C^1 there exists for each ν a number i such that $x_\nu \in O_i'$. Since $|x_\nu - y_\nu| \to 0$, inf dist$(O_i', \partial O_i) > 0$, we have $y_\nu \in O_i$ for large ν. Hence, by virtue of (3.135)

$$|u(x_\nu) - u(y_\nu)| \leq AC|x_\nu - y_\nu|$$

which contradicts (3.136). Thus (3.136) does not hold and we conclude $u \in H^{1,\infty}(\Omega)$.

By virtue of Theorem 3.14 and the closed graph theorem there exist positive constants C, c depending only on m and Ω such that

$$c\|u\|_{m,\infty,\Omega} \leq \|u\|_{m,\infty,\Omega}' \leq C\|u\|_{m,\infty,\Omega}. \tag{3.137}$$

Lemma 3.16 *If $\tilde{u}$ is the function defined by (3.132) for $u \in H^{m,\infty}(R_+^n)$, then $\tilde{u} \in H^{m,\infty}(R^n)$, and there exists a constant C independent of u such that*

$$\|\tilde{u}\|'_{m,\infty} \leq C\|u\|'_{m,\infty,R_+^n}. \tag{3.138}$$

Proof. Since $u \in C^{m-1}(\bar{R}_+^n)$, it follows from Lemma 3.15 that $\tilde{u} \in C^{m-1}(R^n)$. If $x = (x_1, \ldots, x_n), y = (y_1, \ldots, y_n), x_n < 0, y_n \geq 0, |\alpha| \leq m-1$

$$D^\alpha \tilde{u}(x) - D^\alpha \tilde{u}(y) = \sum_{k=1}^{m+1} \lambda_k (-k)^{\alpha_n} \left(D^\alpha u(x_1, \ldots, x_{n-1}, -kx_n) - D^\alpha u(y) \right).$$

The inequality (3.138) follows from this and

$$|-kx_n - y_n| \leq -kx_n + y_n \leq k(-x_n + y_n) = k|x_n - y_n|.$$

Theorem 3.15 *Let m be a nonnegative integer, and $1 \leq p \leq \infty$. If Ω is an open set of R^n uniformly regular of class C^m, then there exists a bounded linear operator E from $W^{m,p}(\Omega)$ to $W^{m,p}(R^n)$ such that for each $u \in W^{m,p}(\Omega)$ the restriction of Eu to Ω coincides with u.*

Proof. First consider the case $1 \leq p < \infty$. In view of Theorem 3.13 it suffices to define Eu for $u \in C_0^\infty(\bar{\Omega})$. Let $\{O_i\}, \{O_i'\}, \{\Phi_i\}, \{\Psi_i\}, N, M$ be as in the definition of uniform regular domains of class C^m, and $\{\zeta_i\}$ be the family of functions defined by (3.130). Put $u_i = \zeta_i u$. Let $v_i, i > 0$, be the functions defined by

$$v_i(y) = \begin{cases} u_i(\Psi_i(y)) & |y| < 1, y_n \geq 0 \\ 0 & |y| \geq 1, y_n \geq 0 \end{cases}.$$

Then, $v_i \in C_0^m(\bar{R}_+^n)$, and by virtue of (3.129)

$$\|v_i\|_{m,p,R_+^n} \leq C\|u_i\|_{m,p,\Omega} \tag{3.139}$$

with a constant C independent of i. If we define the function $\tilde{v}_i$ by

$$\tilde{v}_i(y_1, \ldots, y_{n-1}, y_n) = \begin{cases} v_i(y_1, \ldots, y_{n-1}, y_n) & y_n > 0 \\ \displaystyle\sum_{i=1}^{m+1} \lambda_k v_i(y_1, \ldots, y_{n-1}, -ky_n) & y_n < 0 \end{cases},$$

then in view of Lemma 3.15 $v_i \in C_0^m(R^n)$, $\mathrm{supp}\,\tilde{v}_i \subset \{y; |y| < 1\}$,

$$\|\tilde{v}_i\|_{m,p} \leq C\|v_i\|_{m,p,R_+^n}. \tag{3.140}$$

Next, if we put

$$\tilde{u}_i(x) = \begin{cases} \tilde{v}_i(\Phi_i(x)) & x \in O_i \\ 0 & x \in R^n \setminus O_i \end{cases},$$

then it follows from (3.129) that

$$\|\tilde{u}_i\|_{m,p} \leq C\|\tilde{v}_i\|_{m,p}, \tag{3.141}$$

where C is a constant independent also of i. Let $\tilde{u}_0$ be the extension of u_0 to R^n by 0 outside Ω. Then $\tilde{u}_0 \in C_0^m(R^n)$ and

$$\|\tilde{u}_0\|_{m,p} = \|u_0\|_{m,p,\Omega}. \tag{3.142}$$

Set $\tilde{u} = \sum_{i=0}^{\infty} \tilde{u}_i$. Let $|\alpha| \leq m$. Since for each x only at most $N + 1$ of $D^\alpha u_i(x)$ are different from 0,

$$|D^\alpha \tilde{u}(x)|^p = \left| \sum_{i=0}^{\infty} D^\alpha \tilde{u}_i(x) \right|^p \leq (N+1)^{p-1} \sum_{i=0}^{\infty} |D^\alpha \tilde{u}_i(x)|^p. \tag{3.143}$$

It is easy to show that there exists a constant C independent of i such that

$$\|u_i\|_{m,p,\Omega}^p \leq C \int_{\Omega \cap O_i} \sum_{|\alpha| \leq m} |D^\alpha u(x)|^p dx. \tag{3.144}$$

From (3.139),...,(3.144) it follows that

$$\|\tilde{u}\|_{m,p}^p = \int_{R^n} \sum_{|\alpha| \leq m} |D^\alpha \tilde{u}(x)|^p dx$$

$$\leq (N+1)^{p-1} \sum_{i=0}^{\infty} \int_{R^n} \sum_{|\alpha| \leq m} |D^\alpha \tilde{u}_i(x)|^p dx = (N+1)^{p-1} \sum_{i=0}^{\infty} \|\tilde{u}_i\|_{m,p}^p$$

$$\leq C \sum_{i=0}^{\infty} \|u_i\|_{m,p,\Omega}^p \leq C \sum_{i=0}^{\infty} \int_{\Omega \cap O_i} \sum_{|\alpha| \leq m} |D^\alpha u(x)|^p dx. \tag{3.145}$$

Since $N + 2$ of $\{O_i\}$ have an empty intersection, the last side of (3.145) does not exceed

$$C \int_{\Omega} \sum_{|\alpha| \leq m} |D^\alpha u(x)|^p dx = C\|u\|_{m,p,\Omega}^p.$$

Evidently $\tilde{u}|_\Omega = u$. Thus, by putting $Eu = \tilde{u}$, we complete the proof of the case $1 \leq p < \infty$.

Next suppose $p = \infty$. For $u \in W^{m,\infty}(\Omega)$ let the functions $v_i, \tilde{v}_i, \tilde{u}_i$ be defined as in the case $1 \leq p < \infty$. Since $u \in H^{m,\infty}(\Omega)$ in view of Theorem 3.14, it is clear that $v_i \in H^{m,\infty}(R_+^n)$, and

$$\|v_i\|'_{m,\infty,R_+^n} \leq C\|u_i\|'_{m,\infty,\Omega} \tag{3.146}$$

with a constant C independent of i. By virtue of Lemma 3.16 $\tilde{v}_i \in H^{m,\infty}(R^n)$ and

$$\|\tilde{v}_i\|'_{m,\infty} \leq C\|v_i\|'_{m,\infty,R_+^n}. \tag{3.147}$$

It is evident that $\tilde{u}_i \in H^{m,\infty}(R^n)$ and

$$\|\tilde{u}_i\|'_{m,\infty} \leq C\|\tilde{v}_i\|'_{m,\infty}. \tag{3.148}$$

Put $\tilde{u} = \sum_{i=0}^{\infty} u_i$. Since for $x, y \in R^n, |\alpha| \leq m - 1$

$$|D^\alpha \tilde{u}(x)| \leq (N+1)\sup_i |D^\alpha \tilde{u}_i(x)|,$$

$$|D^\alpha \tilde{u}(x) - D^\alpha \tilde{u}(y)| \leq 2(N+1)\sup_i |D^\alpha \tilde{u}_i(x) - D^\alpha \tilde{u}_i(y)|,$$

we have

$$\|\tilde{u}\|'_{m,\infty} \leq C \sup_i \|\tilde{u}_i\|'_{m,\infty}. \tag{3.149}$$

By virtue of Theorem 3.14, (3.146),(3.147),(3.148),(3.149) we get

$$\|\tilde{u}\|_{m,\infty} \leq C\|\tilde{u}\|'_{m,\infty} \leq C \sup_i \|u_i\|_{m,\infty,\Omega} \leq C\|u\|_{m,\infty,\Omega}.$$

Thus $Eu = \tilde{u}$ satisfies the assertion of the theorem.

3.8   Embedding Theorems

It will be shown that the results of section 3.4 hold for open sets other than R^n.

Theorem 3.16 *Let Ω be an open set of R^n uniformly regular of class C^2. If m, j are integers satisfying $0 \leq j < m$ and $1 \leq p \leq \infty$, then there exists a constant γ such that for $u \in W^{m,p}(\Omega)$*

$$\|u\|_{j,p,\Omega} \leq \gamma \left(|u|_{m,p,\Omega}^{j/m} |u|_{0,p,\Omega}^{1-j/m} + |u|_{0,p,\Omega} \right). \tag{3.150}$$

Proof. We begin with the case $j = m - 1$:

$$\|u\|_{m-1,p,\Omega} \leq \gamma \left(|u|_{m,p,\Omega}^{(m-1)/m} |u|_{0,p,\Omega}^{1/m} + |u|_{0,p,\Omega} \right). \tag{3.151}$$

Let E be the operator of Theorem 3.15 in the case $m = 2$. Then by Theorem 3.7

$$\|Eu\|_{1,p} \leq C \left(|Eu|_{2,p}^{1/2} |Eu|_{0,p}^{1/2} + |Eu|_{0,p} \right).$$

Hence

$$\|u\|_{1,p,\Omega} \leq C \left(\|u\|_{2,p,\Omega}^{1/2} |u|_{0,p,\Omega}^{1/2} + |u|_{0,p,\Omega} \right)$$
$$\leq C \left(|u|_{2,p,\Omega}^{1/2} |u|_{0,p,\Omega}^{1/2} + \|u\|_{1,p,\Omega}^{1/2} |u|_{0,p,\Omega}^{1/2} + |u|_{0,p,\Omega} \right),$$

from which it follows that

$$\|u\|_{1,p,\Omega} \leq C \left(|u|_{2,p,\Omega}^{1/2} |u|_{0,p,\Omega}^{1/2} + |u|_{0,p,\Omega} \right). \tag{3.152}$$

Suppose (3.151) is true with m replaced by $m - 1$:

$$\|u\|_{m-2,p,\Omega} \leq C \left(|u|_{m-1,p,\Omega}^{(m-2)/(m-1)} |u|_{0,p,\Omega}^{1/(m-1)} + |u|_{0,p,\Omega} \right). \tag{3.153}$$

Applying (3.152) to $D^{m-2}u$ we get

$$|u|_{m-1,p,\Omega} \leq C \left(|u|_{m,p,\Omega}^{1/2} |u|_{m-2,p,\Omega}^{1/2} + |u|_{m-2,p,\Omega} \right).$$

Combining this with (3.153)

$$|u|_{m-1,p,\Omega} \leq C \Big(|u|_{m,p,\Omega}^{1/2} |u|_{m-1,p,\Omega}^{(m-2)/2(m-1)} |u|_{0,p,\Omega}^{1/2(m-1)}$$
$$+ |u|_{m,p,\Omega}^{1/2} |u|_{0,p,\Omega}^{1/2} + |u|_{m-1,p,\Omega}^{(m-2)/(m-1)} |u|_{0,p,\Omega}^{1/(m-1)} + |u|_{0,p,\Omega} \Big). \tag{3.154}$$

With the aid of Young's inequality

$$|u|_{m,p,\Omega}^{1/2} |u|_{m-1,p,\Omega}^{(m-2)/2(m-1)} |u|_{0,p,\Omega}^{1/2(m-1)}$$
$$= \left(\epsilon^{-(m-2)/m} |u|_{m,p,\Omega}^{(m-1)/m} |u|_{0,p,\Omega}^{1/m} \right)^{m/2(m-1)} \left(\epsilon |u|_{m-1,p,\Omega} \right)^{(m-2)/2(m-1)}$$
$$\leq \frac{m}{2(m-1)} \epsilon^{-(m-2)/m} |u|_{m,p,\Omega}^{(m-1)/m} |u|_{0,p,\Omega}^{1/m} + \frac{m-2}{2(m-1)} \epsilon |u|_{m-1,p,\Omega},$$
$$|u|_{m,p,\Omega}^{1/2} |u|_{0,p,\Omega}^{1/2} = \left(|u|_{m,p,\Omega}^{(m-1)/m} |u|_0^{1/m} \right)^{m/2(m-1)} |u|_0^{(m-2)/2(m-1)}$$
$$\leq \frac{m}{2(m-1)} |u|_{m,p,\Omega}^{(m-1)/m} |u|_0^{1/m} + \frac{m-2}{2(m-1)} |u|_{0,p,\Omega},$$
$$|u|_{m-1,p,\Omega}^{(m-2)/(m-1)} |u|_{0,p,\Omega}^{1/(m-1)} \leq \frac{m-2}{m-1} \epsilon |u|_{m-1,p,\Omega} + \frac{1}{m-1} \epsilon^{2-m} |u|_{0,p,\Omega}.$$

Combining these three inequalities and (3.154) and letting ϵ be sufficiently small we get

$$|u|_{m-1,p,\Omega} \leq C \left(|u|_{m,p,\Omega}^{(m-1)/m} |u|_{0,p,\Omega}^{1/m} + |u|_{0,p,\Omega} \right). \qquad (3.155)$$

Applying Young's inequality to the first term in the bracket of the right hand side of (3.153) we get

$$\|u\|_{m-2,p,\Omega} \leq C \left(\epsilon |u|_{m-1,p,\Omega} + \epsilon^{2-m} |u|_{0,p,\Omega} + |u|_{0,p,\Omega} \right). \qquad (3.156)$$

The inequality (3.151) follows from (3.155) and (3.156).

Finally we prove (3.150) by induction on j. Suppose (3.150) is true with $j+1$ in place of j:

$$\|u\|_{j+1,p,\Omega} \leq C \left(|u|_{m,p,\Omega}^{(j+1)/m} |u|_{0,p,\Omega}^{1-(j+1)/m} + |u|_{0,p,\Omega} \right). \qquad (3.157)$$

Replacing m by $j+1$ in (3.151)

$$\|u\|_{j,p,\Omega} \leq C \left(|u|_{j+1,p,\Omega}^{j/(j+1)} |u|_{0,p,\Omega}^{1/(j+1)} + |u|_{0,p,\Omega} \right). \qquad (3.158)$$

The combination of (3.157) and (3.158) yields (3.150).

Theorem 3.17 *Let Ω be an open set of R^n uniformly regular of class C^2. Let m, j be integers satisfying $0 \leq j < m$, and let $1 \leq p < \infty$. Suppose that $m - j - n/p$ is not a nonnegative integer. If $j/m \leq a \leq 1, 1/r = j/n + 1/p - am/n \geq 0$, then $W^{m,p}(\Omega) \subset W^{j,r}(\Omega)$ and there exists a constant γ such that for $u \in W^{m,p}(\Omega)$*

$$\|u\|_{j,r,\Omega} \leq \gamma \left(|u|_{m,p,\Omega}^{a} |u|_{0,p,\Omega}^{1-a} + |u|_{0,p,\Omega} \right). \qquad (3.159)$$

Proof. We begin with the case $m = 1$. Let E be the operator of Theorem 3.15 for $m = 1$. Suppose $1 - n/p$ is not a nonnegative integer. In view of Theorem 3.7 if $0 \leq a \leq 1$ and $1/r = 1/p - a/n \geq 0$, then

$$|Eu|_{0,r} \leq C \left(|Eu|_{1,p}^{a} |Eu|_{0,p}^{1-a} + |Eu|_{0,p} \right).$$

Hence

$$|u|_{0,r,\Omega} \leq C \left(\|u\|_{1,p,\Omega}^{a} |u|_{0,p,\Omega}^{1-a} + |u|_{0,p,\Omega} \right) \leq C \left(|u|_{1,p,\Omega}^{a} |u|_{0,p,\Omega}^{1-a} + |u|_{0,p,\Omega} \right).$$

Thus the proof of the present case is complete.

Suppose the assertion of the theorem is true for $m - 1$. We are going to show that it is also true for m. Let j, p, r, a be as in the assumption of the theorem. First we consider the case $j > 0$. If we set $b = (am - 1)/(m - 1)$, then $(j-1)/(m-1) \le b \le 1$. It is easily seen that the hypothesis of the theorem is satisfied with m, j, a replaced by $m-1, j-1, b$. Hence, by induction assumption

$$\|Du\|_{j-1,r,\Omega} \le C\left(|Du|_{m-1,p,\Omega}^{b}|Du|_{0,p,\Omega}^{1-b} + |Du|_{0,p,\Omega}\right), \qquad (3.160)$$

$$\|u\|_{j-1,r,\Omega} \le C\left(|u|_{m-1,p,\Omega}^{b}|u|_{0,p,\Omega}^{1-b} + |u|_{0,p,\Omega}\right). \qquad (3.161)$$

By Theorem 3.16

$$|Du|_{0,p,\Omega} \le |u|_{1,p,\Omega} \le C\left(|u|_{m,p,\Omega}^{1/m}|u|_{0,p,\Omega}^{(m-1)/m} + |u|_{0,p,\Omega}\right).$$

Substituting this in the right hand side of (3.160)

$$\|Du\|_{j-1,r,\Omega} \le C\bigg[|u|_{m,p,\Omega}^{b}\left(|u|_{m,p,\Omega}^{1/m}|u|_{0,p,\Omega}^{(m-1)/m} + |u|_{0,p,\Omega}\right)^{1-b}$$

$$+|u|_{m,p,\Omega}^{1/m}|u|_{0,p,\Omega}^{(m-1)/m} + |u|_{0,p,\Omega}\bigg]$$

$$\le C\bigg(|u|_{m,p,\Omega}^{b+(1-b)/m}|u|_{0,p,\Omega}^{(m-1)(1-b)/m} + |u|_{m,p,\Omega}^{b}|u|_{0,p,\Omega}^{1-b}$$

$$+|u|_{m,p,\Omega}^{1/m}|u|_{0,p,\Omega}^{(m-1)/m} + |u|_{0,p,\Omega}\bigg)$$

Since $b + (1-b)/m = a, 1 - b = m(1-a)/(m-1)$, we get

$$\|Du\|_{j-1,r,\Omega} \le C\bigg(|u|_{m,p,\Omega}^{a}|u|_{0,p,\Omega}^{1-a}$$

$$+|u|_{m,p,\Omega}^{b}|u|_{0,p,\Omega}^{1-b} + |u|_{m,p,\Omega}^{1/m}|u|_{0,p,\Omega}^{(m-1)/m} + |u|_{0,p,\Omega}\bigg). \qquad (3.162)$$

Noting $b \le a, 1/m \le a$ and applying Young's inequality to the right hand sides of

$$|u|_{m,p,\Omega}^{b}|u|_{0,p,\Omega}^{1-b} = \left(|u|_{m,p,\Omega}^{a}|u|_{0,p,\Omega}^{1-a}\right)^{b/a}|u|_{0,p,\Omega}^{(a-b)/a},$$

$$|u|_{m,p,\Omega}^{1/m}|u|_{0,p,\Omega}^{(m-1)/m} = \left(|u|_{m,p,\Omega}^{a}|u|_{0,p,\Omega}^{1-a}\right)^{1/am}|u|_{0,p,\Omega}^{(am-1)/am},$$

we get from (3.162)

$$\|Du\|_{j-1,r,\Omega} \le C\left(|u|_{m,p,\Omega}^{a}|u|_{0,p,\Omega}^{1-a} + |u|_{0,p,\Omega}\right). \qquad (3.163)$$

Similarly from (3.161) we get

$$|u|_{0,r,\Omega} \leq \|u\|_{j-1,r,\Omega}$$
$$\leq C \left[\left(|u|_{m,p,\Omega}^{(m-1)/m} |u|_{0,p,\Omega}^{1/m} + |u|_{0,p,\Omega} \right)^b |u|_{0,p,\Omega}^{1-b} + |u|_{0,p,\Omega} \right]$$
$$\leq C \left(|u|_{m,p,\Omega}^{(m-1)b/m} |u|_{0,p,\Omega}^{b/m+1-b} + |u|_{0,p,\Omega} \right)$$
$$\leq C \left[\left(|u|_{m,p,\Omega}^a |u|_{0,p,\Omega}^{1-a} \right)^{(m-1)b/ma} |u|_{0,p,\Omega}^{(m(a-b)+b)/ma} + |u|_{0,p,\Omega} \right]$$
$$\leq C \left(|u|_{m,p,\Omega}^a |u|_{0,p,\Omega}^{1-a} + |u|_{0,p,\Omega} \right).$$

Combining this with (3.163) we obtain (3.159).

Next we consider the case $j = 0$. What is to be proved is that if $m - n/p$ is not a nonnegative integer, then for $0 \leq a \leq 1, 1/r = 1/p - am/n \geq 0$

$$|u|_{0,r,\Omega} \leq \gamma \left(|u|_{m,p,\Omega}^a |u|_{0,p,\Omega}^{1-a} + |u|_{0,p,\Omega} \right). \tag{3.164}$$

Case $1/m \leq a \leq 1, r < \infty$. Let s be a number such that $1/s = 1/n + 1/r$. Then $1/s = 1/n + 1/p - am/n, p \leq s < r$. Since $m - 1 - n/p$ is not a nonnegative integer, by what is already proved in case $j > 0$

$$\|u\|_{1,s,\Omega} \leq C \left(|u|_{m,p,\Omega}^a |u|_{0,p,\Omega}^{1-a} + |u|_{0,p,\Omega} \right). \tag{3.165}$$

On the other hand from the result of the case $m = 1$ (note that $1 - n/s = -n/r < 0$)

$$|u|_{0,r,\Omega} \leq C\|u\|_{1,s,\Omega}. \tag{3.166}$$

The inequality (3.164) follows from (3.165) and (3.166).

Case $1/m \leq a \leq 1, r = \infty$. In this case $a = n/mp < 1$ since otherwise we would have $m - n/p = 0$ contrary to the hypothesis. Let $1/s = 1/n + 1/p - m/n$. Then $n < s < \infty$. Applying Theorem 3.3 with $1, 0, s, p, n/mp$ as m, j, p, q, a we get

$$|Eu|_{0,\infty} \leq C|Eu|_{1,s}^{n/mp} |Eu|_{0,p}^{1-n/mp},$$

where E is the operator of Theorem 3.15 for $m = 1$. Hence

$$|u|_{0,\infty,\Omega} \leq C\|u\|_{1,s,\Omega}^{n/mp} |u|_{0,p,\Omega}^{1-n/mp}. \tag{3.167}$$

By virtue of the result of the case $j > 0$ which is already proved

$$\|u\|_{1,s,\Omega} \leq C \left(|u|_{m,p,\Omega} + |u|_{0,p,\Omega} \right). \tag{3.168}$$

Combining (3.167) and (3.168) we obtain the desired inequality

$$|u|_{0,\infty,\Omega} \leq C \left(|u|_{m,p,\Omega}^{n/mp} |u|_{0,p,\Omega}^{1-n/mp} + |u|_{0,p,\Omega} \right).$$

Case $0 \leq a < 1/m$. Put $b = am/(m-1)$. Then $0 \leq b < 1$ and $1/r = 1/p - b(m-1)/n$. Hence by induction assumption

$$|u|_{0,r,\Omega} \leq C \left(|u|_{m-1,p,\Omega}^{b} |u|_{0,p,\Omega}^{1-b} + |u|_{0,p,\Omega} \right). \tag{3.169}$$

In view of Theorem 3.16

$$|u|_{m-1,p,\Omega} \leq C \left(|u|_{m,p,\Omega}^{(m-1)/m} |u|_{0,p,\Omega}^{1/m} + |u|_{0,p,\Omega} \right). \tag{3.170}$$

The desired inequality (3.164) follows from (3.169) and (3.170).

Theorem 3.18 *Let Ω be an open set of R^n uniformly regular of class C^2. Let m, j be integers satisfying $0 \leq j < m$ and let $1 \leq p < \infty$. Suppose $m - j - n/p = 0$. Then for any r satisfying $n/(m-j) = p \leq r < \infty$, we have $W^{m,p}(\Omega) \subset W^{j,r}(\Omega)$ and*

$$\|u\|_{j,r,\Omega} \leq \gamma \left(|u|_{m,p,\Omega}^{a} |u|_{0,p,\Omega}^{1-a} + |u|_{0,p,\Omega} \right), \tag{3.171}$$

where a is the number satisfying $1/r = j/n + 1/p - am/n = m/n - am/n$, i.e. $j/m \leq a = 1 - n/mr < 1$.

Proof. The proof of the case $m = 1$ and of the case $j > 0$ under the assumption that the conclusion of the theorem is true for $m-1$ is the same as that of the previous theorem. Supposing that the theorem has been proved for $m-1$ we are going to show that the assertion of the theorem is true for m and $j = 0$:

$$|u|_{0,r,\Omega} \leq \gamma \left(|u|_{m,n/m,\Omega}^{1-n/mr} |u|_{0,n/m,\Omega}^{n/mr} + |u|_{0,n/m,\Omega} \right) \tag{3.172}$$

for $n/m = p \leq r < \infty$. First we consider the case $r \geq n/(m-1)$. Let $1/s = 1/n + 1/r$. Noting $1/m \leq 1 - n/mr$ we apply Theorem 3.17 with $j = 1, p = n/m, a = 1 - n/mr$ to obtain

$$\|u\|_{1,s,\Omega} \leq C \left(|u|_{m,n/m,\Omega}^{1-n/mr} |u|_{0,n/m,\Omega}^{n/mr} + |u|_{0,n/m,\Omega} \right).$$

The inequality (3.172) follows from this inequality and (3.166). The result of the case $n/m \leq r < n/(m-1)$ follows from that of the case $r = n/(m-1)$ and the interpolation inequality

$$|u|_{0,r,\Omega} \leq |u|_{0,n/(m-1),\Omega}^{m-n/r} |u|_{0,n/m,\Omega}^{n/r-m+1}. \tag{3.173}$$

The following two theorems are verified with the aid of Theorem 3.9, Theorem 3.10 and Theorem 3.16.

Theorem 3.19 *Let Ω be an open set of R^n uniformly regular of class C^2. Let m, j be integers satisfying $0 \leq j < m$, and let $1 \leq p \leq \infty$. Suppose $m - j - n/p > 0$. Then for any r satisfying $p \leq r \leq \infty$ we have $W^{m,p}(\Omega) \subset W^{j,r}(\Omega)$ and for any $u \in W^{m,p}(\Omega)$*

$$\|u\|_{j,r,\Omega} \leq \gamma \left(|u|_{m,p,\Omega}^a |u|_{0,p,\Omega}^{1-a} + |u|_{0,p,\Omega} \right), \tag{3.174}$$

where a is a number satisfying $1/r = j/n + 1/p - am/n$.

Theorem 3.20 *Let Ω be an open set of R^n uniformly regular of class C^2. Let m be a positive integer and $1 \leq p < \infty$. Suppose $m - n/p > 0$.*
(i) If n/p is not an integer, then $W^{m,p}(\Omega) \subset B^{m-n/p}(\bar{\Omega})$. Put $l = m - [n/p] - 1$. Then, for each integer j satisfying $0 \leq j \leq l$

$$\|u\|_{j,\infty,\Omega} \leq \gamma \left(|u|_{m,p,\Omega}^{(n+jp)/mp} |u|_{0,p,\Omega}^{(mp-n-jp)/mp} + |u|_{0,p,\Omega} \right). \tag{3.175}$$

For $0 \leq h \leq [n/p] + 1 - n/p$

$$|D^l u|_{h,\infty,\Omega} \leq \gamma \left(|u|_{m,p,\Omega}^{(n+lp+hp)/mp} |u|_{0,p,\Omega}^{(mp-n-lp-hp)/mp} + |u|_{0,p,\Omega} \right). \tag{3.176}$$

Hence for $0 \leq s \leq m - n/p$

$$\|u\|_{s,\infty,\Omega} \leq \gamma \left(|u|_{m,p,\Omega}^{(n+sp)/mp} |u|_{0,p,\Omega}^{(mp-n-sp)/mp} + |u|_{0,p,\Omega} \right). \tag{3.177}$$

In particular

$$\|u\|_{m-n/p,\infty,\Omega} \leq \gamma \|u\|_{m,p,\Omega}.$$

(ii) If n/p is an integer, then for each s satisfying $0 \leq s < m - n/p$ we have $W^{m,p}(\Omega) \subset B^s(\bar{\Omega})$. (3.175) holds for each integer j such that $0 \leq j \leq l = m - n/p - 1$. For $0 \leq h < 1$ (3.176) holds with a constant depending also on h. For $0 \leq s < m - n/p$ (3.177) holds with a constant γ depending also on s.

In order to show only the inclusion relation $W^{m,p}(\Omega) \subset W^{j,r}(\Omega)$ in the above theorems it suffices to assume that Ω is uniformly regular of class C^1 as is shown in the following theorem.

Theorem 3.21 *Suppose that Ω is an open set of R^n uniformly regular of class C^1. Let m, j be integers satisfying $0 \leq j < m$, and let $1 \leq p < \infty$.*

(i) *If $m-j-n/p$ is not a nonnegative integer, $p \leq r \leq \infty$, $j/n+1/p-m/n \leq 1/r$, then $W^{m,p}(\Omega) \subset W^{j,r}(\Omega)$, and there exists a positive constant γ such that for $u \in W^{m,p}(\Omega)$*

$$\|u\|_{j,r,\Omega} \leq \gamma \|u\|_{m,p,\Omega}. \tag{3.178}$$

(ii) *If $m - j - n/p$ is a nonnegative integer, then $W^{m,p}(\Omega) \subset W^{j,r}(\Omega)$ for $p \leq r < \infty$ and (3.171) holds.*

Proof. (i) It suffices to prove the theorem for $j = 0$. We begin with the case $m = 1$. What is to be shown is that if $1 - n/p$ is not a nonnegative integer, $p \leq r \leq \infty$ and $1/p - 1/n \leq 1/r \leq 1/p$, then $W^{1,p}(\Omega) \subset L^r(\Omega)$. In view of Theorem 3.7 we have $W^{1,p}(R^n) \subset L^r(R^n)$. Let E be the operator of Theorem 3.15 for $m = 1$. If $u \in W^{1,p}(\Omega)$, then $Eu \in W^{1,p}(R^n) \subset L^r(R^n)$. Hence $u \in L^r(\Omega)$. Next, suppose that the conclusion of the theorem is true for $m - 1$, i.e. if $m - 1 - n/p$ is not a nonnegative integer, $p \leq s \leq \infty$, $1/p - (m-1)/n \leq 1/s$, then $W^{m-1,p}(\Omega) \subset L^s(\Omega)$. Assume that $m - n/p$ is not a nonnegative integer, $p \leq r \leq \infty$, $1/p - m/n \leq 1/r$. Then, clearly $m - 1 - n/p$ is not a nonnegative integer. Put

$$\frac{1}{s} = \max\left\{\frac{1}{p} - \frac{m-1}{n}, \frac{1}{r}\right\}.$$

(a) If $1/p - (m-1)/n > 1/r$, then $1/s = 1/p - (m-1)/n < 1/p$. By the induction hypothesis we have $W^{m-1,p}(\Omega) \subset L^s(\Omega)$. Hence $W^{m,p}(\Omega) \subset W^{1,s}(\Omega)$. Since $1 - n/s = m - n/p$ is not a nonnegative integer and $1/s - 1/n = 1/p - m/n \leq 1/r < 1/s$, we have $W^{1,s}(\Omega) \subset L^r(\Omega)$ by the result shown in case $m = 1$. Hence $W^{m,p}(\Omega) \subset L^r(\Omega)$.
(b) If $1/p - (m-1)/n \leq 1/r$, then $W^{m-1,p}(\Omega) \subset L^r(\Omega)$ by the induction hypothesis. Hence $W^{m,p}(\Omega) \subset L^r(\Omega)$.
(ii) The conclusion in case $m = 1$ is established with the aid of Theorem 3.8 in place of Theorem 3.7 in the proof of (i). Suppose that the theorem has been proved for $m - 1$. Assume that $m - n/p$ is a nonnegative integer and $p \leq r < \infty$. If $m - 1 - n/p$ is a nonnegative integer, then $W^{m,p}(\Omega) \subset W^{m-1,p}(\Omega) \subset L^r(\Omega)$. If $m - n/p = 0$, then $m - 1 - n/p = -1$ is not a nonnegative integer. Hence in view of the first part (i) we have $W^{m,p}(\Omega) \subset W^{1,n}(\Omega)$. If $r \geq n$, $W^{1,n}(\Omega) \subset L^r(\Omega)$ by Theorem 3.8. Hence $W^{m,p}(\Omega) \subset L^r(\Omega)$. It is evident that this inclusion relation also holds for $n/m = p \leq r < n$ since $W^{m,p}(\Omega) \subset L^p(\Omega)$.

Remark 3. 6 Analogously to Theorem 3.21 it can be shown that a result corresponding to Theorem 3.20 also holds for an open set Ω uniformly reugular of class C^1.

3.9   Additional Topics

Other important materials on Sobolev spaces are Rellich's theorem and traces on the boundary. We state them without proof.

Let Ω be an open subset of R^n uniformly regular of class C^1. If m, j are integers satisfying $0 \leq j < m, 1 \leq p < \infty$ and

$$1 \leq r \leq \infty, \quad \frac{j}{n} + \frac{1}{p} - \frac{m}{n} \leq \frac{1}{r}, \qquad (3.179)$$

then by Theorem 3.21 we have

$$W^{m,p}(\Omega) \subset W^{j,r}(\Omega). \qquad (3.180)$$

If moreover Ω is bounded and the inequality holds in the second inequality of (3.179), then the imbedding (3.180) is compact.

Theorem 3. 22 (Rellich's theorem) *Suppose that Ω is a bounded open set of R^n of class C^1. If m, j are integers satisfying $0 \leq j < m, 1 \leq p < \infty, 1 \leq r \leq \infty, j/n + 1/p - m/n < 1/r$, then the imbedding $W^{m,p}(\Omega) \subset W^{j,r}(\Omega)$ is compact.*

Corollary 3. 2 *Suppose Ω is a bounded open set of R^n of class C^1. If m is a positive integer and $1 \leq p < \infty$, then the imbedding $W^{m,p}(\Omega) \subset W^{m-1,p}(\Omega)$ is compact.*

If Ω is an open set uniformly regular of class C^m, then the "boundary values" $(\partial/\partial\nu)^j u|_{\partial\Omega}, j = 0, \ldots, m - 1$, are defined for functions $u \in W^{m,p}(\Omega)$.

Theorem 3. 23 *Let Ω be an open set of R^n uniformly regular of class $C^m, m > 0$, and $1 \leq p \leq \infty$. Then for $j = 0, 1, \ldots, m - 1$, there exists a bounded linear mapping γ_j from $W^{m,p}(\Omega)$ to $L^p(\partial\Omega)$ such that if $u \in W^{m,p}(\Omega) \cap C^m(\bar{\Omega})$, then $\gamma_j u \in C^{m-j}(\partial\Omega)$ and*

$$(\gamma_j u)(x) = \left(\frac{\partial}{\partial\nu}\right)^j u(x) \qquad (3.181)$$

for $x \in \partial\Omega$, where ν is the outward normal vector of $\partial\Omega$.

Definition 3. 3 *$\gamma_0 u$ is called the trace of u on the boundary $\partial\Omega$.*

Usually we simply write $u, (\partial/\partial\nu)^j u$ in place of $\gamma_0 u, \gamma_j u$ respectively.

Let Ω be an open set of R^n uniformly regular of class C^m, and $1 \leq p < \infty$. We denote by $W^{m-1/p,p}(\partial\Omega)$ the totality of the traces of functions belonging to $W^{m,p}(\Omega)$. The norm of an element g of $W^{m-1/p,p}(\partial\Omega)$ is defined by

$$[g]_{m-1/p,p,\partial\Omega} = \inf\{\|u\|_{m,p,\Omega}; u \in W^{m,p}(\Omega), \gamma_0 u = g\}. \qquad (3.182)$$

Theorem 3. 24 $W^{m-1/p,p}(\partial\Omega)$ *is a Banach space with the norm* (3.182). $C_0^m(\partial\Omega)$ *is dense in* $W^{m-1/p,p}(\partial\Omega)$.

For a nonempty open set Ω of R^n, a nonnegative integer m and $1 \leq p < \infty$, we denote by $W_0^{m,p}(\Omega)$ the closure of $C_0^m(\Omega)$ in $W^{m,p}(\Omega)$.

Theorem 3. 25 *Let Ω be an open set of R^n uniformly regular of class $C^m, m > 0$, and $1 \leq p < \infty$. An element u of $W^{m,p}(\Omega)$ belongs to $W_0^{m,p}(\Omega)$ if and only if*

$$\gamma_0 u = \gamma_1 u = \cdots = \gamma_{m-1} u = 0.$$

If the following conditions are satisfied, then Ω is said to have the *restricted cone property*.

There exist an open covering $\{O_i; i = 1, 2, \ldots\}$ of $\partial\Omega$ and a sequence of open cones $\{C_i; i = 1, 2, \ldots\}$ with vertices at the origine such that $x + C_i \subset \Omega$ for each i and $x \in \Omega \cap O_i$. There exists a positive number r such that for each $x \in \partial\Omega$ the ball $\{y; |y - x| \leq r\}$ is contained in some O_i. There exists a natural number N such that $N + 1$ of O_i has an empty intersection. If we denote the opening and the height of O_i by θ_i and h_i respectively, then $\inf \theta_i > 0, \inf h_i > 0, \sup h_i < \infty$ and $\mathrm{diam} O_i \leq h_i$.

If Ω has the restricted cone property, then imbedding theorems and Rellich's theorem hold if $1 < p < \infty$. The proof is based on Calderón's extension theorem which is established with the aid of Theorem 2.10.

Chapter 4

Elliptic Boundary Value Problems

4.1 Fundamental Solutions of Elliptic Operators

The contents of this chapter are the L^p estimates by S. Agmon, A. Douglis and L. Nirenberg [11] for solutions of general elliptic boundary value problems. We begin with the fundamental solutions of elliptic operators with constant coefficients by F. John [84].

The notations

$$D = (D_1, \ldots, D_n) = (\partial/\partial x_1, \ldots, \partial/\partial x_n),$$
$$D^\alpha = D_1^{\alpha_1} \cdots D_n^{\alpha_n}, \xi^\alpha = \xi_1^{\alpha_1} \cdots \xi_n^{\alpha_n}, |\alpha| = \alpha_1 + \cdots + \alpha_n$$
$$\text{for} \quad \alpha = (\alpha_1, \ldots, \alpha_n)$$

are also used in this section as in the previous section.

A linear differential operator

$$L(x, D) = \sum_{|\alpha| \leq m} a_\alpha(x) D^\alpha \tag{4.1}$$

of order m with coefficients defined in an open set $\Omega \subset R^n$ is called *elliptic* in Ω if

$$\sum_{|\alpha|=m} a_\alpha(x) \xi^\alpha \neq 0 \tag{4.2}$$

for any $x \in \Omega, 0 \neq \xi \in R^n$. A function $K(x, y)$ defined in $\Omega \times \Omega$ possibly except the diagonal $\{(x, x); x \in \Omega\}$ is called a *fundamental solution* of $L(x, D)$

if

$$L(x, D) \int_{\Omega} K(x, y) f(y) dy = f(x) \qquad (4.3)$$

holds for any $f \in C_0^{\infty}(\Omega)$. If $L(x, D) = L(D)$ is an operator in R^n with constant coefficients, a fundamental solution is of the form $K(x - y)$. In this case $K(x)$ is called a fundamental solution of $L(D)$.

As is well known

$$K(x) = \begin{cases} |x|^{2-n}/(2-n)\Omega_n & n > 2 \\ (2\pi)^{-1} \log |x| & n = 2 \end{cases} \qquad (4.4)$$

is a fundamental solution of Δ, where Ω_n is the surface area of the unit sphere in R^n.

F. John [83],[84] constructed a fundamental solution for elliptic operators with smooth coefficients. In this chapter we describe a case of operators with only the highest order part of constant coefficients following [84].

Lemma 4.1 *Let q be a nonnegative integer such that $n + q$ is even. Then if n is odd,*

$$(-1)^{(n-1)/2}|x|^q \Big/ 2^{n+q}\pi^{n/2-1}\Gamma\left(\frac{q+2}{2}\right)\Gamma\left(\frac{n+q}{2}\right), \qquad (4.5)$$

and if n is even,

$$(-1)^{n/2-1}|x|^q \log |x| \Big/ 2^{n+q-1}\pi^{n/2}\Gamma\left(\frac{q+2}{2}\right)\Gamma\left(\frac{n+q}{2}\right) \qquad (4.6)$$

is a fundamental solution of $\Delta^{(n+q)/2}$.

Proof. Suppose first that n is odd. Since

$$\Delta^{(n+q)/2-1}|x|^q = C_{n,q}|x|^{2-n},$$

where

$$C_{n,q} = (-1)^{(n-3)/2}\frac{2^{n+q-2}}{\pi}\Gamma\left(\frac{q+2}{2}\right)\Gamma\left(\frac{n-2}{2}\right)\Gamma\left(\frac{n+q}{2}\right),$$

and (4.4) is a fundamental solution of Δ, we get for $f \in C_0^{\infty}(R^n)$

$$\Delta^{(n+q)/2}\int_{R^n} |x - y|^q f(y) dy = \Delta \int_{R^n} \Delta^{(n+q)/2-1}|x - y|^q f(y) dy$$

$$= C_{n,q}\Delta \int_{R^n} |x - y|^{2-n} f(y) dy = C_{n,q}(2 - n)\Omega_n f(x).$$

Consequently (4.5) is a fundamental solution of $\Delta^{(n+q)/2}$.

Analogously we can show that (4.6) is a fundamental solution of $\Delta^{(n+q)/2}$ in case n is even using

$$\Delta^{(n+q)/2-1}(|x|^q \log |x|)$$
$$= (-1)^{n/2} 2^{n+q-3} \Gamma\left(\frac{q+2}{2}\right) \Gamma\left(\frac{n-2}{2}\right) \Gamma\left(\frac{n+q}{2}\right) |x|^{2-n}$$

if $n > 2$ and

$$\Delta^{q/2}(|x|^q \log |x|) = 2^q \Gamma\left(\frac{q+2}{2}\right)^2 \log |x| + \text{const}$$

if $n = 2$.

In order to construct a fundamental solution for elliptic operators F. John [84] expressed the fundamental solution of $\Delta^{(n+q)/2}$ in the previous lemma in terms of plane waves.

For the logarithm we always take the principal branch in the complex plane slit along the negative real axis.

Lemma 4. 2 *If q is a nonnegative integer such that $n + q$ is even, then*

$$-\frac{1}{(2\pi i)^n q!} \int_{|\xi|=1} (x\xi)^q \log \frac{x\xi}{i} d\sigma_\xi \tag{4.7}$$

is a fundamental solution for $\Delta^{(n+q)/2}$.

Proof. First we show that if q is a nonnegative integer

$$\int_{|\xi|=1} (x\xi)^q \log \frac{x\xi}{i} d\sigma_\xi = \begin{cases} -\dfrac{1}{2\pi i} \displaystyle\int_{|\xi|=1} |x\xi|^q d\sigma_\xi & q \text{ is odd} \\[3mm] \displaystyle\int_{|\xi|=1} |x\xi|^q \log |x\xi| d\sigma_\xi & q \text{ is even} \end{cases} \tag{4.8}$$

This follows from

$$\int_{|\xi|=1} (x\xi)^q \log \frac{x\xi}{i} d\sigma_\xi = \int_{|\xi|=1, x\xi>0} (x\xi)^q \left(\log(x\xi) - \frac{\pi}{2} i\right) d\sigma_\xi$$
$$+ \int_{|\xi|=1, x\xi<0} (x\xi)^q \left(\log |x\xi| + \frac{\pi}{2} i\right) d\sigma_\xi$$
$$= \int_{|\xi|=1, x\xi>0} (x\xi)^q \left(\log(x\xi) - \frac{\pi}{2} i\right) d\sigma_\xi$$
$$+ (-1)^q \int_{|\xi|=1, x\xi>0} (x\xi)^q \left(\log(x\xi) + \frac{\pi}{2} i\right) d\sigma_\xi$$

$$= (1 + (-1)^q) \int_{|\xi|=1, x\xi>0} (x\xi)^q \log(x\xi) d\sigma_\xi$$

$$+ ((-1)^q - 1) \frac{\pi}{2} i \int_{|\xi|=1, x\xi>0} (x\xi)^q d\sigma_\xi$$

$$= \frac{1 + (-1)^q}{2} \int_{|\xi|=1} |x\xi|^q \log|x\xi| d\sigma_\xi + ((-1)^q - 1) \frac{\pi i}{4} \int_{|\xi|=1} |x\xi|^q d\sigma_\xi.$$

Next we show

$$\int_{|\xi|=1} |x\xi|^q d\sigma_\xi = 2\pi^{(n-1)/2} \Gamma\left(\frac{q+1}{2}\right) |x|^q \Big/ \Gamma\left(\frac{n+q}{2}\right), \qquad (4.9)$$

and

$$\int_{|\xi|=1} |x\xi|^q \log|x\xi| d\sigma_\xi$$

$$= 2\pi^{(n-1)/2} \Gamma\left(\frac{q+1}{2}\right) |x|^q \left(\log|x| + c_{n,q}\right) \Big/ \Gamma\left(\frac{n+q}{2}\right), \quad (4.10)$$

where $c_{n,q}$ is a constant depending only on n and q. With the aid of an orthogonal transformation which maps $x\xi$ to $|x|\xi_1$ for a fixed x

$$\int_{|\xi|=1} |x\xi|^q d\sigma_\xi = |x|^q \int_{|\xi|=1} |\xi_1|^q d\sigma_\xi. \qquad (4.11)$$

We obtain (4.9) with the aid of (4.11) and

$$\int_{|\xi|=1} |\xi_1|^q d\sigma_\xi = 2 \int_{|\xi|=1, \xi_1>0} \xi_1^q d\sigma_\xi$$

$$= 2\Omega_{n-1} \int_0^1 t^q (1-t^2)^{(n-3)/2} dt = \Omega_{n-1} \int_0^1 s^{(q-1)/2}(1-s)^{(n-3)/2} ds$$

$$= \Omega_{n-1} B\left(\frac{q+1}{2}, \frac{n-1}{2}\right) = 2\pi^{(n-1)/2} \Gamma\left(\frac{q+1}{2}\right) \Big/ \Gamma\left(\frac{n+q}{2}\right).$$

Analogously (4.10) follows from

$$\int_{|\xi|=1} |x\xi|^q \log|x\xi| d\sigma_\xi = \int_{|\xi|=1} (|x||\xi_1|)^q \log(|x||\xi_1|) d\sigma_\xi$$

$$= |x|^q \left(\log|x| \int_{|\xi|=1} |\xi_1|^q d\sigma_\xi + \int_{|\xi|=1} |\xi_1|^q \log|\xi_1| d\sigma_\xi \right).$$

Now we turn to the proof that (4.7) is a fundamental solution of $\Delta^{(n+q)/2}$. We show it only in the case n is even, since the case n is odd is easier. We denote (4.6),(4.7) by $K_1(x)$, $K(x)$ respectively. In view of (4.8),(4.10) and

$$\Gamma\left(\frac{q+1}{2}\right)\Gamma\left(\frac{q+2}{2}\right) = \frac{\pi^{1/2}q!}{2^q}$$

we have

$$K(x) = K_1(x) + c'_{n,q}|x|^q,$$

where $c'_{n,q}$ is a constant depending only on n and q. By virtue of Lemma 4.1 $K_1(x)$ is a fundamental solution of $\Delta^{(n+q)/2}$. Since $|x|^q$ is a polynomial of degree q, we have $\Delta^{(n+q)/2}|x|^q = 0$. Hence $K(x)$ is a fundamental solution of $\Delta^{(n+q)/2}$.

By virtue of Lemma 4.2 we have for $\phi \in C_0^\infty(R^n)$

$$\phi(x) = -\frac{1}{(2\pi i)^n q!}$$

$$\times \Delta^{(n+q)/2} \int_{R^n} \phi(y) \left(\int_{|\xi|=1} ((x-y)\xi)^q \log \frac{(x-y)\xi}{i} d\sigma_\xi \right) dy. \quad (4.12)$$

Let $L(D) = \sum_{|\alpha|=m} a_\alpha D^\alpha$ be an elliptic operator of order m with constant coefficients consisting only of the principal part. Put $L(\xi) = \sum_{|\alpha|=m} a_\alpha \xi^\alpha$ for $\xi = (\xi_1, \ldots, \xi_n) \in R^n$. Let q be a nonnegative integer such that $n+q$ is even. We are going to show that the function $K(x)$ defined by

$$K(x) = \Delta^{(n+q)/2}W(x), \quad (4.13)$$

$$W(x) = -\frac{1}{(2\pi i)^n(m+q)!} \int_{|\xi|=1} \frac{(x\xi)^{m+q}}{L(\xi)} \log \frac{x\xi}{i} d\sigma_\xi \quad (4.14)$$

is a fundamental solution of $L(D)$. If we put

$$F(s) = s^{m+q} \log \frac{s}{i}$$

for real s, then

$$W(x) = -\frac{1}{(2\pi i)^n(m+q)!} \int_{|\xi|=1} \frac{F(x\xi)}{L(\xi)} d\sigma_\xi. \quad (4.15)$$

Since

$$\left(\frac{d}{ds}\right)^m F(s) = \frac{(m+q)!}{q!} s^q \log \frac{s}{i} + \text{const} s^q,$$

we have

$$L(D)F(x\xi) = L(\xi)\left[\frac{(m+q)!}{q!}(x\xi)^q \log \frac{x\xi}{i} + \text{const}(x\xi)^q \right].$$

Hence,

$$L(D)W(x) = -\frac{1}{(2\pi i)^n q!} \int_{|\xi|=1} (x\xi)^q \log \frac{x\xi}{i} d\sigma_\xi + P(x),$$

where $P(x)$ is a homogeneous polynomial of degree q. Therefore for $\phi \in C_0^\infty(R^n)$

$$L(D) \int_{R^n} K(x-y)\phi(y)dy = L(D)\Delta^{(n+q)/2} \int_{R^n} W(x-y)\phi(y)dy$$

$$= \Delta^{(n+q)/2} \int_{R^n} \left[-\frac{1}{(2\pi i)^n q!} \int_{|\xi|=1} [(x-y)\xi]^q \log \frac{(x-y)\xi}{i} d\sigma_\xi \right.$$

$$\left. + P(x-y) \right] \phi(y) dy$$

$$= -\frac{1}{(2\pi i)^n q!} \Delta^{(n+q)/2} \int_{R^n} \phi(y) \int_{|\xi|=1} [(x-y)\xi]^q \log \frac{(x-y)\xi}{i} d\sigma_\xi dy.$$

By (4.12) the last side is equal to $\phi(x)$. Thus $K(x)$ is a fundamental solution of $L(D)$.

Next we show that $K(x)$ is analytic in $x \neq 0$. Let η be a fixed vector in R^n such that $|\eta| = 1$. For x such that $|x| + x\eta > 0$ put

$$T(\zeta, x) = \zeta + \frac{2\zeta\eta}{|x|} x - \frac{\zeta(x+|x|\eta)}{|x|(|x|+x\eta)}(x+|x|\eta).$$

Since $\xi = T(\zeta, x)$ is an orthogonal transformation for a fixed x and $x\xi = |x|\zeta\eta$,

$$\int_{|\xi|=1} \frac{(x\xi)^{m+q}}{L(\xi)} \log \frac{x\xi}{i} d\sigma_\xi = \int_{|\zeta|=1} \frac{(|x|\zeta\eta)^{m+q}}{L(T(\zeta,x))} \log \left(|x|\frac{\zeta\eta}{i} \right) d\sigma_\zeta$$

$$= |x|^{m+q} \log |x| \int_{|\zeta|=1} \frac{(\zeta\eta)^{m+q}}{L(T(\zeta,x))} d\sigma_\zeta + |x|^{m+q} \int_{|\zeta|=1} \frac{(\zeta\eta)^{m+q}}{L(T(\zeta,x))} \log \frac{\zeta\eta}{i} d\sigma_\zeta.$$

This show that $W(x)$ is analytic in $|x| + x\eta > 0$. Since η is arbitrary, $W(x)$ is, and hence $K(x)$ is analytic in $x \neq 0$.

We denote by $K_q(x)$ the fundamental solution constructed by (4.13),(4.14), and investigate its dependence on q. Letting one of Δ operate under the integral sign

$$K_q(x) = -\frac{1}{(2\pi i)^n (m+q)!} \Delta^{(n+q-2)/2} \int_{|\xi|=1} \frac{1}{L(\xi)}$$

$$\times \left[(m+q)(m+q-1)(x\xi)^{m+q-2} \log \frac{x\xi}{i} + (2m+2q-1)(x\xi)^{m+q-2} \right] d\sigma_\xi$$

$$= K_{q-2}(x) + \psi_q(x),$$

where

$$\psi_q(x) = -\frac{2m+2q-1}{(2\pi i)^n (m+q)!} \Delta^{(n+q-2)/2} \int_{|\xi|=1} \frac{(x\xi)^{m+q-2}}{L(\xi)} d\sigma_\xi. \tag{4.16}$$

If n is odd, then so is $q - 2$. Hence, the integral of (4.16) vanishes since the integrand is an odd function then. If n is even and $m < n$, the right hand side of (4.16) vanishes, since the integral is a polynomial of degree $m+q-2$. If n is even and $m \geq n$, ψ_q is a homogeneous polynomial of degree $m - n$. Therefore, in any case we have

$$D^m K_q(x) = D^m K_{q-2}(x) \tag{4.17}$$

for any derivative D^m of order m.

Theorem 4.1 *The fundamental solution $K(x)$ of $L(D)$ constructed by (4.13), (4.14) has the following form:*

$$K(x) = \psi(x) + p(x) \log |x|,$$

where ψ is a homogeneous function of degree $m - n$, $p(x) \equiv 0$ if either n is odd or $m < n$, and p is a homogeneous polynomial of degree $m - n$ if n is even and $m \geq n$.

Proof. As in the proof of Lemma 4.2

$$\int_{|\xi|=1} \frac{(x\xi)^{m+q}}{L(\xi)} \log \frac{x\xi}{i} d\sigma_\xi$$

$$= \int_{|\xi|=1, x\xi>0} \frac{(x\xi)^{m+q}}{L(\xi)} \left(\log(x\xi) - \frac{\pi}{2} i \right) d\sigma_\xi$$

$$+ \int_{|\xi|=1, x\xi<0} \frac{(x\xi)^{m+q}}{L(\xi)} \left(\log |x\xi| + \frac{\pi}{2} i \right) d\sigma_\xi$$

$$= \frac{1}{2} \int_{|\xi|=1} \frac{|x\xi|^{m+q}}{L(\xi)} \left(\log |x\xi| - \frac{\pi}{2} i \right) d\sigma_\xi$$

$$+ \frac{(-1)^q}{2} \int_{|\xi|=1} \frac{|x\xi|^{m+q}}{L(\xi)} \left(\log |x\xi| + \frac{\pi}{2} i \right) d\sigma_\xi$$

$$= \frac{1 + (-1)^q}{2} \int_{|\xi|=1} \frac{|x\xi|^{m+q}}{L(\xi)} \log |x\xi| d\sigma_\xi$$

$$+((-1)^q - 1)\frac{\pi i}{4}\int_{|\xi|=1}\frac{|x\xi|^{m+q}}{L(\xi)}d\sigma_\xi$$

$$= \begin{cases} -\dfrac{\pi i}{2}\displaystyle\int_{|\xi|=1}\dfrac{|x\xi|^{m+q}}{L(\xi)}d\sigma_\xi & q \text{ is odd} \\ \displaystyle\int_{|\xi|=1}\dfrac{|x\xi|^{m+q}}{L(\xi)}\log|x\xi|d\sigma_\xi & q \text{ is even} \end{cases}.$$

Hence, if n is odd

$$W(x) = \frac{1}{4(2\pi i)^{n-1}(m+q)!}\int_{|\xi|=1}\frac{|x\xi|^{m+q}}{L(\xi)}d\sigma_\xi, \qquad (4.18)$$

and if n is even

$$W(x) = -\frac{1}{(2\pi i)^n(m+q)!}\int_{|\xi|=1}\frac{(x\xi)^{m+q}}{L(\xi)}\log|x\xi|d\sigma_\xi. \qquad (4.19)$$

Therefore, if n is odd, $W(x)$ is homogeneous of degree $m + q$, and hence $K(x)$ is homogeneous of degree $m - n$. Suppose n is even. Then

$$\int_{|\xi|=1}\frac{(x\xi)^{m+q}}{L(\xi)}\log|x\xi|d\sigma_\xi$$
$$= \int_{|\xi|=1}\frac{(x\xi)^{m+q}}{L(\xi)}d\sigma_\xi\log|x| + \int_{|\xi|=1}\frac{(x\xi)^{m+q}}{L(\xi)}\log\left|\frac{x}{|x|}\xi\right|d\sigma_\xi.$$

Hence if we put

$$\psi_1(x) = -\frac{1}{(2\pi i)^n(m+q)!}\int_{|\xi|=1}\frac{(x\xi)^{m+q}}{L(\xi)}\log\left|\frac{x}{|x|}\xi\right|d\sigma_\xi,$$

$$\psi_2(x) = -\frac{1}{(2\pi i)^n(m+q)!}\int_{|\xi|=1}\frac{(x\xi)^{m+q}}{L(\xi)}d\sigma_\xi,$$

we have

$$W(x) = \psi_1(x) + \psi_2(x)\log|x|,$$

$\psi_1(x)$ is homogeneous of degree $m + q$, and $\psi_2(x)$ is a homogeneous polynomial of degree $m + q$. Thus, we conclude the proof of the theorem.

Theorem 4. 2 *The mth order derivatives of the fundamental solution $K(x)$ constructed by (4.13), (4.14) are homogeneous functions of degree $-n$, and satisfy (2.2).*

Proof. The assertion of the theorem is obvious if n is odd in view of Theorems 4.1 and 2.10. If n is even, we may assume $q = 0$ by (4.17), and hence in view of (4.19)

$$K(x) = -\frac{1}{(2\pi i)^n m!}\Delta^{n/2}\int_{|\xi|=1}\frac{(x\xi)^m}{L(\xi)}\log|x\xi|d\sigma_\xi. \qquad (4.20)$$

Since for $|\alpha| = m$

$$D^\alpha[(x\xi)^m \log|x\xi|] = (m!\log|x\xi| + \text{const})\xi^\alpha,$$

we get

$$D^\alpha\int_{|\xi|=1}\frac{(x\xi)^m}{L(\xi)}\log|x\xi|d\xi = \int_{|\xi|=1}(m!\log|x\xi| + \text{const})\frac{\xi^\alpha}{L(\xi)}d\sigma_\xi$$

$$= m!\int_{|\xi|=1}\frac{\xi^\alpha}{L(\xi)}\log|x\xi|d\sigma_\xi + \text{const}$$

$$= m!\int_{|\xi|=1}\frac{\xi^\alpha}{L(\xi)}d\sigma_\xi\log|x| + m!\int_{|\xi|=1}\frac{\xi^\alpha}{L(\xi)}\log\left|\frac{x}{|x|}\xi\right|d\sigma_\xi + \text{const}.$$

With the aid of this relation and (4.20), and noting $\Delta^{n/2}\log|x| = 0$ we obtain

$$D^\alpha K(x) = -\frac{1}{(2\pi i)^n}\Delta^{n/2}\int_{|\xi|=1}\frac{\xi^\alpha}{L(\xi)}\log\left|\frac{x}{|x|}\xi\right|d\sigma_\xi.$$

The right hand side of this equality is obviously homogeneous of degree $-n$, and satisfies (2.2) by Theorem 2.10.

Next, we show that for $u \in C_0^m(R^n)$

$$\int_{R^n}K(x-y)L(D)u(y)dy = u(x). \qquad (4.21)$$

As is easily seen $K(-x)$ is a fundamental solution of $L(-D)$. Hence for $\phi \in C_0^\infty(R^n)$

$$\int_{R^n}\int_{R^n}K(x-y)L(D)u(y)dy\phi(x)dx$$

$$= \int_{R^n}L(D)u(y)\int_{R^n}K(x-y)\phi(x)dxdy$$

$$= \int_{R^n}u(y)L(-D)\int_{R^n}K(x-y)\phi(x)dxdy = \int_{R^n}u(y)\phi(y)dy.$$

This implies that (4.21) is true.

We use the same notation $|\cdot|_{m,p,\Omega}$, etc. as those introduced in section 3.4.

Theorem 4.3 *For a natural number l such that $l \geq m$ and $1 < p < \infty$ there exists a constant $C_{l,p}$ such that for $u \in C_0^l(R^n)$*

$$|u|_{l,p} \leq C_{l,p} |L(D)u|_{l-m,p}. \tag{4.22}$$

Proof. With the aid of Theorem 4.1 and (4.21)

$$D^{m-1}u(x) = \int_{R^n} D_x^{m-1}K(x-y)L(D)u(y)dy.$$

Applying Theorem 2.10 to $D^{m-1}K$ we get the conclusion for $l = m$:

$$|u|_{m,p} \leq C_{m,p} |L(D)u|_{0,p}. \tag{4.23}$$

If $l > m$, the conclusion is obtained by applying (4.23) to $D^{l-m}u$.

4.2 Assumptions and Main Result

Let $L = L(x, D) = \sum_{|\alpha| \leq m} a_\alpha(x)D^\alpha$ be an elliptic operator of order m in an open set Ω of R^n. The principal part of $L(x, D)$ is denoted by $L^0(x, D)$:

$$L^0(x, D) = \sum_{|\alpha|=m} a_\alpha(x)D^\alpha.$$

Let ξ, η be linearly independent real vectors. Owing to the ellipticity of L the polynomial $L^0(x, \xi + \tau\eta)$ of the variable τ has no real roots. In what follows we assume the following conditions.

SMOOTHNESS CONDITION on Ω. Ω is uniformly regular of class C^m.

SMOOTHNESS CONDITION on L. For $|\alpha| = m$ a_α is bounded and uniformly continuous in $\bar{\Omega}$, and for $|\alpha| < m$ a_α is bounded and measurable in Ω.

ELLIPTICITY CONDITION. L is *uniformly elliptic* in Ω, i.e. there exists a positive constant c such that for any $x \in \Omega$ and $\xi \in R^n$

$$\left| L^0(x, \xi) \right| \geq c|\xi|^m. \tag{4.24}$$

ROOT CONDITION. For every pair of linearly independent real vectors ξ, η the polynomial $L^0(x, \xi + \tau\eta)$ of the variable τ has equal number of roots with positive imaginary part and with negative imaginary part.

Due to the Root Condition the order m of L is even, and $L^0(x, \xi + \tau\eta)$ has exactly $m/2$ roots with positive imaginary part. It is easy to verify that if $n \geq 3$ all elliptic operators satisfy the Root Condition, since if τ is a root for ξ, η then $-\tau$ is a root for $-\xi, -\eta$ and a sphere in R^{n-1} is connected.

Let $B_j(x, D) = \sum_{|\beta| \leq m_j} b_{j\beta}(x)D^\beta, j = 1, \ldots, m/2$, be a set of linear differential operators defined on $\partial\Omega$. We assume $m_j < m$ for $j = 1, \ldots, m/2$.

SMOOTHNESS CONDITION on $\{B_j\}_{j=1}^{m/2}$. For $|\beta| \leq m_j, j = 1, \ldots, m/2, b_{j\beta}$ is $m - m_j$ times differentiable on $\partial\Omega$ and the derivatives of order up to $m - m_j$ are all bounded and uniformly continuous on $\partial\Omega$.

COMPLEMENTING CONDITION. Let x be an arbitrary point on $\partial\Omega$ and ν be the outward normal unit vector to $\partial\Omega$ at x. For each tangential vector $\xi \neq 0$ to $\partial\Omega$ at x let $\tau_1(x, \xi), \ldots, \tau_{m/2}(x, \xi)$ be the roots of the polynomial $L^0(x, \xi + \tau\nu)$ with positive imaginary part. Then the polynomials, in τ, $\{B_j^0(x, \xi + \tau\nu)\}_{j=1}^{m/2}$ are linearly independent modulo the polynomial $\prod_{j=1}^{m/2} (\tau - \tau_j(x, \xi))$, i.e. a linear combination of $\{B_j^0(x, \xi + \tau\nu)\}_{j=1}^{m/2}$ is divisible by $\prod_{j=1}^{m/2} (\tau - \tau_j(x, \xi))$ if and only if all the coefficients vanish, where B_j^0 is the principal part of B_j.

Following S. Agmon, A. Douglis and L. Nirenberg [11] it will be shown that under the above conditions the following *a priori* estimate holds:

$$\|u\|_{m,p,\Omega} \leq C_p \left[\|L(\cdot, D)u\|_{0,p,\Omega} \right.$$

$$\left. + \sum_{j=1}^{m/2} [B_j(\cdot, D)u]_{m - m_j - 1/p, p, \partial\Omega} + \|u\|_{0,p,\Omega} \right] \tag{4.25}$$

for $1 < p < \infty$.

4.3 Preliminaries from the Theory of Ordinary Differential Equations

Let $l(\tau) = \sum_{j=0}^m a_{m-j}\tau^j$ be a polynomial of order m. Suppose that $l(\tau)$ has exactly $m/2$ roots with negative real part. Let $b_j(\tau) = \sum_{k=1}^{m_j} b_{jk}\tau^k, j = 1, \ldots, m/2$, be a polynomial of order m_j. We are interested in solutions belonging to $L^2(0, \infty)$ of the initial value problem

$$l\left(\frac{d}{dt}\right)u \equiv \sum_{j=0}^m a_{m-j}\frac{d^j u}{dt^j} = 0, \quad t > 0, \tag{4.26}$$

$$b_j\left(\frac{d}{dt}\right)u\bigg|_{t=0} \equiv \sum_{k=1}^{m_j} b_{jk}\frac{d^k u}{dt^k}\bigg|_{t=0} = g_j, \quad j = 1, \ldots, m/2. \tag{4.27}$$

Let $\lambda_1, \ldots, \lambda_p$ and $\lambda_{p+1}, \ldots, \lambda_{p+q}$ be the distinct roots of $l(\tau)$ with negative and positive real part respectively, and ν_j be the multiplicity of λ_j. The general solution of (4.26) is

$$u(t) = \sum_{j=1}^{p+q} \sum_{k=1}^{\nu_j} c_{j,k} t^{k-1} e^{\lambda_j t}. \tag{4.28}$$

As is easily seen $u(t)$ given by (4.28) belongs to $L^2(0, \infty)$ if and only if $c_{j,k} = 0$ for $j > p$:

$$u(t) = \sum_{j=1}^{p} \sum_{k=1}^{\nu_j} c_{j,k} t^{k-1} e^{\lambda_j t}. \tag{4.29}$$

If we put

$$\hat{l}(\tau) = \prod_{j=1}^{p} (\tau - \lambda_j)^{\nu_j},$$

then (4.29) is a general solution of

$$\hat{l}(d/dt) u(t) = 0. \tag{4.30}$$

If the $m/2$ coefficients $c_{j,k}$ in (4.29) are uniquely determined by the initial conditions (4.27), then the problem (4.30),(4.27) has a unique solution. We wish to express this condition in an easily visible form.

Lemma 4. 3 *Let $p(\tau)$ be a polynomial of order μ whose coefficient of the highest order term is equal to 1:*

$$p(\tau) = \sum_{k=0}^{\mu} \alpha_{\mu-k} \tau^k, \quad \alpha_0 = 1.$$

For $j = 0, 1, \ldots, \mu - 1$ we put

$$p_j(\tau) = \sum_{k=0}^{j} \alpha_{j-k} \tau^k.$$

If γ is a rectifiable Jordan curve enclosing all the roots of $p(\tau)$ in its interior, then for $0 \leq j, k \leq \mu - 1$

$$\frac{1}{2\pi i} \int_{\gamma} \frac{p_{\mu-1-j}(\tau)}{p(\tau)} \tau^k d\tau = \delta_{jk}. \tag{4.31}$$

Proof. If $k < j$, then $p_{\mu-1-j}(\tau)\tau^k$ is a polynomial of order not exceeding $\mu - 2$. Hence, choosing γ to be a circle $|\tau| = R$ and letting $R \to \infty$ we obtain (4.31). If $k = j$,

$$\frac{p_{\mu-1-j}(\tau)}{p(\tau)}\tau^j - \frac{1}{\tau} = \frac{1}{\tau p(\tau)}\left(\sum_{\kappa=0}^{\mu-1-j}\alpha_{\mu-1-j-\kappa}\tau^{\kappa+j+1} - p(\tau)\right)$$

$$= \frac{1}{\tau p(\tau)}\left(\sum_{\kappa=j+1}^{\mu}\alpha_{\mu-\kappa}\tau^\kappa - \sum_{\kappa=0}^{\mu}\alpha_{\mu-\kappa}\tau^\kappa\right) = -\frac{1}{\tau p(\tau)}\sum_{\kappa=0}^{j}\alpha_{\mu-\kappa}\tau^\kappa.$$

Hence, the right hand side of (4.31) is equal to

$$\frac{1}{2\pi i}\int_\gamma \frac{d\tau}{\tau} - \frac{1}{2\pi i}\int_\gamma \frac{1}{\tau p(\tau)}\sum_{\kappa=0}^{j}\alpha_{\mu-\kappa}\tau^\kappa d\tau.$$

The desired result in this case is obtained as in the case $k < j$. If $k > j$,

$$\frac{p_{\mu-1-j}(\tau)}{p(\tau)}\tau^k = \frac{\tau^k}{p(\tau)}\sum_{\kappa=0}^{\mu-1-j}\alpha_{\mu-1-j-\kappa}\tau^\kappa$$

$$= \frac{\tau^{k-j-1}}{p(\tau)}\sum_{\kappa=0}^{\mu-1-j}\alpha_{\mu-1-j-\kappa}\tau^{\kappa+j+1} = \frac{\tau^{k-j-1}}{p(\tau)}\sum_{\kappa=j+1}^{\mu}\alpha_{\mu-\kappa}\tau^\kappa$$

$$= \frac{\tau^{k-j-1}}{p(\tau)}\left(p(\tau) - \sum_{\kappa=0}^{j}\alpha_{\mu-\kappa}\tau^\kappa\right) = \tau^{k-j-1} - \frac{\tau^{k-j-1}}{p(\tau)}\sum_{\kappa=0}^{j}\alpha_{\mu-\kappa}\tau^\kappa.$$

Therefore, the right hand side of (4.31) is equal to

$$-\frac{1}{2\pi i}\int_\gamma \frac{\tau^{k-j-1}}{p(\tau)}\sum_{\kappa=0}^{j}\alpha_{\mu-\kappa}\tau^\kappa d\tau.$$

Hence, we obtain the desired resuslt also in this case as in the case $k < j$.

We set

$$\hat{l}(\tau) = \sum_{k=0}^{m/2}\hat{a}_{m/2-k}\tau^k,$$

and for $j = 0, \ldots, m/2 - 1$

$$\hat{l}_j(\tau) = \sum_{k=0}^{j}\hat{a}_{j-k}\tau^k.$$

In view of Lemma 4.3 we have for $j, k = 1, \ldots, m/2$

$$\frac{1}{2\pi i}\int_\gamma \frac{1}{\hat{l}(\tau)}\hat{l}_{m/2-k}(\tau)\tau^{j-1}d\tau = \delta_{jk}, \tag{4.32}$$

where γ is a rectifiable Jordan curve enclosing all the roots of $\hat{l}$. Consequently the function u defined by

$$u(t) = \frac{1}{2\pi i} \int_\gamma \frac{1}{\hat{l}(\tau)} \sum_{k=1}^{m/2} c_k \hat{l}_{m/2-k}(\tau) e^{t\tau} d\tau \qquad (4.33)$$

is the solution of the initial value problem

$$\hat{l}\left(\frac{d}{dt}\right) u(t) = 0, \quad -\infty < t < \infty, \qquad (4.34)$$

$$\left(\frac{d}{dt}\right)^{j-1} u(0) = c_j, \quad j = 1, \ldots, m/2. \qquad (4.35)$$

Let $b'_j(\tau) = \sum_{i=1}^{m/2} b'_{ji} \tau^{i-1}$ be the remainder when $b_j(\tau)$ is divided by $\hat{l}(\tau)$:

$$b_j(\tau) = p_j(\tau)\hat{l}(\tau) + b'_j(\tau). \qquad (4.36)$$

With the aid of (4.32) and (4.36) we get

$$b_j\left(\frac{d}{dt}\right) u(0) = \frac{1}{2\pi i} \int_\gamma \frac{1}{\hat{l}(\tau)} \sum_{k=1}^{m/2} c_k \hat{l}_{m/2-k}(\tau) b'_j(\tau) d\tau$$

$$= \sum_{k=1}^{m/2} \sum_{i=1}^{m/2} c_k b'_{ji} \frac{1}{2\pi i} \int_\gamma \frac{1}{\hat{l}(\tau)} \hat{l}_{m/2-k}(\tau) \tau^{i-1} d\tau$$

$$= \sum_{k=1}^{m/2} \sum_{i=1}^{m/2} c_k b'_{ji} \delta_{ki} = \sum_{k=1}^{m/2} c_k b'_{jk}.$$

Therefore if $c_1, \ldots, c_{m/2}$ satisfy

$$\sum_{k=1}^{m/2} c_k b'_{jk} = g_j, \quad j = 1, \ldots, m/2, \qquad (4.37)$$

then (4.33) is the solution of (4.30),(4.27). Hence

$$\det\left(b'_{jk}\right) \neq 0 \qquad (4.38)$$

is a necessary and sufficient condition in order that the initial value problem (4.26),(4.27) has a unique solution belonging to $L^2(0, \infty)$ for any $g_1, \ldots, g_{m/2}$. The condition (4.38) is equivalent to the condition that $b'_1(\tau), \ldots, b'_{m/2}(\tau)$ are linearly independent or $b_1(\tau), \ldots, b_{m/2}(\tau)$ are linearly independent modulo the polynomial $\hat{l}(\tau)$, i.e. a linear combination $\lambda_1 b_1(\tau) + \cdots + \lambda_{m/2} b_{m/2}(\tau)$

is divisible by $\hat{l}(\tau)$ if and only if $\lambda_1 = \cdots = \lambda_{m/2} = 0$. Let $\left(b^{jk}\right)$ be the inverse matrix of $\left(b'_{jk}\right)$. Then (4.37) implies $c_k = \sum_{j=1}^{m/2} b^{kj} g_j$. Substituting this into (4.33) we get

$$u(t) = \sum_{j=1}^{m/2} \frac{1}{2\pi i} \int_\gamma \frac{1}{\hat{l}(\tau)} \sum_{k=1}^{m/2} b^{kj} \hat{l}_{m/2-k}(\tau) e^{t\tau} d\tau g_j.$$

If we set $n_j(\tau) = \sum_{k=1}^{m/2} b^{kj} \hat{l}_{m/2-k}(\tau)$, we obtain

$$u(t) = \sum_{j=1}^{m/2} \frac{1}{2\pi i} \int_\gamma \frac{n_j(\tau)}{\hat{l}(\tau)} e^{t\tau} d\tau g_j. \qquad (4.39)$$

It is easy to show

$$\frac{1}{2\pi i} \int_\gamma \frac{n_j(\tau) b_k(\tau)}{\hat{l}(\tau)} d\tau = \delta_{kj}. \qquad (4.40)$$

4.4 Case of Constant Coefficients

In this section we consider the problem in a half space for operators which have only the principal part with constant coefficients:

$$L(D)u(x) \equiv \sum_{|\alpha|=m} a_\alpha D^\alpha u(x) = 0, \quad x \in R^n_+, \qquad (4.41)$$

$$B_j(D)u(x)|_{x_n=0} \equiv \sum_{|\beta|=m_j} b_{j\beta} D^\beta u(x)|_{x_n=0}$$

$$= g_j(x'), \quad j = 1, \ldots, m/2, \quad x' \in R^{n-1}, \qquad (4.42)$$

where

$$R^n_+ = \{x = (x', x_n); x' = (x_1, \ldots, x_{n-1}) \in R^{n-1}, x_n > 0\}.$$

We denote the Fourier transform of functions $g \in L^2(R^{n-1})$ by

$$\hat{g}(\xi') = (2\pi)^{(1-n)/2} \int_{R^{n-1}} e^{-ix'\xi'} g(x') dx',$$

where $\xi' = (\xi_1, \ldots, \xi_{n-1}) \in R^{n-1}$. The partial Fourier transform of functions $u \in L^2(R^n_+)$ with respect to x' is defined by

$$\hat{u}(\xi', x_n) = (2\pi)^{(1-n)/2} \int_{R^{n-1}} e^{-ix'\xi'} u(x', x_n) dx'.$$

If $g_j \in L^2(R^{n-1}), u \in W^{m,2}(R^n_+)$, then the problem (4.41),(4.42) is transformed to

$$L(i\xi', D_n)\hat{u}(\xi', x_n) = 0, \quad x_n > 0, \tag{4.43}$$

$$B_j(i\xi', D_n)\hat{u}(\xi', 0) = \hat{g}_j(\xi'), \quad j = 1, \ldots, m/2. \tag{4.44}$$

Owing to the Root Condition the polynomial $L(\xi', \tau)$ has exactly $m/2$ roots with positive imaginary part, which we denote by $\tau_1(\xi'), \ldots, \tau_{m/2}(\xi')$. We define the following functions.

$$M^+(\xi', \tau) = \prod_{j=1}^{m/2}(\tau - \tau_j(\xi')) = \sum_{k=0}^{m/2} \alpha^+_{m/2-k}(\xi')\tau^k, \tag{4.45}$$

$$M^+_j(\xi', \tau) = \sum_{k=0}^{j} \alpha^+_{j-k}(\xi')\tau^k, \quad j = 0, \ldots, m/2 - 1, \tag{4.46}$$

$$B'_j(\xi', \tau) = \sum_{k=1}^{m/2} b'_{jk}(\xi')\tau^{k-1} \quad \text{is the remainder}$$

of $B_j(\xi', \tau)$ divided by $M^+(\xi', \tau)$,

$(b^{jk}(\xi'))$ is the inverse matrix of $(b'_{jk}(\xi'))$,

$$N_j(\xi', \tau) = \sum_{k=1}^{m/2} b^{kj}(\xi')M^+_{m/2-k}(\xi', \tau). \tag{4.47}$$

The coefficients $\alpha^+_k(\xi'), k = 0, \ldots, m/2$, are analytic functions of $\xi' \neq 0$ as is shown below. For a fixed ξ' let γ be a rectifiable Jordan curve having $\tau_1(\xi'), \ldots, \tau_{m/2}(\xi')$ inside and all the other roots of $L(\xi', \tau)$ outside. Then

$$\frac{1}{2\pi i}\int_\gamma \frac{\tau^k}{L(\xi', \tau)}\frac{\partial}{\partial\tau}L(\xi', \tau)d\tau = \sum_{j=1}^{m/2}\tau_j(\xi')^k \tag{4.48}$$

for $k = 0, 1, \ldots, m/2$. We may assume that when ξ' is in a small neighborhood of each nonzero point, γ can be chosen fixed. Then the left hand side of (4.48) is an analytic function of ξ' in such a small neighborhood. Hence, the elementary symmetric functions $\alpha^+_k(\xi'), k = 0, \ldots, m/2$, of $\tau_1(\xi'), \ldots, \tau_{m/2}(\xi')$ are also analytic.

We apply the result of the previous section to (4.43),(4.44). In place of $l(\tau), b_j(\tau)$ we have $L(i\xi', \tau), B_j(i\xi', \tau)$. The polynomials $\hat{l}(\tau), \hat{l}_j(\tau)$ are replaced by $i^{m/2}M^+(\xi', -i\tau), i^j M^+_j(\xi', -i\tau)$, and $b'_j(\tau)$ is by

$$i^{m_j}B'_j(\xi', -i\tau) = \sum_{k=1}^{m/2} i^{m_j-k+1}b'_{jk}(\xi')\tau^{k-1}.$$

Hence in place of the matrix $\left(b^{jk}\right)$ we have

$$\left(i^{m_j-k+1}b'_{jk}(\xi')\right)^{-1} = \left(i^{j-m_k-1}b^{jk}(\xi')\right),$$

and hence $i^{m/2-m_j-1}N_j(\xi', -i\tau)$ in place of $n_j(\tau)$. Let γ be a rectifiable Jordan curve enclosing all the roots of $M^+(\xi', \tau)$ for a fixed $\xi' \neq 0$. Then $i\gamma$ encloses all the roots of $M^+(\xi', -i\tau)$. Applying (4.39) to the present case we get

$$\hat{u}(\xi', x_n) = \sum_{j=1}^{m/2} \frac{1}{2\pi i} \int_{i\gamma} \frac{i^{m/2-m_j-1}N_j(\xi', -i\tau)}{i^{m/2}M^+(\xi', -i\tau)} e^{x_n\tau} d\tau \hat{g}_j(\xi').$$

By the change of the variable $\tau \to i\tau$

$$\hat{u}(\xi', x_n) = \sum_{j=1}^{m/2} \frac{1}{2\pi i} \int_{\gamma} \frac{N_j(\xi', \tau)}{M^+(\xi', \tau)} e^{ix_n\tau} d\tau \cdot i^{-m_j} \hat{g}_j(\xi'). \qquad (4.49)$$

Corresponding to (4.40) we have

$$\frac{1}{2\pi i} \int_{\gamma} \frac{N_j(\xi', \tau)B_k(\xi', \tau)}{M^+(\xi', \tau)} d\tau = \delta_{jk}. \qquad (4.50)$$

4.5 Poisson Kernels

In this and the following two sections we consider the boundary value problem (4.41),(4.42). Changing the notations we denote by $x = (x_1, \ldots, x_{n-1})$ and t generic points of R^{n-1} and R^1 respectively, and in consequence

$$R_+^n = \{(x, t); x \in R^{n-1}, t > 0\},$$
$$D_i = \frac{\partial}{\partial x_i}, \quad i = 1, \ldots, n-1, \quad D_t = \frac{\partial}{\partial t},$$
$$D_x = (D_1, \ldots, D_{n-1}), \quad D = (D_x, D_t) = (D_1, \ldots, D_{n-1}, D_t).$$

In terms of these notations the problem (4.41),(4.42) is expressed as

$$L(D)u(x, t) = 0, \quad x \in R^{n-1}, t > 0, \qquad (4.51)$$
$$B_j(D)u(x, 0) = g_j(x), \quad x \in R^{n-1}, \quad j = 1, \ldots, m/2. \qquad (4.52)$$

The adjoint variable of $x = (x_1, \ldots, x_{n-1})$ is denoted by $\xi = (\xi_1, \ldots, \xi_{n-1})$. The relation (4.50) is written as

$$\frac{1}{2\pi i} \int_{\gamma} \frac{N_j(\xi, \tau)B_k(\xi, \tau)}{M^+(\xi, \tau)} d\tau = \delta_{jk}, \qquad (4.53)$$

where γ is a Jordan contour enclosing all the roots of $M^+(\xi, \tau)$. If $u(x, t)$ is the solution of (4.51),(4.52) belonging to $W^{m,2}(R_+^n)$, then its partial Fourier transform with respect to x is by (4.49)

$$\hat{u}(\xi, t) = \sum_{j=1}^{m/2} \frac{1}{2\pi i} \int_\gamma \frac{N_j(\xi, \tau)}{M^+(\xi, \tau)} e^{it\tau} d\tau \cdot i^{-m_j} \hat{g}_j(\xi). \qquad (4.54)$$

The functions $K_j, j = 1, \ldots, m/2$, called Poisson kernels, are defined in R_+^n as follows: for $m_j \geq n - 1$

$$K_j(x, t) = \frac{-1}{(2\pi i)^n (m_j - n + 1)!} \int_{|\xi|=1} d\sigma_\xi$$

$$\times \int_\gamma \frac{N_j(\xi, \tau)}{M^+(\xi, \tau)} (x\xi + t\tau)^{m_j - n + 1} \log \frac{x\xi + t\tau}{i} d\tau, \qquad (4.55)$$

and for $m_j < n - 1$

$$K_j(x, t) = (-1)^{n-1-m_j} \frac{(n - m_j - 2)!}{(2\pi i)^n} \int_{|\xi|=1} d\sigma_\xi$$

$$\times \int_\gamma \frac{N_j(\xi, \tau)}{M^+(\xi, \tau)} (x\xi + t\tau)^{m_j - n + 1} d\tau, \qquad (4.56)$$

where $d\sigma_\xi$ is the area element of the unit sphere in R^{n-1} and γ is a Jordan contour in the half plane $\text{Im} \tau > 0$ enclosing all the roots of $M^+(\xi, \tau)$ for all $|\xi| = 1$.

Let $c_{\lambda,\mu}$ be the constant such that

$$\left(\frac{d}{dz} \right)^\lambda \left[z^{\lambda+\mu} \left(\log \frac{z}{i} + c_{\lambda,\mu} \right) \right] = \frac{(\lambda + \mu)!}{\mu!} z^\mu \log \frac{z}{i} \qquad (4.57)$$

for nonnegative integers λ, μ. The existence of such a constant is easily seen. It is also easy to show that for $\mu < 0, \lambda + \mu \geq 0$

$$\left(\frac{d}{dz} \right)^\lambda \left(z^{\lambda+\mu} \log \frac{z}{i} \right) = (-1)^{1+\mu} (\lambda + \mu)! (-1 - \mu)! z^\mu \qquad (4.58)$$

For $j = 1, \ldots, m/2$ and a nonnegative integer q, let $K_{j,q}(x, t)$ be a function defined by: if $m_j \geq n - 1$

$$K_{j,q}(x, t) = \frac{-1}{(2\pi i)^n (m_j + q)!} \int_{|\xi|=1} d\sigma_\xi \qquad (4.59)$$

$$\times \int_\gamma \frac{N_j(\xi, \tau)}{M^+(\xi, \tau)} (x\xi + t\tau)^{m_j + q} \left(\log \frac{x\xi + t\tau}{i} + c_{n+q-1, m_j - n + 1} \right) d\tau,$$

where c_{n+q-1,m_j-n+1} is a constant defined by (4.57), and if $m_j < n - 1$

$$K_{j,q}(x,t) = \frac{-1}{(2\pi i)^n (m_j+q)!} \int_{|\xi|=1} d\sigma_\xi$$
$$\times \int_\gamma \frac{N_j(\xi,\tau)}{M^+(\xi,\tau)} (x\xi + t\tau)^{m_j+q} \log \frac{x\xi + t\tau}{i} d\tau. \qquad (4.60)$$

It is easy to verify that $L(D)K_{j,q}(x,t) = 0$, and if q is a nonnegative integer with the same parity as $n - 1$

$$\Delta_x^{(n+q-1)/2} K_{j,q}(x,t) = K_j(x,t). \qquad (4.61)$$

We continue to use the notation D^k to denote kth derivatives. If $k = 1$ we simply write D.

Lemma 4. 4 Let $F(x,t)$ be a function in $C^2(R_+^n)$ such that

$$|D^k F(x,t)| \le C/t, \quad k = 0, 1, 2, \qquad (4.62)$$

on the hemishpere $\Sigma = \{(x,t) \in R^n; |x|^2 + t^2 = 1, t > 0\}$. Then $F(x,t)$ is bounded on Σ.

Proof. Let $(x,t) \in \Sigma, x \ne 0$. Since

$$DF(0,1) - DF(x,t) = \int_t^1 \frac{\partial}{\partial r} DF\left(\sqrt{1-r^2}\frac{x}{|x|}, r\right) dr$$
$$= \int_t^1 \left[\sum_{i=1}^{n-1} \frac{-r}{\sqrt{1-r^2}} \frac{x_i}{|x|} D_i DF\left(\sqrt{1-r^2}\frac{x}{|x|}, r\right)\right.$$
$$\left. + D_t DF\left(\sqrt{1-r^2}\frac{x}{|x|}, r\right)\right] dr,$$

we get with the aid of (4.62)

$$|DF(0,1) - DF(x,t)| \le \text{const} \int_t^1 \left(\frac{1}{\sqrt{1-r^2}} + \frac{1}{r}\right) dr \le \text{const} \left(1 + \log\frac{1}{t}\right).$$

Hence

$$|DF(x,t)| \le \text{const}(1 + |\log t|). \qquad (4.63)$$

Repeating the same argument using (4.63) instead of (4.62) we obtain

$$|F(0,1) - F(x,t)| \le \text{const},$$

from which the conclusion follows.

Lemma 4. 5 *The kernels $K_{j,q}$ are infinitely differentiable in $\bar{R}^n_+$ except at the origin. For each nonnegative integer l there exists a constant C_l such that*

$$\left|D^l K_{j,q}(x,t)\right| \le C_l \left(|x|^2 + t^2\right)^{(m_j+q-l)/2} \left[1 + \left|\log\left(|x|^2 + t^2\right)^{1/2}\right|\right]. \quad (4.64)$$

If $l \ge m_j + q + 1$, then $D^l K_{j,q}$ is a homogeneous function of order $m_j + q - l$, and the logarithmic term in (4.64) may be omitted.

Proof. Let γ be a closed curve consisting of the arc $\{\tau; |\tau| = C, \mathrm{Im}\,\tau \ge C^{-1}\}$ and the segment joining both of its end points, where C is a large positive constant. Since $N_j(\xi,\tau)/M^+(\xi,\tau)$ is bounded on $\{(\xi,\tau); |\xi| = 1, \tau \in \gamma\}$, it is not difficult to show that (4.64) holds on the unit sphere $\{(x,t); |x|^2+t^2 = 1\}$ if $l \le m_j + q$. From the definition of $K_{j,q}$ it easily follows that (4.64) holds also for general points (x,t). Suppose $l \ge m_j + q + 1$. Since

$$\left(\frac{d}{dz}\right)^l \left[z^{m_j+q}\left(\log\frac{z}{i} + c_{n+q-1,m_j-n+1}\right)\right] = \left(\frac{d}{dz}\right)^l \left(z^{m_j+q}\log\frac{z}{i}\right)$$
$$= (-1)^{l-m_j-q+1}(m_j+q)!\,(l-m_j-q-1)!z^{m_j+q-l},$$

we have

$$D^l K_{j,q}(x,t) = \int_{|\xi|=1}\int_\gamma F(\xi,\tau)(x\xi+t\tau)^{m_j+q-l}d\sigma_\xi d\tau,$$

$$F(\xi,\tau) = \frac{(-1)^{l-m_j-q}}{(2\pi i)^n}(l-m_j-q-1)!\frac{N_j(\xi,\tau)}{M^+(\xi,\tau)}(\xi,\tau)^l,$$

where $(\xi,\tau)^l$ is the value obtained by replacing D_i, D_t by ξ_i, τ in D^l respectively. Hence $D^l K_{j,q}(x,t)$ is homogeneous of degree $m_j + q - l$. The proof will be complete if we show that this is bounded on the hemishere Σ. For that purpose it suffices to show in view of Lemma 4.4 that the inequalities

$$\left|D^k D^l K_{j,q}(x,t)\right| \le C/t, \quad k = 0,1,2$$

holds on Σ. Replacing $k+l$ by l we are going to show that for $l \ge m_j+q+1$

$$\left|D^l K_{j,q}(x,t)\right| \le C/t, \quad (x,t) \in \Sigma. \quad (4.65)$$

If $t \ge 1/2$, then (4.65) clearly holds. Suppose $0 < t < 1/2$. Let ζ be a function in $C^\infty([-1,1])$ such that $\zeta(r) = 1$ for $|r| \le 1/2$ and $\zeta(r) = 0$ for $3/4 \le |r| \le 1$. We write

$$D^l K_{j,q}(x,t) = I_1 + I_2,$$

$$I_1 = \int_{|\xi|=1}\int_\gamma \frac{F(\xi,\tau)}{(x\xi+t\tau)^\mu}\zeta(x\xi)d\tau d\sigma_\xi,$$

$$I_2 = \int_{|\xi|=1}\int_\gamma \frac{F(\xi,\tau)}{(x\xi+t\tau)^\mu}(1-\zeta(x\xi))d\tau d\sigma_\xi,$$

where $\mu = l - m_j - q$. Since $|x\xi + t\tau| \geq c$ for some constant $c > 0$ if $|x\xi| > 1/2, \tau \in \gamma$,

$$|I_2| \leq \text{const} \int_{|\xi|=1} \int_{\gamma} |F(\xi, \tau)| |d\tau| d\sigma_{\xi}. \qquad (4.66)$$

For a fixed $x = (x_1, \ldots, x_{n-1})$ let T_x be an orthogonal transformation which maps $(x_1, \ldots, x_{n-1})$ to $(|x|, 0, \ldots, 0)$. Making the change of the variable $\eta = T_x \xi$ in the integral of I_1

$$I_1 = \int_{|\eta|=1} \int_{\gamma} \frac{F(T_x^{-1}\eta, \tau)}{(|x|\eta_1 + t\tau)^{\mu}} \zeta(|x|\eta_1) \, d\tau d\sigma_{\eta}.$$

Since $0 < t < 1/2$ we have $|x| \geq \sqrt{3/4}$. Hence the integrand of the above integral vanishes for $|\eta_1| > \sqrt{3/4}$. Therefore writing $\eta' = (\eta_2, \ldots, \eta_{n-1})$ we have

$$I_1 = \int_{|\eta'|=1} d\sigma_{\eta'} \int_{-\sqrt{3/4}}^{\sqrt{3/4}} d\eta_1 \int_{\gamma} \frac{F(T_x^{-1}\tilde{\eta}, \tau)\zeta(|x|\eta_1)}{(|x|\eta_1 + t\tau)^{\mu}} (1 - \eta_1^2)^{n/2-2} \, d\tau,$$

where $\tilde{\eta} = (\eta_1, (1 - \eta_1^2)^{1/2}\eta')$. Integrating by parts $\mu - 1$ times with respect to η_1 yields

$$|I_1| = \frac{1}{(\mu-1)!|x|^{\mu-1}} \left| \int_{|\eta'|=1} d\sigma_{\eta'} \int_{-\sqrt{3/4}}^{\sqrt{3/4}} d\eta_1 \right.$$

$$\times \left. \int_{\gamma} (|x|\eta_1 + t\tau)^{-1} \left(\frac{\partial}{\partial \eta_1}\right)^{\mu-1} \left[F\left(T_x^{-1}\tilde{\eta}, \tau\right) \zeta(|x|\eta_1) (1 - \eta_1^2)^{n/2-2} \right] d\tau \right|$$

Noting $|x| \geq \sqrt{3/4}, ||x|\eta_1 + t\tau| > t\text{Im}\tau$ and that $(\partial/\partial\eta_1)^{\mu-1}[\quad]$ is bounded we obtain $|I| \leq C/t$. Combining this with (4.66) we conclude (4.65).

Theorem 4. 4 *For* $g_j \in C_0^{n+l-m_j+1}(R^{n-1}), j = 1, \ldots, m/2, l \geq \max m_j$

$$u(x, t) = \sum_{j=1}^{m/2} \int_{R^{n-1}} K_j(x - y, t)g_j(y)dy = \sum_{j=1}^{m/2} K_j * g_j \qquad (4.67)$$

is a solution of (4.51), (4.52) *belonging to* $C^l(\bar{R}_+^n)$.

Proof. It is evident that the function defined by (4.67) satisfies $L(D)u = 0$ in $t > 0$. Set $u_j = K_j * g_j$ for $j = 1, \ldots, m/2$. We are going to show that

$$B_k(D)u_j(x, 0) = \delta_{kj}g_j(x). \qquad (4.68)$$

Let q be the largest integer with the same parity as $n-1$ and satisfying $q \leq l - m_j + 2$. Since $n + q - 1 \leq n + l - m_j + 1$ we get with the aid of (4.61) and integration by parts

$$u_j(x,t) = \int_{R^{n-1}} \Delta_x^{(n+q-1)/2} K_{j,q}(x-y,t)g_j(y)dy$$

$$= \int_{R^{n-1}} K_{j,q}(x-y,t)\Delta^{(n+q-1)/2}g_j(y)dy. \qquad (4.69)$$

Hence for $0 \leq k \leq l$

$$D^k u_j(x,t) = \int_{R^{n-1}} D^k K_{j,q}(x-y,t)\Delta^{(n+q-1)/2}g_j(y)dy.$$

Since $q = l - m_j + 1$ or $q = l - m_j + 2$, we see that $m_j + q - k > 0$. Therefore $D^k K_{j,q}(x,t)$ is continuous in $\bar{R}_+^n$ by Lemma 4.5, and hence $u_j \in C^l(\bar{R}_+^n)$. By the definition of $K_{j,q}$

$$B_k(D)K_{j,q}(x,t) = \frac{-1}{(2\pi i)^n (m_j + q - m_k)!} \int_{|\xi|=1} d\sigma_\xi$$

$$\times \int_\gamma \frac{N_j(\xi,\tau)}{M^+(\xi,\tau)} B_k(\xi,\tau)(x\xi + t\tau)^{m_j+q-m_k}\left(\log \frac{x\xi + t\tau}{i} + \text{const}\right)d\tau.$$

Hence with the aid of (4.53)

$$B_k(D)K_{j,q}(x,0) = \frac{-1}{(2\pi i)^n (m_j + q - m_k)!} \int_{|\xi|=1} d\sigma_\xi$$

$$\times \int_\gamma \frac{N_j(\xi,\tau)}{M^+(\xi,\tau)} B_k(\xi,\tau)d\tau(x\xi)^{m_j+q-m_k}\left(\log \frac{x\xi}{i} + \text{const}\right)d\tau$$

$$= -\frac{\delta_{jk}}{(2\pi i)^{n-1}q!} \int_{|\xi|=1} (x\xi)^q \left(\log \frac{x\xi}{i} + \text{const}\right)d\sigma_\xi. \qquad (4.70)$$

It follows from (4.69)

$$B_k(D)u_j(x,0) = \int_{R^{n-1}} B_k(D)K_{j,q}(x-y,0)\Delta^{(n+q-1)/2}g_j(y)dy$$

$$= \int_{R^{n-1}} B_k(D)K_{j,q}(y,0)\Delta^{(n+q-1)/2}g_j(x-y)dy$$

$$= \Delta^{(n+q-1)/2} \int_{R^{n-1}} g_j(x-y)B_k(D)K_{j,q}(y,0)dy$$

$$= \Delta^{(n+q-1)/2} \int_{R^{n-1}} g_j(y)B_k(D)K_{j,q}(x-y,0)dy.$$

Therefore in view of (4.70) we see that $B_k(D)u_j(x, 0) = 0$ if $k \neq j$. For $k = j$ by virtue of (4.12)

$$B_j(D)u_j(x, 0) = -\frac{1}{(2\pi i)^{n-1}q!}\Delta^{(n+q-1)/2}\int_{R^{n-1}}g_j(y)dy$$

$$\times \int_{|\xi|=1}((x-y)\xi)^q\left(\log\frac{(x-y)\xi}{i} + \text{const}\right)d\sigma_\xi$$

$$= -\frac{1}{(2\pi i)^{n-1}q!}\Delta^{(n+q-1)/2}\int_{R^{n-1}}g_j(y)$$

$$\times \int_{|\xi|=1}((x-y)\xi)^q\log\frac{(x-y)\xi}{i}d\sigma_\xi dy = g_j(x).$$

The proof of the theorem is complete.

It follows from (4.61) and Lemma 4.5 that

$$\left|D^l K_j(x, t)\right| \leq C_l\left(|x|^2 + t^2\right)^{(m_j-n+1-l)/2}\left[1 + \left|\log\left(|x|^2 + t^2\right)^{1/2}\right|\right].$$
$$(4.71)$$

If $l > m_j - n + 1$, the logarithmic term on the right hand side may be omitted. Next we show that for $x \neq 0$

$$B_k(D)K_j(x, 0) = 0. \qquad (4.72)$$

Let $\phi \in C_0^\infty(R^{n-1})$ and $0 \notin \text{supp}\phi$. By (4.68) we see that

$$\lim_{t\to 0}\int_{R^{n-1}}B_k(D)K_j(x-y, t)\phi(y)dy$$

$$= \lim_{t\to 0}B_k(D)\int_{R^{n-1}}K_j(x-y, t)\phi(y)dy = \delta_{kj}\phi(x).$$

Hence noting $K_j \in C^\infty\left(\bar{R}_+^n \setminus \{0\}\right)$ we get

$$\int_{R^{n-1}}B_k(D)K_j(x-y, 0)\phi(y)dy = 0.$$

Letting $x = 0$ we see that (4.72) holds for $x \neq 0$.

It follows from (4.71) and the mean value theorem that

$$|B_k(D)K_j(x, t)| \leq \text{const} \cdot t\left(|x|^2 + t^2\right)^{(m_j-m_k-n)/2}\left[1 + \left|\log\left(|x|^2 + t^2\right)^{1/2}\right|\right]$$
$$(4.73)$$

In particular

$$|B_j(D)K_j(x, t)| \leq \text{const} \cdot t\left(|x|^2 + t^2\right)^{-n/2}.$$

Lemma 4. 6 *Set $m_0 = \max m_j$. Suppose that for $j = 1, \ldots, m/2$ $g_j \in C^{m+n+1+m_0-m_j}(R^{n-1})$ and for $k = 0, 1, \ldots, m+n+1+m_0-m_j$*

$$D^k g_j(x) = O\left(|x|^{m-n-m_j-k} \log |x|\right)$$

as $|x| \to \infty$. If we set

$$\tilde{u}_j(x, t) = \sum_{k=1}^{m/2} \int_{R^{n-1}} D_x^{m-m_j} B_j(D) K_k(x - y, t) g_k(y) dy,$$

then $\tilde{u}_j \in C^0(\bar{R}_+^n)$, $\tilde{u}_j(x, 0) = D^{m-m_j} g_j(x)$.

Proof. Integrating by parts we get

$$\tilde{u}_j(x, t) = \sum_{k=1}^{m/2} \int_{R^{n-1}} B_j(D) K_k(x - y, t) D^{m-m_j} g_k(y) dy.$$

Let ζ be a function in $C_0^\infty(R^{n-1})$ such that $\zeta(x) = 1$ for $|x| < \rho$ where ρ is a positive constant. If we set

$$w_1(x, t) = \sum_{k=1}^{m/2} \int_{R^{n-1}} K_k(x - y, t) \zeta(y) D^{m-m_j} g_k(y) dy,$$

$$w_2(x, t) = \sum_{k=1}^{m/2} \int_{R^{n-1}} B_j(D) K_k(x - y, t)(1 - \zeta(y)) D^{m-m_j} g_k(y) dy,$$

then

$$\tilde{u}_j(x, t) = B_j(D) w_1(x, t) + w_2(x, t). \tag{4.74}$$

Since

$$\zeta D^{m-m_j} g_k \in C_0^{n+1+m_0-m_k+m_j}(R^{n-1}) \subset C_0^{n+1+m_0-m_k}(R^{n-1}),$$

we see $B_j(D) w_1 \in C^0(\bar{R}_+^n)$ in view of Theorem 4.4, and

$$B_j(D) w_1(x, 0) = \zeta(x) D^{m-m_j} g_j(x).$$

In particular for $|x| < \rho$

$$B_j(D) w_1(x, 0) = D^{m-m_j} g_j(x). \tag{4.75}$$

Suppose $|x| < \rho/2$. If $1 - \zeta(y) \neq 0$, then $|y| \geq \rho$, and hence $|y|/2 \leq |x - y| \leq 3|y|/2$. Therefore, in view of (4.73)

$$|B_j(D) K_k(x - y, t)| \leq \text{const } t|y|^{m_k - m_j - n}(1 + |\log |y||).$$

Consequently

$$|w_2(x,t)| \leq \text{const} \sum_{k=1}^{m/2} t \int_{|y|>\rho} |y|^{m_k-m_j-n}(1+|\log|y||)$$

$$\times |y|^{m_j-m_k-n}(1+|\log|y||)dy = \text{const } t \int_{|y|>\rho} |y|^{-2n}(1+|\log|y||)^2 dy.$$

This shows that $w_2(x,t) \to 0$ as $t \to 0$ for $|x| < \rho/2$. Combining this with (4.74),(4.75) we obtain $\tilde{u}_j(x,t) \to D^{m-m_j}g_j(x)$ in $|x| < \rho/2$. Since ρ is arbitrary we complete the proof.

4.6 Preliminaries on Integral Kernels

Let $K(x,t)$ be the following integral kernel:

$$K(x,t) = \frac{\Omega(x/|x|)}{(|x|^2+t^2)^{(n-1)/2}}, \quad x \in R^{n-1}, \quad t > 0, \qquad (4.76)$$

where $\Omega(\cdot)$ is a bounded measurable function defined on the unit sphere $\Sigma = \{x \in R^{n-1}; |x| = 1\}$ and satisfying

$$\int_\Sigma \Omega(x)d\sigma = 0. \qquad (4.77)$$

We investigate the partial Fourier transform of $K(x,t)$ with respect to x. For $\mu > 0$ set

$$K_\mu(x,t) = \begin{cases} K(x,t) & |x| < \mu \\ 0 & |x| \geq \mu \end{cases}. \qquad (4.78)$$

Using the polar coordinates as in Chapter 2 we write $x = r\sigma, |\xi| = \rho, x\xi = r\rho\cos\phi$. For $t > 0$

$$\int_{R^{n-1}} K_\mu(x,t)e^{-ix\xi}dx$$

$$= \int_\Sigma \Omega(\sigma)d\sigma \int_0^\mu e^{-ir\rho\cos\phi} \frac{r^{n-2}dr}{(r^2+t^2)^{(n-1)/2}}$$

$$= \int_\Sigma \Omega(\sigma)d\sigma \int_0^{\mu\rho} e^{-is\cos\phi} \frac{s^{n-2}ds}{(s^2+t^2\rho^2)^{(n-1)/2}}$$

$$= \int_\Sigma \Omega(\sigma)d\sigma \int_0^{\mu\rho} (e^{-is\cos\phi} - e^{-s}) \frac{s^{n-2}ds}{(s^2+t^2\rho^2)^{(n-1)/2}}. \qquad (4.79)$$

Dividing the inner integral into the real and imaginary parts

$$\int_0^{\mu\rho} \left(e^{-is\cos\phi} - e^{-s}\right) \frac{s^{n-2}ds}{(s^2 + t^2\rho^2)^{(n-1)/2}} = I_1 + I_2 i, \qquad (4.80)$$

$$I_1 = \int_0^{\mu\rho} \left(\cos(s\cos\phi) - e^{-s}\right) \frac{s^{n-2}ds}{(s^2 + t^2\rho^2)^{(n-1)/2}},$$

$$I_2 = -\int_0^{\mu\rho} \sin(s\cos\phi) \frac{s^{n-2}ds}{(s^2 + t^2\rho^2)^{(n-1)/2}}.$$

If $\mu\rho \leq 1$

$$|I_1| \leq \int_0^1 \left|\cos(s\cos\phi) - e^{-s}\right| \frac{ds}{s} \leq B, \qquad (4.81)$$

where B is the constant defined by (2.21). If $\mu\rho > 1$,

$$|I_1| \leq B + \left| \int_1^{\mu\rho} \left(\cos(s\cos\phi) - e^{-s}\right) \frac{s^{n-2}ds}{(s^2 + t^2\rho^2)^{(n-1)/2}} \right|$$

$$\leq B + \int_1^\infty e^{-s} \frac{ds}{s} + \left| \int_1^{\mu\rho} \cos(s\cos\phi) \frac{s^{n-2}ds}{(s^2 + t^2\rho^2)^{(n-1)/2}} \right|. \qquad (4.82)$$

Suppose $\cos\phi > 0$. Making the change of the variable $s \to s/\cos\phi$ and setting $b = t\rho\cos\phi$

$$\int_1^{\mu\rho} \cos(s\cos\phi) \frac{s^{n-2}ds}{(s^2 + t^2\rho^2)^{(n-1)/2}} = \int_{\cos\phi}^{\mu\rho\cos\phi} \cos s \frac{s^{n-2}ds}{(s^2 + b^2)^{(n-1)/2}}. \qquad (4.83)$$

If $\mu\rho\cos\phi \leq 1$, the absolute value of the right hand side of (4.83) does not exceed

$$\int_{\cos\phi}^1 \frac{ds}{s} = \log \frac{1}{\cos\phi}.$$

If $\mu\rho\cos\phi > 1$, we divide the integral of the right hand side of (4.83) into the part over $(\cos\phi, 1)$ and that over $(1, \mu\rho\cos\phi)$. As for the first part

$$\left| \int_{\cos\phi}^1 \cos s \frac{s^{n-2}ds}{(s^2 + b^2)^{(n-1)/2}} \right| \leq \int_{\cos\phi}^1 \frac{ds}{s} = \log \frac{1}{\cos\phi}. \qquad (4.84)$$

Setting $a = \mu\rho\cos\phi$ we write the second part as

$$\int_1^{\mu\rho\cos\phi} \cos s \frac{s^{n-2}ds}{(s^2 + b^2)^{(n-1)/2}} = \int_1^a \cos s \frac{ds}{s + b}$$

$$+ \int_1^a \cos s \left(\frac{s^{n-2}}{(s^2 + b^2)^{(n-1)/2}} - \frac{1}{s + b} \right) ds. \qquad (4.85)$$

By virtue of (2.19),(2.20) the first term of the right hand side of (4.85) is estimated as

$$\left| \int_{1+b}^{a+b} \cos(s-b) \frac{ds}{s} \right| = \left| \cos b \int_{1+b}^{a+b} \cos s \frac{ds}{s} + \sin b \int_{1+b}^{a+b} \sin s \frac{ds}{s} \right|$$

$$\leq 2A + \log\left(1 + \frac{\pi}{2(1+b)}\right) + 2A \leq 4A + \log\left(1 + \frac{\pi}{2}\right).$$

The second term does not exceed

$$\int_0^\infty \left| \frac{s^{n-2}}{(s^2+b^2)^{(n-1)/2}} - \frac{1}{s+b} \right| ds = \int_0^\infty \left| \frac{\tau^{n-2}}{(\tau^2+1)^{(n-1)/2}} - \frac{1}{\tau+1} \right| d\tau,$$

which is easily seen to be convergent. Hence we get

$$\left| \int_1^{\mu\rho} \cos(s\cos\phi) \frac{s^{n-2} ds}{(s^2 + t^2\rho^2)^{(n-1)/2}} \right| \leq \text{const} + \log\frac{1}{\cos\phi}.$$

Estimating analogously in case $\cos\phi < 0$ we obtain

$$|I_1| \leq \text{const} + \log\frac{1}{|\cos\phi|}. \tag{4.86}$$

In case $\cos\phi > 0$

$$I_2 = -\int_0^a \sin s \frac{s^{n-2} ds}{(s^2+b^2)^{(n-1)/2}}$$

$$= -\int_0^a \sin s \frac{ds}{s+b} - \int_0^a \sin s \left(\frac{s^{n-2}}{(s^2+b^2)^{(n-1)/2}} - \frac{1}{s+b}\right) ds,$$

$$\left| \int_0^a \sin s \frac{ds}{s+b} \right| = \left| \int_b^{a+b} \sin(s-b) \frac{ds}{s} \right|$$

$$= \left| \cos b \int_b^{a+b} \sin s \frac{ds}{s} - \sin b \int_b^{a+b} \cos s \frac{ds}{s} \right|$$

$$\leq 4A + |\sin b| \log\left(1 + \frac{\pi}{2b}\right) \leq \text{const},$$

$$\left| \int_0^a \sin s \left(\frac{s^{n-2}}{(s^2+b^2)^{(n-1)/2}} - \frac{1}{s+b}\right) ds \right|$$

$$\leq \int_0^\infty \left| \frac{s^{n-2}}{(s^2+b^2)^{(n-1)/2}} - \frac{1}{s+b} \right| ds = \text{const}.$$

Hence $|I_2| \le$ const. Thus we have shown that

$$\left| \int_0^{\mu\rho} \left(e^{-is\cos\phi} - e^{-s} \right) \frac{s^{n-2}ds}{(s^2 + t^2\rho^2)^{(n-1)/2}} \right| \le \text{const} + \log\frac{1}{|\cos\phi|}. \quad (4.87)$$

Combining this with (4.79)

$$\left| \int_{R^{n-1}} K_\mu(x,t)e^{-ix\xi}dx \right| \le \int_\Sigma |\Omega(\sigma)| \left(\text{const} + \log\frac{1}{|\cos\phi|} \right) d\sigma \le \text{const}. \quad (4.88)$$

Since

$$\lim_{\mu\to\infty} \int_0^{\mu\rho} \left(e^{-is\cos\phi} - e^{-s} \right) \frac{s^{n-2}ds}{(s^2 + t^2\rho^2)^{(n-1)/2}}$$

$$= \lim_{\mu\to\infty} \int_0^{\mu\rho} \left(e^{-is\cos\phi} - e^{-s} \right) \frac{ds}{s}$$

$$+ \lim_{\mu\to\infty} \int_0^{\mu\rho} \left(e^{-is\cos\phi} - e^{-s} \right) \left(\frac{s^{n-2}}{(s^2 + t^2\rho^2)^{(n-1)/2}} - \frac{1}{s} \right) ds$$

$$= \int_0^\infty \left(e^{-is\cos\phi} - e^{-s} \right) \frac{ds}{s}$$

$$+ \int_0^\infty \left(e^{-is\cos\phi} - e^{-s} \right) \left(\frac{s^{n-2}}{(s^2 + t^2\rho^2)^{(n-1)/2}} - \frac{1}{s} \right) ds,$$

exists if $\cos\phi \ne 0$, we conclude with the aid of (4.79),(4.87),(4.88) that

$$\int_{R^{n-1}} K(x,t)e^{-ix\xi}dx = \lim_{\mu\to\infty} \int_{|x|<\mu} K(x,t)e^{-ix\xi}dx$$

exists and is bounded.

It is easily seen that the integral $\int_{R^{n-1}} K(x-y,t)f(y)dy$ is absolutely convergent for $f \in L^2(R^{n-1})$, $t > 0$. Arguing as in section 3 of Chapter 2 we can establish the following theorem.

Theorem 4.5 *For $f \in L^2(R^{n-1})$*

$$(K(\cdot,t)*f)(x) = \int_{R^{n-1}} K(x-y,t)f(y)dy = \lim_{\mu\to\infty} \int_{R^{n-1}} K_\mu(x-y,t)f(y)dy$$

exists in the strong topology of $L^2(R^{n-1})$. The mapping $f \mapsto K(\cdot,t)*f$ is a bounded linear transformation from $L^2(R^{n-1})$ to itself and its norm is bounded in $t > 0$:

$$|K(\cdot,t)*f|_{0,2,R^{n-1}} \le C|f|_{0,2,R^{n-1}}.$$

Next we consider the following more general integral kernel defined in the half space R_+^n:

$$K(x,t) = \Omega\left(\frac{x}{|P|}, \frac{t}{|P|}\right) \Big/ \left(|x|^2 + t^2\right)^{(n-1)/2}, \qquad (4.89)$$

where $P = (x,t)$, $|P| = \left(|x|^2 + t^2\right)^{1/2}$. $\Omega(x,t)$ is a function defined and continuous on the hemisphere $|x|^2 + t^2 = 1, t \geq 0$, and satisfies a uniform Hölder condition at points on the plane $t = 0$: for $|x| = 1, |y|^2 + t^2 = 1, t \geq 0$

$$|\Omega(x,0) - \Omega(y,t)| \leq C_1 \left(|x-y|^2 + t^2\right)^{\rho/2}, \quad C_1 > 0, \quad 0 < \rho \leq 1. \quad (4.90)$$

Furthermore it is assumed that

$$\int_\Sigma \Omega(x,0)d\sigma = 0, \qquad (4.91)$$

where Σ is the unit sphere of R^{n-1}. Set

$$K_1(x,t) = \Omega\left(\frac{x}{|x|}, 0\right) \Big/ \left(|x|^2 + t^2\right)^{(n-1)/2},$$

$$K_2(x,t) = \left[\Omega\left(\frac{x}{|P|}, \frac{t}{|P|}\right) - \Omega\left(\frac{x}{|x|}, 0\right)\right] \Big/ \left(|x|^2 + t^2\right)^{(n-1)/2}.$$

Then

$$K(x,t) = K_1(x,t) + K_2(x,t). \qquad (4.92)$$

Since

$$\left|\frac{x}{|P|} - \frac{x}{|x|}\right|^2 + \left(\frac{t}{|P|}\right)^2 = \frac{(|P| - |x|)^2}{|P|^2} + \frac{t^2}{|P|^2} = \frac{2|P|^2 - 2|x||P|}{|P|^2}$$

$$= 2\frac{|P| - |x|}{|P|} = 2\frac{|P|^2 - |x|^2}{|P|(|P| + |x|)} = \frac{2t^2}{|P|(|P| + |x|)} \leq \frac{2t^2}{|P|^2} = \frac{2t^2}{|x|^2 + t^2},$$

we have

$$|K_2(x,t)| \leq 2^{\rho/2}C_1 \frac{t^\rho}{\left(|x|^2 + t^2\right)^{(n-1+\rho)/2}}.$$

Hence

$$\int_{R^{n-1}} |K_2(x,t)|\,dx \leq 2^{\rho/2}C_1 t^\rho \int_{R^{n-1}} \frac{dx}{\left(|x|^2 + t^2\right)^{(n-1+\rho)/2}} \leq C_2. \quad (4.93)$$

For $f \in L^2(R^{n-1})$ we set

$$u(x,t) = \int_{R^{n-1}} K(x-y,t)f(y)dy.$$

Then

$$u(x, t) = \int_{R^{n-1}} K_1(x - y, t) f(y) dy + \int_{R^{n-1}} K_2(x - y, t) f(y) dy$$
$$= (K_1(\cdot, t) * f)(x) + (K_2(\cdot, t) * f)(x).$$

By virtue of Theorem 4.5 we have

$$|K_1(\cdot, t) * f|_{0,2,R^{n-1}} \leq C |f|_{0,2,R^{n-1}}.$$

With the aid of (4.93)

$$|K_2(\cdot, t) * f|_{0,2,R^{n-1}} \leq |K_2(\cdot, t)|_{0,1,R^{n-1}} |f|_{0,2,R^{n-1}} \leq C_2 |f|_{0,2,R^{n-1}}.$$

Hence we have established the following theorem.

Theorem 4. 6 *Let $K(x, t)$ be the integral kernel of (4.89). Then for $f \in L^2(R^{n-1})$*

$$|K(\cdot, t) * f|_{0,2,R^{n-1}} \leq C |f|_{0,2,R^{n-1}}.$$

Here C is a constant independent of f, t.

For a positive integer k and $1 < p < \infty$ let $W^{k-1/p,p}(R^{n-1})$ be the space of the traces on $\{(x, 0) \in R^n; x \in R^{n-1}\}$ of functions belonging to $W^{k,p}(R^n_+)$ (Chapter 3, section 10). For $f \in W^{k-1/p,p}(R^{n-1})$ set

$$|f|_{k-1/p,p,R^{n-1}} = \inf\{|v|_{k,p,R^n_+}; v \in W^{k,p}(R^n_+), v(\cdot, 0) = f\}. \tag{4.94}$$

Here by definition

$$|v|_{k,p,R^n_+} = \left(\iint_{R^n_+} \sum_{|\alpha|+i=k} |D_x^\alpha D_t^i v(x, t)|^p \, dx dt \right)^{1/p}.$$

Lemma 4. 7 *Let $G(x, t)$ be a measurable function defined in the half space R^n_+. Set $S^+ = \{(x, t) \in R^n; |x|^2 + t^2 = 1, t > 0\}$. Suppose that for some nonnegative measurable function $\Omega(x, t)$ defined on S^+ and satisfying*

$$\int_{S^+} \Omega(x, t) d\sigma_{x,t} < \infty$$

we have

$$|G(x, t)| \leq \frac{\Omega(x/|P|, t/|P|)}{(|x|^2 + t^2)^{n/2}}, \quad P = (x, t).$$

For $v \in L^p(R^n_+), 1 < p < \infty$, set

$$u(x, t) = \iint_{R^n_+} G(x - y, t + s) v(y, s) dy ds.$$

Then $u \in L^p(R_+^n)$ and the following inequality holds:

$$\iint_{R_+^n} |u(x,t)|^p dxdt \leq C \iint_{R_+^n} |v(x,t)|^p dxdt.$$

Proof. Set $S = \{(x,t) \in R^n; |x|^2 + t^2 = 1\}$ and

$$M(x,t) = \begin{cases} \Omega(x/|P|, t/|P|)\left(|x|^2 + t^2\right)^{-n/2} & t > 0 \\ -\Omega(-x/|P|, -t/|P|)\left(|x|^2 + t^2\right)^{-n/2} & t < 0 \end{cases},$$

then $M(x,t)$ is homogeneous of order $-n$ and

$$\int_S |M(x,t)| d\sigma_{x,t} < \infty.$$

If we extend $v(x,t)$ as 0 outside R_+^n to the whole of R^n, then for $t > 0$

$$|u(x,t)| \leq \iint_{R_+^n} M(x-y, t+s)|v(y,s)|dyds$$

$$= \iint_{R^n} M(x-y, t+s)|v(y,s)|dyds \equiv u^*(x,t).$$

By virtue of Lemma 2.1 the above integral exists almost everywhere in R_+^n. Applying Theorem 2.6 we conclude

$$\iint_{R_+^n} |u(x,t)|^p dxdt \leq \iint_{R^n} |u^*(x,t)|^p dxdt \leq \text{const} \iint_{R_+^n} |v(x,t)|^p dxdt.$$

Theorem 4.7 *Let $K(x,t)$ be the integral kernel of (4.89). Suppose moreover that $K(x,t)$ has continuous derivatives bounded on the hemisphere $|x|^2 + t^2 = 1, t \geq 0$. For $f \in W^{1-1/p,p}(R^{n-1}), 1 < p < \infty$, set*

$$u(x,t) = \int_{R^{n-1}} K(x-y, t)f(y)dy.$$

Then for some constant C independent of f the following inequality holds:

$$\left(\iint_{R_+^n} |D_x u(x,t)|^p \, dxdt\right)^{1/p} \leq C|f|_{1-1/p,p,R^{n-1}}. \tag{4.95}$$

Proof. By the definition of the seminorm (4.94) there exists a function $v \in W^{1,p}(R_+^n)$ such that

$$v(x,0) = f(x), \quad |v|_{1,p,R_+^n} \leq 2|f|_{1-1/p,p,R^{n-1}}. \tag{4.96}$$

Since

$$v(x,t) = f(x) + \int_0^t D_s v(x,s)ds,$$

$$\int_{R^{n-1}} \left| \int_0^t D_s v(x,s)ds \right|^p dx \leq \int_{R^{n-1}} t^{p-1} \int_0^t |D_s v(x,s)|^p ds dx$$

$$\leq t^{p-1} \int_{R^n_+} |D_s v(x,s)|^p dx ds \leq t^{p-1} |v|^p_{1,p,R^n_+},$$

we have

$$\left(\int_{R^{n-1}} |v(x,t)|^p dx \right)^{1/p} \leq \left(\int_{R^{n-1}} |f(x)|^p dx \right)^{1/p} + t^{1-1/p} |v|_{1,p,R^n_+}. \quad (4.97)$$

Noting (4.96)

$$u(x,t) = \int_{R^{n-1}} K(x-y,t)v(y,0)dy$$

$$= -\int_{R^{n-1}} \int_0^T D_s\{K(x-y,t+s)v(y,s)\}ds dy$$

$$+ \int_{R^{n-1}} K(x-y,t+T)v(y,T)dy$$

$$= -\int_{R^{n-1}} \int_0^T D_s K(x-y,t+s) \cdot v(y,s)ds dy$$

$$- \int_{R^{n-1}} \int_0^T K(x-y,t+s)D_s v(y,s)ds dy$$

$$+ \int_{R^{n-1}} K(x-y,t+T)v(y,T)dy.$$

By a suitable change of variable we get

$$D_i u(x,t) = -\int_{R^{n-1}} \int_0^T D_s K(x-y,t+s) \cdot D_i v(y,s)ds dy$$

$$- \int_{R^{n-1}} \int_0^T D_i K(x-y,t+s) \cdot D_s v(y,s)ds dy$$

$$+ \int_{R^{n-1}} D_i K(x-y,t+T) \cdot v(y,T)dy. \quad (4.98)$$

Since

$$|D_i K(x,t)| \leq \text{const} \left(|x|^2 + t^2 \right)^{-n/2}, \quad (4.99)$$

it follows that

$$\left(\int_{R^{n-1}} |D_i K(y, t+T)|^{p'} \, dy \right)^{1/p'} \leq \text{const}(t+T)^{-1-(n-1)/p}.$$

Combining this with (4.97) we get

$$\left| \int_{R^{n-1}} D_i K(x-y, t+T) \cdot v(y, T) dy \right| = O\left(T^{-n/p} \right)$$

as $T \to \infty$. Hence letting $T \to \infty$ in (4.98) we find

$$D_i u(x, t) = - \iint_{R_+^n} D_s K(x-y, t+s) \cdot D_i v(t, s) ds dy$$

$$- \iint_{R_+^n} D_i K(x-y, t+s) \cdot D_s v(y, s) ds dy. \qquad (4.100)$$

Analogously to (4.99) we have

$$|D_t K(x, t)| \leq \text{const} \left(|x|^2 + t^2 \right)^{-n/2}.$$

Hence applying Lemma 4.7 we obtain

$$\left(\iint_{R_+^n} |D_i u(x, t)|^p \, dx dt \right)^{1/p} \leq \text{const}|v|_{1, p, R_+^n} \leq \text{const}|f|_{1-1/p, p, R^{n-1}}.$$

Remark 4. 1 In the proof of Theorem 4.7 the condition (4.91) is not necessary. It is shown in [11] that if $D_t^2 K(x, t)$ exists and is bounded on the hemisphere Σ, then the inequality (4.95) also holds for $D_t u$.

We verify that the theorems of this section can be applied to the kernel

$$K(x, t) = D^{m_j + q + n - 1} K_{j, q}(x, t). \qquad (4.101)$$

By Lemma 4.5 this kernel $K(x, t)$ is homogeneous of order $1 - n$. It is easily seen that $K(x, t)$ is expressed in the form (4.89) with a function $\Omega(x, t)$ which is infinitely differentiable on the hemisphere S^+. Since $L(D)$ is elliptic, the coefficient of D_t^m in $L(D)$ is different from 0. Hence the relation $L(D)K(x, t) = 0$ can be rewritten as

$$D_t^m K(x, t) + \sum_{k=1}^{n-1} c_k D_k D^{\alpha(k)} K(x, t) = 0. \qquad (4.102)$$

Here c_k is a constant and $|\alpha(k)| = m - 1$. Since

$$\left| D^{\alpha(k)} K(x, t) \right| \leq \text{const} \left(|x|^2 + t^2 \right)^{(2-m-n)/2}$$

in view of Lemma 4.5, we see

$$\int_{|x| \leq R} D_k D^{\alpha(k)} K(x, t) dx = \int_{|x| = R} D^{\alpha(k)} K(x, t) \frac{x_k}{R} dS \to 0$$

as $R \to \infty$. Hence

$$\int_{R^{n-1}} D_t^m K(x, t) dx = 0. \tag{4.103}$$

Set for $j = 1, 2, \ldots$

$$g_j(t) = \int_{R^{n-1}} D_t^j K(x, t) dx.$$

As is easily seen $g_j(t) = g_j(1) t^{-j}$. Therefore

$$g_j(t) = D_t g_{j-1}(t) = D_t \left(g_{j-1}(1) t^{1-j} \right)$$
$$= (1 - j) g_{j-1}(1) t^{-j} = (1 - j) t^{-1} g_{j-1}(t). \tag{4.104}$$

Since $g_m(t) \equiv 0$ by (4.103), we get $g_1(t) \equiv 0$ from (4.104), or

$$\int_{R^{n-1}} D_t K(x, t) dx = 0. \tag{4.105}$$

If we decompose $K(x, t)$ as (4.92), then $K_2(x, t)$ is absolutely integrable in R^{n-1} with respect to x and the value of the integral does not depend on t. Hence from (4.105) we get

$$\int_{R^{n-1}} D_t K_1(x, t) dx = 0.$$

On the other hand

$$\int_{R^{n-1}} D_t K_1(x, t) dx = \int_{R^{n-1}} \Omega \left(\frac{x}{|x|}, 0 \right) D_t \left(|x|^2 + t^2 \right)^{(1-n)/2} dx$$
$$= (1 - n) t \int_{R^{n-1}} \Omega \left(\frac{x}{|x|}, 0 \right) \left(|x|^2 + t^2 \right)^{-(n+1)/2} dx$$
$$= (1 - n) t \int_\Sigma \Omega(\sigma, 0) d\sigma \int_0^\infty \frac{r^{n-2} dr}{(r^2 + t^2)^{(n+1)/2}}.$$

Hence we conclude

$$\int_\Sigma \Omega(\sigma, 0) d\sigma = 0.$$

4.7 Nonhomogeneous Boundary Conditions

Let u be a function in $C_0^\infty(\bar{R}_+^n)$ satisfying

$$L(D)u(x,t) = f(x,t) \qquad t > 0, \tag{4.106}$$
$$B_j(D)u(x,0) = g_j(x) \qquad j = 1,\ldots,m/2. \tag{4.107}$$

Let $K(x,t)$ be the fundamental solution of $L(D)$ stated in section 1. As in section 7 of Chapter 3 we extend the function f to R^n by

$$f(x,t) = \begin{cases} f(x,t) & t \geq 0 \\ \displaystyle\sum_{k=1}^{N+1} \lambda_k f(x,-kt) & t < 0 \end{cases}. \tag{4.108}$$

Then $f \in C_0^N(R^n)$. Here

$$\sum_{k=1}^{N+1} (-k)^j \lambda_k = 1, \quad j = 0, 1, \ldots, N,$$

and N is taken sufficiently large. If we put

$$v(x,t) = \iint_{R^n} K(x-y, t-s) f(y,s) \, dy \, ds, \tag{4.109}$$

then $v \in C^{N+m-1}(R^n)$ and satisfies $L(D)v = f$. Furthermore, v is infinitely differentiable outside $\operatorname{supp} f$ and by Theorem 4.1

$$D^k v(x,t) = O\left(\left(|x|^2 + t^2\right)^{(m-n-k)/2} \log\left(|x|^2 + t^2\right)^{1/2} \right) \tag{4.110}$$

as $|x|^2 + t^2 \to \infty$. If we set

$$B_j(D)v(x,0) = h_j(x), \tag{4.111}$$

then $h_j \in C^{N+m-1-m_j}(R^{n-1})$. Moreover, h_j is infinitely differentiable outside some bounded set and

$$D^k h_j(x) = O\left(|x|^{m-n-m_j-k} \log|x|\right) \tag{4.112}$$

as $|x| \to \infty$. Since

$$L(D)(u-v) = 0, \qquad t > 0,$$
$$B_j(D)(u-v) = g_j - h_j, \quad j = 1,\ldots,m/2, \quad t = 0,$$

one might expect that the following relation holds:

$$u(x,t) = v(x,t) + \sum_{j=1}^{m/2} \int_{R^{n-1}} K_j(x - y, t) \left(g_j(y) - h_j(y) \right) dy.$$

However, the convergence of the integral of the right hand side is not evident. Therefore, we show the following theorem instead.

Theorem 4.8 *Let u be a function in $C_0^\infty(\bar{R}_+^n)$ satisfying (4.106), (4.107). Then the relation*

$$D^m u(x,t) = D^m v(x,t) + \sum_{j=1}^{m/2} \int_{R^{n-1}} D^m K_j(x - y, t) \left(g_j(y) - h_j(y) \right) dy$$

$$(4.113)$$

holds in the half space $t > 0$. Here v, h_j are functions defined by (4.109), (4.111) respectively.

If we put

$$u - v = w, \quad g_j - h_j = B_j w|_{t=0} = \omega_j, \qquad (4.114)$$

then what is to be shown is

$$D^m w(x,t) = \sum_{j=1}^{m/2} \int_{R^{n-1}} D^m K_j(x - y, t) \omega_j(y) dy. \qquad (4.115)$$

Since in view of Lemma 4.5

$$D^m K_j(x,t) = O\left(\left(|x|^2 + t^2 \right)^{(m_j - m - n + 1)/2} \log \left(|x|^2 + t^2 \right)^{1/2} \right), \quad (4.116)$$

as $|x|^2 + t^2 \to \infty$, and by (4.112)

$$D^k \omega_j(x) = O\left(|x|^{m - n - m_j - k} \log |x| \right) \qquad (4.117)$$

as $|x| \to \infty$ for $k = 1, 2, \ldots, N + m - 1 - m_j$, the integral on the right of (4.115) converges.

In order to prove Theorem 4.8 we prepare the following two lemmas.

Lemma 4.8 *Let u be a function belonging to $C^m(\bar{R}_+^n)$ and satisfying*

$$L(D)u(x,t) = 0, \quad t > 0, \qquad (4.118)$$
$$B_j(D)u(x,0) = 0, \quad j = 1, \ldots, m/2. \qquad (4.119)$$

Suppose that u and its derivatives of order up to m are absolutely integrable on each hyperplane $t = $ const > 0, and those integrals are uniformly convergent in t in every finite interval $0 < \epsilon \le t \le R$. Moreover suppose that u

and its derivatives of order up to m belong to $L^2(R^{n-1})$ on each hyperplane $t = \text{const} \geq 0$ and their norms

$$\left(\int_{R^{n-1}} |D^k u(x,t)|^2 \, dx \right)^{1/2}, \quad 0 \leq k \leq m, \quad t > 0$$

are bounded. If in additon $u \in L^2(R^n_+)$, then $u \equiv 0$.

Proof. Let $\hat{u}(\xi, t)$ be the partial Fourier transform of u with respect to x. The assumptions imply that $\hat{u}$ and its derivatives in t of order up to m are continuous functions of (ξ, t) in $t > 0$. It follows from (4.118) and the hypotheses that

$$L(i\xi, D_t)\hat{u}(\xi, t) = 0 \tag{4.120}$$

for $t > 0$. For almost every ξ we have $\int_0^\infty |\hat{u}(\xi, t)|^2 \, dt < \infty$. Hence by what was stated in section 3 we get for these values of ξ

$$M^+(i\xi, D_t)\,\hat{u}(\xi, t) = 0. \tag{4.121}$$

Since the left hand side of (4.121) is a continuous function of (ξ, t), this holds for every ξ. It also follows that $\hat{u}(\xi, t)$ is infinitely differentiable in $t \geq 0$ for each fixed ξ. If we verify that for almost every ξ

$$B_j(i\xi, D_t)\hat{u}(\xi, 0) = 0, \quad j = 1, \ldots, m/2, \tag{4.122}$$

then we conclude by the argument of section 3 that $\hat{u}(\xi, t) \equiv 0$, and hence $u \equiv 0$. By hypothesis $\{B_j(D)u(\cdot, t)\}$ is a bounded subset of $L^2(R^{n-1})$. Therefore noting that $u \in C^m(\bar{R}^n_+)$ and (4.119) we see that $\{B_j(D)u(\cdot, t)\}$ converges weakly to 0 in $L^2(R^{n-1})$ as $t \to 0$, and hence so does $\{B_j(i\cdot, D_t)\hat{u}(\cdot, t)\}$. Consequently there exists a subsequence $\{t_\nu\}$ tending to 0 such that some sequence of convex combinations of $\{B_j(i\cdot, D_t)\hat{u}(\cdot, t_\nu)\}$ converges to 0 almost everywhere. On the other hand for every fixed ξ we have $B_j(i\xi, D_t)\hat{u}(\xi, t) \to B_j(i\xi, D_t)\hat{u}(\xi, 0)$ as $t \to 0$ since $\hat{u}(\xi, t)$ is a smooth function of $t \geq 0$ for each fixed ξ as was remarked above. Thus (4.122) follows and the proof is complete.

Lemma 4.9 Let $\{w_{i_1,\ldots,i_m}; i_1, \ldots, i_m = 1, \ldots, n\}$ be a set of functions belonging to $C^1(\bar{R}^n_+)$ which are invariant under permutations of $i_1, \ldots, i_m$. If for every $i_1, \ldots, i_{m+1} = 1, \ldots, n$

$$D_{i_1} w_{i_2,i_3,\ldots,i_{m+1}} = D_{i_2} w_{i_1,i_3,\ldots,i_{m+1}},$$

then there exists a function $g \in C^{m+1}(\bar{R}^n_+)$ such that for every $i_1, \ldots, i_m = 1, \ldots, n$

$$w_{i_1,\ldots,i_m} = D_{i_1} \cdots D_{i_m} g.$$

We can choose such a function g so that all of its derivatives of order up to $m-1$ vanish at the origin, and under this condition g is uniquely determined.

Proof. (i) Case $m = 1$. Suppose that $w_1, \ldots, w_n$ are functions in $C^1(\bar{R}^n_+)$ satisfying $D_j w_i = D_i w_j$ for $i, j = 1, \ldots, n$. If we put

$$g(x) = g(x_1, \ldots, x_n) = \sum_{i=1}^{n} \int_0^{x_i} w_i(x_1, \ldots, x_{i-1}, y_i, 0, \ldots, 0) dy_i,$$

then

$$D_j g(x) = w_j(x_1, \ldots, x_{j-1}, x_j, 0, \ldots, 0)$$
$$+ \sum_{i=j+1}^{n} \int_0^{x_i} D_j w_i(x_1, \ldots, x_{i-1}, y_i, 0, \ldots, 0) dy_i$$
$$= w_j(x_1, \ldots, x_{j-1}, x_j, 0, \ldots, 0)$$
$$+ \sum_{i=j+1}^{n} \int_0^{x_i} D_i w_j(x_1, \ldots, x_{i-1}, y_i, 0, \ldots, 0) dy_i$$
$$= w_j(x_1, \ldots, x_{j-1}, x_j, 0, \ldots, 0)$$
$$+ \sum_{i=j+1}^{n} \{w_j(x_1, \ldots, x_{i-1}, x_i, 0, \ldots, 0) - w_j(x_1, \ldots, x_{i-1}, 0, 0, \ldots, 0)\}$$
$$= w_j(x_1, \ldots, x_n).$$

This function g satisfies $g(0) = 0$. It is easy to show that g is uniquely determined by this condition.

(ii) Suppose the assetion of the lemma is true for $m - 1$. Then for each $i = 1, \ldots, n$ there exists a function $g_i \in C^m(\bar{R}^n_+)$ such that

$$w_{i_1, \ldots, i_{m-1}, i} = D_{i_1} \cdots D_{i_{m-1}} g_i$$

and all derivatives of g_i of order up to $m - 2$ vanish at the origin. For $i, j = 1, \ldots, n$

$$D_{i_1} \cdots D_{i_{m-2}} D_j g_i = w_{i_1, \ldots, i_{m-2}, j, i} = w_{i_1, \ldots, i_{m-2}, i, j} = D_{i_1} \cdots D_{i_{m-2}} D_i g_j.$$

Hence if $m = 2$ $D_j g_i = D_i g_j$. If $m \geq 3$, $D_j g_i - D_i g_j$ is a polynomial of degree $m - 3$ at the most, and all of its derivatives of order up to $m - 3$ vanish at the origin. Hence $D_j g_i = D_i g_j$ also holds in this case. By virtue of the result of (i) there exists a function g such that $g_i = D_i g$ and $g(0) = 0$. It is easily seen that this function g is the one sought.

Proof of Theorem 4.8. Let q be an integer with the same parity as $n - 1$ such that $m_j + q \geq 2m + 1$ for $j = 1, \ldots, m/2$, and N an integer satisfying $N \geq m_0 + q - m + n$, where $m_0 = \max_{j=1, \ldots, m/2} m_j$. Then

$$N \geq m + n + 1 \geq m + 3, \quad N \geq m_0 + n + 2. \qquad (4.123)$$

For $m \leq l \leq 2m+1$ we get with the aid of (4.61),(4.64),(4.117)

$$
\int_{R^{n-1}} D^l K_j(x-y,t)\omega_j(y)dy
$$
$$
= \int_{R^{n-1}} D^l \Delta_x^{(n+q-1)/2} K_{j,q}(x-y,t)\omega_j(y)dy
$$
$$
= \int_{R^{n-1}} D^l K_{j,q}(x-y,t) \cdot \Delta^{(n+q-1)/2}\omega_j(y)dy. \qquad (4.124)
$$

Here we note $\omega_j \in C^{N+m-1-m_j}(R^{n-1}) \subset C^{n+q-1}(R^{n-1})$. Since $m_j+q-l \geq 0$, we find with the aid of (4.64) that the last side of (4.124) is continuous in $\bar{R}^n_+$. Hence, if we set

$$
w_{i_1,\ldots,i_m}(x,t) = \sum_{j=1}^{m/2} \int_{R^{n-1}} D_{i_1} \cdots D_{i_m} K_j(x-y,t)\omega_j(y)dy,
$$

then $w_{i_1,\ldots,i_m} \in C^{m+1}(\bar{R}^n_+)$ and we can apply Lemma 4.9 to the set $\{w_{i_1,\ldots,i_m}\}$ to find that there exists a function $g \in C^{2m+1}(\bar{R}^n_+)$ satisfying

$$
D^m g(x,t) = \sum_{j=1}^{m/2} \int_{R^{n-1}} D^m K_j(x-y,t)\omega_j(y)dy. \qquad (4.125)
$$

Hence it suffices to show that

$$
D^m(w-g) = 0. \qquad (4.126)
$$

In view of (4.117),(4.123) we can apply Lemma 4.6 to obtain

$$
D_x^{m-m_j} B_j(D)g(x,0) = D_x^{m-m_j}\omega_j(x). \qquad (4.127)
$$

Suppose that the following statements are true:
(i) for $m \leq l \leq 2m+1$ $D^l g(\cdot,t) \in L^2(R^{n-1})$ and its norm $|D^l g(\cdot,t)|_{0,2,R^{n-1}}$ is bounded in $t \geq 0$,
(ii)

$$
\iint_{R^n_+} |D_x D^m g(x,t)|^2 \, dx dt < \infty,
$$

(iii) for $m+1 \leq l \leq 2m+1$ the integral $\int_{R^{n-1}} |D^l g(x,t)| \, dx$ converges uniformly in any finite interval $0 < \epsilon \leq t \leq R < \infty$.

Then we can prove (4.126) as follows. Since in view of (4.110) we have

$$
|D^l w(x,t)| \leq \text{const} \left(1+|x|^2+t^2\right)^{(m-n-l)/2}
$$

for $m \le l \le 2m + 2(\le N + m - 1)$, the above statements (i),(ii),(iii) holds also for w. If we set $h = w - g$

$$L(D)h(x, t) = 0, \quad t > 0, \tag{4.128}$$
$$D_x^{m-m_j} B_j(D)h(x, 0) = 0, \quad j = 1, \ldots, m/2. \tag{4.129}$$

Hence noting $h \in C^{2m+1}(\bar{R}_+^n)$ we have

$$L(D)D_x^{m+1}h(x, t) = 0, \quad t > 0,$$
$$B_j(D)D_x^{m+1}h(x, 0) = D_x^{m_j+1}D_x^{m-m_j}B_j(D)h(x, 0) = 0, \quad j = 1, \ldots, m/2.$$

We can apply Lemma 4.8 to $D_x^{m+1}h$ to obtain $D_x^{m+1}h = 0$. This implies that $D_x^m h$ depends only on t. However, $D_x^m h \in L^2(R^{n-1})$ on each hyperplane $t = $ const in view of (i). Hence

$$D_x^m h = 0. \tag{4.130}$$

From (4.128) it follows that $D_t D_x^{m-1} L(D)h = 0$. Rewriting this like (4.102) we have

$$0 = D_t D_x^{m-1} \left(D_t^m + \sum_{k=1}^{n-1} c_k D_k D^{\alpha(k)} \right) h$$
$$= D_t^{m+1} D_x^{m-1} h + \sum_{k=1}^{n-1} c_k D_t D^{\alpha(k)} D_x^{m-1} D_k h$$

and using (4.130) we get

$$D_t^{m+1} D_x^{m-1} h = 0. \tag{4.131}$$

Next rewriting $D_t^2 D_x^{m-2} L(D)h = 0$ as

$$0 = D_t^2 D_x^{m-2} \left(D_t^m + \sum_{k=1}^{n-1} c_k D_k D^{\alpha(k)} \right) h$$
$$= D_t^{m+2} D_x^{m-2} h + \sum_{k=1}^{n-1} c_k D_t^2 D^{\alpha(k)} D_x^{m-2} D_k h$$

and using (4.131) we get $D_t^{m+2} D_x^{m-2} h = 0$. Combining this with (4.131) we obtain $D^{m+2} D_x^{m-2} h = 0$. Proceeding in this manner we conclude $D^{2m}h = 0$. This implies that h is a polynomial. However, in view of (i) $D^m h$ belongs to $L^2(R^{n-1})$ on each hyperplane $t = $ const. This implies $D^m h = 0$, and (4.126) is established.

Thus the proof will be complete if we show (i),(ii),(iii). For $m \leq l \leq 2m+1$ we have in view of (4.125)

$$D^l g(x,t) = \sum_{j=1}^{m/2} \int_{R^{n-1}} D^l K_j(x-y,t)\omega_j(y)dy.$$

Hence setting

$$I_j(x,t) = \int_{R^{n-1}} D^l K_j(x-y,t)\omega_j(y)dy,$$

we are going to prove

$$\int_{R^{n-1}} |I_j(x,t)|^2 \, dx \leq \text{const}, \quad t \geq 0, \tag{4.132}$$

$$\int\!\int_{R^n_+} |D_x I_j(x,t)|^2 \, dx dt < \infty, \tag{4.133}$$

$$\int_{R^{n-1}} |DI_j(x,t)| \, dx \quad \text{is uniformly convergent}$$
$$\text{in} \quad 0 < \epsilon \leq t \leq R \quad \text{if} \quad m \leq l \leq 2m. \tag{4.134}$$

In case $l - m_j$ is even, we choose q so that $n + q - 1 - l + m_j$ is even (this q may be different from the one which appeared at the beginning of the proof) we set

$$\tilde{K}(x,t) = D^l \Delta_x^{(n+q-1-l+m_j)/2} K_{j,q}(x,t).$$

The theorems of the previous section can be applied to this kernel according to the remark at the end of the section. With the aid of (4.64),(4.117) we can integrate by parts in

$$I_j(x,t) = \int_{R^{n-1}} D^l \Delta_x^{(n+q-1)/2} K_{j,q}(x-y,t)\omega_j(y)dy$$

to derive

$$I_j(x,t) = \int_{R^{n-1}} \tilde{K}(x-y,t)\tilde{\omega}(y)dy$$

where $\tilde{\omega} = \Delta^{(l-m_j)/2}\omega_j(y)$. Note that $\tilde{\omega} \in C^1(R^{n-1})$ since

$$\omega_j \in C^{N+m-1-m_j}(R^{n-1}) \subset C^{l-m_j+1}(R^{n-1})$$

in view of (4.123). Furthermore by (4.117)

$$\tilde{\omega}(y) = O\left(|y|^{m-n-l}\right), \quad D\tilde{\omega} = O\left(|y|^{m-n-l-1}\right) \tag{4.135}$$

as $|y| \to \infty$. Hence $\tilde{\omega} \in W^{1,2}(R^{n-1})$. Letting ζ be a smooth function such that $\zeta(0) = 1$ and $\zeta(t) = 0$ for $t \geq 1$ set $v_0(y, t) = \tilde{\omega}(y)\zeta(t)$. Then $v_0 \in W^{1,2}(R^n_+)$. This implies $\tilde{\omega} \in W^{1/2,2}(R^{n-1})$. Applying Theorems 4.6, 4.7 we conclude that (4.132),(4.133) are true. In view of Lemma 4.5 we have

$$|D\tilde{K}(x, t)| \leq \text{const} \left(|x|^2 + t^2\right)^{-n/2}.$$

It follows from (4.135) that $\tilde{\omega} \in L^1(R^{n-1})$. Hence

$$\int_{|x|\geq\lambda} |DI_j(x, t)|\, dx = \int_{|x|\geq\lambda} \left|\int_{R^{n-1}} D\tilde{K}(x - y, t)\tilde{\omega}(y)dy\right| dx$$

$$\leq C \int_{|x|\geq\lambda} \int_{R^{n-1}} \left(|x - y|^2 + t^2\right)^{-n/2} |\tilde{\omega}(y)| dy dx$$

$$\leq C \int_{|y|\leq\mu} \int_{|x|\geq\lambda} \left(|x - y|^2 + t^2\right)^{-n/2} dx |\tilde{\omega}(y)| dy$$

$$+ C \int_{|y|\geq\mu} \int_{|x|\geq\lambda} \left(|x - y|^2 + t^2\right)^{-n/2} dx |\tilde{\omega}(y)| dy = I_1 + I_2.$$

By an obvious change of variable

$$I_2 \leq C \int_{|y|\geq\mu} \int_{R^{n-1}} \left(|x - y|^2 + t^2\right)^{-n/2} dx |\tilde{\omega}(y)| dy$$

$$= C t^{-1} \int_{R^{n-1}} \left(|x|^2 + 1\right)^{-n/2} dx \int_{|y|\geq\mu} |\tilde{\omega}(y)| dy.$$

Hence I_2 can be made arbitrarily small uniformly in $t \geq \epsilon$ by letting μ sufficiently large. If $\lambda > \mu$,

$$I_1 \leq C \int_{|x|\geq\lambda} (|x| - \mu)^{-n} dx \int_{R^{n-1}} |\tilde{\omega}(y)| dy.$$

The right hand side is independent of t and tends to 0 as $\lambda \to \infty$. Thus (4.134) is established. In case $l - m_j$ is odd, we set

$$\bar{K}_i(x, t) = D^l D_i \Delta_x^{(n+q-l+m_j-2)/2} K_{j,q}(x, t),$$

$$\tilde{\omega}_i(y) = D_i \Delta_x^{(l-m_j-1)/2} \omega_j(y).$$

Then

$$I_j(x, t) = \sum_{i=1}^{n-1} \int_{R^{n-1}} \bar{K}_i(x - y, t)\tilde{\omega}_i(y)dy.$$

Hence we can show (4.132),(4.133),(4.134) also in this case.

Theorem 4.9 *Let $u \in W^{m,p}(R^n), 1 < p < \infty, l \geq m$. Furthermore suppose that u vanishes outside a bounded set. If we set*

$$L(D)u = f,$$
$$B_j(D)u|_{t=0} = g_j, \quad j = 1, \ldots, m/2,$$

then there exists a constant $C_{l,p}$ which does not depend on u such that

$$|u|_{l,p,R_+^n} \leq C_{l,p} \left[|f|_{l-m,p,R_+^n} + \sum_{j=1}^{m/2} |g_j|_{l-m_j-1/p,p,R^{n-1}} \right].$$

Here the seminorm $| \cdot |_{l-m_j-1/p,p,R^{n-1}}$ is defined by (4.94).

Proof. We may assume $u \in C_0^\infty(\bar{R}_+^n)$. In accordance with Theorem 4.8 we write

$$D^l u(x,t) = D^l v(x,t) + \sum_{j=1}^{m/2} I_j(x,t),$$

$$I_j(x,t) = \int_{R^{n-1}} D^l K_j(x-y,t)\omega_j(y)dy,$$

where v and $\omega_j = g_j - h_j$ are functions defined by (4.109) and (4.114) respectively. In view of Theorem 2.10 and its proof we get

$$|v|_{l,p,R_+^n} \leq C_{l,p}|f|_{l-m,p,R_+^n}.$$

Hence

$$|h_j|_{l-m_j-1/p,p,R^{n-1}} \leq |B_j(D)v|_{l-m_j,p,R_+^n} \leq C_{l,p}|v|_{l,p,R_+^n} \leq C_{l,p}|f|_{l-m,p,R_+^n}.$$

In case $l - m_j$ is even we get with the aid of an integration by parts

$$I_j(x,t) = \sum_{k=1}^{n-1} D_k I_{j,k}(x,t),$$

where

$$I_{j,k}(x,t) = \int_{R^{n-1}} D^l \Delta_x^{(n+q-l+m_j-1)/2} K_{j,q}(x-y,t) D_k \Delta^{(l-m_j-2)/2}\omega_j(y)dy.$$

By virtue of Theorem 4.7

$$\left(\int\!\!\int_{R_+^n} |D_k I_{j,k}(x,t)|^p \, dx dt \right)^{1/p} \leq \text{const} \left| D_k \Delta^{(l-m_j-2)/2}\omega_j \right|_{1-1/p,p,R^{n-1}}$$
$$\leq \text{const}|\omega_j|_{l-m_j-1/p,p,R^{n-1}} = \text{const}|g_j - h_j|_{l-m_j-1/p,p,R^{n-1}}$$
$$\leq \text{const}|g_j|_{l-m_j-1/p,p,R^{n-1}} + \text{const}|f|_{l-m,p,R_+^n}.$$

In case $l - m_j$ is odd we can proceed analogously putting

$$I_{j,k}(x, t) = \int_{R^{n-1}} D^l D_k \Delta_x^{(n+q-l+m_j-2)/2} K_{j,q}(x - y, t) \Delta^{(l-m_j-1)/2} \omega_j(y) dy.$$

4.8 Problems in Uniformly Regular Open Sets

We assume that the assumptions listed in section 4.2 are satisfied by an open set Ω and the operators $L = L(x, D)$ and $\{B_j\}_{j=1}^{m/2} = \{B_j(x, D)\}_{j=1}^{m/2}$. For a positive number R we set

$$\Sigma_R = \{x \in R^n; |x| < R, x_n > 0\}, \quad \sigma_R = \{x \in R^n; |x| < R, x_n = 0\}.$$

For a function u which is a trace on σ_R of a function belonging to $W^{l,p}(\Sigma_R)$ and vanishing near $|x| = R$ we set

$$[u]_{l-1/p,p,\sigma_R} = \inf\{\|v\|_{l,p,R_+^n}; v \in W^{l,p}(R_+^n), v = u \text{ on } \sigma_R\}. \tag{4.136}$$

Lemma 4.10 *Let $L(x, D)$ be an operator of order m defined in Σ_R, and $B_j(x, D)$ be operators of order $m_j, j = 1, \ldots, m/2$, defined on σ_R. Suppose that the assumptions of section 4.2 are satisfied by $\Sigma_R, L, \{B_j\}_{j=1}^{m/2}$. Then for $1 < p < \infty$ there exists a positive number $\rho < R$ such that for $u \in W^{m,p}(\Sigma_R)$ vanishing for $|x| > \rho$ the following inequality holds:*

$$\|u\|_{m,p,\Sigma_R} \leq C \left[\|Lu\|_{0,p,\Sigma_R} + \sum_{j=1}^{m/2} [B_j u]_{m-m_j-1/p,p,\sigma_R} + \|u\|_{0,p,\Sigma_R} \right].$$
$$\tag{4.137}$$

Proof. Let F, G_j be functions defined by

$$L^0(0, D)u = Lu + \left(L^0(0, D) - L^0(x, D)\right) u - L^1(x, D)u = F,$$
$$B_j^0(0, D)u = B_j u + \left(B_j^0(0, D) - B_j^0(x, D)\right) u - B^1(x, D)u = G_j,$$

where L^0, B_j^0 are the principal parts of L, B_j and L^1, B_j^1 are the lower order parts of L, B_j respectively. Then in view of Theorem 4.9

$$|u|_{m,p,R_+^n} \leq C_{l,p} \left[|F|_{0,p,R_+^n} + \sum_{j=1}^{m/2} |G_j|_{m-m_j-1/p,p,R^{n-1}} \right]. \tag{4.138}$$

As is easily seen

$$|F|_{0,p,R_+^n} \leq |Lu|_{0,p,\Sigma_R}$$
$$+ \max_{|\alpha|=m} \max_{x\in\Sigma_\rho} |a_\alpha(x) - a_\alpha(0)| \, |u|_{m,p,\Sigma_R} + C\|u\|_{m-1,p,\Sigma_R}, \quad (4.139)$$

$$|G_j|_{m-m_j-1/p,p,R^{n-1}} \leq [B_j u]_{m-m_j-1/p,p,\sigma_R}$$
$$+ \max_{|\beta|=m_j} \max_{x\in\sigma_\rho} |b_{j\beta}(x) - b_{j\beta}(0)| \, |u|_{m,p,\Sigma_R} + C\|u\|_{m-1,p,\Sigma_R}. \quad (4.140)$$

If we set

$$\epsilon = \max_{|\alpha|=m} \max_{x\in\Sigma_\rho} |a_\alpha(x) - a_\alpha(0)| + \sum_{j=1}^{m/2} \max_{|\beta|=m_j} \max_{x\in\sigma_\rho} |b_{j\beta}(x) - b_{j\beta}(0)|, \quad (4.141)$$

then $\epsilon \to 0$ as $\rho \to 0$. With the aid of Theorem 3.16 and Young's inequality there exists a positive number C_ϵ for each $\epsilon > 0$ such that

$$\|u\|_{m-1,p,\Sigma_R} \leq \epsilon |u|_{m,p,\Sigma_R} + C_\epsilon |u|_{0,p,\Sigma_R}. \quad (4.142)$$

Substituting (4.139),(4.140),(4.141),(4.142) in (4.138)

$$|u|_{m,p,\Sigma_R} = |u|_{m,p,R_+^n} \leq C_{l,p}$$
$$\times \left[|Lu|_{0,p,\Sigma_R} + \sum_{j=1}^{m/2} [B_j u]_{m-m_j-1/p,p,\sigma_R} + 2\epsilon|u|_{m,p,\Sigma_R} + C_\epsilon |u|_{0,p,\Sigma_R} \right].$$

Letting ρ be so small that $2\epsilon C_{l,p} \leq 1/2$ we get

$$|u|_{m,p,\Sigma_R} \leq C_{l,p} \left[|Lu|_{0,p,\Sigma_R} + \sum_{j=1}^{m/2} [B_j u]_{m-m_j-1/p,p,\sigma_R} + |u|_{0,p,\Sigma_R} \right].$$

The desired inequality follows from this inequality and (4.142).

The following lemma is due to Browder [21].

Lemma 4.11 *If Ω is an open set in R^n uniformly regular of class C^m, then there exist a natural number N and positive numbers M, δ_0 such that for $0 < \delta < \delta_0$ there exist an open covering $\{O_i\}_{i=1}^\infty$ of $\bar{\Omega}$ and for each i a homeomorphism Φ_i of class C^m from O_i to the open ball B_δ of radius δ centered at the origin satisfying the following conditions:*
(a) $N+1$ distinct sets of $\{O_i\}$ have an empty intersection;
(b) if O_i and $\partial\Omega$ have a nonempty intersection

$$\Phi_i(O_i \cap \Omega) = \{y; |y| < \delta, y_n > 0\},$$
$$\Phi_i(O_i \cap \partial\Omega) = \{y; |y| < \delta, y_n = 0\};$$

(c) *if $\Phi_{i,k}(x)$, $\Psi_{i,k}(y)$ are the kth components of $\Phi_i(x)$, $\Psi_i(y) = \Phi_i^{-1}(y)$, then for any $i, k, |\alpha| \leq m, x \in O_i, y \in B_\delta$*

$$|D^\alpha \Phi_{i,k}(x)| \leq M, \quad |D^\alpha \Psi_{i,k}(y)| \leq M. \tag{4.143}$$

Proof. Since Ω is uniformly regular of class of C^m, there exists a locally finite sequence $\{O_{i,0}\}_{j=1}^\infty$ of open sets, a homeomorphism $\Phi_{i,0}$ of class C^m from each $O_{i,0}$ to the unit ball of R^n, a natural number N_0 and a positive number M_0 such that the conditions $(1),(2),(3),(4),(5)$ of Definition 3.2 are all satisfied. Replacing M_0 by another constant if necessary we assume that

$$|\partial \Phi_{i,0}(x)/\partial x| \leq M_0, \quad |\partial \Psi_{i,0}(y)/\partial y| \leq M_0$$

also hold, where $\Psi_{i,0} = \Phi_{i,0}^{-1}$. Then

$$|\Phi_{i,0}(x') - \Phi_{i,0}(x)| \leq M_0 |x' - x|, \quad |\Psi_{i,0}(y') - \Psi_{i,0}(y)| \leq M_0 |y' - y|. \tag{4.144}$$

For $0 < d < \min\{M_0/N_0, 1/2\}$ we set $S_d = \{y \in R^n; |y| < 1/2, |y_n| < d\}$. Then $\cup_{i=1}^\infty \Psi_{i,0}(S_d)$ contains the d/M_0 neighborhood of $\partial \Omega$ as is shown as follows. Let $\text{dist}(x, \partial \Omega) < d/M_0$. Since $d/M_0 < N_0^{-1}$ we have $x \in O_{i,0}' = \Psi_{i,0}(\{y; |y| < 1/2\})$ for some i. If we set $y = \Phi_{i,0}(x)$, then

$$|y| < 1/2, \quad |y_n| = |\Phi_{i,0,n}(x)| \leq M_0 \text{dist}(x, \partial \Omega) < d.$$

Hence $y \in S_d$ and $x = \Psi_{i,0}(y) \in \Psi_{i,0}(S_d)$.

Let B_0 be the unit ball and $P_0 = B_0 \cap \{y; y_n = 0\}$, and $B_1 = \{y; |y| < 1/2\}$, $P_1 = B_1 \cap \{y; y_n = 0\}$. Let $0 < d_1 < n^{-1/2} \min\{1/2, M_0/N_0\}$. Set

$$Y' = \{\eta' = (\eta_1, \ldots, \eta_{n-1}, 0); \eta_i \text{ are integers},$$
$$d_1 \eta' = (d_1 \eta_1, \ldots, d_1 \eta_{n-1}, 0) \in P_1\},$$

and for $\eta' \in Y'$

$$S_{\eta'} = \{y; |y - d_1 \eta'| < \sqrt{n} d_1\}.$$

Then as is easily seen $S_{\eta'} \subset B_0$. We show that

$$\cup_{\eta' \in Y'} S_{\eta'} \supset S_{d_1} = \{y \in R^n; |y| < 1/2, |y_n| < d_1\}. \tag{4.145}$$

If $y \in S_{d_1}$, then $y' = \{y_1, \ldots, y_{n-1}, 0\} \in P_1$. Let $d_1 \eta'$ be the nearest point to y' with $\eta' \in Y'$. Then $|y' - d_1 \eta'| < \sqrt{n-1} d_1$. Hence

$$|y - d_1 \eta'| = \{y_n^2 + |y' - d_1 \eta'|^2\}^{1/2} < \{d_1^2 + (n-1)d_1^2\}^{1/2} = \sqrt{n} d_1,$$

or $y \in S_{\eta'}$.

In view of (4.144) we have $\text{diam} \Psi_{i,0}(S_{\eta'}) \leq 2\sqrt{n} d_1 M_0$. We have also

$$\cup_{i=1}^\infty \cup_{\eta' \in Y'} \Psi_{i,0}(S_{\eta'}) \supset \cup_{i=1}^\infty \Psi_{i,0}(S_{d_1}) \supset \text{the } d_1/M_0 \text{ neighborhood of } \partial \Omega.$$

Next, set $d_2 = d_1/(2M_0\sqrt{n})$, and

$$Y = \{\eta = (\eta_1, \dots, \eta_n); \quad \eta_i \text{ are integers}, d_2\eta \in \Omega, \operatorname{dist}(d_2\eta, \partial\Omega) > d_1/M_0\}.$$

Then

$$S_\eta = \{x; |x - d_2\eta| < \sqrt{n}d_2\} \subset \Omega.$$

Let $\{O_i\}$ be the totality of $\Psi_{i,0}(S_{\eta'}), \eta' \in Y', i = 1, 2, \dots$, and $S_\eta, \eta \in Y$. Then there exists a homeomorphism of class C^m from each of O_i to the ball with center the origin and radius $\sqrt{n}d_1$. We conclude the proof without difficulty putting $\delta_0 = \min\{1/2, M_0/N_0\}$ and $\delta = \sqrt{n}d_1$.

Theorem 4.10 *Let Ω be an open set in R^n uniformly regular of class C^m. Suppose that $L(x, D), \{B_j(x, D)\}_{j=1}^{m/2}$ satisfy the conditions listed in section 4.2. Then for each $p \in (1, \infty)$ there exists a positive constant C_p such that the inequality (4.25) holds for $u \in W^{m,p}(\Omega)$:*

$$\|u\|_{m,p,\Omega} \leq C_p\Bigg[\|L(\cdot, D)u\|_{0,p,\Omega}$$

$$+ \sum_{j=1}^{m/p}[B_j(\cdot, D)u]_{m-m_j-1/p,p,\partial\Omega} + \|u\|_{0,p,\Omega}\Bigg].$$

Proof. For $\delta > 0$ let $\{O_i\}, \{\Phi_i\}$ be the families of open sets and mappings satisfying the conditions $(a), (b), (c)$ of Lemma 4.11. We construct the partition of unity $\{\zeta_i\}$ subordinate to the families $\{O_i\}, \{\Phi_i\}$ as was done in section 3.6 with $\{y; |y| < \delta\}$ in place of $\{y; |y| < 1\}$. We set $u_i = \zeta_i u$ for each i. In case $O_i \cap \partial\Omega$ is not empty the images of the operators under the homeomorphism $x = \Psi_i(y)$ satisfy the conditions of section 4.2 in the half ball $\{y; |y| < \delta, y_n > 0\}$. Hence, if δ is sufficiently small we can apply Lemma 4.10 to $u_i(\Psi_i(y))$, and pulling back to the original coodinates x we get

$$\|u_i\|_{m,p,\Omega} \leq C_p\Bigg[\|Lu_i\|_{0,p,\Omega} + \sum_{j=1}^{m/2}[B_ju_i]_{m-m_j-1/p,p,\partial\Omega} + \|u_i\|_{0,p,\Omega}\Bigg].$$

If g_i is a function in $W^{m,p}(\Omega)$ coinciding with $B_j(\cdot, D)u_i$ on $\partial\Omega$, then

$$[B_ju_i]_{m-m_j-1/p,p,\partial\Omega} \leq \|\zeta_ig_i\|_{m-m_j,p,\Omega} + C\sum_{k=1}^{m-1}\|D^ku\|_{L^p(\operatorname{supp}\zeta_i)}.$$

Hence following the proof of Theorem 3.15 we conclude the proof of the present theorem.

Remark 4. 2 In [11] it is assumed that Ω is bounded. However, it is not assumed that $m_j < m$, and is proved that for an integer $l \geq \max_{j=1,\ldots,m/2}\{m, m_j + 1\}$ the estimate

$$\|u\|_{l,p,\Omega} \leq C_{l,p}\left[\|Lu\|_{l-m,p,\Omega} + \sum_{j=1}^{m/2}[B_ju]_{l-m_j-1/p,p,\partial\Omega} + \|u\|_{0,p,\Omega}\right]$$

holds under suitable smoothness conditions on the boundary $\partial\Omega$ and the coefficients of $L(\cdot, D), \{B_j(\cdot, D)\}_{j=1}^{m/2}$.

Chapter 5

Elliptic Boundary Value Problems (Continued)

5.1 Adjoint Boundary Conditions

Let Ω be an open subset of R^n uniformly regular of class C^m. Let $L(x, D)$ be a uniformly elliptic operator of order m with coefficients defined in $\bar{\Omega}$ and satisfying the Root Condition of section 4.2. Let $\{B_j(x, D)\}_{j=1}^{m/2}$ be a set of differential operators with coefficients defined on $\partial\Omega$. Assume that the order m_j of B_j is less than m.

The set $\{B_j\}_{j=1}^{m/2}$ is called *normal* if the following conditions are satisfied:

(i) $\{m_j\}_{j=1}^{m/2}$ are distinct, i.e. $m_j \neq m_k$ if $j \neq k$,

(ii) $\partial\Omega$ is nowhere characteristic with respect to each of B_j, i.e. for any $x \in \partial\Omega$ $B_j^0(x, \nu) \neq 0, j = 1, \ldots, m/2$, where ν is the outward normal vector to $\partial\Omega$ at x and B_j^0 is the principal part of B_j.

Suppose that the coefficients of L and $\{B_j\}_{j=1}^{m/2}$ are sufficiently smooth and $\{B_j\}_{j=1}^{m/2}$ is normal. Let $L'(x, D)$ be the formal adjoint of $L(x, D)$: if

$$L(x, D) = \sum_{|\alpha| \leq m} a_\alpha(x) D^\alpha, \tag{5.1}$$

then

$$L'(x, D)v = \sum_{|\alpha| \leq m} (-D)^\alpha \left(\bar{a}_\alpha(x) v \right). \tag{5.2}$$

In this section following M. Schechter [133] we describe how to construct the adjoint boundary operators.

Theorem 5.1 *Under the hypotheses stated above there exists a normal set of boundary operators $\{B'_j(x, D)\}_{j=1}^{m/2}$ such that a necessary and sufficient condition in order that $u \in C^m(\bar{\Omega})$ satisfies*

$$B_j(x, D)u = 0, \quad j = 1, \ldots, m/2, \quad \text{on} \quad \partial\Omega \tag{5.3}$$

is that the equality

$$(L(x, D)u, v) = (u, L'(x, D)v) \tag{5.4}$$

holds for any $v \in C_0^m(\bar{\Omega})$ satisfying

$$B'_j(x, D)v = 0, \quad j = 1, \ldots, m/2, \quad \text{on} \quad \partial\Omega. \tag{5.5}$$

Conversely, $v \in C_0^m(\bar{\Omega})$ satisfies (5.5) if and only if (5.4) holds for any $u \in C^m(\bar{\Omega})$ satisfying (5.3).

Remark 5.1 In the above theorem $u \in C^m(\bar{\Omega}), v \in C_0^m(\bar{\Omega})$ may be replaced by $u \in C_0^m(\bar{\Omega}), v \in C^m(\bar{\Omega})$.

A normal set of m boundary operators of order less than m is called a *Dirichlet set* of order m. If $\{C_j\}_{j=1}^m$ is a Dirichlet set of order m and μ_j is the order of C_j, then $\{\mu_j\}_{j=1}^m = \{0, 1, \ldots, m-1\}$. If $\partial/\partial\nu$ is the normal derivative, then $\{(\partial/\partial\nu)^{j-1}\}_{j=1}^m$ is a Dirichlet set of order m.

Let a be a point of $\partial\Omega$. We construct the set $\{B'_j\}$ of Theorem 5.1 in some neighborhood of a. By a coordinate transformation we suppose that the part of Ω in the neighborhood is the half ball

$$\Sigma_R = \{x \in R^n; |x| < R, x_n > 0\},$$

and the part of $\partial\Omega$ is the disk

$$\sigma_R = \{x \in R^n; |x| < R, x_n = 0\}.$$

In the following two lemmas we consider only smooth functions in $\bar{\Sigma}_R$.

Lemma 5.1 *Let $\{C_j\}_{j=1}^m$ and $\{C'_j\}_{j=1}^m$ be a couple of Dirichlet sets of order m. If C_j and C'_j are both of order $j-1$ for $j = 1, \ldots, m$, then there exist differential operators Λ_{jk} of order $j-k$ containing only the tangential derivatives $D_1, \ldots, D_{n-1}$ such that*

$$C'_j = \sum_{k=1}^j \Lambda_{jk} C_k \quad \text{on} \quad \sigma_R. \tag{5.6}$$

Λ_{jj} *is a function which vanishes nowhere on σ_R.*

Proof. If the lemma is proved in case $C_j' = D_n^{j-1}, j = 1, \ldots, m$, then we have

$$D_n^{j-1} = \sum_{l=1}^{j} \Lambda_{jl} C_l. \tag{5.7}$$

With some tangential differential operators Γ_{jk} of order at most $j - k$, C_j' is expressed as

$$C_j' = \sum_{k=1}^{j} \Gamma_{jk} D_n^{k-1}, \tag{5.8}$$

Γ_{jj} being a nonvanishing function. It follows from (5.7),(5.8) that

$$C_j' = \sum_{k=1}^{j} \Gamma_{jk} \sum_{l=1}^{k} \Lambda_{kl} C_l = \sum_{l=1}^{j} \sum_{k=l}^{j} \Gamma_{jk} \Lambda_{kl} C_l = \sum_{l=1}^{j} \Theta_{jl} C_l.$$

Here $\Theta_{jl} = \sum_{k=l}^{j} \Gamma_{jk} \Lambda_{kl}$ is a tangential differential operator of order at most $j - l$, and $\Theta_{jj} = \Gamma_{jj} \Lambda_{jj}$ is a nonvanishing function. Hence it suffices to show the lemma only in case $C_j' = D_n^{j-1}, j = 1, \ldots, m$.

Each C_k is expressed as

$$C_k = \sum_{l=1}^{k} \Gamma_{kl} D_n^{l-1},$$

where Γ_{kl} are operators with the same property as those of (5.8). Therefore

$$D_n^{k-1} = \Gamma_{kk}^{-1} C_k - \Gamma_{kk}^{-1} \sum_{l=1}^{k-1} \Gamma_{kl} D_n^{l-1}.$$

Hence, if D_n^{l-1} is expressed as (5.7) for $l = 1, \ldots, k-1$, then so is D_n^{k-1}. If $j = 1$, (5.7) is clear. Hence for any $j = 1, \ldots, m$ D_n^{j-1} is expressed as (5.7).

Lemma 5.2 *If $\{C_j\}_{j=1}^{m}$ is a Dirichlet set and $\{g_j\}_{j=1}^{m}$ is a set of functions defined on σ_R, then there exists a function $v \in C^m(\bar{\Sigma}_R)$ such that $C_j v = g_j, j = 1, \ldots, m$ on σ_R.*

Proof. In view of Lemma 5.1 there exist tangential differential operators $\Gamma_{jk}, \Lambda_{kl}$ such that

$$C_j = \sum_{k=1}^{j} \Gamma_{jk} D_n^{k-1}, \quad D_n^{k-1} = \sum_{l=1}^{k} \Lambda_{kl} C_l.$$

It is easy to show

$$\sum_{k=l}^{j} \Gamma_{jk}\Lambda_{kl} = \delta_{jl}. \tag{5.9}$$

If v is a function defined in $\bar{\Sigma}_R$ satisfying

$$D_n^{k-1}v = \sum_{l=1}^{k} \Lambda_{kl}g_l$$

on σ_R, then with the aid of (5.9) we get that on σ_R

$$C_j v = \sum_{k=1}^{j} \Gamma_{jk}D_n^{k-1}v = \sum_{k=1}^{j}\Gamma_{jk}\sum_{l=1}^{k}\Lambda_{kl}g_l = \sum_{l=1}^{j}\sum_{k=l}^{j}\Gamma_{jk}\Lambda_{kl}g_l = g_j.$$

Proof of Theorem 5.1. If $u, v \in C^m(\bar{\Sigma}_R)$ and v vanishes near $|x| = R$, then by an integration by parts

$$\int_{\Sigma_R} \left[D_n^k u \cdot v - u \cdot (-D_n)^k v\right] dx = -\sum_{i=1}^{k}\int_{\sigma_R} D_n^{i-1}u \cdot (-D_n)^{k-i}v\, dx', \tag{5.10}$$

where $dx' = dx_1 \cdots dx_{n-1}$. We write the operator L as

$$L = \sum_{|\mu|+k\leq m} a_{\mu k}D^\mu D_n^k,$$

where $\mu = (\mu_1, \ldots, \mu_{n-1}, 0)$. With the aid of (5.10)

$(Lu, v) - (u, L'v)$

$$= \sum_{|\mu|+k\leq m}\int_{\Sigma_R}\left[D_n^k u \cdot (-D)^\mu(a_{\mu k}\bar{v}) - u \cdot (-D_n)^k(-D)^\mu(a_{\mu k}\bar{v})\right]dx$$

$$= -\sum_{|\mu|+k\leq m}\sum_{i=1}^{k}\int_{\sigma_R} D_n^{i-1}u \cdot (-D_n)^{k-i}(-D)^\mu(a_{\mu k}\bar{v})\, dx'$$

$$= -\sum_{k=1}^{m}\sum_{i=1}^{k}\int_{\sigma_R} D_n^{i-1}u \cdot \sum_{|\mu|\leq m-k}(-D)^\mu(-D_n)^{k-i}(a_{\mu k}\bar{v})\, dx'$$

$$= \sum_{i=1}^{m}\int_{\sigma_R} D_n^{i-1}u \cdot \overline{N_i v}\, dx', \tag{5.11}$$

where

$$N_i v = -\sum_{k=i}^{m} \sum_{|\mu| \le m-k} (-D)^\mu (-D_n)^{k-i} (\bar{a}_{\mu k} v).$$

N_i is of order at most $m - i$. The ellipticity of L implies $a_{0m} \ne 0$. Hence

$$N_i = (-1)^{m-i-1} \bar{a}_{0m} D_n^{m-i} + \cdots \tag{5.12}$$

is of order $m - i$ and σ_R is not characteristic with respect to N_i. Adding $m/2$ boundary operators $\{B_j\}_{j=m/2+1}^{m}$ to $\{B_j\}_{j=1}^{m/2}$, we obtain a Dirichlet set $\{B_j\}_{j=1}^{m}$. We renumber it so that $\{B_j\}_{j=1}^{m} = \{\tilde{B}_j\}_{j=1}^{m}$ and $\tilde{B}_j$ is of order $j - 1$ for $j = 1, \ldots, m$. We denote the order of B_j, $j = m/2 + 1, \ldots, m$, by m_j. Then $\{m_j\}_{j=1}^{m} = \{0, 1, \ldots, m\}$, and $\tilde{B}_{m_j+1} = B_j$. By virtue of Lemma 5.1 we have

$$D_n^{i-1} = \sum_{l=1}^{i} \Lambda_{il} \tilde{B}_l, \tag{5.13}$$

where Λ_{il} are tangential differential operators of order $i - l$ at the most and Λ_{ii} is a nonvanishing function. Denoting the formal adjoint of Λ_{il} by Λ_{il}', we set

$$C_j = \sum_{i=m-j+1}^{m} \Lambda_{i,m-j+1}' N_i \tag{5.14}$$

for $j = 1, \ldots, m$. C_j is of order at most $j - 1$, and by (5.12)

$$C_j = \Lambda_{m-j+1,m-j+1}' N_{m-j+1} + \cdots = (-1)^j \Lambda_{m-j+1,m-j+1}' \bar{a}_{0m} D_n^{j-1} + \cdots.$$

Hence C_j is of order $j - 1$ and σ_R is not characteristic with respect to C_j. In view of (5.11),(5.13),(5.14)

$$(Lu, v) - (u, L'v) = \sum_{i=1}^{m} \int_{\sigma_R} \sum_{l=1}^{i} \Lambda_{il} \tilde{B}_l u \cdot \overline{N_i v} dx'$$

$$= \sum_{l=1}^{m} \int_{\sigma_R} \tilde{B}_l u \cdot \sum_{i=l}^{m} \overline{\Lambda_{il}' N_i v} dx' = \sum_{l=1}^{m} \int_{\sigma_R} \tilde{B}_l u \cdot \overline{C_{m+1-l} v} dx'.$$

Hence we obtain

$$(Lu, v) - (u, L'v) = \sum_{j=1}^{m} \int_{\sigma_R} B_j u \cdot \overline{C_{m-m_j} v} dx'. \tag{5.15}$$

Suppose that (5.3) holds. If

$$C_{m-m_j} v = 0, \quad j = m/2 + 1, \ldots, m, \quad \text{on} \quad \sigma_R, \tag{5.16}$$

then in view of (5.15) we see that (5.4) holds. Conversely, suppose u is such that (5.4) holds for any v satisfying (5.16). Let ζ be a real valued function belonging to $C_0^\infty(\{x; |x| < R\})$. In view of Lemma 5.2 there exists a function $v \in C^m(\bar{\Sigma}_R)$ satisfying

$$
C_{m-m_j} v = \begin{cases} \zeta^2 B_j u & j = 1, \dots, m/2 \\ 0 & j = m/2 + 1, \dots, m \end{cases} \quad \text{on} \quad \sigma_R.
$$

We can choose v so that $v = 0$ near $|x| = R$. Hence with the aid of (5.15)

$$
\sum_{j=1}^{m/2} \int_{\sigma_R} \zeta^2 |B_j u|^2 dx' = \sum_{j=1}^{m/2} \int_{\sigma_R} B_j u \cdot \overline{C_{m-m_j} v} dx'
$$

$$
= \sum_{j=1}^m \int_{\sigma_R} B_j u \cdot \overline{C_{m-m_j} v} dx' = (Lu, v) - (u, L'v) = 0.
$$

Consequently we get $B_j u = 0, j = 1, \dots, m/2$, on σ_R. Thus $\{B_j'\}_{j=1}^{m/2} = \{C_{m-m_j}\}_{j=m/2+1}^m$ meets our requirement in Σ_R. Returning to the original coordinate system and using the partition of unity we obtain a desired set of boundary operators.

Definition 5.1   Two sets $\{B_j\}_{j=1}^m$ and $\{\tilde{B}_j\}_{j=1}^{\tilde{m}}$ of boundary operators are said to be *equivalent* if the following condition is satisfied:

$$
B_j v = 0, \quad j = 1, \dots, m, \quad \text{on} \quad \partial\Omega
$$

if and only if

$$
\tilde{B}_j v = 0, \quad j = 1, \dots, \tilde{m}, \quad \text{on} \quad \partial\Omega.
$$

Lemma 5.3   *Let $\{B_j\}_{j=1}^m$, $\{\tilde{B}_j\}_{j=1}^{\tilde{m}}$ be normal sets of boundary operators. Denote the orders of B_j, $\tilde{B}_j$ by m_j, $\tilde{m}_j$. If these two sets are equivalent, then $m = \tilde{m}$ and there exist tangential differential operators Λ_{jk} such that*

$$
\tilde{B}_j = \sum_{k=1}^m \Lambda_{jk} B_k, \quad j = 1, \dots, m. \tag{5.17}
$$

The set $\{\Lambda_{jk}\}$ satisfies the following conditions:
(i) Λ_{jk} is of order at most $\tilde{m}_j - m_k$,
(ii) if $\tilde{m}_j = m_k$, then Λ_{jk} is a nowhere vanishing function.

Proof. Let l be a positive integer such that

$$
\{m_j\}_{j=1}^m \cup \{\tilde{m}_j\}_{j=1}^{\tilde{m}} \subset \{0, 1, \dots, l-1\},
$$

and $\{B_j\}_{j=1}^l$ a Dirichlet set containing $\{B_j\}_{j=1}^m$. Then in view of Lemma 5.1 each $\tilde{B}_j$ is expressed as

$$\tilde{B}_j = \sum_{k=1}^l \Lambda_{jk} B_k. \tag{5.18}$$

Denote the order of B_k for $k > m$ by m_k. Then Λ_{jk} is of order at most $\tilde{m}_j - m_k$. For each $j = 1, \ldots, \tilde{m}$ there exists k such that $m_k = \tilde{m}_j$ and for such k Λ_{jk} is a nonvanishing function. If $\Lambda_{ji} \neq 0$ for some $i > m$, then there exists a function g defined on $\partial\Omega$ such that $\Lambda_{ji}g \neq 0$. By Lemma 5.2 there exists a function v such that $B_i v = g$ and $B_k v = 0$ for $k \neq i$ on σ_R. In view of (5.18) $\tilde{B}_j v = \Lambda_{ji}g \neq 0$ on σ_R. Since $B_k v = 0$ for $k = 1, \ldots, m$ on σ_R, this contradicts the equivalence of $\{B_j\}$ and $\{\tilde{B}_j\}$. Hence $\Lambda_{ji} = 0$ for $i > m$. Hence we conclude (5.17) and $\{\tilde{m}_j\}_{j=1}^{\tilde{m}} \subset \{m_j\}_{j=1}^m$. The opposite inclusion is established analogously.

Definition 5.2 Let $\{B_j'(x,D)\}_{j=1}^{m/2}$ be the set of boundary operators of Theorem 5.1. Then $(L'(x,D), \{B_j'(x,D)\}_{j=1}^{m/2}, \Omega)$ is called the *adjoint boundary value problem* of $(L(x,D), \{B_j(x,D)\}_{j=1}^{m/2}, \Omega)$.

It is easy to see that the adjoint boundary value problem of the Dirichlet problem for $L(x,D)$ is the Dirichlet problem for $L'(x,D)$.

Lemma 5.4 *Suppose that $L, \{B_j\}_{j=1}^{m/2}$ satisfies the Complementing Condition and $\{B_j\}_{j=1}^{m/2}$ is normal. If $\{\tilde{B}_j\}_{j=1}^{m/2}$ is a normal set equivalent to $\{B_j\}_{j=1}^{m/2}$, then $L, \{\tilde{B}_j\}_{j=1}^{m/2}$ satisfies the Complementing Condition.*

Proof. By virtue of Lemma 5.3 we have

$$\tilde{B}_j = \sum_{k=1}^{m/2} \Lambda_{jk} B_k.$$

We denote the principal part of $B_j, \tilde{B}_j, \Lambda_{jk}$ by $B_j^0, \tilde{B}_j^0, \Lambda_{jk}^0$. Then

$$\tilde{B}_j^0(x, \xi + \tau\nu) = \sum_{k=1}^{m/2} \Lambda_{jk}^0(x,\xi) B_k^0(x, \xi + \tau\nu),$$

where ξ, ν are a tangential and the outward normal vector to $\partial\Omega$ at $x \in \partial\Omega$ respectively. Since $\{\tilde{B}_j^0(x, \xi + \tau\nu)\}_{j=1}^{m/2}$ is a linearly independent set of polynomials of τ, the matrix $\left(\Lambda_{jk}^0(x,\xi)\right)$ is nonsingular. If

$$\sum_{j=1}^{m/2} c_j \tilde{B}_j^0(x, \xi + \tau\nu) = \sum_{k=1}^{m/2}\sum_{j=1}^{m/2} c_j \Lambda_{jk}^0(x,\xi) B_k^0(x, \xi + \tau\nu) \equiv 0$$

modulo $M^+(x, \xi + \tau\nu)$, then $\sum_{j=1}^{m/2} c_j \Lambda_{jk}^0(x, \xi) = 0, k = 1, \ldots, m/2$. Hence $c_1 = \cdots = c_{m/2} = 0$.

Next we show that if $L, \{B_j\}_{j=1}^{m/2}$ satisfy the Complementing Condition, then so does its adjoint $L', \{B_j'\}_{j=1}^{m/2}$. For complex variables z, ζ we set

$$\sigma_k(z, \zeta) = \sum_{i=1}^{k} z^{k-i} \zeta^{i-1}, \quad \sigma_0(z, \zeta) = 0.$$

For a polynomial $P(z) = \sum_{k=0}^{m} a_k z^k$ of a complex variable z and a complex vector $\omega = (\omega_0, \omega_1, \ldots, \omega_m)$ set

$$P(\omega) = \sum_{k=0}^{m} a_k \omega_k.$$

If we set

$$R(z, \zeta) = P(\sigma(z, \zeta)) = \sum_{k=0}^{m} a_k \sigma_k(z, \zeta), \tag{5.19}$$

as is easily seen

$$R(z, \zeta) = R(\zeta, z), \quad P(z) - P(\zeta) = (z - \zeta) R(z, \zeta). \tag{5.20}$$

If we set

$$R_z^{(k)}(z, \zeta) = \left(\frac{\partial}{\partial z} \right)^k R(z, \zeta), \quad P^k(z) = \left(\frac{d}{dz} \right)^k P(z),$$

then we have

$$(z - \zeta) R_z^{(k)}(z, \zeta) = P^{(k)}(z) - k R_z^{(k-1)}(z, \zeta), \tag{5.21}$$

Lemma 5.5 *Let P_1, P_2 be polynomials of order m_1, m_2 of a complex variable. Set $P = P_1 P_2$ and $m = m_1 + m_2$. Let $\{Q_j\}_{j=1}^{m_1}$ be linearly independent polynomials of order less than m, and $\omega = (\omega_0, \omega_1, \ldots, \omega_{m-1})$ a complex vector. The conditions*

$$P(\sigma(\cdot, \omega)) \equiv 0 \quad modulo \quad P_2, \tag{5.22}$$
$$Q_j(\omega) = 0, \quad j = 1, \ldots, m_1 \tag{5.23}$$

imply $\omega = 0$ if and only if $\{Q_j\}_{j=1}^{m_1}$ is linearly independent modulo P_1.

Proof. By definition

$$P(\sigma(z,w)) = \sum_{k=0}^{m} a_k \sigma_k(z,w) = R(z,w).$$

Let $\{z_i\}$ be the distinct roots of P_2, the multiplicity of z_i being denoted by ν_i. The condition (5.22) holds if and only if

$$R_z^{(k)}(z_i,w) = \left(\frac{\partial}{\partial z}\right)^k R(z,w)\bigg|_{z=z_i} = \left(\frac{\partial}{\partial z}\right)^k P(\sigma(z,w))\bigg|_{z=z_i} = 0 \quad (5.24)$$

for $0 \le k < \nu_i$. As is easily seen

$$R_z^{(k)}(z_i,\zeta) = \frac{k!P(\zeta)}{(\zeta - z_i)^{k+1}} \quad \text{for} \quad 0 \le k < \nu_i.$$

Hence if we put $P_{ik}(\zeta) = P(\zeta)/(\zeta - z_i)^{k+1}$, then P_{ik} is a polynomial of order less than m and $R_z^{(k)}(z_i,\zeta) = k!P_{ik}(\zeta)$. Hence (5.24) is equivalent to

$$P_{ik}(w) = 0, \quad 0 \le k < \nu_i. \tag{5.25}$$

Thus the only vector w satisfying (5.22) and (5.23) is 0 if and only if the m polynomials P_{ik}, Q_j are linearly independent. Set

$$M_{ik}(\zeta) = P_2(\zeta)/(\zeta - z_i)^{k+1} \quad \text{for} \quad 0 \le k < \nu_i.$$

Suppose that $\{Q_j\}_{j=1}^{m_1}$ is linearly independent modulo P_1, and that suppose for some constants a_{ik}, b_j

$$\sum a_{ik} P_{ik} + \sum b_j Q_j \equiv 0.$$

Since $P_{ik} = P_1 M_{ik}$, $\sum b_j Q_j \equiv 0$ modulo P_1. Therefore $b_1 = \cdots = b_{m_1} = 0$. Hence

$$P_1 \sum a_{ik} M_{ik} = \sum a_{ik} P_{ik} \equiv 0,$$

which implies $\sum a_{ik} M_{ik} \equiv 0$. Since $\{M_{ik}\}$ are linearly independent we obtain $a_{ik} = 0$. Thus P_{ik}, Q_j are linearly independent.

Conversely suppose that P_{ik}, Q_j are linearly independent, and that for some constants $b_1, \ldots, b_{m_1}$ and a polynomial H we have

$$\sum b_j Q_j + H P_1 \equiv 0.$$

Since H is of order less than m_2, there exist constants a_{ik} such that $H = \sum a_{ik} M_{ik}$. Hence

$$\sum b_j Q_j + \sum a_{ik} P_{ik} \equiv 0,$$

which implies $b_1 = \cdots = b_{m_1} = 0$.

Theorem 5.2 *Suppose that the assumptions of Theorem 5.1 are satisfied. If L, $\{B_j\}_{j=1}^{m/2}$ satisfies the Complementing Condition, then so does the adjoint L', $\{B_j'\}_{j=1}^{m/2}$. The converse is also true.*

Proof. We consider in the situation of the proof of Theorem 5.1 and follow the notations there. We denote the principal parts of L, L' by L^0, $(L')^0$:

$$L^0 = \sum_{|\mu|+k=m} a_{\mu k} D^\mu D_n^k, \quad (L')^0 = \sum_{|\mu|+k=m} \bar{a}_{\mu k} D^\mu D_n^k.$$

Let $\xi = (\xi_1, \ldots, \xi_{n-1}, 0)$ and $\nu = (0, \ldots, 0, -1)$ be a tangential and the outward normal vector at the origin, and set

$$L^0(\tau) = L^0(0, \xi + \tau\nu) = \sum_{|\mu|+k=m} a_{\mu k}(0)\xi^\mu \tau^k,$$

$$(L')^0(\tau) = (L')^0(0, \xi + \tau\nu) = \sum_{|\mu|+k=m} \bar{a}_{\mu k}(0)\xi^\mu \tau^k.$$

Then $(L')^0(\tau) = \bar{L}^0(\tau)$, where $\bar{L}^0$ is the polynomial whose coefficients are the complex conjugates of the corresponding coefficients of L^0. Let $\tau_1, \ldots, \tau_{m/2}$ and $\tau_{m/2+1}, \ldots, \tau_m$ be the roots of L^0 with positive and negative imaginary parts respectively. Set

$$M^+(\tau) = \prod_{j=1}^{m/2} (\tau - \tau_j), \quad M^-(\tau) = \prod_{j=m/2+1}^{m} (\tau - \tau_j).$$

We write

$$\tilde{B}_l = \sum_{i=1}^{l} \Gamma_{li} D_n^{i-1}, \; l = 1, \ldots, m, \quad D_n^{i-1} = \sum_{k=1}^{i} \Lambda_{ik} \tilde{B}_k, \; i = 1, \ldots, m,$$

where $\{\tilde{B}_l\}_{l=1}^{m}$ be the operators in the proof of Theorem 5.1. Then

$$\sum_{i=k}^{l} \Gamma_{li} \Lambda_{ik} = \delta_{lk}. \tag{5.26}$$

If we set $\gamma_{li} = \Gamma_{li}^0(0, \xi)$, $\lambda_{ik} = \Lambda_{ik}^0(0, \xi)$, then by (5.26) we have

$$\sum_{i=k}^{l} \gamma_{li} \lambda_{ik} = \delta_{lk}. \tag{5.27}$$

The principal parts of N_i, C_j are

$$N_i^0 = (-1)^{i-1} \sum_{k=i}^{m} \sum_{|\mu|=m-k} \bar{a}_{\mu k} D^\mu D_n^{k-i},$$

$$C_j^0 = (-1)^j \sum_{i=m-j+1}^{m} \sum_{k=i}^{m} \sum_{|\mu|=m-k} \bar{a}_{\mu k} \bar{\Lambda}_{i,m-j+1}^0 D^\mu D_n^{k-i}.$$

If we set

$$N_i^0(\tau) = N_i^0(0, \xi + \tau\nu), \quad C_j^0(\tau) = C_j^0(0, \xi + \tau\nu),$$

then

$$N_i^0(\tau) = (-1)^{i-1} \sum_{k=i}^{m} \sum_{|\mu|=m-k} \bar{a}_{\mu k}(0)\xi^\mu \tau^{k-i}, \tag{5.28}$$

$$C_j^0(\tau) = (-1)^j \sum_{i=m-j+1}^{m} (-1)^{i-1} \bar{\lambda}_{i,m-j+1} N_i^0(\tau). \tag{5.29}$$

Suppose that

$$\sum_{j=1}^{m} \beta_j C_{m-j+1}^0(\tau) \equiv 0 \quad \text{modulo} \quad \bar{M}^-(\tau), \tag{5.30}$$

$$\beta_{m_j+1} = 0 \quad \text{for} \quad j = 1, \ldots, m/2. \tag{5.31}$$

If we show that $\beta_1 = \cdots = \beta_m = 0$, then the proof will be complete. For $i = 1, \ldots, m$ set

$$\omega_{i-1} = \sum_{j=1}^{i} (-1)^{j-1} \beta_j \bar{\lambda}_{ij}.$$

From (5.28),(5.29) it follows that

$$\sum_{j=1}^{m} \beta_j C_{m-j+1}^0(\tau) = \sum_{j=1}^{m} \beta_j (-1)^{m-j+1} \sum_{i=j}^{m} (-1)^{i-1} \bar{\lambda}_{ij} N_i^0(\tau)$$

$$= \sum_{i=1}^{m} \sum_{j=1}^{i} (-1)^{j-1} \beta_j \bar{\lambda}_{ij} (-1)^{i-1} N_i^0(\tau) = \sum_{i=1}^{m} (-1)^{i-1} \omega_{i-1} N_i^0(\tau)$$

$$= \sum_{i=1}^{m} \omega_{i-1} \sum_{k=i}^{m} \sum_{|\mu|=m-k} \bar{a}_{\mu k}(0)\xi^\mu \tau^{k-i}$$

$$= \sum_{k=1}^{m} \sum_{|\mu|=m-k} \bar{a}_{\mu k}(0)\xi^\mu \sum_{i=1}^{k} \tau^{k-i} \omega_{i-1}$$

$$= \sum_{|\mu|+k=m} \bar{a}_{\mu k}(0)\xi^{\mu}\sigma_k(\tau,\omega) = \bar{L}^0(\sigma(\tau,\omega)).$$

Combining this with (5.30) we get

$$L^0(\sigma(\tau,\bar{\omega})) \equiv 0 \quad \text{modulo} \quad M^-(\tau). \tag{5.32}$$

On the other hand by (5.27)

$$\tilde{B}_l^0(\bar{\omega}) = \sum_{i=1}^{l} \gamma_{li}\bar{\omega}_{i-1} = \sum_{i=1}^{l} \gamma_{li} \sum_{j=1}^{i} (-1)^{j-1}\bar{\beta}_j \lambda_{ij}$$

$$= \sum_{j=1}^{l} (-1)^{j-1}\bar{\beta}_j \sum_{i=j}^{l} \gamma_{li}\lambda_{ij} = \sum_{j=1}^{l} (-1)^{j-1}\bar{\beta}_j \delta_{jl} = (-1)^{l-1}\bar{\beta}_l.$$

Hence by (5.31) we get for $j = 1, \ldots, m/2$

$$B_j^0(\bar{\omega}) = \tilde{B}_{m_j+1}^0(\bar{\omega}) = (-1)^{m_j}\bar{\beta}_{m_j+1} = 0. \tag{5.33}$$

In view of (5.32),(5.33), Lemma 5.5 and the hypothesis of the theorem we get $\omega_{i-1} = 0$ for $i = 1, \ldots, m$. Since the matrix (λ_{ij}) is nonsingular, we conclude $\beta_1 = \cdots = \beta_m = 0$.

5.2   Existence of Solutions

Let Ω be an open subset of R^n uniformly regular of class C^m, and $L(x, D)$, $\{B_j(x,D)\}_{j=1}^{m/2}$ be operators satisfying the conditions stated in section 4.2. For $1 < p < \infty$ we define the operator A_p as follows:

$$D(A_p) = \{u \in W^{m,p}(\Omega); B_j(\cdot,D)u = 0 \text{ on } \partial\Omega, j = 1, \ldots, m/2\},$$

for $u \in D(A_p)$ $A_p u = L(\cdot,D)u$ in the sense of distributions. $\tag{5.34}$

In view of Theorem 4.10 the following inequality holds:

$$\|u\|_{m,p,\Omega} \leq C_p \left(\|A_p u\|_{0,p,\Omega} + \|u\|_{0,p,\Omega} \right). \tag{5.35}$$

In this section we use the notations of Chapter 3 to denote norms of Sobolev spaces.

Lemma 5. 6 *The operator A_p defined by (5.34) is closed in $L^p(\Omega)$.*

Proof. Suppose that $u_j \in D(A_p)$, $u_j \to u$, $A_p u_j \to f$ in $L^p(\Omega)$. By virtue of (5.35)

$$\|u_j - u_k\|_{m,p,\Omega} \leq C_p \left(\|A_p u_j - A_p u_k\|_{0,p,\Omega} + \|u_j - u_k\|_{0,p,\Omega} \right).$$

Hence $u \in W^{m,p}(\Omega)$, $\|u_j - u\|_{m,p,\Omega} \to 0$. Therefore $u \in D(A_p)$, and $A_p u_j \to A_p u$, $A_p u = f$.

In this section following S. Agmon [9] we show that under some natural assumptions the resolvent set $\rho(A_p)$ of A_p is not empty.

Let θ be an angle such that $0 \leq \theta < 2\pi$. Introducing an auxiliary real variable t we set

$$Q = \Omega \times R = \{(x, t); x \in \Omega, -\infty < t < \infty\}, \tag{5.36}$$
$$\mathcal{L}(x, D_x, D_t) = L(x, D_x) - (-1)^{m/2} e^{i\theta} D_t^m. \tag{5.37}$$

We make the following assumptions:

$$\mathcal{L}(x, D_x, D_t) \quad \text{is elliptic in } \bar{Q}, \tag{5.38}$$
$$\mathcal{L}(x, D_x, D_t), \{B_j(x, D_x)\}_{j=1}^{m/2} \quad \text{satisfyies the}$$
$$\text{Complementing Condition in } \bar{Q}. \tag{5.39}$$

The condition (5.38) holds if and only if

(i) For any $x \in \bar{\Omega}$ and a real vector $\xi \neq 0$ $\arg\{(-1)^{m/2} L^0(x, \xi)\} \neq \theta$.

By the definition of the Complementing Condition the second condition (5.39) is

(ii) Let x be an arbirary point of $\partial\Omega$, ξ be tangential to $\partial\Omega$ at x, ν the outward normal vector to $\partial\Omega$ at x and λ a complex number with argument θ. Suppose moreover $(\xi, \lambda) \neq 0$. Let $\tau_1^+(\xi, \lambda), \ldots, \tau_{m/2}^+(\xi, \lambda)$ be the roots of the polynomial $L^0(x, \xi + \tau\nu) - (-1)^{m/2}\lambda$ with positive imaginary parts. Then the polynomials $\{B_j^0(x, \xi + \tau\nu)\}_{j=1}^{m/2}$ are linearly independent modulo $\prod_{j=1}^{m/2}(\tau - \tau_j^+(\xi, \lambda))$.

Remark 5. 2 If $\lambda = 0$ the condition (ii) reduces to the Complementing Condition for L, $\{B_j\}_{j=1}^{m/2}$. If $\xi = 0$ in the condition (ii), we get that $\tau^{m_j} B_j^0(x, \nu)$ are linearly independent modulo $\prod_{j=1}^{m/2}(\tau - \tau_j^+(x, \nu))$, and hence linearly independent. This implies that $\{B_j\}_{j=1}^{m/2}$ is normal.

The following lemma is a slight extension of Theorem 2.1 of S. Agmon [9].

Lemma 5. 7 *Let Ω be an open set in R^n uniformly regular of class C^m. Suppose that L, $\{B_j\}_{j=1}^{m/2}$ satisfies the conditions of section 4.2 and the assumptions (i),(ii) of this section. Then for any $1 < p < \infty$ there exists a constant C_p such that for any $u \in W^{m,p}(\Omega)$, $g_j \in W^{m-m_j,p}(\Omega), j =$*

$1, \ldots, m/2$, satisfying $B_j(\cdot, D)u = g_j$ on $\partial\Omega$, and a complex number λ satisfying $\arg\lambda = \theta$, $|\lambda| > C_p$ the following inequality holds:

$$\sum_{j=0}^{m} |\lambda|^{(m-j)/m} \|u\|_{j,p,\Omega} \leq C_p \left[\|(L-\lambda)u\|_{0,p,\Omega} \right.$$

$$\left. + \sum_{j=1}^{m/2} |\lambda|^{(m-m_j)/m} \|g_j\|_{0,p,\Omega} + \sum_{j=1}^{m/2} \|g_j\|_{m-m_j,p,\Omega} \right]. \qquad (5.40)$$

Proof. Let ζ be a function in $C^\infty(-\infty, \infty)$ such that $\zeta(t) = 0$ for $|t| > 1$ and $\zeta(t) = 1$ for $|t| < 1/2$. For $r > 0$, $u \in W^{m,p}(\Omega)$ set $v(x, t) = \zeta(t)e^{irt}u(x)$. Then applying Theorem 4.10 to $\mathcal{L}$, $\{B_j\}_{j=1}^{m/2}$ in $\bar{Q}$ we get

$$\|v\|_{m,p,Q} \leq C \left[\|\mathcal{L}v\|_{0,p,Q} + \sum_{j=1}^{m/2} [B_j v]_{m-m_j-1/p,p,\partial Q} + \|v\|_{0,p,Q} \right]. \qquad (5.41)$$

With the aid of Leibnitz' formula

$$\mathcal{L}v(x, t) = \zeta(t)e^{irt} \left(L - r^m e^{i\theta} \right) u(x)$$

$$-(-1)^{m/2} e^{i\theta} \sum_{k=0}^{m-1} \binom{m}{k} \zeta^{(m-k)}(t)(-1)^{k/2} r^k e^{irt} u(x).$$

Hence

$$\|\mathcal{L}v\|_{0,p,Q} \leq \|(L - r^m e^{i\theta})u\|_{0,p,\Omega} + C \sum_{k=0}^{m-1} r^k \|u\|_{0,p,\Omega}. \qquad (5.42)$$

Since $x \in \partial\Omega$ if $(x, t) \in \partial Q$, we have on ∂Q

$$B_j v(x, t) = \zeta(t)e^{irt} B_j u(x) = \zeta(t)e^{irt} g_j(x).$$

Thererfore with the aid of the moment inequality (3.150)

$$[B_j v]_{m-m_j-1/p,p,\partial Q} \leq \|\zeta e^{irt} g_j\|_{m-m_j,p,Q}$$

$$\leq \sum_{k=0}^{m-m_j} r^{m-m_j-k} \|g_j\|_{k,p,\Omega}$$

$$\leq C \left(r^{m-m_j} \|g_j\|_{0,p,\Omega} + \|g_j\|_{m-m_j,p,\Omega} \right). \qquad (5.43)$$

On the other hand

$$\|v\|_{m,p,Q}^p = \sum_{|\alpha|+k \leq m} \int_{-\infty}^{\infty} \int_{\Omega} |D_x^\alpha D_t^k v(x, t)|^p \, dx dt$$

$$\geq \sum_{|\alpha|+k \leq m} \int_{-1/2}^{1/2} \int_{\Omega} |D_t^k e^{irt} D_x^\alpha u(x)|^p \, dx dt$$

$$= \sum_{k=0}^{m} r^{pk} \sum_{|\alpha| \leq m-k} \int_{\Omega} |D^\alpha u(x)|^p \, dx = \sum_{j=0}^{m} r^{p(m-j)} \|u\|_{j,p,\Omega}^p. \quad (5.44)$$

Combining (5.41),(5.42),(5.43),(5.44) we get

$$\sum_{j=0}^{m} r^{m-j} \|u\|_{j,p,\Omega} \leq C \left[\|(L - r^m e^{i\theta})u\|_{0,p,\Omega} + \sum_{k=0}^{m-1} r^k \|u\|_{0,p,\Omega} \right.$$

$$\left. + \sum_{j=1}^{m/2} \left(r^{m-m_j} \|g_j\|_{0,p,\Omega} + \|g_j\|_{m-m_j,p,\Omega} \right) + \|u\|_{0,p,\Omega} \right].$$

If r is sufficiently large, the second and fourth terms in the bracket of the right hand side of this inequality are small compared with the left hand side. Thus putting $\lambda = r^m e^{i\theta}$, we complete the proof.

Letting $g_j = 0$ in Lemma 5.7 we obtain the following theorem.

Theorem 5.3 *Let Ω be an open set of R^n uniformly regular of class C^m. Suppose that L, $\{B_j\}_{j=1}^{m/2}$ satisfy the conditions of section 4.2 and (i),(ii) of this section. Then, for any $1 < p < \infty$ there exists a constant C_p such that if $\arg\lambda = \theta$, $|\lambda| > C_p$, the following inequality holds for any $u \in D(A_p)$:*

$$\sum_{j=0}^{m} |\lambda|^{(m-j)/m} \|u\|_{j,p,\Omega} \leq C_p \|(A_p - \lambda)u\|_{0,p,\Omega}. \quad (5.45)$$

Remark 5.3 In view of the moment inequality (3.150) we see that the inequality (5.45) is equivalent to the following inequality:

$$|\lambda| \|u\|_{0,p,\Omega} + \|u\|_{m,p,\Omega} \leq C_p \|(A_p - \lambda)u\|_{0,p,\Omega} \quad (5.46)$$

replacing C_p by another constant if necessary.

The following definition is due to S. Agmon [9].

Definition 5.3 Let A be a linear closed operator in some Banach space and θ an angle. If for some positive constant C the half line $\{\lambda; \arg\lambda = \theta, |\lambda| > C\}$ is contained in the resolvent set $\rho(A)$ of A and the inequality $\|(A-\lambda)^{-1}\| \leq C/|\lambda|$ holds for λ on this half line, then the half line $\arg\lambda = \theta$ is called a *ray of the minimal growth of the resolvent* of A.

In view of Theorem 5.3 if

$$R(A_p - \lambda) = L^p(\Omega), \quad \arg\lambda = \theta, \quad |\lambda| \geq C_p \qquad (5.47)$$

is shown to hold, we see that $\arg\lambda = \theta$ is a ray of the minimal growth of the resolvent of A_p. In [9] it is announced that the proof of (5.47) will be given somewhere. However, the proof does not seem to have appeared. In Chapter 3 of [149] the proof of (5.47) is given in case Ω is bounded. The outline is as follows. If the coefficients are sufficiently smooth, (5.47) can be shown using M. Schechter's result on the Fredholm alternative between $L - \lambda$, $\{B_j\}$, and its adjoint $L' - \bar{\lambda}, \{B_j'\}$ ([134]). In the general case we approximate the operators L, $\{B_j\}$ by a sequence of operators $L^{(k)}$, $\{B_j^{(k)}\}$ with smooth coefficients. Then (5.47) holds with constant $C_p = C_p^{(k)}$ possibly depending also on k. However, (5.40) holds for $L^{(k)}$, $\{B_j^{(k)}\}$ with constants independent of k. Hence, the inequality

$$\|(A_p^{(k)} - \lambda)^{-1}\| \leq C/|\lambda|$$

holds for $\arg\lambda = \theta$, $|\lambda| > C_p^{(k)}$ with some constant C independent of k, where $A_p^{(k)}$ is the operator defined by (5.34) with $L^{(k)}$, $\{B_j^{(k)}\}$ in place of L, $\{B_j\}$. Consequently with the aid of Neumann series expansion the resolvent $(A_p^{(k)} - \lambda)^{-1}$ can be continued analytically to a fixed sector

$$\Sigma_p = \{\lambda; \theta - \delta < \arg\lambda < \theta + \delta, |\lambda| > C_p\},$$

with uniform estimate there. Thus letting $k \to \infty$ it can be shown that $\rho(A_p)$ contains the set Σ_p.

Definition 5.4 Let $L(x, D)$ be a differential operator of order m in Ω. If for any $x \in \bar{\Omega}$, $\xi \neq 0$

$$\mathrm{Re}\{(-1)^{m/2}L^0(x, \xi)\} > 0,$$

then $L(x, D)$ is called *strongly elliptic*.

For $l = 1, 2, \ldots, (-\Delta)^l$ is strongly elliptic.

Theorem 5.4 *A strongly elliptic operator satisfies the Root Condition.*

Proof. Let $L(x, D)$ be strongly elliptic of order m. Since $L^0(x, -\xi) = (-1)^m L^0(x, \xi)$, m is even. For $0 \leq s \leq 1$

$$L_s(x, D) = sL(x, D) + (1 - s)(-\Delta)^{m/2}$$

is strongly elliptic. If ξ, η are linearly independent real vectors, then the roots of the polynomial $L_s^0(x, \xi + \tau\eta)$ are continuous functions of s, and if $s = 0$ the exactly $m/2$ of them have positive imaginary parts. Hence $L(x, \xi + \tau\eta) = L_1(x, \xi + \tau\eta)$ has exactly $m/2$ roots with positive imaginary parts.

That $L(x, D)$ is strongly elliptic is equivalent to saying that for $\theta \in [\pi/2, 3\pi/2]$

$$\mathcal{L}_\theta(x, D_x, D_t) = L(x, D_x) - (-1)^{m/2} e^{i\theta} D_t^m \qquad (5.48)$$

is elliptic. In what follows in this section we assume that

(iii) $L(x, D)$ is strongly elliptic in $\bar\Omega$,

(iv) for any $\theta \in [\pi/2, 3\pi/2]$ $\mathcal{L}_\theta(x, D_x, D_t)$, $\{B_j(x, D_x)\}_{j=1}^{m/2}$ satisfy the Complementing Condition in $\bar{Q}$.

From Theorem 5.3 we obtain the following theorem.

Theorem 5.5 *Let Ω be an open set of R^n uniformly regular of class C^m. Assume that $L(x, D)$, $\{B_j(x, D)\}_{j=1}^{m/2}$ satisfy the conditions listed in section 4.2 and (iii),(iv) of this section. Then for each $1 < p < \infty$ there exists a positive constant C_p such that for any $u \in W^{m,p}(\Omega)$, $g_j \in W^{m-m_j,p}(\Omega)$ satisfying $B_j(x, D)u = g_j, j = 1, \ldots, m/2$, on $\partial\Omega$, and a complex number λ satisfying $\mathrm{Im}\lambda \leq 0$, $|\lambda| > C_p$ the following inequality holds:*

$$\sum_{j=0}^{m} |\lambda|^{(m-j)/m} \|u\|_{j,p,\Omega} \leq C_p \Bigg[\|(L(\cdot, D) - \lambda)u\|_{0,p,\Omega}$$

$$+ \sum_{j=1}^{m/2} |\lambda|^{(m-m_j)/m} \|g_j\|_{0,p,\Omega} + \sum_{j=1}^{m/2} \|g_j\|_{m-m_j,p,\Omega} \Bigg]. \qquad (5.49)$$

Especially if $u \in D(A_p)$ we have the following inequality

$$|\lambda| \|u\|_{0,p,\Omega} + \|u\|_{m,p,\Omega} \leq C_p \|(A_p - \lambda)u\|_{0,p,\Omega}. \qquad (5.50)$$

If the assumptions of Theorem 5.5 are satisfied and furthermore if for some λ satisfying $\mathrm{Im}\lambda \leq 0$, $|\lambda| > C_p$ (5.47) holds, then $-A_p$ generates an analytic semigroup in $L^p(\Omega)$. Note that $D(A_p)$ is dense in $L^p(\Omega)$ since it contains $C_0^m(\Omega)$. By what was stated after Definition 5.3 $-A_p$ generates an analytic semigroup if Ω is bounded.

We make the following assumption in order to consider the adjoint boundary value problem.

(v) The coefficients of the lower order terms of the formal adjoint $L'(x, D)$ of $L(x, D)$ are bounded and measurable.

Namely the formal adjoint of L constructed regardless of the differentiabilty of the coefficients of L satisfies the Smoothness Condition of section 4.2. Note that the coefficients of the highest order terms of L' are the complex conjugates of those of the highest order terms of L, and hence uniformly continuous in $\bar{\Omega}$.

Next assume the adjoint boundary operators $\{B'_j\}_{j=1}^{m/2}$ constructed in section 5.1 also satisfy the conditions in section 4.2.

(vi) There exists a normal set $\{B'_j\}_{j=1}^{m/2}$ of boundary operators satisfying the following conditions. Let m'_j be the order of B'_j. Then $m'_j < m$ and the coefficients of B'_j have bounded and uniformly continuous derivatives up to order $m - m'_j$ on $\partial\Omega$. Let $v \in W^{m,2}(\Omega)$. Then $(Lu, v) = (u, L'v)$ holds for any $u \in W^{m,2}(\Omega)$ satisfying $B_j(\cdot, D)u = 0, j = 1, \ldots, m/2$, on $\partial\Omega$ if and only if v satisfies $B'_j(\cdot, D)v = 0, j = 1, \ldots, m/2$, on $\partial\Omega$, and the assertion also holds interchanging $u, \{B_j\}$ and $v, \{B'_j\}$.

The adjoint boundary value problem of

$$(L(x, D_x) - (-1)^{m/2} e^{i\theta} D_t^m, \{B_j(x, D_x)\}_{j=1}^{m/2}, Q)$$

is

$$(L'(x, D_x) - (-1)^{m/2} e^{-i\theta} D_t^m, \{B'_j(x, D_x)\}_{j=1}^{m/2}, Q). \qquad (5.51)$$

Hence (5.51) satisfies the Complementing Condition for any $\theta \in [\pi/2, 3\pi/2]$. The operator A'_p is defined just as A_p was defined:

$$D(A'_p) = \{u \in W^{m,p}(\Omega); B'_j(\cdot, D)u = 0, j = 1, \ldots, m/2, \text{ on } \partial\Omega\},$$
$$\text{for } u \in D(A'_p) \; A'_p u = L'(\cdot, D)u \text{ in the sense of distributions.} \qquad (5.52)$$

By virtue of Theorem 5.5 the following statement holds. For $1 < p < \infty$ set $p' = p/(p-1)$. If we replace C_p by another constant if necessary, then for any $u \in D(A'_{p'})$ and λ satisfying $\text{Im}\lambda \leq 0, |\lambda| > C_p$

$$|\lambda|\|u\|_{0,p',\Omega} + \|u\|_{m,p',\Omega} \leq C_p\|(A'_p - \lambda)u\|_{0,p,\Omega}. \qquad (5.53)$$

In what follows we assume that the coefficients of B_j and B'_j are extended to the whole of $\bar{\Omega}$ so that they belong to $B^{m-m_j}(\bar{\Omega})$ and $B^{m-m'_j}(\bar{\Omega})$ respectively for each $j = 1, \ldots, m/2$.

The following definition is also due to F. E. Browder [21].

Definition 5. 5 Let Ω be an open set in R^n. If for any $a \in \partial\Omega$ the part of Ω, $\partial\Omega$ in some neighborhood of a is expressed as

$$x_i > \psi(x_1, \ldots, x_{i-1}, x_{i+1}, \ldots, x_n), \quad x_i = \psi(x_1, \ldots, x_{i-1}, x_{i+1}, \ldots, x_n)$$

respectively for some $i = 1, \ldots, n$ and a C^{2m} function ψ, then Ω is called an open set *locally regular of class* C^{2m}.

It is possible to show the following result following the idea of Browder [21]. In addition to the assumptions made so far we assume that Ω is locally regular of class C^{2m}. Then for $\operatorname{Im}\lambda \leq 0$, $|\lambda| > C_p$

$$R(A_p - \lambda) = L^p(\Omega), \quad R(A'_{p'} - \bar{\lambda}) = L^{p'}(\Omega). \tag{5.54}$$

The proof of this statement is very lengthy, and so we state only the outline. Since by Theorem 5.5 $A_p - \lambda$ has a continuous inverse, $R(A_p - \lambda)$ is closed. Hence it suffices to show that $R(A_p - \lambda)$ is dense. This is the same for $R(A'_{p'} - \bar{\lambda})$. If we approximate the coefficients of L, $\{B_j\}$ by smooth functions, then the inequality (5.40) holds with common constants for this approximating sequence. Hence it suffices to consider only the case of smooth coefficients. Let p be in a compact subinterval of $(1, \infty)$ containing 2. For a complex number λ satisfying $\operatorname{Im}\lambda \leq 0$ and with sufficiently large absolute value we replace L by $L - \lambda$ and show

$$R(A_p) = L^p(\Omega). \tag{5.55}$$

In this case for $u \in W^{m,p}(\Omega)$, $g_j \in W^{m-m_j,p}(\Omega)$ such that $B_j u = g_j$ on $\partial\Omega$ for $j = 1, \ldots, m/2$ the inequality

$$\|u\|_{m,p,\Omega} \leq C \left(\|Lu\|_{0,p,\Omega} + \sum_{j=1}^{m/2} \|g_j\|_{m-m_j,p,\Omega} \right) \tag{5.56}$$

holds. As for the adjoint problem we have for $u \in D(A'_{p'})$

$$\|u\|_{m,p',\Omega} \leq C\|A'_{p'}u\|_{0,p',\Omega}. \tag{5.57}$$

Lemma 5. 8 *Let u be a function in $L^p(\Omega)$. Assume that for some neighborhood U of each point of $\bar{\Omega}$ $u \in W^{m,p}(U)$ and $Lu \in L^p(\Omega)$. If moreover u satisfies the boundary conditions $B_j u|_{\partial\Omega} = 0, j = 1, \ldots, m/2$, then $u \in W^{m,p}(\Omega)$.*

Browder [21: Theorem 3] proved this lemma in the case of Dirichlet boundary conditions. The proof remains valid in the general case owing to Theorem 4.10.

Let $f \in C_0^\infty(\Omega)$. In view of (5.57) A'_2 has a continuous inverse. Hence by Theorem 1.3 $R((A'_2)^*) = L^2(\Omega)$. Therefore there exists $u \in L^2(\Omega)$ such that $f = (A'_2)^* u$. This function u is a weak solution of the boundary value problem

$$Lu = f \text{ in } \Omega, \quad B_j u = 0, \ j = 1, \ldots, m/2, \text{ on } \partial\Omega. \tag{5.58}$$

By virtue of a regularity result of weak solutions of elliptic boundary value problems (Schechter [134] in case Ω is bounded or Browder [21] in case of Dirichlet boundary conditions in an unbounded domain) we have $u \in C^m(\bar{\Omega})$. Hence u is a classical solution of (5.58). Applying Lemma 5.8 we get $u \in W^{m,2}(\Omega)$, and hence $u \in D(A_2), f = A_2 u$. Thus (5.55) is established for $p = 2$.

Next suppose that $1/2 < 1/p \leq 1/2 + 1/n$. Let ϕ be a function in $C_0^m(R^n)$ such that $\phi(x) = 1$ for $|x| \leq 1$ and $\phi(x) = 0$ for $|x| \geq 2$, and set $\phi_R(x) = \phi(x/R)$ for $R \geq 1$. Let $f \in L^p(\Omega) \cap L^2(\Omega)$. By what was shown just above there exists $u \in D(A_2)$ such that $f = A_2 u$. Applying (5.56) to $\phi_R u$ with $g_j = B_j(\phi_R u) - \phi_R B_j u$ we get

$$\|\phi_R u\|_{m,p,\Omega} \leq C \left(\|L(\phi_R u)\|_{0,p,\Omega} + \sum_{j=1}^{m/2} \|B_j(\phi_R u) - \phi_R B_j u\|_{m-m_j,p,\Omega} \right).$$
(5.59)

By Leibnitz' formula

$$L(\phi_R u) = \phi_R f + \sum_{|\alpha|+|\beta| \leq m, \alpha \neq 0} c_{\alpha\beta} D^\alpha \phi_R \cdot D^\beta u,$$

$$B_j(\phi_R u) - \phi_R B_j u = \sum_{|\alpha|+|\beta| \leq m_j, \alpha \neq 0} c'_{\alpha\beta} D^\alpha \phi_R \cdot D^\beta u.$$

With the aid of Hölder's inequality

$$\|c_{\alpha\beta} D^\alpha \phi_R \cdot D^\beta u\|_{0,p,\Omega} \leq C \left(\int_\Omega |D^\alpha \phi_R \cdot D^\beta u|^p \, dx \right)^{1/p}$$

$$\leq C \left(\int_{R^n} |D^\alpha \phi_R|^{2p/(2-p)} \, dx \right)^{(2-p)/2p} \left(\int_\Omega |D^\beta u|^2 \, dx \right)^{1/2}$$

$$= C R^{n(1/p-1/2)-|\alpha|} \left(\int_{R^n} |D^\alpha \phi|^{2p/(2-p)} \, dx \right)^{(2-p)/2p} \|D^\beta u\|_{0,2,\Omega}.$$

Noting $n(1/p - 1/2) - |\alpha| \leq 1 - |\alpha| \leq 0$ we get

$$\|c_{\alpha\beta} D^\alpha \phi_R \cdot D^\beta u\|_{0,p,\Omega} \leq C \|u\|_{m-1,2,\Omega}.$$
(5.60)

Analogously

$$\|B_j(\phi_R u) - \phi_R B_j u\|_{m-m_j,p,\Omega} \leq C \|u\|_{m-1,2,\Omega}.$$
(5.61)

Substituting (5.60),(5.61) in (5.59) we get

$$\|\phi_R u\|_{m,p,\Omega} \leq C \left(\|f\|_{0,p,\Omega} + \|u\|_{m-1,2,\Omega} \right).$$
(5.62)

Letting $R \to \infty$ we get $u \in W^{m,p}(\Omega)$, and hence $u \in D(A_p), f = A_p u$. With the aid of the same argument we show that (5.55) is true for p satisfying

$1/p_1 < 1/p \leq 1/p_1 + 1/n$ with some p_1 satisfying $1/2 < 1/p_1 \leq 1/2 + 1/n$. Continuing this process we can verify that (5.55) holds for $1 < p \leq 2$.

Next, if $1/2 - m/n \leq 1/p, 2 < p < \infty$, then in view of Theorem 3.19 $W^{m,2}(\Omega) \subset L^p(\Omega)$. Hence if for $f \in C_0^\infty(\Omega)$ $f = A_2 u$, then $u \in L^p(\Omega) \cap C^m(\bar{\Omega})$. By virtue of Lemma 5.7 $u \in W^{m,p}(\Omega)$, and hence $u \in D(A_p), f = A_p u$. If $m < n/2$, set $1/p_1 = 1/2 - m/n$, and we can prove by the same argument as above that for p satisfying $1/p_1 - m/n \leq 1/p \leq 1/p_1$ (5.55) holds. Continuing this process we can show that (5.55) is true for $2 < p < \infty$.

In view of (5.50),(5.53),(5.54) we see that

$$\rho(A_p) \supset \{\lambda; \mathrm{Im}\lambda \leq 0, |\lambda| > C_p\}, \tag{5.63}$$

$$\rho(A'_{p'}) \supset \{\lambda; \mathrm{Im}\lambda \leq 0, |\lambda| > C_p\}. \tag{5.64}$$

Since as is seen without difficulty $(A_p u, v) = (u, A'_{p'} v)$ for $u \in D(A_p), v \in D(A'_{p'})$, we have

$$(A_p)^* \supset A'_{p'}. \tag{5.65}$$

If u is an arbitrary element of $D((A_p)^*)$ and $\mathrm{Im}\lambda \leq 0, |\lambda| > C_p$, then in view of (5.64) there exists a function $v \in D(A'_{p'})$ such that

$$((A_p)^* - \bar{\lambda})u = (A'_{p'} - \bar{\lambda})v. \tag{5.66}$$

Combining (5.65),(5.66) we get

$$((A_p)^* - \bar{\lambda})u = ((A_p)^* - \bar{\lambda})v. \tag{5.67}$$

Since $\lambda \in \rho(A_p)$ implies $\bar{\lambda} \in \rho((A_p)^*)$, we obtain from (5.67) that $u = v \in D(A'_{p'})$. Thus we conclude

$$(A_p)^* = A'_{p'}. \tag{5.68}$$

Especially if L is formally self-adjoint, i.e. $L = L'$, and if $\{B'_j\}$ is equivalent to $\{B_j\}$, then A_2 is self-adjoint.

Summing up we have established the following theorem.

Theorem 5. 6 *If in addition to the hypothesis of theorem 5.5 Ω is locally regular of class C^{2m}, then $-A_p$, $-A'_{p'}$ generate analytic semigroups $\exp(-tA_p)$, $\exp(-tA'_{p'})$ in $L^p(\Omega)$, $L^{p'}(\Omega)$ respectively, and*

$$\exp(-tA_p)^* = \exp(-tA'_{p'}). \tag{5.69}$$

5.3   Estimates of the Kernels of $exp(-tA_p)$ and $(A_p - \lambda)^{-1}$

In this section under the assumptions of Theorem 5.6 we obtain the estimates of the kernels of $\exp(-tA_p)$ and $(A_p - \lambda)^{-1}$. We follow the idea of R. Beals [19] of establishing the asymptotic behavior of the resolvent kernel of A_2 in case A_2 is self-adjoint under mild smoothness assumptions on the coefficients. We also follow L. Hörmander's idea of using exponential functions to derive sharp interior estimates of resolvent kernels ([75]). For direct construction of the Green function for the parabolic initial boundary value problem with coefficients depending also on t see S. D. Èĭdel'man and S. D. Ivasišen [57], S. D. Ivasišen [78],[79] and V. A. Solonnikov [143].

If the assumptions (i),(ii) of section 5.2 are satisfied, then the same assumptions are also satisfied for angles sufficiently close to θ. Hence, we assume that (5.49) holds for λ satisfying $\arg\lambda \notin (-\theta_0, \theta_0)$ for some $\theta_0 \in (0, \pi/2)$.

Lemma 5.9   *If $f \in L^p(\Omega) \cap L^q(\Omega)$, $\arg\lambda \notin (-\theta_0, \theta_0)$, $|\lambda| > \max\{C_p, C_q\}$, then*

$$(A_p - \lambda)^{-1}f = (A_q - \lambda)^{-1}f. \tag{5.70}$$

Proof. Assume $p < q$. Replacing L by $L - \lambda$ we show

$$A_p^{-1}f = A_q^{-1}f. \tag{5.71}$$

As was stated in the proof of (5.54) the inequality (5.56) holds. Suppose $1/p - 1/n \leq 1/q$. Let ϕ be a function in $C_0^\infty(R^n)$ such that $\phi(x) = 1$ for $|x| \leq 1$ and $\phi(x) = 0$ for $|x| \geq 2$. Set $\phi_R(x) = \phi(x/R)$ for $R \geq 1$. Set $u = A_q^{-1}f$. If we apply (5.56) to $\phi_R u$, then by the same calculation used in deriving (5.62) we get

$$\|\phi_R u\|_{m,p,\Omega} \leq C(\|f\|_{0,p,\Omega} + \|u\|_{m-1,q,\Omega}).$$

Letting $R \to \infty$ we obtain $u \in W^{m,p}(\Omega)$, and hence $u \in D(A_p)$. Thus (5.71) is verified. In the general case we choose $p_1, \ldots, p_{k-1}$ so that $p = p_0 < p_1 < \cdots < p_k = q$, $1/p_{i-1} - 1/n \leq 1/p_i$. Then $f \in L^{p_i}(\Omega)$, $i = 0, 1, \ldots, k$. By the same argument as above we can show

$$A_p^{-1}f = A_{p_1}^{-1}f = \cdots = A_{p_{k-1}}^{-1}f = A_q^{-1}f.$$

Lemma 5.10   *Let Ω be a nonempty open subset of R^n, and S, T bounded linear operators from $L^2(\Omega)$ to itself. Suppose that $R(S) \subset B(\bar\Omega)$, $R(T^*) \subset B(\bar\Omega)$, where $B(\bar\Omega)$ is the set of bounded and continuous functions in $\bar\Omega$, and* that $\{Sg; \|g\|_{0,2,\Omega} \leq 1\}$ and $\{T^*h; \|h\|_{0,2,\Omega} \leq 1\}$ are equicontinuous in $\bar{\Omega}$. Then, ST is an integral operator with kernel $K(x,y)$:

$$(STf)(x) = \int_\Omega K(x,y)f(y)dy. \tag{5.72}$$

$K(x,y)$ is bounded, uniformly continuous in $\Omega \times \Omega$ and satisfies

$$|K(x,y)| \leq \|S\|_{B(L^2,L^\infty)}\|T^*\|_{B(L^2,L^\infty)}. \tag{5.73}$$

Proof. By assumption there exists a strictly increasing continuous function ω defined in $[0,\infty)$ such that $\omega(0) = 0$ and

$$|(Sg)(x) - (Sg)(x')| \leq \|g\|_{0,2,\Omega}\omega(|x - x'|), \tag{5.74}$$
$$|(T^*h)(y) - (T^*h)(y')| \leq \|h\|_{0,2,\Omega}\omega(|y - y'|). \tag{5.75}$$

Since for $f \in L^2(\Omega), x \in \Omega$

$$|(STf)(x)| \leq \|STf\|_{0,\infty,\Omega} \leq \|S\|_{B(L^2,L^\infty)}\|Tf\|_{0,2,\Omega}$$
$$\leq \|S\|_{B(L^2,L^\infty)}\|T\|_{B(L^2,L^2)}\|f\|_{0,2,\Omega}, \tag{5.76}$$

there exists a function $K(x,y)$ such that $K(x,\cdot) \in L^2(\Omega)$ and

$$(STf)(x) = \int_\Omega K(x,y)f(y)dy, \tag{5.77}$$

$$\left(\int_\Omega |K(x,y)|^2 dy\right)^{1/2} \leq \|S\|_{B(L^2,L^\infty)}\|T\|_{B(L^2,L^2)}. \tag{5.78}$$

Since for $f \in L^2(\Omega) \cap L^1(\Omega), g \in L^2(\Omega)$

$$|(Tf,g)| = |(f,T^*g)| \leq \|f\|_{0,1,\Omega}\|T^*g\|_{0,\infty,\Omega}$$
$$\leq \|f\|_{0,1,\Omega}\|T^*\|_{B(L^2,L^\infty)}\|g\|_{0,2,\Omega},$$

we get

$$\|Tf\|_{0,2,\Omega} \leq \|T^*\|_{B(L^2,L^\infty)}\|f\|_{0,1,\Omega}. \tag{5.79}$$

Hence

$$|(STf)(x)| \leq \|S\|_{B(L^2,L^\infty)}\|T^*\|_{B(L^2,L^\infty)}\|f\|_{0,1,\Omega}. \tag{5.80}$$

From (5.77),(5.80) it follows that

$$\operatorname{ess\,sup}_{y\in\Omega}|K(x,y)| \leq \|S\|_{B(L^2,L^\infty)}\|T^*\|_{B(L^2,L^\infty)} \equiv \gamma. \tag{5.81}$$

In view of (5.74),(5.79) for $x, x' \in \Omega$

$$\left|\int_\Omega (K(x,y) - K(x',y))f(y)dy\right| = |(STf)(x) - (STf)(x')|$$
$$\leq \|Tf\|_{0,2,\Omega}\omega(|x - x'|) \leq \|T^*\|_{B(L^2,L^\infty)}\|f\|_{0,1,\Omega}\omega(|x - x'|).$$

Hence

$$\text{ess sup}_{y \in \Omega} |K(x, y) - K(x', y)| \leq \|T^*\|_{B(L^2, L^\infty)} \omega(|x - x'|). \tag{5.82}$$

Let $\{x_j\}$ be a countable dense subset of Ω. Then, in view of (5.81),(5.82) there exists a null set $N \subset \Omega$ such that for $y \notin N$

$$|K(x_j, y)| \leq \gamma, \ |K(x_j, y) - K(x_k, y)| \leq \|T^*\|_{B(L^2, L^\infty)} \omega(|x_j - x_k|). \tag{5.83}$$

Let x be an arbitrary point of Ω. By virtue of (5.83) for a subsequence $\{x_{j_l}\}$ tending to x the limit

$$\tilde{K}(x, y) = \lim_{l \to \infty} K(x_{j_l}, y)$$

exists for any $y \notin N$. It is obvious that the limit does not depend on the choice of $\{x_{j_l}\}$. For $x, x' \in \Omega, y \notin N$ we have

$$|\tilde{K}(x, y)| \leq \gamma, \ |\tilde{K}(x, y) - \tilde{K}(x', y)| \leq \|T^*\|_{B(L^2, L^\infty)} \omega(|x - x'|). \tag{5.84}$$

It is easily seen that $\tilde{K}(x, y)$ is measurable in $\Omega \times \Omega$, and by (5.78)

$$\left(\int_\Omega |\tilde{K}(x, y)|^2 dy \right)^{1/2} \leq \|S\|_{B(L^2, L^\infty)} \|T\|_{B(L^2, L^2)}. \tag{5.85}$$

Let $f \in L^2(\Omega) \cap L^1(\Omega)$ and $x \in \Omega$. Let again $\{x_{j_l}\}$ be a sequence tending to x. Then, in view of (5.83) we get from (5.77) with x_{j_l} in place of x

$$(STf)(x) = \int_\Omega \tilde{K}(x, y) f(y) dy. \tag{5.86}$$

By virtue of (5.85) this holds for any $f \in L^2(\Omega)$.

Let $f \in L^2(\Omega), g \in L^2(\Omega) \cap L^1(\Omega)$. Then

$$|(f, S^*g)| = |(Sf, g)| \leq \|Sf\|_{0,\infty,\Omega} \|g\|_{0,1,\Omega}$$
$$\leq \|S\|_{B(L^2, L^\infty)} \|f\|_{0,2,\Omega} \|g\|_{0,1,\Omega}.$$

Hence the restriction of S^* to $L^2(\Omega) \cap L^1(\Omega)$ can be extended continuously to a bounded linear operator S' from $L^1(\Omega)$ to $L^2(\Omega)$, and for any $g \in L^1(\Omega)$

$$\|S'g\|_{0,2,\Omega} \leq \|S\|_{B(L^2, L^\infty)} \|g\|_{0,1,\Omega}. \tag{5.87}$$

As is easily seen for any $f \in L^2(\Omega), g \in L^1(\Omega)$

$$(f, S'g) = (Sf, g). \tag{5.88}$$

If $f \in L^2(\Omega) \cap L^1(\Omega), g \in L^1(\Omega)$, then with the aid of (5.84),(5.86),(5.88)

$$(f, T^*S'g) = (Tf, S'g) = (STf, g)$$
$$= \int_\Omega \int_\Omega \tilde{K}(x, y)f(y)dy\overline{g(x)}dx = \int_\Omega f(y)\int_\Omega \tilde{K}(x, y)\overline{g(x)}dxdy.$$

Hence if we set $\overline{\tilde{K}(x, y)} = K'(x, y)$, then we have

$$(T^*S'g)(y) = \int_\Omega K'(x, y)g(x)dx \tag{5.89}$$

almost everywhere in Ω. Therefore, if $\{g_j\}$ is a countable dense subset of $L^1(\Omega)$, then there exists a null set $N' \supset N$ such that for any $y \notin N'$ and j

$$(T^*S'g_j)(y) = \int_\Omega K'(x, y)g_j(x)dx.$$

Since T^* is a bounded linear operator from $L^2(\Omega)$ to $B(\bar{\Omega})$ by the hypothesis and the closed graph theorem, we see noting (5.84) that (5.89) holds for any $g \in L^1(\Omega)$ and $y \notin N'$.

Let $y, y' \notin N'$ and $g \in L^1(\Omega)$. Then it follows from (5.89),(5.75),(5.87) that

$$\left|\int_\Omega (K'(x, y) - K'(x, y'))g(x)dx\right| = |(T^*S'g)(y) - (T^*S'g)(y')|$$
$$\leq \|S'g\|_{0,2,\Omega}\omega(|y - y'|) \leq \|S\|_{B(L^2,L^\infty)}\|g\|_{0,1,\Omega}\omega(|y - y'|).$$

Hence noting that $\tilde{K}(x, y)$ is continuous in x we obtain

$$\sup_{x\in\Omega}|\tilde{K}(x, y) - \tilde{K}(x, y')| = \sup_{x\in\Omega}|K'(x, y) - K'(x, y')| \leq \|S\|_{B(L^2,L^\infty)}\omega(|y-y'|).$$

Therefore $\tilde{K}(x, y)$ is extended to the whole of $\Omega \times \Omega$ as a continuous function, and

$$|\tilde{K}(x, y) - \tilde{K}(x', y)| \leq \|T^*\|_{B(L^2,L^\infty)}\omega(|x - x'|),$$
$$|\tilde{K}(x, y) - \tilde{K}(x, y')| \leq \|S\|_{B(L^2,L^\infty)}\omega(|y - y'|).$$

Therefore rewriting $\tilde{K}(x, y)$ as $K(x, y)$ we see that (5.72) holds. Using (5.80)

$$\left|\int_\Omega K(x, y)f(y)dy\right| = |(STf)(x)| \leq \|S\|_{B(L^2,L^\infty)}\|T^*\|_{B(L^2,L^\infty)}\|f\|_{0,1,\Omega},$$

which implies (5.73).

For an n-dimensional complex vector $\eta = (\eta_1, \ldots, \eta_n)$ the operators A_p^η, $A_p'^\eta$ are defined as follows.

$$D(A_p^\eta) = \{u \in W^{m,p}(\Omega); B_j(\cdot, D+\eta)u = 0, j = 1, \ldots, m/2, \text{ on } \partial\Omega\},$$
$$\text{(5.90)}$$
for $u \in D(A_p^\eta)$ $A_p^\eta u = L(\cdot, D+\eta)u$ in the sense of distributions,

$$D(A_p'^\eta) = \{u \in W^{m,p}(\Omega); B_j'(\cdot, D+\eta)u = 0, j = 1, \ldots, m/2, \text{ on } \partial\Omega\},$$
$$\text{(5.91)}$$
for $u \in D(A_p'^\eta)$ $A_p'^\eta u = L'(\cdot, D+\eta)u$ in the sense of distributions.

Since

$$e^{-x\eta} D_j(e^{x\eta} u(x)) = (D_j + \eta_j)u(x),$$

we have

$$e^{-x\eta} L(x, D)(e^{x\eta} u(x)) = L(x, D+\eta)u(x), \qquad \text{(5.92)}$$
$$e^{-x\eta} B_j(x, D)(e^{x\eta} u(x)) = B_j(x, D+\eta)u(x). \qquad \text{(5.93)}$$

From the proof of Theorem 5.1 we see that there exist normal sets of boundary operators $\{C_j(x, D)\}_{j=1}^{m/2}$, $\{C_j'(x, D)\}_{j=1}^{m/2}$ such that

$$\int_\Omega \left(L(x, D)u \cdot \bar{v} - u \cdot \overline{L'(x, D)v} \right) dx$$

$$= \sum_{j=1}^{m/2} \int_{\partial\Omega} B_j(x, D)u \cdot \overline{C_j(x, D)v} dS$$

$$+ \sum_{j=1}^{m/2} \int_{\partial\Omega} C_j'(x, D)u \cdot \overline{B_j'(x, D)v} dS. \qquad \text{(5.94)}$$

By virtue of (5.92),(5.93),(5.94)

$$\int_\Omega \left(L(x, D+\eta)u \cdot \bar{v} - u \cdot \overline{L'(x, D-\bar{\eta})v} \right) dx$$

$$= \int_\Omega \left(e^{-x\eta} L(x, D)(e^{x\eta} u) \cdot \bar{v} - u \cdot \overline{e^{x\bar{\eta}} L'(x, D)(e^{-x\bar{\eta}} v)} \right) dx$$

$$= \int_\Omega \left(L(x, D)(e^{x\eta} u) \cdot \overline{e^{-x\bar{\eta}} v} - e^{x\eta} u \cdot \overline{L'(x, D)(e^{-x\bar{\eta}} v)} \right) dx$$

$$= \sum_{j=1}^{m/2} \int_{\partial\Omega} B_j(x, D)(e^{x\eta} u) \cdot \overline{C_j(x, D)(e^{-x\bar{\eta}} v)} dS$$

$$+ \sum_{j=1}^{m/2} \int_{\partial\Omega} C_j'(x, D)(e^{x\eta} u) \cdot \overline{B_j'(x, D)(e^{-x\bar{\eta}} v)} dS$$

$$= \sum_{j=1}^{m/2} \int_{\partial\Omega} B_j(x, D+\eta)u \cdot \overline{C_j(x, D-\bar{\eta})v} dS$$

$$+ \sum_{j=1}^{m/2} \int_{\partial\Omega} C_j'(x, D+\eta)u \cdot \overline{B_j'(x, D-\bar{\eta})v} dS.$$

This shows that the adjoint of $L(x, D+\eta), \{B_j(x, D+\eta)\}_{j=1}^{m/2}$ is $L'(x, D-\bar{\eta}), \{B_j'(x, D-\bar{\eta})\}_{j=1}^{m/2}$.

Lemma 5.11 *For $1 < p < \infty$ there exists a positive number δ such that for $\arg \lambda \notin (-\theta_0, \theta_0), |\lambda| > C_p, |\eta| \le \delta |\lambda|^{1/m}, u \in W^{m,p}(\Omega), h_j \in W^{m-m_j, p}(\Omega),$ $B_j(x, D+\eta)u = h_j, j = 1, \ldots, m/2$ on $\partial\Omega$ the following inequality holds:*

$$\sum_{k=0}^{m} |\lambda|^{(m-k)/m} \|u\|_{k,p,\Omega} \le C_p' \bigg[\|(L(x, D+\eta) - \lambda)u\|_{0,p,\Omega}$$

$$+ \sum_{j=1}^{m/2} |\lambda|^{(m-m_j)/m} \|h_j\|_{0,p,\Omega} + \sum_{j=1}^{m/2} \|h_j\|_{m-m_j,p,\Omega} \bigg]. \qquad (5.95)$$

Especially for $u \in D(A_p^\eta)$

$$|\lambda| \|u\|_{0,p,\Omega} + \|u\|_{m,p,\Omega} \le C_p' \|(A_p^\eta - \lambda)u\|_{0,p,\Omega}. \qquad (5.96)$$

Analogously, for $u \in D(A_{p'}'^{\bar{\eta}})$

$$|\lambda| \|u\|_{0,p,\Omega} + \|u\|_{m,p,\Omega} \le C_p' \|(A_{p'}'^{\bar{\eta}} - \bar{\lambda})u\|_{0,p,\Omega}. \qquad (5.97)$$

Proof. Set

$$g_j = (B_j(x, D) - B_j(x, D+\eta))u + h_j.$$

Then $g_j \in W^{m-m_j, p}(\Omega)$ and $g_j = B_j(x, D)u$ on $\partial\Omega$. We have

$$\|(L(x, D) - \lambda)u\|_{0,p,\Omega}$$
$$\le \|(L(x, D+\eta) - \lambda)u\|_{0,p,\Omega} + \|(L(x, D+\eta) - L(x, D))u\|_{0,p,\Omega}$$
$$\le \|(L(x, D+\eta) - \lambda)u\|_{0,p,\Omega} + C \sum_{k=0}^{m-1} |\eta|^{m-k} \|u\|_{k,p,\Omega},$$

$$\|g_j\|_{0,p,\Omega} \le C \sum_{k=0}^{m_j-1} |\eta|^{m_j-k} \|u\|_{k,p,\Omega} + \|h_j\|_{0,p,\Omega},$$

$$\|g_j\|_{m-m_j,p,\Omega} \le C \sum_{k=m-m_j}^{m-1} |\eta|^{m-k} \|u\|_{k,p,\Omega} + \|h_j\|_{m-m_j,p,\Omega}.$$

Hence with the aid of (5.49)

$$\sum_{k=0}^{m} |\lambda|^{(m-k)/m} \|u\|_{k,p,\Omega} \le C_p' \Big[\|(L(x, D + \eta) - \lambda)u\|_{0,p,\Omega}$$

$$+ \sum_{k=0}^{m-1} |\eta|^{m-k} \|u\|_{k,p,\Omega} + \sum_{j=1}^{m/2} |\lambda|^{(m-m_j)/m} \sum_{k=0}^{m_j-1} |\eta|^{m_j-k} \|u\|_{k,p,\Omega}$$

$$+ \sum_{j=1}^{m/2} |\lambda|^{(m-m_j)/m} \|h_j\|_{0,p,\Omega} + \sum_{j=1}^{m/2} \|h_j\|_{m-m_j,p,\Omega} \Big].$$

If $0 < \delta \le 1, |\eta| \le \delta |\lambda|^{1/m}$, then

$$\sum_{k=0}^{m} |\lambda|^{(m-k)/m} \|u\|_{k,p,\Omega}$$

$$\le C_p' \Big[\|(L(x, D + \eta) - \lambda)u\|_{0,p,\Omega} + \delta \sum_{k=0}^{m-1} |\lambda|^{(m-k)/m} \|u\|_{k,p,\Omega}$$

$$+ \sum_{j=1}^{m/2} |\lambda|^{(m-m_j)/m} \delta \sum_{k=0}^{m_j-1} |\lambda|^{(m_j-k)/m} \|u\|_{k,p,\Omega}$$

$$+ \sum_{j=1}^{m/2} |\lambda|^{(m-m_j)/m} \|h_j\|_{0,p,\Omega} + \sum_{j=1}^{m/2} \|h_j\|_{m-m_j,p,\Omega} \Big]$$

$$\le C_p' \Big[\|(L(x, D + \eta) - \lambda)u\|_{0,p,\Omega} + \delta \sum_{k=0}^{m-1} |\lambda|^{(m-k)/m} \|u\|_{k,p,\Omega}$$

$$+ \sum_{j=1}^{m/2} |\lambda|^{(m-m_j)/m} \|h_j\|_{0,p,\Omega} + \sum_{j=1}^{m/2} \|h_j\|_{m-m_j,p,\Omega} \Big].$$

If δ is sufficiently small, the inequality (5.95) follows from this inequality.

Similarly to (5.68) we have

$$(A_p^\eta)^* = A'^{\bar\eta}_{p'}. \tag{5.98}$$

Hence with the aid of Theorem 1.3 we obtain the following lemma.

Lemma 5.12 *Let $1 < p < \infty$ and δ be as in Lemma 5.11. Then if $\arg \lambda \notin (-\theta_0, \theta_0), |\lambda| > C_p, |\eta| \le \delta |\lambda|^{1/m}$, then $\lambda \in \rho(A_p^\eta), \bar\lambda \in \rho(A'^{\bar\eta}_{p'})$, and for any $f \in L^p(\Omega)$ or $f \in L^{p'}(\Omega)$*

$$\|(A_p^\eta - \lambda)^{-1} f\|_{0,p,\Omega} \le C' \|f\|_{0,p,\Omega} / |\lambda|, \tag{5.99}$$

$$\|(A_p^\eta - \lambda)^{-1}f\|_{m,p,\Omega} \le C_p'\|f\|_{0,p,\Omega}, \tag{5.100}$$

$$\|(A_{p'}'^{\bar\eta} - \bar\lambda)^{-1}f\|_{0,p',\Omega} \le C_p'\|f\|_{0,p',\Omega}/|\lambda|, \tag{5.101}$$

$$\|(A_{p'}'^{\bar\eta} - \bar\lambda)^{-1}f\|_{m,p',\Omega} \le C_p'\|f\|_{0,p',\Omega}. \tag{5.102}$$

If η is pure imaginary, then

$$(A_p^\eta - \lambda)^{-1}f = e^{-x\eta}(A_p - \lambda)^{-1}(e^{x\eta}f). \tag{5.103}$$

Proof. That $\lambda \in \rho(A_p^\eta), \bar\lambda \in \rho(A_{p'}'^{\bar\eta})$ and (5.99)-(5.102) follow from (5.96)-(5.98) and Theorem 1.3. Noting that if η is pure imaginary, $u \in W^{m,p}(\Omega)$ if and only if $e^{x\eta}u \in W^{m,p}(\Omega)$, we can easily show (5.103) with the aid of (5.92),(5.93).

Remark 5. 4 If Ω is bounded, (5.103) holds for an arbitrary complex vector η.

Lemma 5. 13 *Let $L(x,D), \{B_j(x,D)\}_{j=1}^{m/2}$ be operators satisfying the conditions of section 4.2. Suppose in addition that $\{B_j\}_{j=1}^{m/2}$ is normal and for any $u \in W^{m,p}(\Omega)$ the following inequality holds:*

$$\|u\|_{m,p,\Omega} \le C\left[\|L(x,D)u\|_{0,p,\Omega} + \sum_{j=1}^{m/2}[B_j(x,D)u]_{m-m_j-1/p,p,\partial\Omega}\right]. \tag{5.104}$$

Moreover suppose that $0 \in \rho(A_p)$ for the operator A_p defined by (5.34). Then for any $f \in L^p(\Omega), g_j \in W^{m-m_j-1/p,p}(\partial\Omega), j = 1, \ldots, m/2$, the solution of the boundary value problem

$$L(x,D)u = f \quad in \quad \Omega, \tag{5.105}$$

$$B_j(x,D)u = g_j, \; j = 1, \ldots, m/2, \; on \; \partial\Omega \tag{5.106}$$

exists and is unique.

Proof. In view of Theorem 3.27 there exits a sequence $\{g_{j,\nu}\} \subset C_0^m(\partial\Omega)$ such that $g_{j,\nu} \to g_j$ in $W^{m-m_j-1/p,p}(\partial\Omega)$. By virtue of Lemma 5.2 there exists a function $w_{j,\nu} \in C_0^m(\bar\Omega)$ satisfying $B_j(x,D)w_{j,\nu} = g_{j,\nu}$ on $\partial\Omega$. If we set

$$u_\nu = A_p^{-1}(f - Lw_{j,\nu}) + w_{j,\nu},$$

then $u_\nu \in W^{m,p}(\Omega)$, $Lu_\nu = f$ in Ω, $B_ju_\nu = g_{j,\nu}, j = 1, \ldots, m/2$ on $\partial\Omega$. Applying (5.104) to $u_\nu - u_\mu$ we have

$$\|u_\nu - u_\mu\|_{m,p,\Omega} \le C\sum_{j=1}^{m/2}[g_{j,\nu} - g_{j,\mu}]_{m-m_j-1/p,p,\partial\Omega}.$$

Hence $\{u_\nu - u_\mu\}$ is a Cauchy sequence in $W^{m,p}(\Omega)$. The limit is obviously the unique solution of (5.105),(5.106).

Lemma 5.14 *Let $1 < p < \infty$, λ, η be as in Lemma 5.12. Then, for any $f \in L^p(\Omega), g_j \in W^{m-m_j-1/p,p}(\partial\Omega), j = 1, \ldots, m/2$, the boundary value problem*

$$(L(x, D + \eta) - \lambda)u = f \quad \text{in} \quad \Omega, \tag{5.107}$$

$$B_j(x, D + \eta)u = g_j, \ j = 1, \ldots, m/2, \ \text{on} \ \partial\Omega. \tag{5.108}$$

has a unique solution.

Proof. In view of Lemma 5.11 there exists a constant $C_{p,\lambda}$ depending on p, λ such that for $u \in W^{m,p}(\Omega), h_j \in W^{m-m_j,p}(\Omega)$ satisfying $B_j(x, D + \eta) = h_j$ on $\partial\Omega$

$$\|u\|_{m,p,\Omega} \le C_{p,\lambda} \left[\|(L(x, D + \eta) - \lambda)u\|_{0,p,\Omega} + \sum_{j=1}^{m/2} \|h_j\|_{m-m_j,p,\Omega} \right].$$

Replacing the right hand side by the greatest lower bound with respect to h_j yields

$$\|u\|_{m,p,\Omega} \le C_{p,\lambda} \Bigg[\|(L(x, D + \eta) - \lambda)u\|_{0,p,\Omega}$$

$$+ \sum_{j=1}^{m/2} [B_j(x, D + \eta)u]_{m-m_j-1/p,p,\partial\Omega} \Bigg].$$

Hence using Lemma 5.13 we conclude that the solution of (5.107),(5.108) exists and is unique.

Lemma 5.15 *Under the assumptions of Lemma 5.11 $(A_p^\eta - \lambda)^{-1}$ is a holomorphic function of η in $|\eta| \le \delta|\lambda|^{1/m}$.*

Proof. For $f \in L^p(\Omega)$ set $u_\eta = (A_p^\eta - \lambda)^{-1}f$. Then

$$(L(x, D + \eta) - \lambda)u_\eta = f \quad \text{in} \quad \Omega, \tag{5.109}$$

$$B_j(x, D + \eta)u_\eta = 0, \ j = 1, \ldots, m/2, \ \text{on} \ \partial\Omega. \tag{5.110}$$

If we formally differentiate both sides of (5.109),(5.101) with respect to η_1 and set

$$g(x) = -(\partial L/\partial \xi_1)(x, D + \eta)u_\eta(x),$$
$$g_j(x) = -(\partial B_j/\partial \xi_1)(x, D + \eta)u_\eta(x),$$

then we obtain

$$(L(x, D + \eta) - \lambda)\partial u_\eta(x)/\partial \eta_1 = g(x) \quad \text{in} \quad \Omega,$$
$$B_j(x, D + \eta)\partial u_\eta(x)/\partial \eta_1 = g_j(x), \; j = 1, \ldots, m/2, \text{ on } \partial\Omega.$$

Evidently $g \in W^{1,p}(\Omega) \subset L^p(\Omega), g_j \in W^{m-m_j+1,p}(\Omega) \subset W^{m-m_j,p}(\Omega)$.
Hence in view of Lemma 5.14 the solution of the boundary value problem

$$(L(x, D + \eta) - \lambda)v = g \quad \text{in} \quad \Omega,$$
$$B_j(x, D + \eta)v = g_j, \; j = 1, \ldots, m/2, \text{ on } \partial\Omega,$$

exists. If we set $u_{\eta,h} = (u_\zeta - u_\eta)/h$, where $\zeta = (\eta_1 + h, \eta_2, \ldots, \eta_n)$, then

$$(L(x, D + \eta) - \lambda)(u_{\eta,h} - v) = -h^{-1}(L(x, D + \zeta) - L(x, D + \eta))u_\zeta - g$$

in Ω, and

$$B_j(x, D + \eta)(u_{\eta,h} - v) = -h^{-1}(B_j(x, D + \zeta) - B_j(x, D + \eta))u_\zeta - g_j$$

on $\partial\Omega$. Thus applying (5.95) to $u_{\eta,h} - v$ and letting $h \to 0$ we obtain $\|u_{\eta,h} - v\|_{m,p,\Omega} \to 0$.

We choose natural numbers l, s and exponents $2 = q_1 < \cdots < q_s < q_{s+1} = \infty, 2 = r_1 < \cdots < r_{l-s} < r_{l-s+1} = \infty$ as follows:
(i) in case $m > n/2, l = 2, s = 1$, i.e. $2 = q_1 < q_2 = \infty, \; 2 = r_1 < r_2 = \infty$;
(ii) in case $m < n/2, s > n/2m, l - s > n/2m$,

$$\frac{1}{q_j} - \frac{1}{q_{j+1}} < \frac{m}{n}, \; j = 1, \ldots, s - 1, \qquad \frac{1}{q_{s-1}} > \frac{m}{n} > \frac{1}{q_s}, \tag{5.111}$$

$$\frac{1}{r_j} - \frac{1}{r_{j+1}} < \frac{m}{n}, \; j = 1, \ldots, l - s - 1, \qquad \frac{1}{r_{l-s-1}} > \frac{m}{n} > \frac{1}{r_{l-s}}, \tag{5.112}$$

and $m - n/q_s, m - n/r_{l-s}$ are not integers;
(iii) in case $m = n/2, l = 4, s = 2, 2 = q_1 < q_2 < q_3 = \infty, 2 = r_1 < r_2 < r_3 = \infty$.
Set

$$a_j = (n/m)(1/q_j - 1/q_{j+1}), \quad j = 1, \ldots, s.$$

Then $0 < a_j < 1$ and in view of Theorems 3.19, 3.20 we have for $j = 1, \ldots, s - 1$

$$\|u\|_{0,q_{j+1},\Omega} \le C\|u\|_{m,q_j,\Omega}^{a_j}\|u\|_{0,q_j,\Omega}^{1-a_j}, \tag{5.113}$$

and for $0 \le h \le m - n/q_s$

$$\|u\|_{h,\infty,\Omega} \le C\|u\|_{m,q_s,\Omega}^{a_s+h/m}\|u\|_{0,q_s,\Omega}^{1-a_s-h/m}. \tag{5.114}$$

Replacing L by $L+\lambda_0$ for some sufficiently large positive number λ_0 if necessary we assume that (5.49) holds for $p = q_1, \ldots, q_s, r_1, \ldots, r_{l-s}$ and for any λ satisfying $\arg \lambda \notin (-\theta_0, \theta_0)$. Set $C_0 = \max\{C_p; p = q_1, \ldots, q_s, r_1, \ldots, r_{l-s}\}$, and denote by δ the minimum value of δ of Lemma 5.11 for $p = q_1, \ldots, q_s$, $r_1, \ldots, r_{l-s}$. Let $\lambda_j, j = 1, \ldots, l$, be complex numbers such that $\arg \lambda_j \notin (-\theta_0, \theta_0)$, and η a complex vector such that $|\eta| \leq \delta \min\{|\lambda_1|^{1/m}, \ldots, |\lambda_l|^{1/m}\}$. Set

$$S = (A_2^\eta - \lambda_s)^{-1}(A_2^\eta - \lambda_{s-1})^{-1} \cdots (A_2^\eta - \lambda_2)^{-1}(A_2^\eta - \lambda_1)^{-1}, \quad (5.115)$$
$$T = (A_2^\eta - \lambda_{s+1})^{-1}(A_2^\eta - \lambda_{s+2})^{-1} \cdots (A_2^\eta - \lambda_{l-1})^{-1}(A_2^\eta - \lambda_l)^{-1}. \quad (5.116)$$

By (5.111) and Theorem 3.17

$$R((A_2^\eta - \lambda_1)^{-1}) \subset W^{m,2}(\Omega) = W^{m,q_1}(\Omega) \subset L^{q_2}(\Omega).$$

Hence in view of Lemma 5.9 $(A_2^\eta - \lambda_2)^{-1}$ on the right hand side of (5.115) may be replaced by $(A_{q_2}^\eta - \lambda_2)^{-1}$. Repeating this yields

$$S = (A_{q_s}^\eta - \lambda_s)^{-1}(A_{q_{s-1}}^\eta - \lambda_{s-1})^{-1} \cdots (A_{q_2}^\eta - \lambda_2)^{-1}(A_2^\eta - \lambda_1)^{-1}. \quad (5.117)$$

In view of Theorem 3.20 and (5.111)

$$R((A_{q_s}^\eta - \lambda_s)^{-1}) \subset W^{m,q_s}(\Omega) \subset B^{m-n/q_s}(\bar{\Omega}).$$

Therefore

$$R(S) \subset B^{m-n/q_s}(\bar{\Omega}). \quad (5.118)$$

By virtue of (5.98) we get

$$T^* = (A'^{\bar{\eta}}_2 - \bar{\lambda}_l)^{-1}(A'^{\bar{\eta}}_2 - \bar{\lambda}_{l-1})^{-1} \cdots (A'^{\bar{\eta}}_2 - \bar{\lambda}_{s+2})^{-1}(A'^{\bar{\eta}}_2 - \bar{\lambda}_{s+1})^{-1}.$$

Analogously to (5.117),(5.118) we obtain

$$T^* = (A'^{\bar{\eta}}_{r_{l-s}} - \bar{\lambda}_l)^{-1} \cdots (A'^{\bar{\eta}}_{r_2} - \bar{\lambda}_{s+2})^{-1}(A'^{\bar{\eta}}_2 - \bar{\lambda}_{s+1})^{-1}, \quad (5.119)$$

$$R(T^*) \subset B^{m-n/r_{l-s}}(\bar{\Omega}). \quad (5.120)$$

In view of Lemma 5.12

$$\|(A_{q_j}^\eta - \lambda_j)^{-1}f\|_{0,q_j,\Omega} \leq C\|f\|_{0,q_j,\Omega}/|\lambda_j|, \quad (5.121)$$

$$\|(A_{q_j}^\eta - \lambda_j)^{-1}f\|_{m,q_j,\Omega} \leq C\|f\|_{0,q_j,\Omega} \quad (5.122)$$

for $j = 1, \ldots, s$. By virtue of (5.113),(5.121),(5.122) we get

$$\|(A_{q_j}^\eta - \lambda_j)^{-1}f\|_{0,q_{j+1},\Omega}$$
$$\leq C\|(A_{q_j}^\eta - \lambda_j)^{-1}f\|_{m,q_j,\Omega}^{a_j}\|(A_{q_j}^\eta - \lambda_j)^{-1}f\|_{0,q_j,\Omega}^{1-a_j}$$
$$\leq C\|f\|_{0,q_j,\Omega}^{a_j}\left(\|f\|_{0,q_j,\Omega}/|\lambda_j|\right)^{1-a_j} = C|\lambda_j|^{a_j-1}\|f\|_{0,q_j,\Omega}$$

for $j = 1, \ldots, s - 1$. Hence

$$\|Sf\|_{0,q_s,\Omega} \leq C|\lambda_s|^{-1} \prod_{j=1}^{s-1} |\lambda_j|^{a_j-1} \|f\|_{0,2,\Omega}, \tag{5.123}$$

$$\|Sf\|_{m,q_s,\Omega} \leq C \prod_{j=1}^{s-1} |\lambda_j|^{a_j-1} \|f\|_{0,2,\Omega}. \tag{5.124}$$

It follows from (5.114),(5.123),(5.124) that for $0 \leq h \leq m - n/q_s$

$$\|Sf\|_{h,\infty,\Omega} \leq C\|Sf\|_{m,q_s,\Omega}^{a_s+h/m} \|Sf\|_{0,q_s,\Omega}^{1-a_s-h/m}$$

$$\leq C|\lambda_s|^{h/m} \prod_{j=1}^{s} |\lambda_j|^{a_j-1} \|f\|_{0,2,\Omega}. \tag{5.125}$$

Analogously for $0 \leq h' \leq m - n/r_{l-s}$

$$\|T^* f\|_{h',\infty,\Omega} \leq C|\lambda_l|^{h'/m} \prod_{j=s+1}^{l} |\lambda_j|^{a_j-1} \|f\|_{0,2,\Omega}, \tag{5.126}$$

where $a_j = (n/m)(1/r_{j-s} - 1/r_{j-s+1})$ for $j = s + 1, \ldots, l$. Therefore for $0 < h < \min\{m - n/q_s, 1\}, 0 < h' < \min\{m - n/r_{l-s}, 1\}$

$$|(Sg)(x) - (Sg)(x')| \leq C|\lambda_s|^{h/m} \prod_{j=1}^{s} |\lambda_j|^{a_j-1} \|g\|_{0,2,\Omega} |x - x'|^h, \tag{5.127}$$

$$|(T^*h)(y) - (T^*h)(y')| \leq C|\lambda_l|^{h'/m} \prod_{j=s+1}^{l} |\lambda_j|^{a_j-1} \|h\|_{0,2,\Omega} |y - y'|^{h'}. \tag{5.128}$$

Owing to (5.118),(5.120),(5.127),(5.128) we can apply Lemma 5.10 to the operator

$$ST = (A_2^\eta - \lambda_1)^{-1}(A_2^\eta - \lambda_2)^{-1} \cdots (A_2^\eta - \lambda_l)^{-1}$$

to conclude that ST is an integral operator whose kernel we denote by $K_{\lambda_1,\ldots,\lambda_l}^\eta(x,y)$. In view of (5.73) and (5.125),(5.126) with $h = 0, h' = 0$ respectively

$$|K_{\lambda_1,\ldots,\lambda_l}^\eta(x,y)| \leq C \prod_{j=1}^{l} |\lambda_j|^{a_j-1}. \tag{5.129}$$

If η is pure imaginary, we have by Lemma 5.12

$$STf = e^{-x\eta}(A_2 - \lambda_1)^{-1}(A_2 - \lambda_2)^{-1} \cdots (A_2 - \lambda_l)^{-1}(e^{x\eta}f).$$

Hence if we denote the kernel of $(A_2 - \lambda_1)^{-1} \cdots (A_2 - \lambda_l)^{-1}$ by $K_{\lambda_1,\ldots,\lambda_l}(x, y)$, then we have

$$K_{\lambda_1,\ldots,\lambda_l}(x, y) = e^{(x-y)\eta} K^\eta_{\lambda_1,\ldots,\lambda_l}(x, y) \qquad (5.130)$$

for a pure imaginary η. By Lemma 5.15 ST is a holomorphic function of η in $|\eta| \leq \delta \min\{|\lambda_1|^{1/m}, \ldots, |\lambda_l|^{1/m}\}$, and hence so is $K^\eta_{\lambda_1,\ldots,\lambda_l}(x, y)$ as is easily seen using Morera's theorem. Therefore (5.130) holds also if η is not pure imaginary. In what follows we consider only real η. Owing to Remark 5.4 if Ω is bounded, we need not consider nonreal η. It follows from (5.129),(5.130) that

$$|K_{\lambda_1,\ldots,\lambda_l}(x, y)| \leq C e^{(x-y)\eta} \prod_{j=1}^l |\lambda_j|^{a_j - 1}.$$

Minimizing the right hand side of this inequality with respect to η we obtain

$$|K_{\lambda_1,\ldots,\lambda_l}(x, y)|$$

$$\leq C \exp\left[-\delta \min\left(|\lambda_1|^{1/m}, \ldots, |\lambda_l|^{1/m}\right) |x - y|\right] \prod_{j=1}^l |\lambda_j|^{a_j - 1}$$

$$\leq C \sum_{k=1}^l \exp\left(-\delta |\lambda_k|^{1/m} |x - y|\right) \prod_{j=1}^l |\lambda_j|^{a_j - 1}. \qquad (5.131)$$

The following lemma is easily shown.

Lemma 5.16 *For any positive number a there exists a constant C such that $t^a \leq C e^t$ for any $t \geq 0$.*

Let Γ be a smooth contour running from $\infty e^{-i\theta_0}$ to $\infty e^{i\theta_0}$ in the closed angular domain $\Sigma = \{\lambda; \arg \lambda \notin (-\theta_0, \theta_0)\}$. Then

$$\exp(-tA_2) = \frac{1}{2\pi i} \int_\Gamma e^{-\lambda t} (A_2 - \lambda)^{-1} d\lambda. \qquad (5.132)$$

Hence,

$$\exp(-lt A_2) = [\exp(-tA_2)]^l$$

$$= \frac{1}{(2\pi i)^l} \int_\Gamma \cdots \int_\Gamma e^{-\lambda_1 t - \cdots - \lambda_l t} (A_2 - \lambda_1)^{-1} \cdots (A_2 - \lambda_l)^{-1} d\lambda_1 \cdots d\lambda_l.$$

Consequently, $\exp(-tA_2)$ has a kernel $G(x, y, t)$, and

$$G(x, y, lt) = \frac{1}{(2\pi i)^l} \int_\Gamma \cdots \int_\Gamma e^{-\lambda_1 t - \cdots - \lambda_l t} K_{\lambda_1,\ldots,\lambda_l}(x, y) d\lambda_1 \cdots d\lambda_l \qquad (5.133)$$

holds for $x, y \in \Omega, |\arg t| \leq \pi/2 - \theta_0$. For fixed x, y, t we deform the integral contour Γ to the following path $\Gamma_{x,y,t}$:

$$\Gamma_{x,y,t} = \{\lambda; |\arg \lambda| = \theta_0, |\lambda| \geq a\} \cup \{\lambda; \lambda = ae^{i\theta}, |\theta| \geq \theta_0\},$$

where

$$a = \epsilon \frac{|x-y|^{m/(m-1)}}{|t|^{m/(m-1)}} = \frac{\epsilon\rho}{|t|}, \qquad \rho = \frac{|x-y|^{m/(m-1)}}{|t|^{1/(m-1)}}, \qquad \epsilon > 0. \qquad (5.134)$$

If we put $\lambda = re^{\pm i\theta_0}$ for $|\arg \lambda| = \theta_0$, we get

$$\mathrm{Re}\lambda t = \mathrm{Re}\lambda \cdot \mathrm{Re}\,t - \mathrm{Im}\lambda \cdot \mathrm{Im}\,t$$
$$= r\mathrm{Re}\,t \left(\cos\theta_0 \mp \sin\theta_0 \frac{\mathrm{Im}\,t}{\mathrm{Re}\,t}\right) \geq r\mathrm{Re}\,t \left(\cos\theta_0 - \sin\theta_0 \frac{|\mathrm{Im}\,t|}{\mathrm{Re}\,t}\right).$$

Hence if t is in the closed sector

$$\frac{|\mathrm{Im}\,t|}{\mathrm{Re}\,t} \leq (1 - \epsilon_0)\frac{\cos\theta_0}{\sin\theta_0}, \qquad (5.135)$$

where $0 < \epsilon_0 < 1$, then $\mathrm{Re}\lambda t \geq r\mathrm{Re}\,t \cdot \epsilon_0 \cos\theta_0$. Therefore there exists a positive constant c such that for t in the sector (5.135) and $\lambda = re^{\pm i\theta_0}$

$$\mathrm{Re}\lambda t \geq cr|t|. \qquad (5.136)$$

From (5.131),(5.133) it follows that

$$|G(x, y, lt)| \leq C \int_{\Gamma_{x,y,t}} \cdots \int_{\Gamma_{x,y,t}} e^{-\mathrm{Re}\lambda_1 t - \cdots - \mathrm{Re}\lambda_l t}$$
$$\times \sum_{k=1}^{l} \exp\left(-\delta|\lambda_k|^{1/m}|x-y|\right) \prod_{j=1}^{l} |\lambda_j|^{a_j - 1}|d\lambda_1 \cdots d\lambda_l|. \qquad (5.137)$$

The term with $k = 1$ in the right hand side of (5.137) is

$$C \int_{\Gamma_{x,y,t}} \exp\left(-\mathrm{Re}\lambda_1 t - \delta|\lambda_1|^{1/m}|x-y|\right) |\lambda_1|^{a_1 - 1}|d\lambda_1|$$
$$\times \prod_{j=2}^{l} \int_{\Gamma_{x,y,t}} e^{-\mathrm{Re}\lambda_j t}|\lambda_j|^{a_j - 1}|d\lambda_j|. \qquad (5.138)$$

Using (5.136) we get for $j = 2, \ldots, l$

$$\int_{\Gamma_{x,y,t}} e^{-\mathrm{Re}\lambda_j t}|\lambda_j|^{a_j - 1}|d\lambda_j|$$
$$= \int_{|\lambda_j|=a} + \int_{|\lambda_j|\geq a} \leq e^{a|t|}a^{a_j - 1}2\pi a + 2\int_a^\infty e^{-cr|t|}r^{a_j - 1}dr.$$

Using the definition (5.134) of a, ρ and with the change of the variable $s = r|t|$ in the last integral we get with the aid of Lemma 5.16

$$\int_{\Gamma_{x,y,t}} e^{-\mathrm{Re}\lambda_j t} |\lambda_j|^{a_j-1} |d\lambda_j|$$

$$\leq 2\pi \left(\frac{\epsilon\rho}{|t|}\right)^{a_j} e^{\epsilon\rho} + \frac{2}{|t|^{a_j}} \int_0^\infty e^{-cs} s^{a_j-1} ds \leq C\frac{e^{2\epsilon\rho}}{|t|^{a_j}}. \qquad (5.139)$$

Next if we put

$$\int_{\Gamma_{x,y,t}} \exp\left(-\mathrm{Re}\lambda_1 t - \delta|\lambda_1|^{1/m}|x-y|\right) |\lambda_1|^{a_1-1} |d\lambda_1|$$

$$= \int_{|\lambda_1|=a} + \int_{|\lambda_1|>a} = I_1 + I_2, \qquad (5.140)$$

then

$$I_1 \leq 2\pi a^{a_1} \exp(a|t| - \delta a^{1/m}|x-y|)$$
$$= 2\pi \left(\frac{\epsilon\rho}{|t|}\right)^{a_1} \exp(\epsilon\rho - \delta\epsilon^{1/m}\rho) \leq \frac{C}{|t|^{a_1}} \exp(2\epsilon\rho - \delta\epsilon^{1/m}\rho), \qquad (5.141)$$

$$I_2 \leq 2 \int_a^\infty \exp(-cr|t| - \delta r^{1/m}|x-y|) r^{a_1-1} dr \qquad (5.142)$$

Making the change of the variable $r = as$ in the right hand side of (5.142) and substituting (5.134)

$$I_2 \leq 2a^{a_1} \int_1^\infty s^{a_1-1} \exp(-c\epsilon\rho s - \delta\epsilon^{1/m}\rho s^{1/m}) ds$$

$$\leq 2a^{a_1} \exp(-\delta\rho\epsilon^{1/m}) \int_1^\infty s^{a_1-1} e^{-c\epsilon\rho s} ds$$

$$\leq 2 \left(\frac{\epsilon\rho}{|t|}\right)^{a_1} \exp(-\delta\rho\epsilon^{1/m}) \frac{1}{(\epsilon\rho)^{a_1}} \int_0^\infty \xi^{a_1-1} e^{-c\xi} d\xi$$

$$= C|t|^{-a_1} \exp(-\delta\rho\epsilon^{1/m}). \qquad (5.143)$$

From (5.140),(5.141),(5.143) it follows that

$$\int_{\Gamma_{x,y,t}} \exp\left(-\mathrm{Re}\lambda_1 t - \delta|\lambda_1|^{1/m}|x-y|\right) |\lambda_1|^{a_1-1} |d\lambda_1|$$

$$\leq C|t|^{-a_1} \exp(2\epsilon\rho - \delta\epsilon^{1/m}\rho). \qquad (5.144)$$

In view of (5.139),(5.144) and $\sum_{j=1}^l a_j = n/m$ (5.138) does not exceed

$$C|t|^{-n/m} \exp\left[(2l\epsilon - \delta\epsilon^{1/m})\rho\right].$$

The terms with $k = 2, \ldots, l$ in the right hand side of (5.138) are estimated similarly. Hence, choosing ϵ so small that $2l\epsilon - \delta\epsilon^{1/m} < 0$ we find that there exist positive constant C, c such that in the sector (5.135) the following inequality holds:

$$|G(x, y, t)| \leq \frac{C}{|t|^{n/m}} \exp\left(-c\frac{|x - y|^{m/(m-1)}}{|t|^{1/(m-1)}}\right). \qquad (5.145)$$

By an analogous argument we can show that for $|\alpha| < m$

$$|D_x^\alpha G(x, y, t)| \leq \frac{C}{|t|^{(n+|\alpha|)/m}} \exp\left(-c\frac{|x - y|^{m/(m-1)}}{|t|^{1/(m-1)}}\right). \qquad (5.146)$$

The outline of the proof of (5.146) is as follows. If we choose $q_1, \ldots, q_s$ so that $q_s > n$ in addition to (i),(ii),(iii), then in view of Theorem 3.20 and (5.114) we have

$$R(S) \subset B^{m-n/q_s}(\bar{\Omega}) \subset B^{m-1}(\bar{\Omega}),$$

and

$$\|u\|_{k,\infty,\Omega} \leq C\|u\|_{m,q_s,\Omega}^{a_s+k/m}\|u\|_{0,q_s,\Omega}^{1-a_s-k/m}, \quad k = 0, \ldots, m - 1. \qquad (5.147)$$

Hence by (5.114),(5.123),(5.124) we have for $|\alpha| < m$

$$\|D_x^\alpha Sf\|_{0,\infty,\Omega} \leq \|Sf\|_{|\alpha|,\infty,\Omega} \leq C|\lambda_s|^{|\alpha|/m}\prod_{j=1}^{s}|\lambda_j|^{a_j-1}\|f\|_{0,2,\Omega}. \qquad (5.148)$$

Since $D_x^\alpha K_{\lambda_1,\ldots,\lambda_l}^\eta(x, y)$ is the kernel of $D_x^\alpha ST$ we get from (5.126) with $h' = 0$, (5.148) and Lemma 5.10

$$|D_x^\alpha K_{\lambda_1,\ldots,\lambda_l}^\eta(x, y)| \leq C|\lambda_s|^{|\alpha|/m}\prod_{j=1}^{l}|\lambda_j|^{a_j-1}.$$

Therefore

$$|D_x^\alpha K_{\lambda_1,\ldots,\lambda_l}(x, y)| = \left|D_x^\alpha\left[e^{(x-y)\eta}K_{\lambda_1,\ldots,\lambda_l}^\eta(x, y)\right]\right|$$

$$= \left|\sum_{\beta \leq \alpha}\binom{\alpha}{\beta}D_x^{\alpha-\beta}e^{(x-y)\eta} \cdot D_x^\beta K_{\lambda_1,\ldots,\lambda_l}^\eta(x, y)\right|$$

$$\leq C\sum_{\beta \leq \alpha}|\eta|^{|\alpha-\beta|}e^{(x-y)\eta}|\lambda_s|^{|\beta|/m}\prod_{j=1}^{l}|\lambda_j|^{a_j-1}$$

$$\leq C\sum_{\beta \leq \alpha}|\lambda_s|^{|\alpha|/m}e^{(x-y)\eta}\prod_{j=1}^{l}|\lambda_j|^{a_j-1}. \qquad (5.149)$$

Following the argument by which we derived (5.145) we can show that (5.146) holds. Noting that $\bar{G}(y, x, t)$ is the kernel of $\exp(-tA_2')$ we can establish an analogous estimate for $D_y^\beta G(x, y, t), |\beta| < m$:

$$|D_y^\beta G(x, y, t)| \le \frac{C}{|t|^{(n+|\beta|)/m}} \exp\left(-c\frac{|x-y|^{m/(m-1)}}{|t|^{1/(m-1)}}\right). \qquad (5.150)$$

Since $\exp(-2tA_2) = \{\exp(-tA_2)\}^2$, we have

$$G(x, y, 2t) = \int_\Omega G(x, z, t)G(z, y, t)dz. \qquad (5.151)$$

Using

$$|x-z|^{m/(m-1)} + |y-z|^{m/(m-1)} \ge 2^{-1/(m-1)}|x-y|^{m/(m-1)},$$
$$|x-z|^{m/(m-1)} + |y-z|^{m/(m-1)} \ge |x-z|^{m/(m-1)}$$

we obtain

$$\int_\Omega \exp\left(-c\frac{|x-z|^{m/(m-1)}}{|t|^{1/(m-1)}}\right) \exp\left(-c\frac{|y-z|^{m/(m-1)}}{|t|^{1/(m-1)}}\right) dz$$

$$= \int_\Omega \left\{\exp\left[-\frac{c}{2|t|^{1/(m-1)}}\left(|x-z|^{m/(m-1)} + |y-z|^{m/(m-1)}\right)\right]\right\}^2 dz$$

$$\le \exp\left(-\frac{c}{2^{m/(m-1)}}\frac{|x-y|^{m/(m-1)}}{|t|^{1/(m-1)}}\right) \int_{R^n} \exp\left(-\frac{c}{2}\frac{|x-z|^{m/(m-1)}}{|t|^{1/(m-1)}}\right) dz$$

$$= |t|^{n/m} \exp\left(-\frac{c}{2^{m/(m-1)}}\frac{|x-y|^{m/(m-1)}}{|t|^{1/(m-1)}}\right) \int_{R^n} \exp\left(-\frac{c}{2}|\xi|^{m/(m-1)}\right) d\xi.$$
$$(5.152)$$

From (5.146),(5.150),(5.151),(5.152) it follows that for $|\alpha| < m, |\beta| < m$

$$|D_x^\alpha D_y^\beta G(x, y, t)| \le \frac{C}{|t|^{(n+|\alpha|+|\beta|)/m}} \exp\left(-c\frac{|x-y|^{m/(m-1)}}{|t|^{1/(m-1)}}\right) \qquad (5.153)$$

with another positive constant c. Since

$$\frac{d}{dt}\exp(-tA_2) = -\frac{1}{2\pi i}\int_\Gamma \lambda e^{-\lambda t}(A_2 - \lambda)^{-1}d\lambda,$$

we can verify similarly

$$\left|\frac{\partial}{\partial t}G(x, y, t)\right| \le \frac{C}{|t|^{n/m+1}} \exp\left(-c\frac{|x-y|^{m/(m-1)}}{|t|^{1/(m-1)}}\right). \qquad (5.154)$$

Next we estimate the kernel of $(A_2 - \lambda)^{-1}$. Letting $0 < \theta_1 < \pi/2 - \theta_0, 0 < \epsilon_1 < 1$ we consider in the region

$$\text{Im}\lambda > 0, \quad \frac{|\text{Re}\lambda|}{\text{Im}\lambda} \leq (1 - \epsilon_1)\frac{\sin\theta_1}{\cos\theta_1}. \tag{5.155}$$

For $t = re^{i\theta_1}$

$$\text{Re}\lambda t = \text{Re}\lambda \cdot \text{Re}t - \text{Im}\lambda \cdot \text{Im}t$$
$$= \text{Re}\lambda \cdot r\cos\theta_1 - \text{Im}\lambda \cdot \sin\theta_1 \leq -\epsilon_1 r\sin\theta_1 \cdot \text{Im}\lambda.$$

Hence there exists a positive constant c such that

$$\text{Re}\lambda t \leq -cr|\lambda|. \tag{5.156}$$

Therefore if λ is in the region (5.155) we can integrate along the half line $t = re^{i\theta_1}, 0 < r < \infty$, in the following integral

$$(A_2 - \lambda)^{-1} = \int e^{-\lambda t}\exp(-tA_2)dt. \tag{5.157}$$

If we denote the kernel of $(A_2 - \lambda)^{-1}$ by $K_\lambda(x, y)$, then by (5.157)

$$K_\lambda(x, y) = \int_{t=re^{i\theta_1}} e^{\lambda t}G(x, y, t)dt. \tag{5.158}$$

In view of (5.145),(5.156)

$$|K_\lambda(x, y)| \leq \int_0^\infty e^{-cr|\lambda|}\frac{C}{r^{n/m}}\exp\left(-c\frac{|x - y|^{m/(m-1)}}{r^{1/(m-1)}}\right)dr. \tag{5.159}$$

In case $n > m$ changing the variable as $r = |x - y|^m s$ and putting $h = (|\lambda|^{1/m}|x - y|)^{1-m}$

$$\int_0^\infty r^{-n/m}e^{-cr|\lambda|}\exp(-c|x - y|^{m/(m-1)}r^{-1/(m-1)})dr$$

$$= |x - y|^{m-n}\int_0^\infty s^{-n/m}\exp(-cs^{-1/(m-1)} - c|\lambda||x - y|^m s)ds$$

$$\leq |x - y|^{m-n}\int_0^h s^{-n/m}\exp(-cs^{-1/(m-1)})ds$$

$$+|x - y|^{m-n}\int_h^\infty s^{-n/m}\exp(-ch^{-m/(m-1)}s)ds = I_1 + I_2.$$

If $|\lambda|^{1/m}|x - y| < 1$, using

$$\int_0^h s^{-n/m}\exp(-cs^{-1/(m-1)})ds \leq \int_0^\infty s^{-n/m}\exp(-cs^{-1/(m-1)})ds < \infty,$$

and $e^{-|\lambda|^{1/m}|x-y|} > e^{-1}$, we get

$$\int_0^h s^{-n/m} \exp(-cs^{-1/(m-1)})ds \le C \exp(-|\lambda|^{1/m}|x-y|).$$

If $|\lambda|^{1/m}|x-y| \ge 1$, noting $h \le 1$ and using Lemma 5.16 we have for some positive constant c'

$$\int_0^h s^{-n/m} \exp(-cs^{-1/(m-1)})ds$$

$$\le C \int_0^h \exp(-c's^{-1/(m-1)})ds \le Ch \exp(-c'h^{-1/(m-1)})$$

$$\le C \exp(-c'h^{-1/(m-1)}) = C \exp(-c'|\lambda|^{1/m}|x-y|).$$

Consequently for some constants C, c

$$I_1 \le C|x-y|^{m-n} \exp(-c|\lambda|^{1/m}|x-y|). \tag{5.160}$$

Again with the aid of Lemma 5.16

$$\int_h^\infty s^{-n/m} \exp(-ch^{-m/(m-1)}s)ds$$

$$\le \exp(-ch^{-1/(m-1)}) \int_h^\infty s^{-n/m}ds$$

$$= \frac{m}{n-m} h^{-(n-m)/m} \exp(-ch^{-1/(m-1)})$$

$$\le C \exp(-c''h^{-1/(m-1)}) = C \exp(-c''|\lambda|^{1/m}|x-y|). \tag{5.161}$$

Therefore we have shown that if $n > m$ the following inequality holds:

$$|K_\lambda(x,y)| \le C|x-y|^{m-n} \exp(-c|\lambda|^{1/m}|x-y|). \tag{5.162}$$

In case $n = m$ with the aid of the change of the variable $r = |x-y|^m s$

$$\int_0^\infty r^{-1} e^{-cr|\lambda|} \exp(-c|x-y|^{m/(m-1)}r^{-1/(m-1)})dr$$

$$= \int_0^\infty s^{-1} \exp(-cs^{-1/(m-1)}) \exp(-ch^{-m/(m-1)}s)ds. \tag{5.163}$$

If $|\lambda|^{1/m}|x-y| < 1$, put $b = (|\lambda||x-y|^m)^{-1} = h^{m/(m-1)}$. Then $b > 1$. The right hand side of (5.163) does not exceed

$$\int_0^b s^{-1} \exp(-cs^{-1/(m-1)})ds + \int_b^\infty s^{-1} \exp(-cb^{-1}s)ds$$

$$\leq \int_0^1 s^{-1} \exp(-cs^{-1/(m-1)})ds + \int_1^b s^{-1}ds + \int_1^\infty e^{-cs}\frac{ds}{s}$$

$$\leq C\left[1 + \log(|\lambda|^{1/m}|x-y|)^{-1}\right].$$

If $|\lambda|^{1/m}|x-y| \geq 1$, $h \leq 1$. The left hand side of (5.163) does not exceed

$$\int_0^h s^{-1}\exp(-cs^{-1/(m-1)})ds$$

$$+ \int_h^\infty s^{-1}\exp(-cs^{-1/(m-1)})\exp(-ch^{-m/(m-1)}s)ds$$

$$\leq \exp(-2^{-1}ch^{-1/(m-1)})\int_0^h s^{-1}\exp(-2^{-1}cs^{-1/(m-1)})ds$$

$$+ \exp(-2^{-1}ch^{-1/(m-1)})$$

$$\times \left[\int_h^1 s^{-1}\exp(-cs^{-1/(m-1)})ds + \int_1^\infty s^{-1}\exp(-2^{-1}cs)ds\right]$$

$$\leq C\exp(-2^{-1}c|\lambda|^{1/m}|x-y|).$$

Therefore if $n = m$ we have

$$|K_\lambda(x,y)| \leq C\exp(-c|\lambda|^{1/m}|x-y|)\left[1 + \log^+(|\lambda|^{1/m}|x-y|)^{-1}\right]. \quad (5.164)$$

Finally in case $n < m$ making the change of the variable $r = s/|\lambda|$ and setting $\tilde{h} = |\lambda|^{1/m}|x-y|$

$$\int_0^\infty r^{-n/m}e^{-cr|\lambda|}\exp(-c|x-y|^{m/(m-1)}r^{-1/(m-1)})dr$$

$$= |\lambda|^{n/m-1}\int_0^\infty s^{-n/m}e^{-cs}\exp(-c\tilde{h}^{m/(m-1)}s^{-1/(m-1)})ds$$

$$\leq |\lambda|^{n/m-1}\exp(-c\tilde{h})\int_0^{\tilde{h}} s^{-n/m}e^{-cs}ds$$

$$+ |\lambda|^{n/m-1}e^{-c\tilde{h}/2}\int_{\tilde{h}}^\infty s^{-n/m}e^{-cs/2}ds \leq C|\lambda|^{n/m-1}e^{-c\tilde{h}/2}.$$

Hence if $n < m$ we have

$$|K_\lambda(x,y)| \leq C|\lambda|^{n/m-1}\exp(-c|\lambda|^{1/m}|x-y|). \quad (5.165)$$

Analogously in the region $\text{Im}\lambda < 0, |\text{Re}\lambda|/|\text{Im}\lambda| \leq (1-\epsilon_1)\sin\theta_1/\cos\theta_1$ we deform the integral path in the integral (5.158) to the half line $t = re^{-i\theta_1}, 0 < r < \infty$, to obtain the same estimates as (5.162),(5.164),(5.165).

In the set $-\mathrm{Re}\lambda/|\mathrm{Im}\lambda| > (1-\epsilon_1)\sin\theta_1/\cos\theta_1$ we can establish the estimates of the same form integrating along the positive real axis in the integral (5.158). If we use (5.153) we can obtain the estimates of $D_x^\alpha D_y^\beta K_\lambda(x,y), |\alpha| < m, |\beta| < m$. It is clear that $G(x,y,t), K_\lambda(x,y)$ are also the kernels of $\exp(-tA_p), (A_p - \lambda)^{-1}$ respectively. Thus recalling that we replaced L by $L + \lambda_0$ we have proved the following theorem.

Theorem 5.7 *Suppose that the hypothesis of Theorem 5.6 are satisfied. Then the kernels $G(x,y,t), K_\lambda(x,y)$ of $\exp(-tA_p), (A_p - \lambda)^{-1}, 1 < p < \infty$, exist, and there exist positive constants C, c, a real number ω and an angle $\theta_0 \in (0, \pi/2)$ such that*
(i) for $|\alpha| < m, |\beta| < m, (x,y) \in \Omega \times \Omega, |\arg t| < \pi/2 - \theta_0$,

$$|D_x^\alpha D_y^\beta G(x,y,t)| \le \frac{C}{|t|^{(n+|\alpha|+|\beta|)/m}} \exp\left(-c\frac{|x-y|^{m/(m-1)}}{|t|^{1/(m-1)}}\right) e^{\omega|t|},$$

$$(5.166)$$

$$\left|\frac{\partial}{\partial t}G(x,y,t)\right| \le \frac{C}{|t|^{n/m+1}} \exp\left(-c\frac{|x-y|^{m/(m-1)}}{|t|^{1/(m-1)}}\right) e^{\omega|t|}. \qquad (5.167)$$

(ii) for $\arg\lambda \notin (-\theta_0, \theta_0), |\lambda| > C, |\alpha| < m, |\beta| < m, (x,y) \in \Omega \times \Omega$

$$|D_x^\alpha D_y^\beta K_\lambda(x,y)| \le Ce^{-c|\lambda|^{1/m}|x-y|}$$

$$\times \begin{cases} |x-y|^{m-n-|\alpha|-|\beta|} & m < n + |\alpha| + |\beta| \\ |\lambda|^{(n+|\alpha|+|\beta|)/m-1} & m > n + |\alpha| + |\beta| \\ \left[1 + \log^+(|\lambda|^{-1/m}|x-y|^{-1})\right] & m = n + |\alpha| + |\beta| \end{cases}. \quad (5.168)$$

Since for $|\arg t| < \pi/2 - \theta_0, |\arg s| < \pi/2 - \theta_0$

$$\exp(-(t+s)A_p) = \exp(-tA_p)\exp(-sA_p),$$

we have

$$G(x,y,t+s) = \int_\Omega G(x,z,t)G(z,y,s)dz. \qquad (5.169)$$

5.4 Boundary Value Problems in L^1

Suppose that the assumptions of Theorem 5.6 are satisfied. It will be shown that the realization $-A_1$ of the operator $-L$ in $L^1(\Omega)$ under the boundary conditions $B_j u|_{\partial\Omega} = 0, j = 1, \ldots, m/2$, generates an analytic semigroup in $L^1(\Omega)$. In this connection we mention H. Amann [13] and D. Guidetti [72]. Amann obtained the desired result for second order operators using the theory of dual semigroups. Guidetti established directly the estimate of the resolvent kernel for systems without considering the adjoint problem.

Set

$$H(x,t) = \exp\left(-c\frac{|x|^{m/(m-1)}}{|t|^{1/(m-1)}}\right)$$

so that by Theorem 5.7

$$|G(x,y,t)| \le \frac{Ce^{\omega|t|}}{|t|^{n/m}}H(x-y,t). \tag{5.170}$$

With the aid of the change of the variable $x = |t|^{1/m}y$ we get

$$\int_{R^n} H(x,t)ds = |t|^{n/m}\int_{R^n}\exp(-c|y|^{m/(m-1)})dy. \tag{5.171}$$

For $f \in L^1(\Omega)$ set

$$(G(t)f)(x) = \int_\Omega G(x,y,t)f(y)dy. \tag{5.172}$$

In view of (5.170),(5.171) we have for $|\arg t| < \pi/2 - \theta_0$

$$\int_\Omega |(G(t)f)(x)|dx \le \int_\Omega\int_\Omega \frac{Ce^{\omega|t|}}{|t|^{n/m}}H(x-y,t)dx|f(y)|dy$$

$$\le \frac{Ce^{\omega|t|}}{|t|^{n/m}}\int_{R^n}H(x,t)dx\int_\Omega|f(y)|dy$$

$$= Ce^{\omega|t|}\int_{R^n}\exp(-c|y|^{m/(m-1)})dy|f|_{0,1,\Omega}. \tag{5.173}$$

Hence $G(t)$ is a bounded linear operator from $L^1(\Omega)$ to itself. By virtue of (5.169) we have

$$G(t+s) = G(t)G(s), \quad |\arg t| < \pi/2 - \theta_0, \quad |\arg s| < \pi/2 - \theta_0. \tag{5.174}$$

Let $f \in C_0(\Omega)$ and $f(x) = 0$ for $|x| > N$. Then

$$\int_{|x|>N+1}|(G(t)f)(x)|dx \le \int_{|x|>N+1}\int_\Omega \frac{Ce^{\omega|t|}}{|t|^{n/m}}H(x-y,t)|f(y)|dydx$$

$$\le |f|_{0,\infty,\Omega}\frac{Ce^{\omega|t|}}{|t|^{n/m}}\int_{|x|>N+1}\int_{|y|<N}H(x-y,t)dydx$$

$$\le |f|_{0,\infty,\Omega}\frac{Ce^{\omega|t|}}{|t|^{n/m}}\int_{|y|<N}\int_{|x-y|>1}H(x-y,t)dxdy$$

$$\le |f|_{0,\infty,\Omega}Ce^{\omega|t|}\int_{|y|<N}dy\int_{|x|>|t|^{-1/m}}\exp(-c|x|^{m/(m-1)})dx.$$

Therefore

$$\lim_{t\to 0}\int_{|x|>N+1}|(G(t)f)(x)-f(x)|dx = \lim_{t\to 0}\int_{|x|>N+1}|(G(t)f)(x)|dx = 0.$$
$$(5.175)$$

On the other hand with the aid of Hölder's inequality

$$\int_{|x|\le N+1}|(G(t)f)(x)-f(x)|dx$$
$$\le |\{x\in\Omega; |x|\le N+1\}|^{1-1/p}|G(t)f-f|_{0,p,\Omega} \to 0 \quad (5.176)$$

as $t\to 0$, where $1<p<\infty$. From (5.175),(5.176) it follows that

$$\lim_{t\to 0}|G(t)f-f|_{0,1,\Omega} \to 0. \qquad (5.177)$$

Since $C_0(\Omega)$ is dense in $L^1(\Omega)$ and $\|G(t)\|_{\mathcal{L}(L^1,L^1)}$ is uniformly bounded if t is bounded in view of (5.173), we conclude that (5.177) holds for every $f\in L^1(\Omega)$. By virtue of (5.167) $G(t)f$ is differentiable in t and

$$|(d/dt)G(t)f|_{0,1,\Omega} \le Ce^{\omega t}|f|_{0,1,\Omega}/t, \quad t>0,$$

for any $f\in L^1(\Omega)$. Thus we have shown that $G(t)$ is an analytic C_0-semigroup in $L^1(\Omega)$. The infinitesimal generator of $G(t)$ is denoted by A_1:

$$G(t) = \exp(-tA_1). \qquad (5.178)$$

If κ is a complex number, the kernel of $\exp(-t(A_p+\kappa)) = e^{-\kappa t}\exp(-tA_p)$ is $e^{-\kappa t}G(x,y,t)$. Hence replacing $L(x,D)$ by $L(x,D)+\kappa$ for a sufficiently large positive number κ, we assume that (5.166),(5.167) hold for $\omega<0$. Then, in view of Lemma 1.2 we have for every $1<p<\infty, f\in L^p(\Omega), t>0$

$$|\exp(-tA_p)f|_{0,p,\Omega} \le \frac{Ce^{\omega t}}{t^{n/m}}\int_{R^n}H(x,t)dx|f|_{0,p,\Omega}$$
$$= Ce^{\omega t}\int_{R^n}\exp(-c|x|^{m/(m-1)})dx|f|_{0,p,\Omega}, \qquad (5.179)$$

and hence $0\in\rho(A_p)$. Similarly by (5.173) $0\in\rho(A_1)$. By Theorem 5.7 the kernel $K(x,y) = K_0(x,y)$ of A_p^{-1} exists and there exist positive constants C,c such that for $|\alpha|<m, |\beta|<m$

$$|D_x^\alpha D_y^\beta K(x,y)| \le Ce^{-c|x-y|}$$
$$\times \begin{cases} |x-y|^{m-n-|\alpha|-|\beta|} & m-n-|\alpha|-|\beta|<0 \\ 1 & m-n-|\alpha|-|\beta|>0 \\ 1+\log^+|x-y|^{-1} & m-n-|\alpha|-|\beta|=0 \end{cases}. \quad (5.180)$$

Since $A_1^{-1} = \int_0^\infty G(t)dt$, $K(x,y)$ is also the kernel of A_1^{-1}.

Theorem 5. 8 *The domain $D(A_1)$ of A_1 coincides with the totality of functions u satisfying the following three conditions:*
(i) $u \in W^{m-1,q}(\Omega)$ for any q satisfying $1 \leq q < n/(n-1)$;
(ii) $L(x,D)u \in L^1(\Omega)$ in the sense of distributions;
(iii) for any p such that $0 < (n/m)(1 - 1/p) < 1$ and $v \in D(A'_{p'})$

$$(Lu, v) = (u, L'v). \tag{5.181}$$

Furthermore, for $u \in D(A_1)$, $A_1 u = L(x,D)u$ and if $m_j < m-1$, $B_j(x,D)u = 0$ on $\partial\Omega$.

Proof. First we verify that both sides of (5.181) are meaningful. Since $m - n/p' > 0$, we have $v \in W^{m,p'}(\Omega) \subset L^\infty(\Omega)$ by Theorem 3.19. Therefore the left hand side makes sense. As for the right side it suffices to show that u satisfying (i) belongs to $L^p(\Omega)$. If $n \geq m$, set $1/q = 1/r + (m-1)/n$ for r satisfying $\{1 - (m-1)/n\}^{-1} \leq r < n/(n-m)$. Then $1 \leq q < n/(n-1)$. Since $m - 1 - n/q < 0$, $W^{m-1,q}(\Omega) \subset L^r(\Omega)$ in view of Theorem 3.17. Hence $u \in L^r(\Omega)$ for any $1 \leq r < n/(n-m)$. Since $1 < p < n/(n-m)$, we conclude $u \in L^p(\Omega)$. If $n < m$, we have $W^{m-1,q}(\Omega) \subset W^{n,q}(\Omega) \subset L^r(\Omega)$ for any $q \leq r < \infty$ by Theorem 3.18 or 3.19. Hence, $u \in L^r(\Omega)$ for any $1 \leq r < \infty$. Especially $u \in L^p(\Omega)$.

Suppose $u \in D(A_1)$. Set $A_1 u = f$. Then

$$u(x) = \int_\Omega K(x,y)f(y)dy. \tag{5.182}$$

By virtue of (5.180) for $|\alpha| = m - 1$

$$|D_x^\alpha K(x,y)| \leq C|x-y|^{1-n}e^{-c|x-y|}.$$

Hence, if $1 \leq q < n/(n-1)$ we get by Lemma 1.2 that

$$\left(\int_\Omega |D^\alpha u(x)|^q dx\right)^{1/q} \leq C \left[\int_{R^n} \left(|x|^{1-n}e^{-c|x|}\right)^q dx\right]^{1/q} |f|_{0,1,\Omega} < \infty.$$

Therefore

$$D^\alpha u \in L^q(\Omega), \quad |\alpha| = m - 1. \tag{5.183}$$

Let

$$0 \leq (n/m)(1 - 1/p) < 1. \tag{5.184}$$

If $m < n$, we have in view of (5.180)

$$|K(x,y)| \leq C|x-y|^{m-n}e^{-c|x-y|}.$$

Since $1 \le p < n/(n-m)$,

$$\int_{R^n} \left(|x|^{m-n} e^{-c|x|} \right)^p dx < \infty.$$

Hence, we get $u \in L^p(\Omega)$ applying Lemma 1.2. If $m > n$, $|K(x,y)| \le Ce^{-c|x-y|}$. Therefore we readily obtain $u \in L^p(\Omega)$. If $m = n$, we have

$$|K(x,y)| \le Ce^{-c|x-y|}(1 + \log^+ |x-y|^{-1}),$$

$$\int_{R^n} \left[e^{-c|x|}(1 + \log^+ |x|^{-1}) \right]^p dx < \infty,$$

and we see $u \in L^p(\Omega)$ also in this case. Consequently $u \in L^p(\Omega)$ in any case. Since (5.184) is equivalent to $1 \le p < n/(n-m)$ if $m < n$ and to $1 \le p < \infty$ if $m \ge n$, we see that $u \in L^q(\Omega)$ for $1 \le q < n/(n-1)$, and hence combining this with (5.183) conclude that u satisfies (i).

Next suppose that $0 < (n/m)(1 - 1/p) < 1$. Then $u \in L^p(\Omega)$ as was shown above. Hence

$$G(t)A_1 u = -\frac{d}{dt}G(t)u = -\frac{d}{dt}e^{-tA_p}u = A_p e^{-tA_p}u = LG(t)u. \qquad (5.185)$$

As $t \to 0$ $G(t)u \to u$, $LG(t)u = G(t)A_1 u \to A_1 u$ in $L^1(\Omega)$. Therefore, $Lu = A_1 u \in L^1(\Omega)$ in the sense of distributions, and u satisfies (ii). If $v \in D(A'_{p'}), 0 < (n/m)(1 - 1/p) < 1$,

$$(G(t)A_1 u, v) = (A_p e^{-tA_p}u, v) = (e^{-tA_p}u, A'_{p'}v). \qquad (5.186)$$

Since $v \in W^{m,p'}(\Omega) \subset L^\infty(\Omega)$ as was remarked at the beginning of the proof, we obatin (5.181) letting $t \to 0$ in (5.186). Thus we have proved that $u \in D(A_1)$ implies that u satisfies (i),(ii),(iii). Conversely suppose that u satisfies (i),(ii),(iii). If v is the solution of $A_1 v = Lu$, v also satisfies (i),(ii),(iii), and $L(u - v) = 0$. Let g be an arbitrary element of $L^{p'}(\Omega), 0 < (n/m)(1 - 1/p) < 1$ and w the solution of $A'_{p'}w = g$, then by (iii)

$$0 = (L(u-v), w) = (u - v, L'w) = (u - v, g).$$

Hence $u = v \in D(A_1)$.

Suppose $u \in D(A_1), 1 \le q < n/(n-1)$. If $|\alpha| \le m - 1, 0 < s \le 1$,

$$\left[\int_{R^n} \left(\frac{C}{s^{(n+|\alpha|)/m}} H(x,s) \right)^q dx \right]^{1/q} \le C_q s^{n/mq - (n+|\alpha|)/m},$$

$$\frac{n}{mq} - \frac{n+|\alpha|}{m} > -1.$$

Hence as $t \to 0$

$$\|G(t)u - u\|_{m-1,q,\Omega} = \left\| \int_0^t G(s)A_1 u ds \right\|_{m-1,q,\Omega}$$

$$\leq C_q t^{n/mq - (n-1)/m} |A_1 u|_{0,1,\Omega} \to 0.$$

Therefore, if $m_j < m - 1$ we have $B_j u = 0$ on $\partial\Omega$.

5.5 Boundary Value Problems in Spaces of Continuous Functions

In this section we assume that Ω is a bounded open subset of R^n of class C^2. We consider elliptic boundary value problems in the space $C(\bar{\Omega})$. The related results are H. B. Stewart [144],[145].

Let

$$L = -\sum_{i,j=1}^n \frac{\partial}{\partial x_j} \left(a_{ij} \frac{\partial}{\partial x_j} \right) + \sum_{i=1}^n b_i \frac{\partial}{\partial x_i} + c$$

be an elliptic differential operator of second order satisfying the assumptions of the previous section 5.4. Let A be the operator defined by

$$D(A) = \{u \in W^{2,p}(\Omega) \text{ for any } p \in (1, \infty), u = 0 \text{ on } \partial\Omega, Lu \in C(\bar{\Omega})\},$$

or

$$D(A) = \{u \in W^{2,p}(\Omega) \text{ for any } p \in (1, \infty), \partial u / \partial n_L = 0 \text{ on } \partial\Omega, Lu \in C(\bar{\Omega})\},$$

and

$$Au = Lu \quad \text{for} \quad u \in D(A)$$

according as we are concerned with the Dirichlet or Neumann boundary condition respectively. Let A_p be the operator defined by (5.34) for the operator L and the boundary conditions mentioned above. Then by virtue of Sobolev's imbedding theorem we have

$$D(A) = [\cap_{1 < p < \infty} D(A_p)] \cap \{u \in C^1(\bar{\Omega}); Lu \in C(\bar{\Omega})\}.$$

Let $G(x, y, t)$ and $K_\lambda(x, y)$ be the kernels of the semigroup $\exp(-tA_p)$ and the resolvent $(A_p - \lambda)^{-1}$ respectively. Then by Theorem 5.7 there exist positive constants C, c such that

$$|G(x, y, t)| \leq \frac{C}{t^{n/2}} \exp\left(-c\frac{|x - y|^2}{t}\right) e^{\omega t} \tag{5.187}$$

$$|K_\lambda(x,y)| \le Ce^{-c|\lambda|^{1/2}|x-y|}$$

$$\times \begin{cases} |x-y|^{2-n} & n > 2, \\ \left[1 + \log^+(|\lambda|^{-1/2}|x-y|^{-1})\right] & n = 2, \\ |\lambda|^{-1/2} & n = 1 \end{cases} \qquad (5.188)$$

for $x, y \in \Omega, t > 0, \lambda \in \Sigma$, where $\Sigma = \{\lambda; \theta_0 < \arg \lambda < 2\pi - \theta_0, |\lambda| \ge C_1\}, 0 < \theta_0 < \pi/2$.

Remark 5. 5 In the above the factor $e^{\omega t}$ and the restriction $|\lambda| \ge C_1$ would be removed, but we are not interested in this problem here.

Let f be an arbitrary element of $C(\bar\Omega)$. Since Ω is bounded, $f \in L^p(\Omega)$ for any $p \in (1, \infty)$. For $\lambda \in \Sigma$ we put

$$u(x) = \int_\Omega K_\lambda(x,y)f(y)dy. \qquad (5.189)$$

Then $u \in D(A_p)$ and $(A_p - \lambda)u = f$. Hence $Lu = f + \lambda u \in C(\bar\Omega)$ owing to Sobolev's imbedding theorem. Consequently $u \in D(A)$ and

$$(A - \lambda)u = f. \qquad (5.190)$$

Conversely if u is an element of $D(A)$ such that $(A - \lambda)u = f$. Then $(A_p - \lambda)u = f$ for $1 < p < \infty$. Therefore (5.189) is a unique element satifying (5.190). Hence we conclude $\rho(A) \supset \Sigma$. By virtue of (5.188)

$$|u(x)| \le C \int_\Omega |x-y|^{2-n} e^{-c|\lambda|^{1/2}|x-y|} dy |f|_{0,\infty,\Omega}$$

$$\le C \int_{R^n} |y|^{2-n} e^{-c|\lambda|^{1/2}|y|} dy |f|_{0,\infty,\Omega} \le C|f|_{0,\infty,\Omega}/|\lambda|$$

if $n > 2$. If $n = 1, 2$, we can prove the same result. Therefore

$$|(A - \lambda)^{-1}f|_{0,\infty,\Omega} \le C|f|_{0,\infty,\Omega}/|\lambda|.$$

Consequently $-A$ generates an analytic semigroup $\exp(-tA)$ in $C(\bar\Omega)$. The kernel of $\exp(-tA)$ is

$$\frac{1}{2\pi i} \int_\Omega e^{-\lambda t} K_\lambda(x,y)d\lambda = G(x,y,t).$$

In the case of the Dirichlet boundary condition $D(A)$ is not dense in $C(\bar\Omega)$ since

$$D(A) \subset C_0(\bar\Omega) = \{u \in C(\bar\Omega); u = 0 \text{ on } \partial\Omega\}.$$

If $f \in C_0^2(\Omega)$, then $f \in D(A_p)$ for any $1 < p < \infty$. Therefore choosing p so that $p > n/2$

$$| \exp(-tA)f - f|_{0,\infty,\Omega} = | \exp(-tA_p)f - f|_{0,\infty,\Omega}$$
$$\leq C|A_p(\exp(-tA_p)f - f)|_{0,p,\Omega} = C| \exp(-tA_p)A_p f - A_p f|_{0,p,\Omega} \to 0$$

as $t \to 0$. Since $C_0^2(\Omega)$ is dense in $C_0(\bar{\Omega})$, we see that for any $f \in C_0(\bar{\Omega})$ $\exp(-tA)f \to f$ in $C(\bar{\Omega})$ as $t \to 0$.

In the case of the Neumann boundary condition we use the fact

$$\{u \in C^2(\bar{\Omega}); \partial u/\partial n_L = 0 \text{ on } \partial\Omega\} \quad \text{is dense in} \quad C(\bar{\Omega}) \tag{5.191}$$

without proof. If f belongs to the space (5.191), then $f \in D(A_p)$ for any $1 < p < \infty$. Hence we can show as in the case of the Dirichlet boundary condition that for any $f \in C(\bar{\Omega})$, $\exp(-tA)f \to f$ in $C(\bar{\Omega})$ as $t \to 0$. Therefore, $D(A)$ is dense in $C(\bar{\Omega})$.

Theorem 5.9 *The operator A defined at the beginning of this section generates an analytic semigroup in $C(\bar{\Omega})$. In the case of the Dirichlet boundary condition $\overline{D(A)} = C_0(\bar{\Omega})$, and in the case of the Neumann boundary condition $D(A)$ is dense in $C(\bar{\Omega})$.*

Remark 5.6 We can prove an analogous result for operators of arbitrary order.

Remark 5.7 In the case of the Neumann boundary condition we can prove that $D(A)$ is dense in $C(\bar{\Omega})$ also in the following manner. Owing to H. Triebel [154: Theorems 1.15.3, 4.9.1] and T. Seeley [135],[136] we have $D(A_p^\theta) = H_p^{2\theta}(\Omega)$ if $0 < 2\theta < 1 + 1/p$, where $H_p^{2\theta}(\Omega)$ is defined through complex interpolation (A. P. Calderón [26]) between the Sobolev spaces $W^{k,p}(\Omega)$, $k = 0, 1, 2, \ldots$. In view of Sobolev's imbedding theorem $H_p^{2\theta}(\Omega) \subset C(\bar{\Omega})$ if $1/p < 2\theta/n$. Therefore, first choosing p so large that $n/p < 1 + 1/p$ and then θ so that $n/p < 2\theta < 1 + 1/p$ we have

$$D(A_p^\theta) = H_p^{2\theta}(\Omega) \subset C(\bar{\Omega}).$$

If $f \in D(A_p^\theta)$, then

$$| \exp(-tA)f - f|_{0,\infty,\Omega} \leq C|A_p^\theta(\exp(-tA_p)f - f)|_{0,p,\Omega}$$
$$= C| \exp(-tA_p)A_p^\theta f - A_p^\theta f|_{0,p,\Omega} \to 0$$

as $t \to 0$. Since $C^2(\bar{\Omega}) \subset W^{2,p}(\Omega) \subset H_p^{2\theta}(\Omega) = D(A_p^\theta)$ and $C^2(\bar{\Omega})$ is dense in $C(\bar{\Omega})$, $D(A_p^\theta)$ is also dense in $C(\bar{\Omega})$. Hence we conclude $\exp(-tA)f \to f$ in $C(\bar{\Omega})$ for any $f \in C(\bar{\Omega})$.

5.6   Example of a Higher Order Operator

In this section we give an example of 6th order elliptic operator which is not variational and generates an analytic semigroup.
  Let

$$\mathcal{B}_0 = -\sum_{i,j=1}^{n} a_{ij}(x)\frac{\partial^2}{\partial x_i \partial x_j} + \sum_{i=1}^{n} b_i(x)\frac{\partial}{\partial x_i} + c(x), \qquad (5.192)$$

$$\mathcal{B}_1 = \sum_{|\alpha|\le 4} a_\alpha(x)D_x^\alpha \qquad (5.193)$$

be elliptic operators of second and fourth oder respectively with smooth coefficients in a bounded domain Ω in R^n with smooth boundary. Here we assume $A(x) = (a_{ij}(x))$ is a real, positive definite symmetric matrix for each $x \in \bar{\Omega}$, and $\sum_{|\alpha|=4} a_\alpha(x)\xi^\alpha > 0$ for $0 \ne \xi \in R^n$. For the sake of simplicity we assume that $a_\alpha(x), |\alpha| = 4$, are real valued and the principal part of $\mathcal{B}_1$ is formally selfadjoint. Let B_0, B_1 be the realizations of $\mathcal{B}_0, \mathcal{B}_1$ in $L^p(\Omega), 1 < p < \infty$, under the Dirichlet boundary conditions. We will show that any half line starting from the origin except the positive real axis is a ray of minimal growth of the resolvents of $B_0 B_1$ and $B_1 B_0$. This is an extension of what is given in A. Favini and H. Tanabe [62] where it is assumed that $\mathcal{B}_1 = \mathcal{B}_0^2$, with proof almost unaltered. It can be seen that these operators are not variational by considering the case $\mathcal{B}_0 = -\Delta, \mathcal{B}_1 = \Delta^2$. By taking the adjoint it suffices to consider only $B_1 B_0$. Then the boudary conditions are $u = \mathcal{B}_0 u = (\partial/\partial\nu)\mathcal{B}_0 u = 0$ on $\partial\Omega$.
  Let $x \in \partial\Omega$ be fixed. Let ν be the outward unit vector at x, $\xi \in R^n$ be tangential to $\partial\Omega$ at x, and $r \ge 0, 0 < \theta < 2\pi, r^2 + |\xi|^2 > 0$. Let s_1, s_2, s_3 be the roots of the polynomial

$$\sum_{|\alpha|=4} a_\alpha(x)(s\nu + \xi)^\alpha \cdot (A(x)(s\nu + \xi), s\nu + \xi) - re^{i\theta} \qquad (5.194)$$

with positive imaginary parts. In what follows we write A for $A(x)$. What is to be shown is that the polynomials

$$1, \quad (A(s\nu + \xi), s\nu + \xi), \quad s\,(A(s\nu + \xi), s\nu + \xi)$$

are linearly independent modulo $(s - s_1)(s - s_2)(s - s_3)$, namely

$$c_1 + c_2\,(A(s\nu + \xi), s\nu + \xi) + c_3 s\,(A(s\nu + \xi), s\nu + \xi)$$
$$\equiv c_3(A\nu, \nu)(s - s_1)(s - s_2)(s - s_3) \qquad (5.195)$$

if and only if $c_1 = c_2 = c_3 = 0$. Comparing the coefficients of both sides of (5.195) we obtain

$$
\begin{aligned}
c_1 + c_2(A\xi, \xi) &= -c_3(A\nu, \nu)s_1 s_2 s_3, \\
2c_2(A\nu, \xi) + c_3(A\xi, \xi) &= c_3(A\nu, \nu)(s_2 s_3 + s_3 s_1 + s_1 s_2), \\
c_2(A\nu, \nu) + 2c_3(A\nu, \xi) &= -c_3(A\nu, \nu)(s_1 + s_2 + s_3).
\end{aligned}
\tag{5.196}
$$

Calculating the determinant of the system of linear equations (5.196) for the unknowns c_1, c_2, c_3 we reduce the problem to showing

$$
\begin{aligned}
4(A\nu, \xi)^2 &+ 2(A\nu, \xi)(A\nu, \nu)(s_1 + s_2 + s_3) \\
&- (A\nu, \nu)(A\xi, \xi) + (A\nu, \nu)^2(s_2 s_3 + s_3 s_1 + s_1 s_2) \neq 0.
\end{aligned}
\tag{5.197}
$$

We use the following classical theorem on polynomials by Hermit and Biehler (T. Takagi [147: Chapter 2]).

Lemma 5.17 *Let $f(z)$ be a polynomial whose roots have imaginary parts of the same sign. Decomposing the coefficients of $f(z)$ into the sum of the real and imaginary parts we write $f(z) = g(z) + ih(z)$. Then the roots of $g(z)$ and $h(z)$ are real distinct and separate one another.*

Proof. We denote the degree of $f(z)$ by n, and suppose that the imaginary parts of the roots $\alpha_1, \ldots, \alpha_n$ of $f(z)$ are all positive. Let $\bar{f}(z)$ be the polynomial whose coefficients are the complex conjugates of the corresponding coefficients of $f(z)$. Then we have

$$
g(z) = (f(z) + \bar{f}(z))/2, \quad h(z) = (f(z) - \bar{f}(z))/2i.
$$

Let $f(z) = a_0(z - \alpha_1) \cdots (z - \alpha_n)$. Then $\bar{f}(z) = \bar{a}_0(z - \bar{\alpha}_1) \cdots (z - \bar{\alpha}_n)$. Put

$$
Z = \frac{(z - \alpha_1) \cdots (z - \alpha_n)}{(z - \bar{\alpha}_1) \cdots (z - \bar{\alpha}_n)}, \quad \kappa = -\frac{\bar{a}_0}{a_0}.
$$

Then $g(z) = 0$ and $h(z) = 0$ are equivalent to $Z = \kappa$ and $Z = -\kappa$ respectively, and $|\kappa| = 1$. Let $\mathrm{Im}\,\alpha > 0$. Then when z goes along the real axis from $-\infty$ to ∞, $\dfrac{z - \alpha}{z - \bar{\alpha}}$ goes along the unit circle from 1 to 1 counterclockwisely. Hence when z goes along the real axis from $-\infty$ to ∞, Z goes along the unit circle n times counterclockwisely. When Z reaches κ, $g(z) = 0$ holds, and when Z reaches $-\kappa$, $h(z) = 0$ holds. This is repeated n times.

Remark 5.8 Either $g(z)$ or $h(z)$ may be of degree $n - 1$. In this case letting one of the roots of $g(z)$ or $h(z)$ be ∞, the conclusion of the theorem holds.

Suppose that (5.197) is false. Then we have

$$4(A\nu, \xi)^2 + 2(A\nu, \xi)(A\nu, \nu)(s_1 + s_2 + s_3)$$
$$-(A\nu, \nu)(A\xi, \xi) + (A\nu, \nu)^2(s_2 s_3 + s_3 s_1 + s_1 s_2) = 0. \quad (5.198)$$

We apply the theorem of Hermit-Biehler to the polynomial

$$f(z) = (z - s_1)(z - s_2)(z - s_3).$$

Then

$$g(z) = z^3 - \mathrm{Re}(s_1 + s_2 + s_3) \cdot z^2$$
$$+ \mathrm{Re}(s_2 s_3 + s_3 s_1 + s_1 s_2) \cdot z - \mathrm{Re} s_1 s_2 s_3, \quad (5.199)$$
$$h(z) = -\mathrm{Im}(s_1 + s_2 + s_3) \cdot z^2$$
$$+ \mathrm{Im}(s_2 s_3 + s_3 s_1 + s_1 s_2) \cdot z - \mathrm{Im} s_1 s_2 s_3. \quad (5.200)$$

In view of Lemma 5.17 and Remark 5.8 we have

$$g(z) = (z - \alpha_1)(z - \alpha_2)(z - \alpha_3), \quad (5.201)$$
$$h(z) = -\mathrm{Im}(s_1 + s_2 + s_3) \cdot (z - \beta_1)(z - \beta_2)$$

(note $\mathrm{Im}(s_1 + s_2 + s_3) \neq 0$) for some $\alpha_1, \alpha_2, \alpha_3, \beta_1, \beta_2$ satsifying

$$\alpha_1 < \beta_1 < \alpha_2 < \beta_2 < \alpha_3. \quad (5.202)$$

It follows from (5.201),(5.202) that

$$g'\left(\frac{\alpha_1 + \alpha_2}{2}\right) = -\left(\frac{\alpha_1 - \alpha_2}{2}\right)^2 < 0, \; g'\left(\frac{\alpha_2 + \alpha_3}{2}\right) = -\left(\frac{\alpha_2 - \alpha_3}{2}\right)^2 < 0.$$

Hence $g'(z) < 0$ for $\dfrac{\alpha_1 + \alpha_2}{2} < z < \dfrac{\alpha_2 + \alpha_3}{2}$. Since (5.202) implies

$\dfrac{\alpha_1 + \alpha_2}{2} < \dfrac{\beta_1 + \beta_2}{2} < \dfrac{\alpha_2 + \alpha_3}{2}$, we have $g'\left(\dfrac{\beta_1 + \beta_2}{2}\right) < 0$. From (5.200),

(5.198) it follows that

$$\frac{\beta_1 + \beta_2}{2} = \frac{\mathrm{Im}(s_2 s_3 + s_3 s_1 + s_1 s_2)}{2\mathrm{Im}(s_1 + s_2 + s_3)} = -\frac{(A\nu, \xi)}{(A\nu, \nu)},$$

and hence

$$g'\left(-\frac{(A\nu, \xi)}{(A\nu, \nu)}\right) < 0. \quad (5.203)$$

On the other hand differentiating both sides of (5.199), substituting $z = -\dfrac{(A\nu, \xi)}{(A\nu, \nu)}$, using the assumption (5.198) and Schwarz' inequality we get

$$g'\left(-\frac{(A\nu, \xi)}{(A\nu, \nu)}\right) = \frac{(A\xi, \xi)(A\nu, \nu) - (A\nu, \xi)^2}{(A\nu, \nu)^2} \geq 0.$$

This contradicts (5.203).

Chapter 6

Parabolic Evolution Equations

This chapter is devoted to the study of evolution equations of parabolic type

$$du(t)/dt = A(t)u(t) + f(t), \quad t \in (0, T], \tag{6.1}$$
$$u(0) = x, \tag{6.2}$$

in a Banach space X. The main results are due to P. Acquistapace and B. Terreni [3],[6],[8] and P. Acquistapace [1]. "Parabolic" means that $A(t)$ generates an analytic semigroup for each t. It is not assumed that $A(t)$ is densely defined. A. Yagi [161] gave another proof using fractional powers of $A(t)$. We also note that the result of A. Yagi [159] which is a very general one in the case where $A(t)$ is densely defined is independent of the above results of [1],[3],[6],[8].

6.1 Equations with Coefficients Differentiable in t

We denote the resolvent $(\lambda - A)^{-1}$ of an operator A by $R(\lambda, A)$. We state the assumptions of this section:

(P1) There exists an angle $\theta_0 \in (\pi/2, \pi]$ such that

(i) $\rho(A(t)) \supset \Sigma = \{\lambda; |\arg \lambda| < \theta_0\} \cup \{0\}$ for each $t \in [0, T]$,

(ii) there exists a positive constant M such that

$$\|R(\lambda, A(t))\| \le M/|\lambda|, \quad \lambda \in \Sigma, \quad t \in [0, T]. \tag{6.3}$$

(P2) For each $\lambda \in \Sigma$ the operator valued function $R(\lambda, A(t))$ of t is continuously differentiable in $[0, T]$ in the uniform operator topology, and there exist positive constants L_1 and $\rho \in (0, 1]$ such that

$$\left\| \frac{\partial}{\partial t} R(\lambda, A(t)) \right\| \leq \frac{L_1}{|\lambda|^\rho}, \quad \lambda \in \Sigma, \quad t \in [0, T]; \tag{6.4}$$

(P3) There exist positive constants L_2 and $\alpha \in (0, 1]$ such that

$$\left\| \frac{d}{dt} A(t)^{-1} - \frac{d}{dt} A(\tau)^{-1} \right\| \leq L_2 |t - \tau|^\alpha, \quad t, \tau \in [0, T]. \tag{6.5}$$

In view of Theorem 1.6 $A(t)$ generates an analytic semigroup $\{e^{\tau A(t)}; \tau \geq 0\}$ under the assumption (P1). The following equalities hold:

$$e^{\tau A(t)} = \frac{1}{2\pi i} \int_\Gamma e^{\tau \lambda} R(\lambda, A(t)) d\lambda, \quad \tau > 0, \quad t \in [0, T], \tag{6.6}$$

$$A(t) e^{\tau A(t)} = \frac{1}{2\pi i} \int_\Gamma \lambda e^{\tau \lambda} R(\lambda, A(t)) d\lambda, \quad \tau > 0, \ t \in [0, T], \tag{6.7}$$

where Γ is a smooth path connecting $\infty e^{-i\theta}$ and $\infty e^{i\theta}$, $\pi/2 < \theta < \theta_0$, in Σ.

The norm of X is denoted by $\| \cdot \|$. We denote by $C([0, T]; X)$ the Banach space consisting of all continuous functions defined in the interval $[0, T]$ and taking values in X with norm

$$\|u\|_{C([0,T];X)} = \sup_{t \in [0,T]} \|u(t)\|.$$

For $\alpha \in (0, 1)$ $C^\alpha([0, T]; X)$ is the set of functions belonging to $C([0, T]; X)$ and Hölder continuous with exponent α with norm

$$\|u\|_{C^\alpha([0,T];X)} = \|u\|_{C([0,T];X)} + \sup_{t,s \in [0,T], t \neq s} \frac{\|u(t) - u(s)\|}{|t - s|^\alpha},$$

and $C^1([0, T]; X)$ is the set of all continuously differentiable functions defined in $[0, T]$ and with values in X with norm

$$\|u\|_{C^1([0,T];X)} = \|u\|_{C([0,T];X)} + \|u'\|_{C([0,T];X)}.$$

For $\alpha \in (0, 1)$

$$C^{1+\alpha}([0, T]; X) = C^1([0, T]; X) \cap \{u; u' \in C^\alpha([0, T]; X)\}$$

with norm

$$\|u\|_{C^{1+\alpha}([0,T];X)} = \|u\|_{C^1([0,T];X)} + \|u'\|_{C^\alpha([0,T];X)}.$$

In addition we denote by

$$C((0,T];X), \quad C^\alpha((0,T];X) \quad C^1((0,T];X) \quad C^{1+\alpha}((0,T];X)$$

the set of functions belonging to

$$C([\epsilon,T];X), \quad C^\alpha([\epsilon,T];X), \quad C^1([\epsilon,T];X), \quad C^{1+\alpha}([\epsilon,T];X)$$

for each $\epsilon > 0$.

For $1 \le p \le \infty$, $L^p(0,T;X)$ is the set of all functions u with values in X such that u is strongly measurable in $(0,T)$ and $\|u(t)\|^p$ is integrable in $(0,T)$ if $p < \infty$ and essentially bounded in $(0,T)$ if $p = \infty$. The norm of $L^p(0,T;X)$ is defined by

$$\|u\|_{L^p(0,T;X)} = \begin{cases} \left(\int_0^T \|u(t)\|^p dt \right)^{1/p} & p < \infty, \\ \operatorname{ess\,sup}_{t \in [0,T]} \|u(t)\| & p = \infty. \end{cases}$$

The set of all bounded linear operators from a Banach space X into itself is denoted by $\mathcal{L}(X)$. The norm of $\mathcal{L}(X)$ is also denoted by $\|\cdot\|$. For two real numbers α, β we write $\alpha \wedge \beta = \min\{\alpha, \beta\}$.

Let $f \in C([0,T];X)$ and $x \in \overline{D(A(0))}$.

Definition 6. 1 u is a *classical solution* of (6.1),(6.2) if $u \in C([0,T];X) \cap C^1((0,T];X), u(t) \in D(A(t))$ for any $t \in (0,T]$ and

$$du(t)/dt = A(t)u(t) + f(t) \quad t \in (0,T], \quad u(0) = x.$$

Definition 6. 2 u is a *strict solution* of (6.1),(6.2) if $u \in C^1([0,T];X), u(t) \in D(A(t))$ for any $t \in [0,T]$ and

$$du(t)/dt = A(t)u(t) + f(t) \quad t \in [0,T], \quad u(0) = x.$$

Remark 6. 1 In addition to strict and classical solutions in the above definitions solutions in the following sense called *strong solutions* are considered in Acquistapace and Terreni [3]: $u \in C([0,T];X)$ and there exists a sequence $\{u_n\} \subset C([0,T];X)$ such that $u_n(t) \in D(A(t))$ for any $t \in [0,T]$, $A(\cdot)u_n(\cdot) \in C([0,T];X)$ and
(i) $u_n \to u$ in $C([0,T];X)$,
(ii) $u_n' - A(\cdot)u_n = f_n \in C([0,T];X)$ and $f_n \to f$ in $C([0,T];X)$,
(iii) $u_n(0) \to x$ in X.
Hypothesis (P3) is not necessary so long as strong solutions are concerned. It is used only to prove the existence and regularity of strict and classical solutions.

6.2 Preliminaries(1)

The following lemma is easily proved as in section 3 of Chapter 1.

Lemma 6.1 *Under the assumptions* (P1),(P2) *we have*
(i) $\|e^{\tau A(t)}\| \leq C, \quad t, \tau \in [0, T],$
(ii) $\|A(t)e^{\tau A(t)}\| \leq C/\tau, \quad t \in [0, T], \quad \tau \in (0, T].$

We define an operator valued function $P(t, s)$, $0 \leq s < t \leq T$, by

$$P(t, s) = \frac{1}{2\pi i} \int_{\Gamma} e^{\lambda(t-s)} \frac{\partial}{\partial t} R(\lambda, A(t)) d\lambda = \left(\frac{\partial}{\partial t} + \frac{\partial}{\partial s} \right) e^{(t-s)A(t)}. \quad (6.8)$$

Arguing as in the proof of (1.11) we can show that the following inequality holds with some positive constant K:

$$\|P(t, s)\| \leq \frac{K}{(t-s)^{1-\rho}}, \quad 0 \leq s < t \leq T. \quad (6.9)$$

Lemma 6.2 *Under the hypothses* (P1),(P2) $P(t, s)$ *is continuous in* $0 \leq s < t \leq T$ *in the norm topology of* $\mathcal{L}(X)$.

Proof. The assertion follows since the integral (6.8) converges in the norm topology of $\mathcal{L}(X)$ uniformly in $\{(t, s); t, s \in [0, T], t - s \geq \epsilon\}$ for any $\epsilon \in (0, T)$.

Lemma 6.3 *Under the hypothesis* (P1),(P2),(P3) *we have for* $\lambda \in \Sigma, t, \tau \in [0, T]$

$$\left\| \frac{\partial}{\partial t} R(\lambda, A(t)) - \frac{\partial}{\partial \tau} R(\lambda, A(\tau)) \right\| \leq C(|t - \tau|^{\alpha} + |\lambda|^{1-\rho}|t - \tau|). \quad (6.10)$$

Proof. Since

$$\frac{\partial}{\partial t} R(\lambda, A(t)) = -A(t)R(\lambda, A(t))\frac{d}{dt}A(t)^{-1} \cdot A(t)R(\lambda, A(t)),$$

we have

$$\frac{\partial}{\partial t} R(\lambda, A(t)) - \frac{\partial}{\partial \tau} R(\lambda, A(\tau))$$

$$= -[A(t)R(\lambda, A(t)) - A(\tau)R(\lambda, A(\tau))]\frac{d}{dt}A(t)^{-1} \cdot A(t)R(\lambda, A(t))$$

$$-A(\tau)R(\lambda, A(\tau))\left[\frac{d}{dt}A(t)^{-1} - \frac{d}{d\tau}A(\tau)^{-1} \right] A(t)R(\lambda, A(t))$$

$$-A(\tau)R(\lambda, A(\tau))\frac{d}{dt}A(\tau)^{-1} \cdot [A(t)R(\lambda, A(t)) - A(\tau)R(\lambda, A(\tau))].$$

Combining this with

$$\|A(t)R(\lambda, A(t)) - A(\tau)R(\lambda, A(\tau))\| = \|\lambda[R(\lambda, A(t)) - R(\lambda, A(\tau))]\|$$
$$= \left\|\lambda \int_\tau^t \frac{\partial}{\partial \sigma} R(\lambda, A(\sigma)) d\sigma\right\| \le C|t - \tau||\lambda|^{1-\rho},$$

we complete the proof.

Lemma 6.4 *Under the hypothses* (P1),(P2),(P3) *we have for* $0 \le s < \tau < t \le T$ *and* $0 < \delta \le \alpha \wedge \rho$

$$\|P(t, s) - P(\tau, s)\|$$
$$\le C[(t - \tau)^\alpha (t - s)^{-1} + (t - \tau)(t - s)^{-1}(\tau - s)^{\rho-1}] \qquad (6.11)$$
$$\le C(t - \tau)^\delta (\tau - s)^{\alpha \wedge \rho - \delta - 1}. \qquad (6.12)$$

Proof. Since

$$P(t, s) - P(\tau, s) = \frac{1}{2\pi i} \int_\Gamma e^{\lambda(t-s)} \left[\frac{\partial}{\partial t} R(\lambda, A(t)) - \frac{\partial}{\partial \tau} R(\lambda, A(\tau))\right] d\lambda$$
$$+ \frac{1}{2\pi i} \int_{\tau-s}^{t-s} \int_\Gamma \lambda e^{\lambda \sigma} \frac{\partial}{\partial \tau} R(\lambda, A(\tau)) d\lambda d\sigma,$$
$$\left\|\int_\Gamma \lambda e^{\lambda \sigma} \frac{\partial}{\partial \tau} R(\lambda, A(\tau)) d\lambda\right\| \le C\sigma^{\rho-2}, \qquad (6.13)$$

we get with the aid of Lemma 6.3

$$\|P(t, s) - P(\tau, s)\|$$
$$\le C\left[\frac{(t - \tau)^\alpha}{t - s} + \frac{t - \tau}{(t - s)^{2-\rho}} + \int_{\tau-s}^{t-s} \sigma^{\rho-2} d\sigma\right]. \qquad (6.14)$$

Using (6.14) and

$$\frac{t - \tau}{(t - s)^{2-\rho}} < \int_{\tau-s}^{t-s} \sigma^{\rho-2} d\sigma = \frac{1}{1-\rho}\left[(\tau - s)^{\rho-1} - (t - s)^{\rho-1}\right]$$
$$= \frac{1}{1-\rho}(\tau - s)^{\rho-1}\left[1 - \left(\frac{\tau - s}{t - s}\right)^{1-\rho}\right]$$
$$< \frac{1}{1-\rho}(\tau - s)^{\rho-1}\left(1 - \frac{\tau - s}{t - s}\right) = \frac{1}{1-\rho}(\tau - s)^{\rho-1}\frac{t - \tau}{t - s}$$

we obtain (6.11). The inequality (6.12) is an easy consequence of (6.11).

Proposition 6.1 *If the assumptions* (P1),(P2) *are satisfied, then* $P(\cdot, 0)x \in C((0, T]; X) \cap L^1(0, T; X)$ *for* $x \in X$. *If moreover the assumption* (P3) *is satisfied, then* $P(\cdot, 0)x \in C^\alpha((0, T]; X)$.

Proof. The assertion follows from Lemma 6.2, (6.9) and Lemma 6.4.

Proposition 6. 2 *Suppose that the assumptions* (P1),(P2) *are satisfied. Then, we have*
(i) *The operator valued function* $e^{tA(t)}$ *is continuously differentiable in* $(0,T]$ *in the norm topology of* $\mathcal{L}(X)$ *and*

$$(d/dt)e^{tA(t)} = A(t)e^{tA(t)} + P(t,0). \tag{6.15}$$

(ii) *Let* $x \in X$. *Then* $e^{tA(t)}x \in C([0,T];X)$ *if and only if* $x \in \overline{D(A(0))}$. *In this case* $e^{tA(t)}x|_{t=0} = x$.
(iii) *If moreover* (P3) *is satisfied, then* $e^{tA(t)}x \in C^{1+\alpha}((0,T];X)$.

Proof. The assertion (i) follows from

$$\frac{d}{dt}e^{tA(t)} = \frac{d}{dt}\left[\frac{1}{2\pi i}\int_\Gamma e^{\lambda t}R(\lambda, A(t))d\lambda\right]$$

$$= \frac{1}{2\pi i}\int_\Gamma \lambda e^{\lambda t}R(\lambda, A(t))d\lambda + \frac{1}{2\pi i}\int_\Gamma e^{\lambda t}\frac{\partial}{\partial t}R(\lambda, A(t))d\lambda$$

$$= A(t)e^{tA(t)} + P(t,0).$$

The assertion (ii) follows from

$$e^{tA(t)}x - x = (e^{tA(t)} - e^{tA(0)})x + e^{tA(0)}x - x,$$

$$\|e^{tA(t)} - e^{tA(0)}\| = \left\|\frac{1}{2\pi i}\int_\Gamma e^{\lambda t}(R(\lambda, A(t)) - R(\lambda, A(0)))d\lambda\right\|$$

$$\leq C\int_\Gamma e^{\mathrm{Re}\lambda t}\frac{t}{|\lambda|^\rho}|d\lambda| \leq Ct^\rho$$

and Theorem 1.7.
(iii) By Proposition 6.1 $P(\cdot,0)x \in C^\alpha((0,T];X)$. Hence in view of (6.15) it suffices to show that $A(t)e^{tA(t)}x \in C^\alpha((0,T];X)$. For $0 \leq \tau < t \leq T$

$$\|A(t)e^{tA(t)} - A(\tau)e^{\tau A(\tau)}\|$$

$$\leq \|A(t)e^{tA(t)} - A(\tau)e^{tA(\tau)}\| + \|A(\tau)e^{tA(\tau)} - A(\tau)e^{\tau A(\tau)}\|$$

$$= \left\|\frac{1}{2\pi i}\int_\Gamma \lambda e^{\lambda t}(R(\lambda, A(t)) - R(\lambda, A(\tau)))d\lambda\right\| + \left\|\int_\tau^t \frac{\partial}{\partial s}A(\tau)e^{sA(\tau)}ds\right\|$$

$$\leq C\int_\Gamma |\lambda|e^{\mathrm{Re}\lambda t}\frac{t-\tau}{|\lambda|^\rho}|d\lambda| + C\int_\tau^t \frac{ds}{s^2} \leq C\left[(t-\tau)t^{\rho-2} + (t-\tau)(t\tau)^{-1}\right].$$

Thus the proof is complete.

Next we investigate the operator P defined by

$$(Pf)(t) = \int_0^t P(t,s)f(s)ds. \tag{6.16}$$

Proposition 6.3 *Suppose that the assumptions* (P1),(P2) *are satisfied.*
(i) $P \in \mathcal{L}(L^1(0,T;X)) \cap \mathcal{L}(C([0,T];X))$.
(ii) *If* $f \in C((0,T];X) \cap L^1(0,T;X)$, *then* $Pf \in C((0,T];X)$.
Suppose moreover that (P3) *is satisfied.*
(iii) *If* $f \in C((0,T];X) \cap L^1(0,T;X)$, *then* $Pf \in C^\delta((0,T];X)$ *for any*
$\delta \in (0,\alpha) \cap (0,\rho]$.

Proof. (i) Using (6.9) and Fubini's theorem we easily obtain

$$\|Pf\|_{L^1(0,T;X)} \leq \frac{KT^\rho}{\rho}\|f\|_{L^1(0,T;X)}.$$

That $P \in \mathcal{L}(C([0,T];X))$ is an easy consequence of (6.9) and the continuity
of $P(t,s)$ in $0 \leq s < t \leq T$.
(ii) For $t > \tau > \epsilon > 0$

$$\|(Pf)(t) - (Pf)(\tau)\| = \left\| \int_\tau^t P(t,s)f(s)ds \right.$$

$$+ \int_\epsilon^\tau (P(t,s) - P(\tau,s))f(s)ds + \int_0^\epsilon (P(t,s) - P(\tau,s))f(s)ds \Bigg\|$$

$$\leq C(t-\tau)^\rho \sup_{\epsilon \leq s \leq T} \|f(s)\| + \int_\epsilon^\tau \|P(t,s) - P(\tau,s)\|ds \sup_{\epsilon \leq s \leq T} \|f(s)\|$$

$$+ \sup_{0 \leq s \leq \epsilon} \|P(t,s) - P(\tau,s)\| \int_0^\epsilon \|f(s)\|ds. \tag{6.17}$$

Each term of the last side tends to 0 as $t \to \tau$ or $\tau \to t$.
(iii) With the aid of (6.17), Lemma 6.4 and its proof

$$\|(Pf)(t) - (Pf)(\tau)\| \leq C(t-\tau)^\rho \sup_{\epsilon \leq s \leq T} \|f(s)\|$$

$$+ C \int_\epsilon^\tau \left[(t-\tau)^\alpha (t-s)^{-1} + \int_{\tau-s}^{t-s} \sigma^{\rho-2}d\sigma \right] ds \sup_{\epsilon \leq s \leq T} \|f(s)\|$$

$$+ C \sup_{0 \leq s \leq \epsilon} \left[\frac{(t-\tau)^\alpha}{t-s} + \frac{t-\tau}{t-s}(\tau-s)^{\rho-1} \right] \int_0^\epsilon \|f(s)\|ds.$$

Since

$$\int_\epsilon^\tau \left[(t-\tau)^\alpha (t-s)^{-1} + \int_{\tau-s}^{t-s} \sigma^{\rho-2}d\sigma \right] ds$$

$$= (t-\tau)^\alpha \log \frac{t-\epsilon}{t-\tau} + \frac{1}{\rho(1-\rho)}[(\tau-\epsilon)^\rho + (t-\tau)^\rho - (t-\epsilon)^\rho]$$

$$< (t-\tau)^\alpha \log \frac{t-\epsilon}{t-\tau} + \frac{1}{\rho(1-\rho)}(t-\tau)^\rho,$$

we obtain

$$\|(Pf)(t) - (Pf)(\tau)\| \le C(t - \tau)^\rho \sup_{\epsilon \le s \le T} \|f(s)\|$$

$$+ C(t - \tau)^\alpha \log \frac{t - \epsilon}{t - \tau} \sup_{\epsilon \le s \le T} \|f(s)\|$$

$$+ C \left[(t - \tau)^\alpha (t - \epsilon)^{-1} + (t - \tau)(t - \epsilon)^{-1}(\tau - \epsilon)^{\rho - 1} \right] \int_0^\epsilon \|f(s)\| ds,$$

which implies the desired result.

Let Q be the operator defined by

$$Q = (1 + P)^{-1}. \tag{6.18}$$

Since P is an integral operator of Volterra type, the inverse of $1 + P$ exists and is given by

$$(Qf)(t) = f(t) + \int_0^t R(t, s) f(s) ds, \tag{6.19}$$

where

$$R(t, s) = \sum_{n=1}^\infty (-1)^n P_n(t, s), \tag{6.20}$$

$$P_1(t, s) = P(t, s), \tag{6.21}$$

$$P_n(t, s) = \int_s^t P(t, \tau) P_{n-1}(\tau, s) d\tau, \quad n = 2, 3, \ldots. \tag{6.22}$$

By induction we can easily show that

$$\|P_n(t, s)\| \le \frac{(K\Gamma(\rho))^n}{\Gamma(n\rho)} (t - s)^{n\rho - 1}, \tag{6.23}$$

$$P_n(t, s) = \int_s^t P_{n-1}(t, \tau) P(\tau, s) d\tau, \quad n = 2, 3, \ldots, \tag{6.24}$$

where K is the constant in (6.9). By virtue of (6.23) the series (6.20) is uniformly convergent in any compact set of $\{(t, s); 0 \le s < t \le T\}$ in the norm topology of $\mathcal{L}(X)$. Hence $R(t, s)$ is continuous in $0 \le s < t \le T$ in the norm topology, and satisfies

$$\|R(t, s)\| \le \frac{C}{(t - s)^{1-\rho}}. \tag{6.25}$$

With the aid of (6.22),(6.24) it is easy to show that

$$P(t, s) + R(t, s) + \int_s^t P(t, \tau)R(\tau, s)d\tau = 0, \tag{6.26}$$

$$P(t, s) + R(t, s) + \int_s^t R(t, \tau)P(\tau, s)d\tau = 0. \tag{6.27}$$

In view of (6.25) and the continuity of $R(t, s)$ in (t, s) the assertions of Proposition 6.3 (i),(ii) hold with $P(t, s)$ replaced by $R(t, s)$.

Proposition 6. 4 *Under the assumptions* (P1),(P2) *we have*
(i) $Q \in \mathcal{L}(L^1(0, T; X)) \cap \mathcal{L}(C([0, T]; X))$.
(ii) *If $f \in C((0, T]; X) \cap L^1(0, T; X)$, then $Qf \in C((0, T]; X)$.*
Under the assumptions (P1),(P2),(P3) *we have*
(iii) *If $f \in L^1(0, T; X) \cap C^\delta((0, T]; X)$, $\delta \in (0, \alpha) \cap (0, \rho]$, then $Qf \in C^\delta((0, T]; X)$.*

Proof. The assertions (i),(ii) are consequences of Proposition 6.3 (i),(ii) with $R(t, s)$ in place of $P(t, s)$. We now prove (iii). If we set $g = Qf = f + Rf$, where $(Rf)(t) = \int_0^t R(t, s)f(s)ds$, then $Rf = -Pg$. In view of Proposition 6.3 (ii) with P replaced by R we get $Rf \in C((0, T]; X)$. Hence $g \in C((0, T]; X)$. Clearly $g \in L^1(0, T; X)$. Hence $Pg \in C^\delta((0, T]; X)$ in view of Proposition 6.3 (iii). Thus we conclude

$$Qf = f + Rf = f - Pg \in C^\delta((0, T]; X).$$

Let T be the operator defined by

$$(Tf)(t) = \int_0^t e^{(t-s)A(t)} f(s)ds. \tag{6.28}$$

Proposition 6. 5 *Suppose that the assumptions* (P1),(P2) *are satisfied.*
(i) *If $f \in L^1(0, T; X)$, then $Tf \in C([0, T]; X)$ and $(Tf)(0) = 0$.*
(ii) *If $f \in L^1(0, T; X) \cap C^\delta((0, T]; X)$, $\delta \in (0, 1]$, then $(Tf)(t) \in D(A(t))$ for $t \in (0, T]$ and*

$$A(t)(Tf)(t) = \int_0^t A(t)e^{(t-s)A(t)}(f(s) - f(t))ds + (e^{tA(t)} - 1)f(t). \tag{6.29}$$

(iii) *If $f \in L^1(0, T; X) \cap C^\delta((0, T]; X), \delta \in (0, 1]$, then $Tf \in C^1((0, T]; X)$ and*

$$\frac{d}{dt}(Tf)(t) = \int_0^t A(t)e^{(t-s)A(t)}(f(s) - f(t))ds + e^{tA(t)}f(t) + \int_0^t P(t, s)f(s)ds. \tag{6.30}$$

If moreover (P3) *is satisfied, we have also*

(iv)   *If* $f \in L^1(0,T;X) \cap C^\delta((0,T];X)$, $\delta \in (0,\alpha) \cap (0,\rho]$, *then* $Tf \in C^{1+\delta}((0,T];X)$.

Proof. The assertion (i) is a simple consequence of Lebesgue's dominated convergence theorem.

(ii) For $0 < \epsilon < t$

$$\int_0^{t-\epsilon} e^{(t-s)A(t)} f(s)ds = \int_0^{t-\epsilon} e^{(t-s)A(t)}(f(s) - f(t))ds$$

$$+ \int_0^{t-\epsilon} A(t)e^{(t-s)A(t)}A(t)^{-1}f(t)ds$$

$$= \int_0^{t-\epsilon} e^{(t-s)A(t)}(f(s) - f(t))ds + (e^{tA(t)} - e^{\epsilon A(t)})A(t)^{-1}f(t).$$

Letting $\epsilon \to 0$ we get

$$(Tf)(t) = \int_0^t e^{(t-s)A(t)}(f(s) - f(t))ds + (e^{tA(t)} - 1)A(t)^{-1}f(t). \quad (6.31)$$

Since

$$\|A(t)e^{(t-s)A(t)}(f(s) - f(t))\| \leq \begin{cases} C_\epsilon(t-s)^{\delta-1} & \epsilon < s < t \\ \dfrac{C}{t-\epsilon}\|f(s) - f(t)\| & 0 < s < \epsilon \end{cases},$$

each term of the right hand side of (6.31) belongs to $D(A(t))$ and (6.29) holds.

(iii) Suppose $0 < \epsilon < \tau < t \leq T$. Then

$$(Tf)(t) - (Tf)(\tau) = \int_\tau^t e^{(t-s)A(t)} f(s)ds$$

$$+ \int_0^\tau (e^{(t-s)A(t)} - e^{(\tau-s)A(\tau)})f(s)ds$$

$$= \int_\tau^t e^{(t-s)A(t)}(f(s) - f(t))ds + \int_\tau^t e^{(t-s)A(t)} f(t)ds$$

$$+ \int_0^\tau (e^{(t-s)A(t)} - e^{(\tau-s)A(t)})f(s)ds$$

$$+ \int_0^\tau (e^{(\tau-s)A(t)} - e^{(\tau-s)A(\tau)})f(s)ds$$

$$= \int_\tau^t e^{(t-s)A(t)}(f(s) - f(t))ds + (e^{(t-\tau)A(t)} - 1)A(t)^{-1}f(t)$$

$$+ \int_0^\tau (e^{(t-s)A(t)} - e^{(\tau-s)A(t)})(f(s) - f(\tau))ds$$

$$+ \int_0^\tau (e^{(t-s)A(t)} - e^{(\tau-s)A(t)})f(\tau)ds$$

$$+ \int_0^\tau (e^{(\tau-s)A(t)} - e^{(\tau-s)A(\tau)})f(s)ds.$$

The fourth term in the last side is equal to

$$(e^{tA(t)} - e^{(t-\tau)A(t)} - e^{\tau A(t)} + 1)A(t)^{-1}f(\tau)$$
$$= (e^{(t-\tau)A(t)} - 1)(e^{\tau A(t)} - 1)A(t)^{-1}f(\tau)$$
$$= (e^{(t-\tau)A(t)} - 1)(e^{\tau A(t)} - 1)A(t)^{-1}(f(\tau) - f(t))$$
$$+ (e^{(t-\tau)A(t)} - 1)(e^{\tau A(t)} - e^{tA(t)})A(t)^{-1}f(t)$$
$$+ (e^{(t-\tau)A(t)} - 1)(e^{tA(t)} - 1)A(t)^{-1}f(t).$$

Hence

$$\frac{(Tf)(t) - (Tf)(\tau)}{t - \tau} = \frac{1}{t-\tau}\int_\tau^t e^{(t-s)A(t)}(f(s) - f(t))ds$$

$$+ \frac{e^{(t-\tau)A(t)} - 1}{t - \tau}A(t)^{-1}f(t)$$

$$+ \int_0^\tau \frac{e^{(t-s)A(t)} - e^{(\tau-s)A(t)}}{t - \tau}(f(s) - f(\tau))ds$$

$$+ \frac{e^{(t-\tau)A(t)} - 1}{t - \tau}(e^{\tau A(t)} - 1)A(t)^{-1}(f(\tau) - f(t))$$

$$+ \frac{e^{(t-\tau)A(t)} - 1}{t - \tau}(e^{\tau A(t)} - e^{tA(t)})A(t)^{-1}f(t)$$

$$+ \frac{e^{(t-\tau)A(t)} - 1}{t - \tau}(e^{tA(t)} - 1)A(t)^{-1}f(t)$$

$$+ \int_0^\tau \frac{e^{(\tau-s)A(t)} - e^{(\tau-s)A(\tau)}}{t - \tau}f(s)ds = \sum_{i=1}^7 I_i.$$

It is easily seen that I_1, I_4, I_5 go to 0 as $\tau \to t$, and by Lemma 1.5 (i)

$$I_2 + I_6 = \frac{e^{(t-\tau)A(t)} - 1}{t - \tau}A(t)^{-1}e^{tA(t)}f(t) \to e^{tA(t)}f(t).$$

By Lebesgue's dominated convergence theorem $I_7 \to \int_0^t P(t,s)f(s)ds$. Finally we investigate I_3. We have for $0 < \epsilon < t - \eta < \tau < t \leq T$

$$I_3 - \int_0^t A(t)e^{(t-s)A(t)}(f(s) - f(t))ds$$

$$= \int_0^{t-\eta} \left[\frac{e^{(t-s)A(t)} - e^{(\tau-s)A(t)}}{t - \tau} - A(t)e^{(t-s)A(t)} \right] (f(s) - f(\tau))ds$$

$$+ \int_0^{t-\eta} A(t)e^{(t-s)A(t)}ds(f(t) - f(\tau))$$

$$+ \int_{t-\eta}^{\tau} \frac{e^{(t-s)A(t)} - e^{(\tau-s)A(t)}}{t - \tau}(f(s) - f(\tau))ds$$

$$- \int_{t-\eta}^{t} A(t)e^{(t-s)A(t)}(f(s) - f(t))ds. \tag{6.32}$$

Since

$$\|e^{(t-s)A(t)} - e^{(\tau-s)A(t)}\|$$

$$= \left\| \int_{\tau}^{t} A(t)e^{(\sigma-s)A(t)}d\sigma \right\| \le C \int_{\tau}^{t} \frac{d\sigma}{\sigma - s} = C \log \frac{t - s}{\tau - s}, \tag{6.33}$$

$$\log \frac{t - s}{\tau - s} = \log \left(1 + \frac{t - \tau}{\tau - s} \right) \le \frac{t - \tau}{\tau - s}, \tag{6.34}$$

the norm of the third term in the right hand side of (6.32) does not exceed

$$C_\epsilon \int_{t-\eta}^{\tau} (\tau - s)^{\delta-1}ds = C_\epsilon(\tau - t + \eta)^\delta \le C_\epsilon \eta^\delta.$$

The norm of the last term is also dominated by $C_\epsilon \eta^\delta$. The second term is equal to

$$(e^{tA(t)} - e^{\eta A(t)})(f(t) - f(\tau)),$$

hence its norm is dominated by $C_\epsilon(t - \tau)^\delta$. The integrand of the first term tends to 0 as $\tau \to t$ for each fixed s. Its norm does not exceed

$$C \left(\frac{1}{\tau - s} + \frac{1}{t - s} \right) \|f(s) - f(\tau)\| \le \frac{C}{\tau - s} \left(\|f(s)\| + \sup_{\epsilon \le \tau \le T} \|f(\tau)\| \right).$$

Hence by Lebesque's dominated convergence theorem the first term goes to 0 as $\tau \to t$ for each fixed η. Thus

$$I_3 \to \int_0^t A(t)e^{(t-s)A(t)}(f(s) - f(t))ds$$

as $\tau \to t$. Summing up we have established that Tf has a left derivative which is equal to the right hand side of (6.30).

We prove now the continuity of the left derivative of Tf. For $0 < \tau <$

$t \leq T$ we have

$$\int_0^t A(t)e^{(t-s)A(t)}(f(s) - f(t))ds - \int_0^\tau A(\tau)e^{(\tau-s)A(\tau)}(f(s) - f(\tau))ds$$

$$= \int_\tau^t A(t)e^{(t-s)A(t)}(f(s) - f(t))ds$$

$$+ \int_0^\tau \left[A(t)e^{(t-s)A(t)}(f(s) - f(t)) - A(\tau)e^{(\tau-s)A(\tau)}(f(s) - f(\tau)) \right] ds$$

$$= \int_\tau^t A(t)e^{(t-s)A(t)}(f(s) - f(t))ds$$

$$+ \int_0^\tau \left[(A(t)e^{(t-s)A(t)} - A(\tau)e^{(t-s)A(\tau)})(f(s) - f(t)) \right.$$

$$+ (A(\tau)e^{(t-s)A(\tau)} - A(\tau)e^{(\tau-s)A(\tau)})(f(s) - f(\tau))$$

$$\left. + A(\tau)e^{(t-s)A(\tau)}(f(\tau) - f(t)) \right] ds$$

$$= \int_\tau^t A(t)e^{(t-s)A(t)}(f(s) - f(t))ds$$

$$+ \int_0^\tau (A(t)e^{(t-s)A(t)} - A(\tau)e^{(t-s)A(\tau)})(f(s) - f(t))ds$$

$$+ \int_0^\tau (A(\tau)e^{(t-s)A(\tau)} - A(\tau)e^{(\tau-s)A(\tau)})(f(s) - f(\tau))ds$$

$$+ (e^{tA(\tau)} - e^{(t-\tau)A(\tau)})(f(\tau) - f(t)). \tag{6.35}$$

Using

$$\|A(t)e^{(t-s)A(t)} - A(\tau)e^{(t-s)A(\tau)}\| \leq C(t - \tau)(t - s)^{\rho-2},$$

$$\|A(\tau)e^{(t-s)A(\tau)} - A(\tau)e^{(\tau-s)A(\tau)}\| \leq C \frac{t - \tau}{(t - s)(\tau - s)}$$

we see that for $0 < \epsilon < \tau < t \leq T$ the norm of the left hand side of (6.35) does not exceed

$$C_\epsilon \int_\tau^t (t - s)^{\delta-1}ds + C_\epsilon \int_\epsilon^\tau (t - \tau)(t - s)^{\rho+\delta-2}ds$$

$$+ C \int_0^\epsilon (t - \tau)(t - s)^{\rho-2}\|f(s) - f(t)\|ds + C_\epsilon \int_\epsilon^\tau \frac{t - \tau}{t - s}(\tau - s)^{\delta-1}ds$$

$$+ C \int_0^\epsilon \frac{t - \tau}{(t - s)(\tau - s)}\|f(s) - f(\tau)\|ds + C_\epsilon(t - \tau)^\delta. \tag{6.36}$$

Using $t - \tau \leq (t - \tau)^\delta(t - s)^{1-\delta}$ in the second term of (6.36) and

$$\int_\epsilon^\tau \frac{(\tau - s)^{\delta-1}}{t - s}ds < \int_0^\tau \frac{s^{\delta-1}}{t - \tau + s}ds$$

$$= \int_0^{\tau/(t-\tau)} \frac{\sigma^{\delta-1}}{1+\sigma} d\sigma (t-\tau)^{\delta-1} < \int_0^\infty \frac{\sigma^{\delta-1}}{1+\sigma} d\sigma (t-\tau)^{\delta-1} \quad (6.37)$$

in the fourth term of (6.36) we get

$$\left\| \int_0^t A(t) e^{(t-s)A(t)} (f(s) - f(t)) ds - \int_0^\tau A(\tau) e^{(\tau-s)A(\tau)} (f(s) - f(\tau)) ds \right\|$$

$$\leq C_\epsilon (t-\tau)^\delta + C \frac{t-\tau}{(t-\epsilon)^{2-\rho}} \int_0^\epsilon \|f(s) - f(t)\| ds$$

$$+ C \frac{t-\tau}{(t-\epsilon)(\tau-\epsilon)} \int_0^\epsilon \|f(s) - f(\tau)\| ds. \quad (6.38)$$

On the other hand we have

$$\|e^{tA(t)} f(t) - e^{\tau A(\tau)} f(\tau)\| \leq \|e^{tA(t)} (f(t) - f(\tau))\|$$

$$+ \|(e^{tA(t)} - e^{tA(\tau)}) f(\tau)\| + \|(e^{tA(\tau)} - e^{\tau A(\tau)}) f(\tau)\|$$

$$\leq C_\epsilon (t-\tau)^\delta + C(t-\tau) t^{\rho-1} \|f(\tau)\| + C \log \frac{t}{\tau} \|f(\tau)\|. \quad (6.39)$$

Thus the left derivative of Tf is continuous, and applying Lemma 1.3 we complete the proof of (iii).

(iv) The assertion is a direct consequence of (6.38),(6.39) and Proposition 6.3 (iii).

6.3   Classical Solutions

Using the results in the previous section we can prove the existence of classical solutions of (6.1),(6.2).

Theorem 6. 1 *Suppose that the assumptions* (P1),(P2),(P3) *are satisfied. If* $x \in \overline{D(A(0))}$ *and* $f \in C([0,T]; X) \cap C^\sigma((0,T]; X)$, $\sigma \in (0,1]$, *then the function* u *defined by*

$$u(t) = e^{tA(t)} x + \int_0^t e^{(t-s)A(t)} \left[(1+P)^{-1} (f - P(\cdot, 0))x \right] (s) ds \quad (6.40)$$

is a classical solution of (6.1),(6.2). *Moreover* $u \in C^{1+\sigma \wedge \delta}((0,T], X)$ *for* $\delta \in (0,\alpha) \cap (0,\rho]$.

Proof. In view of Proposition 6.2 (ii) $e^{tA(t)} x \in C([0,T]; X)$ and $e^{tA(t)} x|_{t=0} = x$. Since $P(\cdot, 0)x \in L^1(0,T; X)$ by Proposition 6.1, we have $f - P(\cdot, 0)x \in L^1(0,T; X)$. Hence by Proposition 6.4 (i)

$$g \equiv (1+P)^{-1}(f - P(\cdot, 0)x) \in L^1(0,T; X). \quad (6.41)$$

Therefore by Proposition 6.5 (i) we have $Tg \in C([0,T]; X)$ and $(Tg)(0) = 0$. Thus $u \in C([0,T]; X)$ and $u(0) = x$. Let $\delta \in (0, \alpha) \cap (0, \rho]$. In view of Proposition 6.1

$$f - P(\cdot, 0)x \in C^{\sigma \wedge \alpha}((0,T]; X) \subset C^{\sigma \wedge \delta}((0,T]; X).$$

Hence by Proposition 6.4 (iii)

$$g \in C^{\sigma \wedge \delta}((0,T]; X). \tag{6.42}$$

Therefore applying Proposition 6.5 (ii) we get $(Tg)(t) \in D(A(t))$ for $t \in (0,T]$. Thus $u(t) \in D(A(t))$ for $t \in (0,T]$ and

$$A(t)u(t) = A(t)e^{tA(t)}x$$
$$+ \int_0^t A(t)e^{(t-s)A(t)}(g(s) - g(t))ds + (e^{tA(t)} - 1)g(t).$$

By virtue of Proposition 6.2 (iii) $e^{tA(t)}x \in C^{1+\alpha}((0,T]; X)$. From (6.41),(6.42) and Proposition 6.5 (iv) it follows that $Tg \in C^{1+\sigma \wedge \delta}((0,T]; X)$, and by Proposition 6.5 (iii)

$$\frac{d}{dt}(Tg)(t) = \int_0^t A(t)e^{(t-s)A(t)}(g(s) - g(t))ds + e^{tA(t)}g(t) + \int_0^t P(t,s)g(s)ds$$

for $t \in (0,T]$. Hence $u \in C^{1+\sigma \wedge \delta}((0,T]; X)$ and

$$du(t)/dt - A(t)u(t) = P(t,0)x + \int_0^t P(t,s)g(s)ds + g(t) = f(t)$$

since $(1 + P)g = f - P(\cdot, 0)x$. Thus the proof is complete.

Next we prove the uniqueness of a classical solution. We set

$$\tilde{P}(t,s) = \frac{1}{2\pi i}\int_\Gamma e^{\lambda(t-s)}\frac{\partial}{\partial s}R(\lambda, A(s))d\lambda = \left(\frac{\partial}{\partial t} + \frac{\partial}{\partial s}\right)e^{(t-s)A(s)}$$

for $0 \le s < t \le T$. Then $\tilde{P}(t,s)$ has the same property as $P(t,s)$. Therefore for the operator $\tilde{P}$ defined by

$$(\tilde{P}f)(t) = \int_0^t \tilde{P}(t,s)f(s)ds$$

the assertions of Propositions 6.3, 6.4 hold. Similarly for the operator $\tilde{T}$ defined by

$$(\tilde{T}f)(t) = \int_0^t e^{(t-s)A(s)}f(s)ds$$

we have the assertions of Proposition 6.5.

Theorem 6. 2 *Suppose that the assumptions* (P1),(P2) *are satisfied. If* u *is a classical solution of* (6.1),(6.2) *for* $x \in \overline{D(A(0))}$ *and* $f \in C([0, T]; X)$, *then we have*

$$u = (1 - \tilde{P})^{-1} \left(e^{\cdot A(0)} x + \tilde{T} f \right). \qquad (6.43)$$

Hence a classical solution is uniquely determined by x *and* f.

Proof. Let u be a classical solution of (6.1),(6.2). For a fixed $t \in (0, T]$ set $v(s) = e^{(t-s)A(s)} u(s)$ for $0 \le s \le t$. Then

$$v'(s) = e^{(t-s)A(s)} u'(s) - e^{(t-s)A(s)} A(s) u(s) + \tilde{P}(t, s) u(s)$$
$$= e^{(t-s)A(s)} f(s) + \tilde{P}(t, s) u(s).$$

Integrating both sides from 0 to $t - \epsilon, 0 < \epsilon < t$, we obtain

$$v(t - \epsilon) - v(0) = \int_0^{t-\epsilon} e^{(t-s)A(s)} f(s) ds + \int_0^{t-\epsilon} \tilde{P}(t, s) u(s) ds. \qquad (6.44)$$

Writing

$$v(t - \epsilon) = e^{\epsilon A(t-\epsilon)} u(t - \epsilon) = e^{\epsilon A(t-\epsilon)} (u(t - \epsilon) - u(t))$$
$$+ (e^{\epsilon A(t-\epsilon)} - e^{\epsilon A(t)}) u(t) + e^{\epsilon A(t)} u(t),$$

we investigate the behavior of each term of the right hand side as $\epsilon \to 0$. In view of Lemma 6.1 the first term goes to 0. As in the proof of Proposition 6.2 (ii) we have

$$\|e^{\epsilon A(t-\epsilon)} - e^{\epsilon A(t)}\| \le C \epsilon^\rho.$$

Hence the second term also goes to 0. Since $u(t) \in D(A(t))$ the third term tends to $u(t)$ by Theorem 1.7. Hence $v(t - \epsilon) \to u(t)$. Letting $\epsilon \to 0$ in (6.44) and noting $v(0) = e^{tA(0)} x$ we obtain

$$u(t) = e^{tA(0)} x + (\tilde{T} f)(t) + (\tilde{P} u)(t),$$

from which (6.43) follows.

Remark 6. 2 If we put the initial condition at $t = s \in (0, T)$ instead of $t = 0$, then the classical solution is expressed as

$$u(t) = e^{(t-s)A(t)} x + \int_s^t e^{(t-\tau)A(t)} \left[(1 + P_s)^{-1} (f - P(\cdot, s)x) \right] (\tau) d\tau, \quad (6.45)$$

where

$$(P_s f)(t) = \int_s^t P(t, \tau) f(\tau) d\tau.$$

By virtue of (6.26) or (6.27)

$$\left((1+P_s)^{-1}f\right)(t) = f(t) + \int_s^t R(t,\tau)f(\tau)d\tau.$$

Hence

$$\left[(1+P_s)^{-1}(f - P(\cdot,s)x)\right](t)$$
$$= f(t) - P(t,s)x + \int_s^t R(t,\tau)(f(\tau) - P(\tau,s)x)d\tau$$
$$= f(t) + \int_s^t R(t,\tau)f(\tau)d\tau + R(t,s)x.$$

Substituting this in (6.45) we obtain

$$u(t) = e^{(t-s)A(t)}x$$
$$+ \int_s^t e^{(t-\tau)A(t)}\left[f(\tau) + \int_s^\tau R(\tau,\sigma)f(\sigma)d\sigma + R(\tau,s)x\right]d\tau$$
$$= e^{(t-s)A(t)}x + \int_s^t e^{(t-\tau)A(t)}R(\tau,s)xd\tau$$
$$+ \int_s^t e^{(t-\tau)A(t)}f(\tau)d\tau + \int_s^t \int_\sigma^t e^{(t-\tau)A(t)}R(\tau,\sigma)d\tau f(\sigma)d\sigma$$
$$= U(t,s)x + \int_s^t U(t,\tau)f(\tau)d\tau,$$

where

$$U(t,s) = e^{(t-s)A(t)} + \int_s^t e^{(t-\tau)A(t)}R(\tau,s)d\tau$$

is the *evolution operator* or *fundamental solution* constructed by T. Kato and H. Tanabe [93] in case where $A(t)$ is densely defined. If $f(t) \equiv 0$, then following the proof of Theorem 6.1

$$u'(t) = A(t)e^{(t-s)A(t)}x$$
$$+ \int_s^t A(t)e^{(t-\tau)A(t)}(R(\tau,s)x - R(t,s)x)d\tau + (e^{(t-s)A(t)} - 1)R(t,s)x.$$

Hence we can show following the argument of [93] that for $x \in \overline{D(A(s))}$

$$\left\|\frac{\partial}{\partial t}U(t,s)x\right\| = \|A(t)U(t,s)x\| \le \frac{C}{t-s}\|x\|.$$

By Remark 1.1 if X is reflexive, then $A(t)$ is densely defined. Therefore Theorems 6.1 and 6.2 reduce to Theorems 3.1 and 4.2 of [93] except the precise Hölder continuity of the derivative of the solution.

6.4   Preliminaries(2)

In this section we establish some preliminary results which will be needed in the next section to show that the classical solution constructed in Theorem 6.1 is a strict solution under some additional conditions on the data.

Proposition 6. 6 *Suppopse that the assumptions* (P1),(P2) *are satisfied. Let* $x \in D(A(0))$. *Then*
(i) $P(\cdot, 0)x \in L^{\infty}(0, T; X)$ *and as* $t \to 0$

$$P(t, 0)x = (1 - e^{tA(0)}) \cdot \frac{d}{dt}A(t)^{-1}\Big|_{t=0} A(0)x + o(1);$$

(ii) *As* $t \to 0$

$$\frac{e^{tA(t)} - 1}{t}x = \frac{e^{tA(0)} - 1}{t}x - (e^{tA(0)} - 1) \cdot \frac{d}{dt}A(t)^{-1}\Big|_{t=0} A(0)x + o(1);$$

(iii) *As* $t \to 0$

$$A(t)e^{tA(t)}x = e^{tA(0)}A(0)x - tA(0)e^{tA(0)} \cdot \frac{d}{dt}A(t)^{-1}\Big|_{t=0} A(0)x + o(1);$$

(iv) *If* (P3) *is also satisfied, then as* $t - \tau \to +0$

$$P(t, 0)x - P(\tau, 0)x$$
$$= -\left[e^{tA(\tau)} - e^{\tau A(\tau)}\right] \cdot \frac{d}{dt}A(t)^{-1}\Big|_{t=0} A(0)x + O\left((t - \tau)^{\rho \wedge \alpha}\right).$$

Proof. (i) Differentiating both sides of

$$R(\lambda, A(t))A(t)^{-1} = \lambda^{-1}R(\lambda, A(t)) + \lambda^{-1}A(t)^{-1}$$

with respect to t we get

$$\frac{\partial}{\partial t}R(\lambda, A(t)) \cdot A(t)^{-1}$$
$$= \frac{1}{\lambda}\frac{\partial}{\partial t}R(\lambda, A(t)) + \frac{1}{\lambda}\frac{d}{dt}A(t)^{-1} - R(\lambda, A(t))\frac{d}{dt}A(t)^{-1}. \quad (6.46)$$

Multiplying both sides of this equality by $e^{\lambda t}$ and integrating along Γ we obtain

$$P(t, 0)A(t)^{-1} = \frac{1}{2\pi i}\int_{\Gamma}\frac{e^{\lambda t}}{\lambda}\frac{\partial}{\partial t}R(\lambda, A(t))d\lambda + (1 - e^{tA(t)})\frac{d}{dt}A(t)^{-1}.$$

Consequently

$$P(t,0)x = P(t,0)(A(0)^{-1} - A(t)^{-1})A(0)x + P(t,0)A(t)^{-1}A(0)x$$

$$= P(t,0)(A(0)^{-1} - A(t)^{-1})A(0)x + \frac{1}{2\pi i}\int_\Gamma \frac{e^{\lambda t}}{\lambda}\frac{\partial}{\partial t}R(\lambda, A(t))A(0)xd\lambda$$

$$+(1 - e^{tA(t)})\frac{d}{dt}A(t)^{-1}A(0)x.$$

Combining this with (6.4),(6.5),(6.9) and

$$\|e^{sA(t)} - e^{sA(0)}\| \le Cts^{\rho-1} \tag{6.47}$$

we obtain the desired result.
(ii) By a standard calculation

$$\frac{e^{tA(t)} - 1}{t}x = \frac{e^{tA(0)} - 1}{t}x - (e^{tA(0)} - 1)\cdot\frac{d}{dt}A(t)^{-1}\Big|_{t=0}A(0)x$$

$$+\frac{e^{tA(t)} - 1}{t}\left[A(0)^{-1} - A(t)^{-1} + t\cdot\frac{d}{dt}A(t)^{-1}\Big|_{t=0}\right]A(0)x$$

$$+\left[\frac{e^{tA(t)} - 1}{t}A(t)^{-1} - \frac{e^{tA(0)} - 1}{t}A(0)^{-1}\right]A(0)x$$

$$-(e^{tA(t)} - e^{tA(0)})\cdot\frac{d}{dt}A(t)^{-1}\Big|_{t=0}A(0)x. \tag{6.48}$$

With the aid of (6.47) we get

$$\|(e^{tA(t)} - 1)A(t)^{-1} - (e^{tA(0)} - 1)A(0)^{-1}\|$$

$$=\left\|\int_0^t (e^{sA(t)} - e^{sA(0)})ds\right\| \le Ct^{\rho+1}. \tag{6.49}$$

Hence the norms of the fourth and fifth terms of the right hand side of (6.48) do not exceed Ct^ρ. Evidently the third term goes to 0 as $t \to 0$. Thus the proof of (ii) is complete.
(iii) A standard calculation shows

$$A(t)e^{tA(t)}x = e^{tA(0)}A(0)x - tA(0)e^{tA(0)}\cdot\frac{d}{dt}A(t)^{-1}\Big|_{t=0}A(0)x$$

$$+tA(t)e^{tA(t)}\left[\frac{A(0)^{-1} - A(t)^{-1}}{t} + \frac{d}{dt}A(t)^{-1}\Big|_{t=0}\right]A(0)x$$

$$+(e^{tA(t)} - e^{tA(0)})A(0)x$$

$$-\left(tA(t)e^{tA(t)} - tA(0)e^{tA(0)}\right)\cdot\frac{d}{dt}A(t)^{-1}\Big|_{t=0}A(0)x. \tag{6.50}$$

With the aid of Lemma 6.1 we see that the third term of the right hand side of (6.50) goes to 0 as $t \to 0$. By (6.47) the fourth term is $O(t^\rho)$. Analogously it is easily seen that the last term is also $O(t^\rho)$. Thus the proof of (iii) is complete.

(iv) For $0 \leq \tau < t \leq T$

$$P(t,0)x - P(\tau,0)x = \frac{1}{2\pi i} \int_\Gamma e^{\lambda t} \left[\frac{\partial}{\partial t} R(\lambda, A(t)) - \frac{\partial}{\partial \tau} R(\lambda, A(\tau)) \right] x d\lambda$$

$$+ \frac{1}{2\pi i} \int_\Gamma \int_\tau^t \lambda e^{\lambda \sigma} d\sigma \cdot \frac{\partial}{\partial \tau} R(\lambda, A(\tau)) x d\lambda. \qquad (6.51)$$

With the aid of (6.46)

$$\left[\frac{\partial}{\partial t} R(\lambda, A(t)) - \frac{\partial}{\partial \tau} R(\lambda, A(\tau)) \right] A(0)^{-1}$$

$$= \frac{\partial}{\partial t} R(\lambda, A(t)) \cdot (A(0)^{-1} - A(t)^{-1}) + \frac{\partial}{\partial t} R(\lambda, A(t)) \cdot A(t)^{-1}$$

$$- \frac{\partial}{\partial \tau} R(\lambda, A(\tau)) \cdot (A(0)^{-1} - A(\tau)^{-1}) - \frac{\partial}{\partial \tau} R(\lambda, A(\tau)) \cdot A(\tau)^{-1}$$

$$= \left[\frac{\partial}{\partial t} R(\lambda, A(t)) - \frac{\partial}{\partial \tau} R(\lambda, A(\tau)) \right] (A(0)^{-1} - A(t)^{-1})$$

$$+ \frac{\partial}{\partial \tau} R(\lambda, A(\tau))(A(\tau)^{-1} - A(t)^{-1}) + \frac{1}{\lambda} \left[\frac{\partial}{\partial t} R(\lambda, A(t)) - \frac{\partial}{\partial \tau} R(\lambda, A(\tau)) \right]$$

$$+ \frac{1}{\lambda} \left[\frac{d}{dt} A(t)^{-1} - \frac{d}{d\tau} A(\tau)^{-1} \right] - [R(\lambda, A(t)) - R(\lambda, A(\tau))] \frac{d}{d\tau} A(\tau)^{-1}$$

$$- R(\lambda, A(t)) \left[\frac{d}{dt} A(t)^{-1} - \frac{d}{d\tau} A(\tau)^{-1} \right],$$

$$\frac{\partial}{\partial \tau} R(\lambda, A(\tau)) A(0)^{-1} = \frac{\partial}{\partial \tau} R(\lambda, A(\tau))(A(0)^{-1} - A(\tau)^{-1})$$

$$+ \frac{1}{\lambda} \frac{\partial}{\partial \tau} R(\lambda, A(\tau)) + \frac{1}{\lambda} \frac{d}{d\tau} A(\tau)^{-1}$$

$$- R(\lambda, A(\tau)) \left[\frac{d}{d\tau} A(\tau)^{-1} - \frac{d}{dt} A(t)^{-1} \Big|_{t=0} \right] - R(\lambda, A(\tau)) \cdot \frac{d}{dt} A(t)^{-1} \Big|_{t=0}.$$

Substituting these in (6.51) we get

$$P(t,0)x - P(\tau,0)x$$

$$= \frac{1}{2\pi i} \int_\Gamma e^{\lambda t} \left\{ \left[\frac{\partial}{\partial t} R(\lambda, A(t)) - \frac{\partial}{\partial \tau} R(\lambda, A(\tau)) \right] (A(0)^{-1} - A(t)^{-1}) \right.$$

$$\left. + \frac{\partial}{\partial \tau} R(\lambda, A(\tau))(A(\tau)^{-1} - A(t)^{-1}) + \frac{1}{\lambda} \left[\frac{\partial}{\partial t} R(\lambda, A(t)) - \frac{\partial}{\partial \tau} R(\lambda, A(\tau)) \right] \right.$$

$$+\frac{1}{\lambda}\left[\frac{d}{dt}A(t)^{-1} - \frac{d}{d\tau}A(\tau)^{-1}\right] - [R(\lambda, A(t)) - R(\lambda, A(\tau))]\frac{d}{d\tau}A(\tau)^{-1}$$

$$-R(\lambda, A(t))\left[\frac{d}{dt}A(t)^{-1} - \frac{d}{d\tau}A(\tau)^{-1}\right]\Big\}A(0)xd\lambda$$

$$+\frac{1}{2\pi i}\int_\Gamma \int_\tau^t \lambda e^{\lambda\sigma} d\sigma \Big\{ \frac{\partial}{\partial\tau}R(\lambda, A(\tau))(A(0)^{-1} - A(\tau)^{-1})$$

$$+\frac{1}{\lambda}\frac{\partial}{\partial\tau}R(\lambda, A(\tau)) + \frac{1}{\lambda}\frac{\partial}{\partial\tau}A(\tau)^{-1}$$

$$-R(\lambda, A(\tau))\left[\frac{d}{d\tau}A(\tau)^{-1} - \frac{d}{dt}A(t)^{-1}\Big|_{t=0}\right]\Big\}A(0)xd\lambda$$

$$-\left[e^{tA(\tau)} - e^{\tau A(\tau)}\right]\cdot\frac{d}{dt}A(t)^{-1}\Big|_{t=0} A(0)x.$$

Using this it is easy to show that

$$P(t,0)x - P(\tau,0)x$$

$$= O\left((t-\tau)^\rho\right) + O\left((t-\tau)^\alpha\right) - \left[e^{tA(\tau)} - e^{\tau A(\tau)}\right]\cdot\frac{d}{dt}A(t)^{-1}\Big|_{t=0} A(0)x.$$

Proposition 6. 7 *Suppose that the assumptions* (P1),(P2),(P3) *are satisfied. If* $f \in L^\infty(0, T; X)$, *then* $Pf \in C^\delta([0, T]; X)$ *for any* $\delta \in (0, \alpha) \cap (0, \rho]$.

Proof. Letting $\epsilon = 0$ in the proof of Proposition 6.3 (iii) we obtain

$$\|(Pf)(t) - (Pf)(\tau)\| \leq C\left[(t-\tau)^\rho + (t-\tau)^\alpha \log\frac{t}{t-\tau}\right]\sup_{0\leq s\leq T}\|f(s)\|,$$

from which the conclusion readily follows.

6.5   Strict Solutions

In this section it is shown that under some additional conditions on the initial value x and the inhomogeneous term f the classical solution constructed in Theorem 6.1 is a strict solution. The results of this and next sections are due to P. Acquistapace and B. Terreni [3: section 5].

Theorem 6. 3 *Suppose that the assumptions* (P1),(P2),(P3) *are satisfied. Let* $x \in D(A(0)), f \in C^\sigma([0, T]; X), \sigma \in (0, 1],$ *and suppose that*

$$A(0)x + f(0) - \frac{d}{dt}A(t)^{-1}\Big|_{t=0} A(0)x \in \overline{D(A(0))}. \tag{6.52}$$

Then the function u *defined by* (6.40) *is a unique strict solution of* (6.1),(6.2). *Moreover,* $u \in C^{1+\sigma\wedge\delta}((0, T]; X)$ *for each* $\delta \in (0, \alpha) \cap (0, \rho]$.

Proof. In order to show that u is a strict solution it suffices to verify that $u'(0)$ exists and $u' \in C([0,T]; X)$. Put $g = (1+P)^{-1}(f - P(\cdot, 0)x)$. Then by (6.40)

$$\frac{u(t) - x}{t} = \frac{e^{tA(t)} - 1}{t} x + \frac{1}{t}(Tg)(t). \tag{6.53}$$

Since

$$g = f - P(\cdot, 0)x - Pg, \tag{6.54}$$

we have

$$Tg = T(f - f(0)) - TP(\cdot, 0)x - TPg + Tf(0). \tag{6.55}$$

Since $f \in C^\sigma([0,T]; X)$

$$(T(f - f(0)))(t) = O(t^{\sigma+1}). \tag{6.56}$$

In view of Proposition 6.6 (i)

$$P(\cdot, 0)x \in L^\infty(0, T; X), \tag{6.57}$$

which implies $f - P(\cdot, 0)x \in L^\infty(0, T; X)$. Hence by Proposition 6.4 (i) $g \in L^\infty(0, T; X)$. Therefore

$$(Pg)(t) = O(t^\rho), \tag{6.58}$$

and hence

$$(TPg)(t) = O(t^{\rho+1}). \tag{6.59}$$

By virtue of Proposition 6.6 (i)

$$(TP(\cdot, 0)x)(t) = \int_0^t e^{(t-s)A(t)} P(s, 0)x\, ds$$

$$= \int_0^t (e^{(t-s)A(t)} - e^{(t-s)A(0)}) P(s, 0)x\, ds$$

$$+ \int_0^t e^{(t-s)A(0)} \left[(1 - e^{sA(0)}) \cdot \left. \frac{d}{dt} A(t)^{-1} \right|_{t=0} A(0)x + O(s^{\rho \wedge \alpha}) \right] ds.$$

Since $P(\cdot, 0)x \in L^\infty(0, T; X)$ and by (6.47)

$$\|e^{(t-s)A(t)} - e^{(t-s)A(0)}\| \leq Ct(t-s)^{\rho-1},$$

we have

$$(TP(\cdot, 0)x)(t)$$

$$= \int_0^t (e^{(t-s)A(0)} - e^{tA(0)}) \cdot \left. \frac{d}{dt} A(t)^{-1} \right|_{t=0} A(0)x\, ds + O(t^{\rho \wedge \alpha+1})$$

$$= (e^{tA(0)} - 1)A(0)^{-1} \cdot \left. \frac{d}{dt} A(t)^{-1} \right|_{t=0} A(0)x$$

$$- te^{tA(0)} \cdot \left. \frac{d}{dt} A(t)^{-1} \right|_{t=0} A(0)x + O(t^{\rho \wedge \alpha+1}). \tag{6.60}$$

Using (6.49)

$$(Tf(0))(t) = \int_0^t e^{(t-s)A(t)} f(0)ds = (e^{tA(t)} - 1)A(t)^{-1}f(0)$$

$$= (e^{tA(0)} - 1)A(0)^{-1}f(0) + O(t^{\rho+1}). \tag{6.61}$$

From (6.53),(6.55),(6.56),(6.59),(6.60),(6.61) and Proposition 6.6 (ii) it follows that

$$\frac{u(t) - x}{t} = \frac{e^{tA(0)} - 1}{t} x - (e^{tA(0)} - 1) \cdot \frac{d}{dt} A(t)^{-1} \Big|_{t=0} A(0)x$$

$$- \frac{e^{tA(0)} - 1}{t} A(0)^{-1} \cdot \frac{d}{dt} A(t)^{-1} \Big|_{t=0} A(0)x$$

$$+ e^{tA(0)} \cdot \frac{d}{dt} A(t)^{-1} \Big|_{t=0} A(0)x + \frac{e^{tA(0)} - 1}{t} A(0)^{-1} f(0) + o(1)$$

$$= \frac{e^{tA(0)} - 1}{t} x + \frac{d}{dt} A(t)^{-1} \Big|_{t=0} A(0)x$$

$$- \frac{e^{tA(0)} - 1}{t} A(0)^{-1} \cdot \frac{d}{dt} A(t)^{-1} \Big|_{t=0} A(0)x + \frac{e^{tA(0)} - 1}{t} A(0)^{-1} f(0) + o(1)$$

$$= \frac{e^{tA(0)} - 1}{t} A(0)^{-1} \left(A(0)x + f(0) - \frac{d}{dt} A(t)^{-1} \Big|_{t=0} A(0)x \right)$$

$$+ \frac{d}{dt} A(t)^{-1} \Big|_{t=0} A(0)x + o(1) \to A(0)x + f(0)$$

as $t \to 0$ by (6.52) and Lemma 1.5. Hence we have established that $u'(0)$ exists and

$$u'(0) = A(0)x + f(0). \tag{6.62}$$

Next we show that u' is continuous at $t = 0$. As was shown in the proof of Theorem 6.1

$$u'(t) = A(t)e^{tA(t)}x + P(t,0)x + \int_0^t A(t)e^{(t-s)A(t)}(g(s) - g(t))ds$$

$$+ e^{tA(t)}g(t) + \int_0^t P(t,s)g(s)ds. \tag{6.63}$$

Since $g \in L^\infty(0, T; X), Pg \in C^\delta([0, T]; X), \delta \in (0, \alpha) \cap (0, \rho]$ by Proposition 6.7. It is easily seen that

$$\|A(t)e^{(t-s)A(t)} - A(s)e^{(t-s)A(s)}\| \le C(t-s)^{\rho-1}.$$

Using these and (6.54),(6.57) and applying Proposition 6.6 (iv) we get

$$\int_0^t A(t)e^{(t-s)A(t)}(g(s) - g(t))ds$$

$$= -\int_0^t A(t)e^{(t-s)A(t)}\left[(Pg)(s) - (Pg)(t)\right]ds$$

$$+ \int_0^t A(t)e^{(t-s)A(t)}(f(s) - f(t))ds$$

$$- \int_0^t \left[A(t)e^{(t-s)A(t)} - A(s)e^{(t-s)A(s)}\right][P(s,0)x - P(t,0)x]ds$$

$$- \int_0^t A(s)e^{(t-s)A(s)}\left[(e^{tA(s)} - e^{sA(s)}) \cdot \frac{d}{dt}A(t)^{-1}\Big|_{t=0} A(0)x\right.$$

$$\left. + O\left((t-s)^{\rho\wedge\alpha}\right)\right]ds = O(t^\delta) + O(t^\sigma) + O(t^\rho)$$

$$+ \int_0^t A(0)(e^{tA(0)} - e^{(2t-s)A(0)}) \cdot \frac{d}{dt}A(t)^{-1}\Big|_{t=0} A(0)x\,ds$$

$$+ \int_0^t \left[A(s)(e^{tA(s)} - e^{(2t-s)A(s)}) - A(0)(e^{tA(0)} - e^{(2t-s)A(0)})\right]$$

$$\cdot \frac{d}{dt}A(t)^{-1}\Big|_{t=0} A(0)x\,ds + O(t^{\rho\wedge\alpha}). \tag{6.64}$$

Since

$$\|A(s)(e^{tA(s)} - e^{(2t-s)A(s)}) - A(0)(e^{tA(0)} - e^{(2t-s)A(0)})\|$$
$$\leq \|A(s)e^{tA(s)} - A(0)e^{tA(0)}\| + \|A(s)e^{(2t-s)A(s)} - A(0)e^{(2t-s)A(0)}\|$$
$$\leq Cst^{\rho-2} + Cs(2t-s)^{\rho-2} \leq Cst^{\rho-2},$$

the norm of the fifth term of the last side of (6.64) is $O(t^\rho)$. Hence

$$\int_0^t A(t)e^{(t-s)A(t)}(g(s) - g(t))ds = o(1)$$

$$+ \left[tA(0)e^{tA(0)} - e^{tA(0)}(e^{tA(0)} - 1)\right] \cdot \frac{d}{dt}A(t)^{-1}\Big|_{t=0} A(0)x. \tag{6.65}$$

Next, by (6.47),(6.54),(6.57),(6.58) and Proposition 6.6 (i) we have

$$e^{tA(t)}g(t) = e^{tA(t)}(-(Pg)(t) + f(t) - P(t,0)x)$$
$$= -e^{tA(t)}(Pg)(t) + e^{tA(t)}(f(t) - f(0)) + (e^{tA(t)} - e^{tA(0)})f(0)$$
$$+ e^{tA(0)}f(0) - (e^{tA(t)} - e^{tA(0)})P(t,0)x - e^{tA(0)}P(t,0)x$$

$$= e^{tA(0)} f(0) + e^{tA(0)} (e^{tA(0)} - 1) \cdot \frac{d}{dt} A(t)^{-1} \Big|_{t=0} A(0)x$$

$$+ O(t^{\rho \wedge \alpha}) + O(t^\sigma). \tag{6.66}$$

From (6.63),(6.65),(6.66),(6.58) and Proposition 6.6 (i),(iii) it follows that

$$u'(t) = e^{tA(0)} \left[A(0)x + f(0) - \frac{d}{dt} A(t)^{-1} \Big|_{t=0} A(0)x \right]$$

$$+ \frac{d}{dt} A(t)^{-1} \Big|_{t=0} A(0)x + o(1) \to A(0)x + f(0) = u'(0).$$

6.6   Maximal Regularity

If A is a linear closed not necessarily densely defined operator which generates an analytic semigroup, then the interpolation space $(D(A), X)_{1-\theta,\infty}$, $0 < \theta < 1$, between $D(A)$ and X is defined as in the case of densely defined operators:

$$(D(A), X)_{1-\theta,\infty} = \{u(0); t^{1-\theta} u(t), t^{1-\theta} Au(t), t^{1-\theta} u'(t) \in L^\infty(0, T; X)\}.$$

It is clear that

$$D(A) \subset (D(A), X)_{1-\theta,\infty} \subset \overline{D(A)}. \tag{6.67}$$

Let A' be the restriction of A to the subspace

$$Z = \{x \in D(A); Ax \in \overline{D(A)}\}.$$

If $x \in D(A)$, then as $0 < \lambda \to \infty$

$$\|\lambda(\lambda - A)^{-1}x - x\| = \|(\lambda - A)^{-1}Ax\| \le C\|Ax\|/\lambda \to 0. \tag{6.68}$$

Since $\lambda(\lambda - A)^{-1}$ is uniformly bounded as $\lambda \to +\infty$, we see from (6.68) that $\lambda(\lambda - A)^{-1}x \to x$ as $\lambda \to +\infty$ if $x \in \overline{D(A)}$. Furthermore if $x \in D(A)$, then $\lambda(\lambda - A)^{-1}x \in Z$ since

$$A(\lambda(\lambda - A)^{-1}x) = \lambda^2(\lambda - A)^{-1}x - \lambda x \in \overline{D(A)}.$$

Hence $D(A') = Z$ is dense in $\overline{D(A)}$. Therefore A' generates an analytic C_0-semigroup in $\overline{D(A)}$ which is a Banach space with norm of X.

Lemma 6. 5 $(D(A'), \overline{D(A)})_{1-\infty,\theta} = (D(A), X)_{1-\theta,\infty}.$

Proof. It is evident that $(D(A'), \overline{D(A)})_{1-\theta,\infty} \subset (D(A), X)_{1-\theta,\infty}$. Conversely suppose that $x \in (D(A), X)_{1-\theta,\infty}$. Then there exists a function u with values in X such that

$$t^{1-\theta} u(t), t^{1-\theta} Au(t), t^{1-\theta} u'(t) \in L^\infty(0, \infty; X), u(0) = x.$$

If we put $w(t) = (1 - tA)^{-1}u(t)$, then $w(t) \in D(A')$ for $t > 0$,

$$\|t^{1-\theta}w(t)\| \leq Ct^{1-\theta}\|u(t)\|, \quad \|t^{1-\theta}Aw(t)\| \leq Ct^{1-\theta}\|Au(t)\|,$$
$$\|t^{1-\theta}w'(t)\| = t^{1-\theta}\|(1 - tA)^{-1}u'(t) + (1 - tA)^{-2}Au(t)\|$$
$$\leq Ct^{1-\theta}\|u'(t)\| + Ct^{1-\theta}\|Au(t)\|,$$
$$\|w(t) - x\| \leq \|(1 - tA)^{-1}(u(t) - x)\| + \|(1 - tA)^{-1}x - x\|$$
$$\leq C\|u(t) - x\| + \|(1 - tA)^{-1}x - x\| \to 0$$

as $t \to 0$ since $x \in \overline{D(A)}$ in view of (6.67). Hence $x \in (D(A'), \overline{D(A)})_{1-\theta,\infty}$.

In what follows we write $D_A(\theta, \infty)$ instead of $(D(A), X)_{1-\theta,\infty}$. Since A' is densely defined, the characterizations of $D_A(\theta, \infty)$ stated in Chapter 1 holds for A owing to Lemma 6.5:

$$D_A(\theta, \infty) = \{x \in X; \sup_{t>0} t^{-\theta}\|e^{tA}x - x\| < \infty\}, \tag{6.69}$$

$$= \{x \in X; \sup_{\lambda>0} \lambda^{\theta}\|AR(\lambda, A)x\| < \infty\} \tag{6.70}$$

$$= \{x \in X; \sup_{\lambda \in \Sigma} |\lambda|^{\theta}\|AR(\lambda, A)x\| < \infty\}. \tag{6.71}$$

Now we return to the situation where the assumptions (P1),(P2) or (P1), (P2),(P3) are satisfied.

Lemma 6.6 *Suppose that the assumptions* (P1),(P2) *are satisfied. If* $x \in D_{A(0)}(\beta, \infty), \beta \in (0, 1),$ *then*

$$\|A(\tau)e^{tA(\tau)}x\| \leq C_x t^{\beta-1} \tag{6.72}$$

for $0 \leq \tau \leq t \leq T, 0 < t,$ *where* C_x *is a constant depending on* x.

Proof. We have

$$A(\tau)e^{tA(\tau)}x = A(\tau)e^{tA(\tau)}[x - e^{tA(0)}x$$
$$+ (A(0)^{-1} - A(\tau)^{-1})A(0)e^{tA(0)}x] + e^{tA(\tau)}A(0)e^{tA(0)}x.$$

Hence (6.72) follows from Lemma 6.1, (6.69) and the inequality

$$\|A(0)e^{tA(0)}x\| \leq C_x t^{\beta-1}$$

which is a consequence of (6.71).

Propostion 6.8 *Suppose that the assumptions* (P1),(P2) *are satisfied. If* $\beta \in (0, \rho],$ *then* $e^{tA(t)}x \in C^{\beta}([0, T]; X)$ *if and only if* $x \in D_{A(0)}(\beta, \infty)$.

Proof. Let $x \in D_{A(0)}(\beta, \infty)$ and $0 \leq \tau < t \leq T$. We have

$$e^{tA(t)}x - e^{\tau A(\tau)}x = (e^{tA(t)} - e^{tA(\tau)})x$$
$$+ \left[(e^{tA(\tau)} - e^{\tau A(\tau)}) - (e^{tA(0)} - e^{\tau A(0)}) \right] x$$
$$+ (e^{tA(0)} - e^{\tau A(0)})x. \tag{6.73}$$

The first term of the right hand side of (6.73) is equal to

$$\int_\tau^t \frac{\partial}{\partial \sigma} e^{tA(\sigma)} x \, d\sigma = \int_\tau^t \frac{\partial}{\partial \sigma} \frac{1}{2\pi i} \int_\Gamma e^{\lambda t} R(\lambda, A(\sigma)) x \, d\lambda \, d\sigma$$
$$= \frac{1}{2\pi i} \int_\Gamma e^{\lambda t} \int_\tau^t \frac{\partial}{\partial \sigma} R(\lambda, A(\sigma)) x \, d\sigma \, d\lambda. \tag{6.74}$$

With the aid of (6.46)

$$\frac{\partial}{\partial \sigma} R(\lambda, A(\sigma))x = \frac{\partial}{\partial \sigma} R(\lambda, A(\sigma)) \Big[x - e^{\sigma A(0)} x$$
$$+ (A(0)^{-1} - A(\sigma)^{-1}) A(0) e^{\sigma A(0)} x + A(\sigma)^{-1} A(0) e^{\sigma A(0)} x \Big]$$
$$= \frac{\partial}{\partial \sigma} R(\lambda, A(\sigma)) \Big[x - e^{\sigma A(0)} x + (A(0)^{-1} - A(\sigma)^{-1}) A(0) e^{\sigma A(0)} x \Big]$$
$$+ \left[\frac{1}{\lambda} \frac{\partial}{\partial \sigma} R(\lambda, A(\sigma)) + \frac{1}{\lambda} \frac{d}{d\sigma} A(\sigma)^{-1} - R(\lambda, A(\sigma)) \frac{d}{d\sigma} A(\sigma)^{-1} \right] A(0) e^{\sigma A(0)} x.$$

Substituting this in (6.74)

$$(e^{tA(t)} - e^{tA(\tau)})x$$
$$= \frac{1}{2\pi i} \int_\Gamma e^{\lambda t} \int_\tau^t \frac{\partial}{\partial \sigma} R(\lambda, A(\sigma)) \Big[x - e^{\sigma A(0)} x$$
$$+ (A(0)^{-1} - A(\sigma)^{-1}) A(0) e^{\sigma A(0)} x \Big] d\sigma \, d\lambda$$
$$+ \frac{1}{2\pi i} \int_\Gamma e^{\lambda t} \int_\tau^t \left[\frac{1}{\lambda} \frac{\partial}{\partial \sigma} R(\lambda, A(\sigma)) + \frac{1}{\lambda} \frac{d}{d\sigma} A(\sigma)^{-1} \right.$$
$$\left. - R(\lambda, A(\sigma)) \frac{d}{d\sigma} A(\sigma)^{-1} \right] A(0) e^{\sigma A(0)} x \, d\sigma \, d\lambda. \tag{6.75}$$

Similarly applying the same argument to the second term of (6.73) which is equal to

$$\int_0^\tau \frac{\partial}{\partial r} (e^{tA(r)} - e^{\tau A(r)}) x \, dr$$

$$= \int_0^\tau \frac{\partial}{\partial r} \frac{1}{2\pi i} \int_\Gamma (e^{\lambda t} - e^{\lambda \tau}) R(\lambda, A(r)) x \, d\lambda \, dr$$

$$= \frac{1}{2\pi i} \int_\Gamma \int_\tau^t \lambda e^{\lambda \sigma} \int_0^\tau \frac{\partial}{\partial r} R(\lambda, A(r)) x \, dr \, d\sigma \, d\lambda,$$

we obtain

$$\left[(e^{tA(\tau)} - e^{\tau A(\tau)}) - (e^{tA(0)} - e^{\tau A(0)}) \right] x$$

$$= \frac{1}{2\pi i} \int_\Gamma \int_\tau^t \lambda e^{\lambda \sigma} \int_0^\tau \frac{\partial}{\partial r} R(\lambda, A(r)) \left[x - e^{\tau A(0)} x \right.$$

$$+ (A(0)^{-1} - A(r)^{-1}) A(0) e^{\tau A(0)} x \Big] dr \, d\sigma \, d\lambda$$

$$+ \frac{1}{2\pi i} \int_\Gamma \int_\tau^t \lambda e^{\lambda \sigma} \int_0^\tau \left[\frac{1}{\lambda} \frac{\partial}{\partial r} R(\lambda, A(r)) + \frac{1}{\lambda} \frac{d}{dr} A(r)^{-1} \right.$$

$$- R(\lambda, A(r)) \frac{d}{dr} A(r)^{-1} \Big] A(0) e^{\tau A(0)} x \, dr \, d\sigma \, d\lambda. \tag{6.76}$$

By (6.69) and Lemma 6.6 the norm of the first term of the right hand side of (6.75) is dominated by

$$Ct^{\rho-1} \int_\tau^t \sigma^\beta \, d\sigma \leq Ct^\rho \int_\tau^t \sigma^{\beta-1} d\sigma \leq Ct^\rho (t^\beta - \tau^\beta) \leq Ct^\rho (t - \tau)^\beta,$$

and that of the second term by $C(t - \tau)^\beta$. The norm of the first term of the right hand side of (6.76) does not exceed

$$C\tau^{\beta+1} \int_\tau^t \sigma^{\rho-2} d\sigma \leq C\tau^\rho \int_\tau^t \sigma^{\beta-1} d\sigma \leq C\tau^\rho (t - \tau)^\beta,$$

and the second term does not exceed

$$C\tau^\beta \log \frac{t}{\tau} = C\tau^\beta \log \left(1 + \frac{t - \tau}{\tau} \right) \leq C\tau^\beta \left(\frac{t - \tau}{\tau} \right)^\beta \Big/ \beta = C(t - \tau)^\beta / \beta.$$

Finally by Lemma 6.6

$$\| e^{tA(0)} x - e^{\tau A(0)} x \| = \left\| \int_\tau^t A(0) e^{sA(0)} x \, ds \right\| \leq C \int_\tau^t s^{\beta-1} ds \leq C(t - \tau)^\beta.$$

Summing up we obtain $e^{tA(t)} x \in C^\beta([0, T]; X)$.

Conversely suppose that $e^{tA(t)} x \in C^\beta([0, T]; X)$. Then by Proposition 6.2 (ii) we have $x \in \overline{D(A(0))}$ and $e^{tA(t)} x|_{t=0} = x$. From this and (6.47) we obtain

$$e^{tA(0)} x - x = (e^{tA(0)} - e^{tA(t)}) x + (e^{tA(t)} x - x) = O(t^\rho) + O(t^\beta).$$

Hence we conclude $x \in D_{A(0)}(\beta, \infty)$ with the aid of (6.69).

Remark 6. 3 The "if" part of Proposition 6.8 holds for $\beta \in (0, 1]$.

Proposition 6. 9 *Suppose that the assumptions* (P1),(P2),(P3) *are satisfied. If* $f \in C^\delta([0,T];X)$, $\delta \in (0,\alpha) \cap (0,\rho]$, *then*

$$Qf = (1+P)^{-1}f \in C^\delta([0,T];X).$$

Proof. $Qf = f + Rf = f - P(f + Rf)$ and by Proposition 6.7 $P(f + Rf) \in C^\delta([0,T];X)$. Hence we conclude $Qf \in C^\delta([0,T];X)$.

Proposition 6. 10 *Suppose that the assumptions* (P1),(P2),(P3) *are satisfied. If* $f \in C^\delta([0,T];X)$, $\delta \in (0,\alpha) \cap (0,\rho]$ *and* $f(0) = 0$, *then* $Tf \in C^{1+\delta}([0,T];X)$ *and* $(Tf)'(0) = 0$.

Proof. The assertion is obtained by letting $\epsilon = 0$ in (6.38),(6.39) in the proof of Proposition 6.5 (iii), noting

$$\|f(\tau)\| \le C\tau^\delta, \quad \log\frac{t}{\tau} = \log\left(1 + \frac{t-\tau}{\tau}\right) \le \frac{1}{\delta}\left(\frac{t-\tau}{\tau}\right)^\delta,$$

and using Proposition 6.7.

Theorem 6. 4 *Suppose that the assumptions* (P1),(P2),(P3) *are satisfied. Let* $x \in D(A(0))$, $f \in C^\delta([0,T];X)$, $\delta \in (0,\alpha) \cap (0,\rho]$, *and suppose that* u *is a strict solution of* (6.1),(6.2). *Then* $u \in C^{1+\delta}([0,T];X)$ *if and only if the following condition is satisfied:*

$$A(0)x + f(0) - \frac{d}{dt}A(t)^{-1}\Big|_{t=0}A(0)x \in D_{A(0)}(\delta,\infty).$$

Proof. Let z be the solution of

$$z'(t) = A(t)z(t) + A(0)x + f(0) - \frac{d}{dt}A(t)^{-1}\Big|_{t=0}A(0)x,$$

$$z(0) = 0.$$

By Theorem 6.3 z is a strict solution, and is expressed as

$$z(t) = \int_0^t e^{(t-s)A(t)}h(s)ds, \tag{6.77}$$

where h is the solution of

$$h(t) + \int_0^t P(t,s)h(s)ds = A(0)x + f(0) - \frac{d}{dt}A(t)^{-1}\Big|_{t=0}A(0)x.$$

We set

$$w(t) = u(t) - A(t)^{-1}A(0)x - z(t). \tag{6.78}$$

Then w is a strict solution of

$$w'(t) = A(t)w(t) + f(t) - f(0) - \left[\frac{d}{dt}A(t)^{-1} - \frac{d}{dt}A(t)^{-1}\Big|_{t=0}\right]A(0)x,$$

$$w(0) = 0,$$

and represented by

$$w(t) = \int_0^t e^{(t-s)A(t)}k(s)ds,$$

where

$$k(t) + \int_0^t P(t,s)k(s)ds$$

$$= f(t) - f(0) - \left[\frac{d}{dt}A(t)^{-1} - \frac{d}{dt}A(t)^{-1}\Big|_{t=0}\right]A(0)x.$$

In view of Proposition of 6.9 $k \in C^\delta([0,T]; X)$, $k(0) = 0$. Hence by Proposition 6.10 $w \in C^{1+\delta}([0,T]; X)$. Therefore in view of (6.78) $u \in C^{1+\delta}([0,T]; X)$ if and only if $z \in C^{1+\delta}([0,T]; X)$. By Proposition 6.9 $h \in C^\delta([0,T]; X)$. By (6.77) and Proposition 6.5 (iii)

$$\frac{d}{dt}z(t) = \frac{d}{dt}(T(h-h(0)))(t) + \frac{d}{dt}(Th(0))(t)$$

$$= \frac{d}{dt}(T(h-h(0)))(t) + e^{tA(t)}h(0) + (Ph(0))(t). \tag{6.79}$$

By Proposition 6.10 the first term of the right hand side of (6.79) belongs to $C^\delta([0,T]; X)$. By virtue of Proposition 6.7 the last term also belongs to $C^\delta([0,T]; X)$. Hence $z \in C^{1+\delta}([0,T]; X)$ if and only if $e^{tA(t)}h(0) \in C^\delta([0,T]; X)$. Therefore using Proposition 6.8 we conclude that $u \in C^{1+\delta}([0,T]; X)$ if and only if

$$A(0)x + f(0) - \frac{d}{dt}A(t)^{-1}\Big|_{t=0}A(0)x = h(0) \in D_{A(0)}(\delta, \infty).$$

6.7   An Example

Following Acquistapace and Terreni [3] we state an example satisfying the assumptions (P1),(P2),(P3) of section 6.1.

Let $X = C([0, 1])$ with norm $\|u\| = \sup_{x \in [0,1]} |u(x)|$. We define the operator $A(t)$ by

$$D(A(t)) = \{u \in C^2([0, 1]); u(0) = 0, \alpha(t)u(1) + \beta(t)u'(1) = 0\},$$

$$A(t)u = u'' \quad \text{for} \quad u \in D(A(t)). \tag{6.80}$$

Here $\alpha(\cdot)$ and $\beta(\cdot)$ are real valued functions in $C^1([0, T])$ satisfying

$$\alpha(t) \geq 0, \quad \beta(t) \geq 0, \quad \alpha(t) + \beta(t) > 0 \quad \text{for} \quad t \in [0, T]. \tag{6.81}$$

Proposition 6.11   *We have*

(i) $\overline{D(A(t))} = \begin{cases} \{u \in C([0, 1]); u(0) = u(1) = 0\} & \beta(t) = 0, \\ \{u \in C([0, T]); u(0) = 0\} & \beta(t) \neq 0. \end{cases}$

In particular $D(A(t))$ is not dense in X.

(ii) $\sigma(A(t)) \subset (-\infty, 0]$. *If* $0 < \theta < \pi$ *and* $\lambda \in \Sigma_\theta = \{\lambda \in \mathbf{C} \setminus \{0\}; |\arg \lambda| \leq \theta\}$, *then* $\lambda \in \rho(A(t))$ *and*

$$\|R(\lambda, A(t))\| \leq \left(1 + \tan^2 \frac{\theta}{2}\right) \Big/ |\lambda|. \tag{6.82}$$

Proof. (i) The assertion is easily shown, but for the sake of completeness we give the proof of the latter case $\beta(t) \neq 0$. We write α, β short for $\alpha(t), \beta(t)$ for a fixed t. Suppose $u \in C([0, 1])$ and $u(0) = 0$. Then for $\epsilon > 0$ there exists a function $w \in C^2([0, T])$ such that $w(0) = 0$ and $\sup_{x \in [0,1]} |u(x) - w(x)| \leq \epsilon/2$. Let δ be a positive number satisfying

$$\delta < 1, \quad \delta \left| \frac{\alpha}{\beta} w(1) + w'(1) \right| \leq \frac{\epsilon}{2},$$

and ϕ a function in $C^2(-\infty, \infty)$ such that $\phi(1) = 1, 0 \leq \phi(x) \leq 1$ for $x \in (-\infty, \infty)$ and $\phi(x) = 0$ for $|x - 1| \geq \delta$. Set

$$v(x) = w(x) + \phi(x)(1 - x)\left(\frac{\alpha}{\beta} w(1) + w'(1)\right).$$

Then $v \in D(A(t))$ and

$$|v(x) - w(x)| \leq \phi(x)|1 - x|\left|\frac{\alpha}{\beta} w(1) + w'(1)\right| \leq \delta \left|\frac{\alpha}{\beta} w(1) + w'(1)\right| \leq \frac{\epsilon}{2}.$$

Hence $\|u - v\| \leq \epsilon$.

(ii) Let $\lambda \in \Sigma_\theta, 0 < \theta < \pi$. Then for $f \in X$ the boundary value problem

$$\lambda u - u'' = f, \qquad x \in [0, 1],$$
$$u(0) = 0, \quad \alpha(t)u(1) + \beta(t)u'(1) = 0$$

has the unique solution

$$u(x;t) = \int_0^1 K(x, y; t) f(y) dy, \qquad (6.83)$$

where taking the branch $\operatorname{Re}\sqrt{\lambda} \geq 0$

$$K(x, y; t)$$
$$= \begin{cases} \dfrac{\sinh \sqrt{\lambda} y}{\sqrt{\lambda}} \dfrac{\alpha(t) \sinh \sqrt{\lambda}(1-x) + \sqrt{\lambda}\beta(t) \cosh \sqrt{\lambda}(1-x)}{\alpha(t) \sinh \sqrt{\lambda} + \sqrt{\lambda}\beta(t) \cosh \sqrt{\lambda}} & y \leq x, \\[3mm] \dfrac{\sinh \sqrt{\lambda} x}{\sqrt{\lambda}} \dfrac{\alpha(t) \sinh \sqrt{\lambda}(1-y) + \sqrt{\lambda}\beta(t) \cosh \sqrt{\lambda}(1-y)}{\alpha(t) \sinh \sqrt{\lambda} + \sqrt{\lambda}\beta(t) \cosh \sqrt{\lambda}} & y \geq x. \end{cases}$$

Clearly $u(\cdot; t) = R(\lambda, A(t))f$ and in view of (6.83) we have

$$\|R(\lambda, A(t))f\| \leq \|f\| \sup_{x \in [0,1]} \int_0^1 |K(x, y; t)| dy. \qquad (6.84)$$

If we set $\rho = \operatorname{Re}\sqrt{\lambda}, \sigma = \operatorname{Im}\sqrt{\lambda}$, then $\rho > 0$ and

$$|K(x, y; t)|$$
$$\leq \begin{cases} \dfrac{\cosh \rho y}{|\sqrt{\lambda}|} \cosh \rho(1-x) \dfrac{\alpha(t) + |\sqrt{\lambda}|\beta(t)}{|\alpha(t) \sinh \sqrt{\lambda} + \sqrt{\lambda}\beta(t) \cosh \sqrt{\lambda}|} & y \leq x, \\[3mm] \dfrac{\cosh \rho x}{|\sqrt{\lambda}|} \cosh \rho(1-y) \dfrac{\alpha(t) + |\sqrt{\lambda}|\beta(t)}{|\alpha(t) \sinh \sqrt{\lambda} + \sqrt{\lambda}\beta(t) \cosh \sqrt{\lambda}|} & y \geq x. \end{cases}$$

Hence

$$\int_0^1 |K(x, y; t)| dy$$
$$\leq |\sinh \rho x \cosh \rho(1-x) + \cosh \rho x \sinh \rho(1-x)|$$
$$\times \frac{\alpha(t) + |\sqrt{\lambda}|\beta(t)}{\rho|\sqrt{\lambda}||\alpha(t) \sinh \sqrt{\lambda} + \sqrt{\lambda}\beta(t) \cosh \sqrt{\lambda}|}$$
$$= \frac{\sinh \rho}{\rho|\sqrt{\lambda}|} \frac{\alpha(t) + |\sqrt{\lambda}|\beta(t)}{|\alpha(t) \sinh \sqrt{\lambda} + \sqrt{\lambda}\beta(t) \cosh \sqrt{\lambda}|}.$$

With the aid of a direct calculation we get

$$|\alpha(t) \sinh \sqrt{\lambda} + \sqrt{\lambda}\beta(t) \cosh \sqrt{\lambda}|$$
$$= \left| \frac{e^\rho}{2} \{[(\alpha + \rho\beta) \cos \sigma - \sigma\beta \sin \sigma] + i[(\alpha + \rho\beta) \sin \sigma + \sigma\beta \cos \sigma]\} \right.$$

$$+\frac{e^{-\rho}}{2}\{[(\rho\beta-\alpha)\cos\sigma+\sigma\beta\sin\sigma]+i[(\alpha-\rho\beta)\sin\sigma+\sigma\beta\cos\sigma]\}\bigg|$$

$$\geq\bigg|\frac{e^{\rho}}{2}\{[(\alpha+\rho\beta)\cos\sigma-\sigma\beta\sin\sigma]+i[(\alpha+\rho\beta)\sin\sigma+\sigma\beta\cos\sigma]\}\bigg|$$

$$-\bigg|\frac{e^{-\rho}}{2}\{[(\rho\beta-\alpha)\cos\sigma+\sigma\beta\sin\sigma]+i[(\alpha-\rho\beta)\sin\sigma+\sigma\beta\cos\sigma]\}\bigg|$$

$$=\frac{e^{\rho}}{2}\left[(\alpha+\rho\beta)^2+\sigma^2\beta^2\right]^{1/2}-\frac{e^{-\rho}}{2}\left[(\alpha-\rho\beta)^2+\sigma^2\beta^2\right]^{1/2}$$

$$\geq\frac{1}{2}\left(e^{\rho}-e^{-\rho}\right)\left[(\alpha+\rho\beta)^2+\sigma^2\beta^2\right]^{1/2}\geq\sinh\rho\cdot(\alpha+\rho\beta).$$

Hence we have

$$|\alpha(t)\sinh\sqrt{\lambda}+\sqrt{\lambda}\beta(t)\cosh\sqrt{\lambda}|\geq\sinh\rho\cdot(\alpha(t)+\rho\beta(t)). \tag{6.85}$$

Therefore we obtain

$$\int_0^1|K(x,y;t)|dy\leq\frac{\alpha(t)+|\sqrt{\lambda}|\beta(t)}{\rho|\sqrt{\lambda}|(\alpha(t)+\rho\beta(t))}. \tag{6.86}$$

Since $|\arg\sqrt{\lambda}|\leq\theta/2<\pi/2$, we have

$$0<\rho\leq|\sqrt{\lambda}|\leq\left(1+\tan^2\frac{\theta}{2}\right)^{1/2}\rho. \tag{6.87}$$

With the aid of (6.86),(6.87) we obtain

$$\int_0^1|K(x,y;t)|dy\leq\left(1+\tan^2\frac{\theta}{2}\right)\frac{1}{|\lambda|}.$$

Combining this and (6.84) we conclude (6.82).

Proposition 6. 12 *If* $|\lambda|\geq\epsilon>0$ *and* $\lambda-\epsilon\in\Sigma_\theta$, *then* $R(\lambda,A(\cdot))\in$ $C^1([0,T];\mathcal{L}(X))$ *and satisfies*

$$\left\|\frac{\partial}{\partial t}R(\lambda,A(t))\right\|\leq\frac{C_\epsilon}{|\lambda|^{1/2}}. \tag{6.88}$$

Proof. With the aid of a direct calculation we find

$$[R(\lambda,A(t))f](x)=\frac{\sinh\sqrt{\lambda}x}{\sqrt{\lambda}}$$

$$\times\frac{\alpha(t)\int_0^1\sinh\sqrt{\lambda}(1-y)f(y)dy+\sqrt{\lambda}\beta(t)\int_0^1\cosh\sqrt{\lambda}(1-y)f(y)dy}{\alpha(t)\sinh\sqrt{\lambda}+\sqrt{\lambda}\beta(t)\cosh\sqrt{\lambda}}$$

$$-\frac{1}{\sqrt{\lambda}}\int_0^x\sinh\sqrt{\lambda}(x-y)f(y)dy.$$

Hence we have

$$\left[\frac{\partial}{\partial t} R(\lambda, A(t)) f \right](x) = \sinh \sqrt{\lambda} x$$

$$\times \frac{\alpha(t)\beta'(t) - \alpha'(t)\beta(t)}{(\alpha(t)\sinh\sqrt{\lambda} + \sqrt{\lambda}\beta(t)\cosh\sqrt{\lambda})^2} \int_0^1 f(y)\sinh\sqrt{\lambda}y\,dy. \quad (6.89)$$

Since

$$|\sinh\sqrt{\lambda}x| = \left(\sinh^2\rho x + \sin^2\sigma x\right)^{1/2},$$

we have $|\sinh\sqrt{\lambda}x| \le \sinh\rho + 1$ for $0 \le x \le 1$, and

$$\int_0^1 |\sinh\sqrt{\lambda}y|dy \le \int_0^1 (\sinh\rho y + 1)dy = \frac{1}{\rho}(\cosh\rho - 1) + 1.$$

Hence recalling (6.85) we obtain

$$\left| \left[\frac{\partial}{\partial t} R(\lambda, A(t)) f \right](x) \right|$$

$$\le (\sinh\rho + 1)\frac{C}{\sinh^2\rho \cdot (\alpha(t) + \rho\beta(t))^2} \left[\frac{1}{\rho}(\cosh\rho - 1) + 1 \right] \|f\|.$$

If $|\lambda| \ge \epsilon, \lambda - \epsilon \in \Sigma_\theta$, then ρ is bounded away from 0, and hence so are $\sinh\rho, \alpha(t) + \rho\beta(t), \sinh\rho/\cosh\rho$. Thus, using also (6.87), we conclude (6.88).

Proposition 6.13 *If* $\alpha, \beta \in C^{1+h}([0,T]), h \in (0,1)$, *then* $R(1, A(\cdot)) \in C^{1+h}([0,T]; \mathcal{L}(X))$.

Proof. The assertion is an easy consequence of (6.89).

Owing to Propositions 6.11, 6.12, 6.13 the results of the previous sections are applied to the problem

$$u_t(x,t) = u_{xx}(x,t) + f(x,t), \qquad (x,t) \in [0,1] \times [0,T],$$
$$u(0,t) = 0, \qquad t \in [0,T],$$
$$\alpha(t)u(1,t) + \beta(t)u_x(1,t) = 0, \qquad t \in [0,T],$$
$$u(x,0) = \phi(x), \qquad x \in [0,1],$$

where $f \in C([0,1] \times [0,T]), \phi \in C([0,1])$. Concerning the condition (6.52) of Theorem 6.3 there exists the following result.

Proposition 6.14 *Let $\phi \in D(A(0))$ and $f \in C([0,T]; X)$. Then, a necessary and sufficient condition in order that the condition (6.52) holds with $A(0)$ replaced by $A(0) - 1$ is that*

$$f(0,0) + \phi''(0) = 0 \qquad\qquad\qquad\qquad\text{if } \beta(0) \neq 0,$$

$$f(0,0) + \phi''(0) = f(0,1) + \phi''(1) + \frac{\beta'(0)}{\alpha(0)}\phi'(1) = 0 \qquad \text{if } \beta(0) = 0.$$

Proof. We examine the condition

$$(A(0) - 1)\phi + f(0, \cdot) + \frac{d}{dt}R(1, A(t))\bigg|_{t=0} (A(0) - 1)\phi \in \overline{D(A(0))}. \quad (6.90)$$

In view of (6.89) and the boundary condition $\phi(0) = 0$ the value of the function of (6.90) at $x = 0$ is equal to $\phi''(0) + f(0,0)$. If $\beta(0) = 0$, then the value of the same function at $x = 1$ is

$$\phi''(1) + f(0,1) + \frac{\beta'(0)}{\alpha(0)\sinh 1}\int_0^1 (\phi''(y) - \phi(y))\sinh y\, dy, \qquad (6.91)$$

since $\phi(1) = 0$. Integrating by parts twice we obtain

$$\int_0^1 (\phi''(y) - \phi(y))\sinh y\, dy = \phi'(1)\sinh 1.$$

Hence (6.91) is equal to

$$\phi''(1) + f(0,1) + \frac{\beta'(0)}{\alpha(0)}\phi'(1).$$

Consequently the result follows from Proposition 6.11 (i).

Remark 6.4 The spaces $D_{A(t)}(\theta, \infty)$ are characterized as follows: if $0 < \theta < 1/2$

$$D_{A(t)}(\theta, \infty) = C^{2\theta}([0,1]) \cap \overline{D(A(t))},$$

if $1/2 < \theta < 1$

$$D_{A(t)}(\theta, \infty) = \{u \in C^{2\theta}([0,1]); u(0) = \alpha(t)u(1) + \beta(t)u'(1) = 0\},$$

if $\theta = 1/2$

$$D_{A(t)}(1/2, \infty)$$

$$= \begin{cases} \{u \in C^{*,1}([0,1]); u(0) = 1, \ \sup_{x \in [0,1)} |u(x) - u(1)|/|x-1| < \infty\} \\ \hspace{7cm} \beta(t) > 0, \\ \{u \in C^{*,1}([0,1]); u(0) = u(1) = 0\}, \hspace{1.5cm} \beta(t) = 0, \end{cases}$$

where $C^{*,1}([0,1])$ is the *Zygmund class* of functions defined by

$$C^{*,1}([0,1])$$
$$= \left\{ u \in C([0,1]); \sup_{x,y\in[0,1],x\neq y} \frac{|u(x) + u(y) - 2u((x+y)/2)|}{|x-y|} < \infty \right\}.$$

The proofs are found in P. Acquistapace and B. Terreni [4],[5], G. Da Prato and P. Grisvard [39] and A. Lunardi [101]. See also P. Acquistapace and B. Terreni [7].

6.8   Equations with Coefficients Hölder Continuous in t

In the papers [1],[6],[8] P. Acquistapace and B. Terreni proved under very general hypotheses the existence and uniqueness of the solutions of (6.1), (6.2). It is also shown that the hypothses are independent of those of section 6.1. One of the hypotheses reads

(P4) There exist a positive constant L, a positive integer k and real numbers $\alpha_1, \ldots, \alpha_k, \beta_1, \ldots, \beta_k$ with $0 \leq \beta_i < \alpha_i \leq 2$ such that

$$\|A(t)R(\lambda, A(t))(A(t)^{-1} - A(s)^{-1})\| \leq L \sum_{i=1}^{k}(t-s)^{\alpha_i}|\lambda|^{\beta_i - 1} \qquad (6.92)$$

for $\lambda \in \Sigma = \{\lambda; |\arg \lambda| < \theta_0\} \setminus \{0\}$ and $0 \leq s < t \leq T$. Set

$$\delta = \min\{\min_{i=1,\ldots,k}(\alpha_i - \beta_i), 1\}. \qquad (6.93)$$

In [1],[6] it is shown that the evolution operator $U(t,s)$ of (6.1),(6.2) is given by

$$U(t,s) = e^{(t-s)A(s)} + \int_s^t Z(\tau, s)d\tau, \qquad (6.94)$$

$$Z(t,s) = A(t)e^{(t-s)A(t)} - A(s)e^{(t-s)A(s)}$$
$$+ \int_s^t R(t,\tau)(A(\tau)e^{(\tau-s)A(\tau)} - A(s)e^{(\tau-s)A(s)})d\tau$$
$$+ \int_s^t (R(t,\tau) - R(t,s))A(s)e^{(\tau-s)A(s)}d\tau$$
$$+ R(t,s)(e^{(t-s)A(s)} - 1), \qquad (6.95)$$

where $R(t, s)$ is the solution of the integral equation

$$R(t, s) = Q(t, s) + \int_s^t R(t, \tau)Q(\tau, s)d\tau$$

$$= Q(t, s) + \int_s^t Q(t, \tau)R(\tau, s)d\tau \qquad (6.96)$$

and $Q(t, s)$ is the operator valued function

$$Q(t, s) = A(t)^2 e^{(t-s)A(t)} \left(A(t)^{-1} - A(s)^{-1}\right). \qquad (6.97)$$

By a formal calulation we explain a general idea which suggests that $U(t, s)$ given above is an evolution operator of (6.1),(6.2).

Let u be a strict solution of

$$du(t)/dt = A(t)u(t), \quad t \in [0, T].$$

For a fixed $0 \le s < t \le T$ set

$$v(\sigma) = e^{(t-\sigma)A(t)}u(\sigma), \quad \sigma \in [s, t].$$

Then

$$v'(\sigma) = e^{(t-\sigma)A(t)}A(\sigma)u(\sigma) - A(t)e^{(t-\sigma)A(t)}u(\sigma)$$

$$= A(t)e^{(t-\sigma)A(t)}(A(t)^{-1} - A(\sigma)^{-1})A(\sigma)u(\sigma). \qquad (6.98)$$

Integration from s to t yields

$$u(t) = e^{(t-s)A(t)}u(s) + \int_s^t A(t)e^{(t-\sigma)A(t)}(A(t)^{-1} - A(\sigma)^{-1})A(\sigma)u(\sigma)d\sigma.$$

Letting $A(t)$ operate on both sides

$$A(t)u(t) = A(t)e^{(t-s)A(t)}u(s) + \int_s^t Q(t, \sigma)A(\sigma)u(\sigma)d\sigma. \qquad (6.99)$$

From (6.96),(6.99) we conclude

$$\int_s^t Q(t, \tau)A(\tau)u(\tau)d\tau = \int_s^t R(t, \tau)A(\tau)e^{(\tau-s)A(\tau)}u(s)d\tau.$$

Substituting this in the right hand side of (6.99) and rewriting the result appropriately

$$A(t)u(t) = A(t)e^{(t-s)A(t)}u(s) + \int_s^t R(t, \tau)A(\tau)e^{(\tau-s)A(\tau)}u(s)d\tau$$

$$= A(s)e^{(t-s)A(s)}u(s) + (A(t)e^{(t-s)A(t)} - A(s)e^{(t-s)A(s)})u(s)$$
$$+ \int_s^t R(t,\tau)(A(\tau)e^{(\tau-s)A(\tau)} - A(s)e^{(\tau-s)A(s)})u(s)d\tau$$
$$+ \int_s^t (R(t,\tau) - R(t,s))A(s)e^{(\tau-s)A(s)}u(s)d\tau$$
$$+ R(t,s)(e^{(t-s)A(s)} - 1)u(s)$$
$$= A(s)e^{(t-s)A(s)}u(s) + Z(t,s)u(s).$$

Integrating the above from s to t we conclude

$$u(t) = e^{(t-s)A(s)}u(s) + \int_s^t Z(\tau,s)u(s)d\tau = U(t,s)u(s). \tag{6.100}$$

Thus one might expect that $U(t,s)$ defined by (6.94)-(6.97) would be a desired evolution operator.

6.9   Preliminary Lemmas and Remarks

Lemma 6.7 *For $0 \le s < t \le T, \lambda \in \Sigma$*

$$\|R(\lambda, A(t)) - R(\lambda, A(s))\| \le C \sum_{i=1}^k (t-s)^{\alpha_i}|\lambda|^{\beta_i - 1}. \tag{6.101}$$

Proof. This is a simple consequence of (6.92) and

$$R(\lambda, A(t)) - R(\lambda, A(s))$$
$$= A(t)R(\lambda, A(t))(A(s)^{-1} - A(t)^{-1})A(s)R(\lambda, A(s)). \tag{6.102}$$

Lemma 6.8 *For a nonnegative integer m, a positive number σ and $0 \le s < t \le T$,*

$$\|A(t)^m e^{\sigma A(t)} - A(s)^m e^{\sigma A(s)}\| \le C \sum_{i=1}^k \frac{(t-s)^{\alpha_i}}{\sigma^{m+\beta_i}}, \tag{6.103}$$

$$\|A(t)^m e^{\sigma A(t)} - A(s)^m e^{\sigma A(s)}\|_{\mathcal{L}(D(A(s)),X)} \le C \sum_{i=1}^k \frac{(t-s)^{\alpha_i}}{\sigma^{m+\beta_i-1}}. \tag{6.104}$$

Proof. These are simple consequences of

$$A(t)^m e^{\sigma A(t)} - A(s)^m e^{\sigma A(s)}$$
$$= \frac{1}{2\pi i} \int_\Gamma \lambda^m e^{\lambda\sigma} (R(\lambda, A(t)) - R(\lambda, A(s)))d\lambda \tag{6.105}$$

and Lemma 6.7.

Corollary 6.1 *For* $0 \leq s < t \leq T$

$$\|A(t)e^{(t-s)A(t)} - A(s)e^{(t-s)A(s)}\| \leq C\sum_{i=1}^{k}(t-s)^{\alpha_i-\beta_i-1} \leq C(t-s)^{\delta-1}.$$

$$(6.106)$$

Recall that δ is the number defined by (6.93).

Lemma 6.9 (i) *For* $0 \leq s < t \leq T$

$$\|Q(t,s)\| \leq C\sum_{i=1}^{k}(t-s)^{\alpha_i-\beta_i-1} \leq C(t-s)^{\delta-1}. \qquad (6.107)$$

(ii) *For* $0 \leq s < \tau < t \leq T$ *and* $\eta \in (0, \delta)$

$$\|Q(t,s) - Q(t,\tau)\| \leq C_\eta(\tau-s)^\eta(t-\tau)^{\delta-\eta-1}. \qquad (6.108)$$

Proof. (i) (6.107) readily follows from (6.92) and

$$Q(t,s) = \frac{1}{2\pi i}\int_\Gamma \lambda e^{\lambda(t-s)}A(t)R(\lambda, A(t))\left(A(t)^{-1} - A(s)^{-1}\right)d\lambda.$$

(ii) Using

$$\begin{aligned}
Q(t,s) &- Q(t,\tau) \\
&= (A(t)^2 e^{(t-s)A(t)} - A(\tau)^2 e^{(t-s)A(\tau)})(A(\tau)^{-1} - A(s)^{-1}) \\
&+ A(\tau)^2 e^{(t-s)A(\tau)}(A(\tau)^{-1} - A(s)^{-1}) \\
&+ (A(t)^2 e^{(t-s)A(t)} - A(t)^2 e^{(t-\tau)A(t)})\left(A(t)^{-1} - A(\tau)^{-1}\right) \\
&= I_1 + I_2 + I_3,
\end{aligned}$$

where

$$\begin{aligned}
I_1 &= \frac{1}{2\pi i}\int_\Gamma \lambda^2 e^{\lambda(t-s)}A(t)R(\lambda, A(t)) \\
&\quad \times (A(\tau)^{-1} - A(t)^{-1})A(\tau)R(\lambda, A(\tau))(A(\tau)^{-1} - A(s)^{-1})d\lambda, \\
I_2 &= \frac{1}{2\pi i}\int_\Gamma \lambda e^{\lambda(t-s)}A(\tau)R(\lambda, A(\tau))(A(\tau)^{-1} - A(s)^{-1})d\lambda, \\
I_3 &= \frac{1}{2\pi i}\int_\Gamma \lambda(e^{\lambda(t-s)} - e^{\lambda(t-\tau)})A(t)R(\lambda, A(t))(A(t)^{-1} - A(\tau)^{-1})d\lambda \\
&= \frac{1}{2\pi i}\int_{t-\tau}^{t-s}\int_\Gamma \lambda^2 e^{\lambda\sigma}A(t)R(\lambda, A(t))(A(t)^{-1} - A(\tau)^{-1})d\lambda d\sigma,
\end{aligned}$$

we get

$$\|Q(t,s) - Q(t,\tau)\|$$

$$\leq C\sum_{i=1}^{k} \frac{(\tau-s)^{\alpha_i}}{(t-s)^{\beta_i+1}}\left[1 + \sum_{i=1}^{k}\frac{(t-\tau)^{\alpha_i}}{(t-s)^{\beta_i}}\right] + C\sum_{i=1}^{k}\int_{t-\tau}^{t-s}\frac{(t-\tau)^{\alpha_i}}{\sigma^{\beta_i+2}}d\sigma.$$

Noting

$$\frac{(t-\tau)^{\alpha_i}}{(t-s)^{\beta_i}} \leq (t-s)^{\alpha_i-\beta_i}, \quad \frac{(\tau-s)^{\alpha_i}}{(t-s)^{\beta_i+1}} \leq \frac{(\tau-s)^{\alpha_i-\beta_i}}{t-s} \leq C\frac{(\tau-s)^\delta}{t-s},$$

$$\int_{t-\tau}^{t-s}\frac{(t-\tau)^{\alpha_i}}{\sigma^{\beta_i+2}}d\sigma \leq (t-\tau)^{\alpha_i-\beta_i}\int_{t-\tau}^{t-s}\frac{d\sigma}{\sigma^2}$$

$$= \frac{(t-\tau)^{\alpha_i-\beta_i-1}(\tau-s)}{t-s} \leq C\frac{(t-\tau)^{\delta-1}(\tau-s)}{t-s},$$

we can easily show (6.108).

Lemma 6.10 (i) *For* $0 \leq s < t \leq T$

$$\|R(t,s)\| \leq C(t-s)^{\delta-1}. \tag{6.109}$$

(ii) *For* $0 \leq s < \tau < t \leq T$ *and* $\eta \in (0,\delta)$

$$\|R(t,s) - R(t,\tau)\| \leq C_\eta(\tau-s)^\eta(t-\tau)^{\delta-\eta-1}. \tag{6.110}$$

Proof. (i) (6.109) is easily obtained by estimating each term of the successive approximation series for the solution of the integral equation (6.96).
(ii) This is a simple consequence of (6.96) and (6.108).

It was pointed out in P. Acquistapace and B. Terreni [8] that the sets of the assumptions (P1),(P2),(P3) and (P1),(P2),(P4) are independent of each other. In order to show that it suffices to verify that (P1),(P2),(P3) do not imply (P4).

Lemma 6.11 *Suppose in addition to the assumptions* (P1),(P2),(P4) *that* $A(\cdot)^{-1} \in C^1([0,T];\mathcal{L}(X))$. *Then we have* $R\left(\dfrac{d}{dt}A(t)^{-1}\right) \subset \overline{D(A(t))}$ *for* $t \in [0,T]$.

Proof. For $u \in X$ and $0 \leq t < t+h \leq T$ we have

$$\frac{A(t+h)^{-1} - A(t)^{-1}}{h}u$$

$$= \frac{1}{h}\left[R\left(\frac{1}{h}, A(t+h)\right) - R\left(\frac{1}{h}, A(t)\right)\right]\frac{A(t+h)^{-1} - A(t)^{-1}}{h}u$$

$$+\frac{1}{h}R\left(\frac{1}{h},A(t)\right)\left[\frac{A(t+h)^{-1}-A(t)^{-1}}{h}-\frac{d}{dt}A(t)^{-1}\right]u$$

$$+\frac{1}{h}R\left(\frac{1}{h},A(t)\right)\frac{d}{dt}A(t)^{-1}u$$

$$-A(t+h)R\left(\frac{1}{h},A(t+h)\right)\frac{A(t+h)^{-1}-A(t)^{-1}}{h}u=\sum_{i=1}^{4}I_i.$$

In view of Lemma 6.7 and the assumption (P4)

$$\|I_1\|\le C\sum_{i=1}^{k}h^{\alpha_i-\beta_i},\quad \|I_4\|\le\sum_{i=1}^{k}h^{\alpha_i-\beta_i}.$$

Clearly $I_2\to 0$ as $h\to 0$. Hence

$$\frac{d}{dt}A(t)^{-1}u=\lim_{h\to 0}I_3\in\overline{D(A(t))}.$$

In the example of section 6.7 let $\alpha(t)=1,\beta(t)=t$. Then

$$[A(t)^{-1}f](x)=\int_0^x(x-y)f(y)dy-\left[\int_0^1 f(y)dy-\frac{1}{1+t}\int_0^1 yf(y)dy\right]x.$$

Therefore

$$\left[\frac{d}{dt}A(t)^{-1}f\right](x)=-\frac{1}{(1+t)^2}\int_0^1 yf(y)dy\cdot x.\qquad(6.111)$$

Suppose that the assumption (P4) holds. Then by Lemma 6.11 we have

$$R\left(\frac{d}{dt}A(t)^{-1}\bigg|_{t=0}\right)\subset\overline{D(A(0))}=\{u\in C([0,T]);u(0)=u(1)=0\}.$$

However, if we take $f(x)\equiv 1$ in (6.111), we have

$$\left[\frac{d}{dt}A(t)^{-1}\bigg|_{t=0}f\right](x)=-\frac{x}{2},$$

and the function does not belong to $\overline{D(A(0))}$, which is a contradiction. Therefore (P4) does not hold.

6.10 Representation of Solutions by Fundamental Solution

Lemmas 6.7, 6.8, 6.9, 6.10 show that all the integrals in (6.94),(6.95),(6.96) are convergent. Hence the operator valued function $U(t,s)$ is well defined by these equalities and (6.97), and we have

$$\|U(t,s)\| \le C, \quad \left\|\frac{\partial}{\partial t}U(t,s)\right\| \le \frac{C}{t-s}, \quad \|Z(t,s)\| \le C(t-s)^{\delta-1}. \quad (6.112)$$

For $r \in [0,T)$ and $g \in C([r,T];X)$ set

$$(Q_r g)(t) = \int_r^t Q(t,s)g(s)ds. \quad (6.113)$$

By virtue of Lemma 6.9 (i) $Q_r \in \mathcal{L}(C([r,T];X))$ and the inverse of $I - Q_r$ is given by

$$((I - Q_r)^{-1}f)(t) = f(t) + \int_r^t R(t,\tau)f(\tau)d\tau. \quad (6.114)$$

If $r = 0$, we write $Q_r = Q$. Using (6.114) $Z(t,s)$ is also given by

$$Z(t,s) = \left[(1 - Q_s)^{-1}(A(\cdot)e^{(\cdot-s)A(\cdot)} - A(s)e^{(\cdot-s)A(s)})\right](t)$$
$$+ \int_s^t (R(t,\tau) - R(t,s))A(s)e^{(\tau-s)A(s)}d\tau$$
$$+ R(t,s)(e^{(t-s)A(s)} - 1). \quad (6.115)$$

Lemma 6.12 Let $f \in C([0,T];X)$ and $x \in D(A(0))$. If a strict solution u of (6.1),(6.2) exists, then $A(0)x + f(0) \in \overline{D(A(0))}$.

Proof. By a direct calculation we have

$$\frac{u(t)-x}{t} = \frac{1}{t}\left[R\left(\frac{1}{t},A(t)\right) - R\left(\frac{1}{t},A(0)\right)\right]\frac{u(t)-x}{t}$$
$$+ \frac{1}{t}R\left(\frac{1}{t},A(0)\right)\left[\frac{u(t)-x}{t} - (A(0)x + f(0))\right]$$
$$+ \frac{1}{t}R\left(\frac{1}{t},A(0)\right)(A(0)x + f(0)) - \frac{1}{t}R\left(\frac{1}{t},A(t)\right)(A(t)u(t) - A(0)x)$$
$$- \frac{1}{t}A(t)R\left(\frac{1}{t},A(t)\right)(A(t)^{-1} - A(0)^{-1})A(0)x = \sum_{i=1}^{5}I_i.$$

With the aid of Lemma 6.7 we get

$$\|I_1\| \le C \sum_{i=1}^{k} t^{\alpha_i - \beta_i} \sup_{t \in [0,T]} \left\| \frac{u(t) - x}{t} \right\|,$$

and by (6.92)

$$\|I_5\| \le \sum_{i=1}^{k} t^{\alpha_i - \beta_i} \|A(0)x\|.$$

Clearly $I_2 \to 0$, $I_4 \to 0$ as $t \to 0$. Hence

$$A(0)x + f(0) = u'(0) = \lim_{t \to 0} \frac{u(t) - x}{t}$$
$$= \lim_{t \to 0} \frac{1}{t} R\left(\frac{1}{t}, A(0)\right)(A(0)x + f(0)) \in \overline{D(A(0))}.$$

The following theorem is due to P. Acquistapace and B. Terreni [6].

Theorem 6.5 (i) *Let* $f \in C([0,T]; X), x \in D(A(0))$ *and suppose that* $A(0)x + f(0) \in \overline{D(A(0))}$. *If* u *is a strict solution of* (6.1),(6.2), *then* u *is given by*

$$u(t) = U(t,0)x + \int_0^t U(t,s)f(s)ds, \quad t \in [0,T]. \tag{6.116}$$

(ii) *Let* $f \in C([0,T]; X)$ *and* $x \in \overline{D(A(0))}$. *If* u *is a classical solution of the problem* (6.1),(6.2), *then* (6.116) *holds for* $t \in (0,T]$.

Consequently a classical solution, and hence a strict solution also, is uniquely determined by the inhomogeneous term f *and the initial value* x.

Proof. (i) Let u be a strict solution of (6.1),(6.2). For $0 < t \le T$ set

$$v(s) = e^{(t-s)A(t)}u(s), \quad s \in [0,t].$$

Differentiating both sides we get this time instead of (6.98)

$$v'(s) = A(t)e^{(t-s)A(t)}(A(t)^{-1} - A(s)^{-1})A(s)u(s) + e^{(t-s)A(t)}f(s). \tag{6.117}$$

Let $0 < \epsilon < 1$. Integrating both sides of (6.117) from 0 to $t - \epsilon t$ we get

$$v(t - \epsilon t) - v(0)$$
$$= \int_0^{t-\epsilon t} A(t)e^{(t-s)A(t)}(A(t)^{-1} - A(s)^{-1})A(s)u(s)ds$$
$$+ \int_0^{t-\epsilon t} e^{(t-s)A(t)}f(s)ds. \tag{6.118}$$

Letting $A(t)$ operate on both sides of (6.118) and using the definition of $v(\cdot)$ we get

$$A(t)e^{\epsilon tA(t)}u(t-\epsilon t) - A(t)e^{tA(t)}x$$
$$= \int_0^{t-\epsilon t} Q(t,s)A(s)u(s)ds + \int_0^{t-\epsilon t} A(t)e^{(t-s)A(t)}f(s)ds.$$

We rewrite this as follows:

$$A(t)u(t) - \int_0^t Q(t,s)A(s)u(s)ds = G_\epsilon(t), \qquad (6.119)$$

where

$$G_\epsilon(t) = -\int_{t-t\epsilon}^t Q(t,s)A(s)u(s)ds + A(t)u(t)$$
$$-A(t)e^{\epsilon tA(t)}u(t-t\epsilon) + A(t)e^{tA(t)}x + A(t)\int_0^{t-t\epsilon} e^{(t-s)A(t)}f(s)ds$$
$$= -\int_{t-\epsilon t}^t Q(t,s)A(s)u(s)ds$$
$$-(e^{\epsilon tA(t)} - 1 - \epsilon t A(t)e^{\epsilon tA(t)})(A(t)u(t) + f(t)) + (e^{\epsilon tA(t)} - 1)f(t)$$
$$-A(t)e^{\epsilon tA(t)}(u(t-\epsilon t) - u(t) + \epsilon tu'(t)) + (A(t)e^{tA(t)} - A(0)e^{tA(0)})x$$
$$+A(0)e^{tA(0)}x + \int_0^{t-\epsilon t}(A(t)e^{(t-s)A(t)} - A(s)e^{(t-s)A(s)})f(s)ds$$
$$+\int_0^{t-\epsilon t} A(s)e^{(t-s)A(s)}f(s)ds = \sum_{i=1}^8 I_i.$$

In view of (6.113) with $r = 0$ we obtain from (6.119)

$$A(t)u(t) = [(I-Q)^{-1}G_\epsilon](t),$$

and hence

$$u'(t) = [(I-Q)^{-1}G_\epsilon](t) + f(t).$$

Integrating this from ϵt to t

$$u(t) = u(\epsilon t) + \int_{\epsilon t}^t f(s)ds + \int_{\epsilon t}^t [(I-Q)^{-1}G_\epsilon](\tau)d\tau. \qquad (6.120)$$

Using (6.114),(6.109),(6.107) we get

$$\int_{\epsilon t}^t \|[(1-Q)^{-1}(I_1 + I_2 + I_4)](\tau)\|d\tau \le C\int_0^t \|I_1(\tau) + I_2(\tau) + I_4(\tau)\|d\tau,$$

$$\|I_1(\tau)\| \le \int_{\tau-\epsilon\tau}^{\tau} \|Q(\tau,s)A(s)u(s)\| ds \le C(\epsilon\tau)^\delta \sup_{s\in[0,T]} \|A(s)u(s)\|,$$

$$\|I_2(\tau)\| \le C\|A(\tau)u(\tau) + f(\tau)\|.$$

By the proof of Lemma 6.12 we see that $A(t)u(t) + f(t) \in \overline{D(A(t))}$. Using this and Lemma 1.5 (ii) we see that $\|I_2(\tau)\| \to 0$ as $\epsilon \to 0$. Noting

$$u(t-\epsilon t) - u(t) + \epsilon t u'(t) = \epsilon t \int_0^1 (u'(t) - u'(t-\epsilon\sigma t))d\sigma,$$

we obtain

$$\|I_4(\tau)\| = \left\| \epsilon t A(t) e^{\epsilon t A(t)} \int_0^1 (u'(t) - u'(t-\epsilon\sigma t))d\sigma \right\|$$

$$\le C \int_0^1 \|u'(t) - u'(t-\epsilon\sigma t)\| d\sigma \to 0$$

as $\epsilon \to 0$, and

$$\|I_4(\tau)\| \le C \sup_{s\in[0,T]} \|u'(s)\|.$$

Therefore by Lebesque's dominated convergence theorem we find that

$$\int_{\epsilon t}^{t} \left\| \left[(1-Q)^{-1}(I_1+I_2+I_4) \right](\tau) \right\| d\tau \to 0. \tag{6.121}$$

Since $\|I_5\| \le Ct^{\delta-1}\|x\|$ by Corollary 6.1, we have

$$\lim_{\epsilon\to0} \int_{\epsilon t}^{t} \left[(I-Q)^{-1}I_5 \right](\tau)d\tau$$

$$= \int_0^t \left[(I-Q)^{-1}(A(\cdot)e^{\cdot A(\cdot)} - A(0)e^{\cdot A(0)})x \right](\tau)d\tau. \tag{6.122}$$

Similarly

$$\lim_{\epsilon\to0} \int_{\epsilon t}^{t} \left[(I-Q)^{-1}I_7 \right](\tau)d\tau = \int_0^t \left[(I-Q)^{-1} \right.$$

$$\left. \left(\int_0^\cdot (A(\cdot)e^{(\cdot-s)A(\cdot)} - A(s)e^{(\cdot-s)A(s)})f(s)ds \right) \right](\tau)d\tau. \tag{6.123}$$

Using (6.114) in the right hand side of (6.123) we find that

$$\lim_{\epsilon\to0} \int_{\epsilon t}^{t} \left[(I-Q)^{-1}I_7 \right](\tau)d\tau$$

$$= \int_0^t \left[\int_0^\tau (A(\tau)e^{(\tau-s)A(\tau)} - A(s)e^{(\tau-s)A(s)})f(s)ds \right.$$
$$\left. + \int_0^\tau R(\tau,\sigma) \int_0^\sigma (A(\sigma)e^{(\sigma-s)A(\sigma)} - A(s)e^{(\sigma-s)A(s)})f(s)dsd\sigma \right]d\tau$$

$$= \int_0^t \int_0^\tau (A(\tau)e^{(\tau-s)A(\tau)} - A(s)e^{(\tau-s)A(s)})f(s)dsd\tau$$
$$+ \int_0^t \int_0^\tau \int_s^\tau R(\tau,\sigma)(A(\sigma)e^{(\sigma-s)A(\sigma)} - A(s)e^{(\sigma-s)A(s)})d\sigma f(s)dsd\tau$$

$$= \int_0^t \int_s^t (A(\tau)e^{(\tau-s)A(\tau)} - A(s)e^{(\tau-s)A(s)})d\tau f(s)ds$$
$$+ \int_0^t \int_s^t \int_s^\tau R(\tau,\sigma)(A(\sigma)e^{(\sigma-s)A(\sigma)} - A(s)e^{(\sigma-s)A(s)})d\sigma d\tau f(s)ds$$

$$= \int_0^t \int_s^t \left[(I-Q_s)^{-1}(A(\cdot)e^{(\cdot-s)A(\cdot)} - A(s)e^{(\cdot-s)A(s)}) \right](\tau)d\tau f(s)ds.$$

$$\text{(6.124)}$$

Since

$$\int_{\epsilon t}^t \left[(I-Q)^{-1}I_6 \right](\tau)d\tau$$

$$= \int_{\epsilon t}^t \left[A(0)e^{\tau A(0)}x + \int_0^\tau R(\tau,s)A(0)e^{sA(0)}xds \right]d\tau$$

$$= (e^{tA(0)} - e^{\epsilon tA(0)})x + \int_{\epsilon t}^t \int_0^\tau (R(\tau,s) - R(\tau,0))A(0)e^{sA(0)}xdsd\tau$$

$$+ \int_{\epsilon t}^t R(\tau,0)(e^{\tau A(0)} - 1)xd\tau,$$

we see

$$\lim_{\epsilon \to 0} \int_{\epsilon t}^t \left[(I-Q)^{-1}I_6 \right](\tau)d\tau$$

$$= (e^{tA(0)} - 1)x + \int_0^t \int_0^\tau (R(\tau,s) - R(\tau,0))A(0)e^{sA(0)}xdsd\tau$$

$$+ \int_0^t R(\tau,0)(e^{\tau A(0)} - 1)xd\tau. \qquad \text{(6.125)}$$

Finally as for I_3, I_8 we have

$$\int_{\epsilon t}^t \left[(I-Q)^{-1}(I_3 + I_8) \right](\tau)d\tau$$

$$= \int_{\epsilon t}^{t} \left\{ (I - Q)^{-1} \left[(e^{\epsilon \cdot A(\cdot)} - 1) f + \int_{0}^{\cdot - \epsilon \cdot} A(s) e^{(\cdot - s) A(s)} f(s) ds \right] \right\} (\tau) d\tau$$

$$= \int_{\epsilon t}^{t} \left[(I - Q)^{-1} (e^{\epsilon \cdot A(\cdot)} - 1) f \right] (\tau) d\tau$$

$$+ \int_{\epsilon t}^{t} \int_{0}^{\tau - \epsilon \tau} A(s) e^{(\tau - s) A(s)} f(s) ds d\tau$$

$$+ \int_{\epsilon t}^{t} \int_{0}^{\tau} R(\tau, \sigma) \int_{0}^{\sigma - \epsilon \sigma} A(s) e^{(\sigma - s) A(s)} f(s) ds d\sigma d\tau$$

$$= \int_{\epsilon t}^{t} \left[(I - Q)^{-1} (e^{\epsilon \cdot A(\cdot)} - 1) f \right] (\tau) d\tau$$

$$+ \int_{\epsilon t}^{t} \int_{0}^{\tau - \epsilon \tau} A(s) e^{(\tau - s) A(s)} f(s) ds d\tau$$

$$+ \int_{\epsilon t}^{t} \int_{0}^{\tau} \int_{0}^{\sigma - \epsilon \sigma} (R(\tau, \sigma) - R(\tau, s)) A(s) e^{(\sigma - s) A(s)} f(s) ds d\sigma d\tau$$

$$+ \int_{\epsilon t}^{t} \int_{0}^{\tau} \int_{0}^{\sigma - \epsilon \sigma} R(\tau, s) A(s) e^{(\sigma - s) A(s)} f(s) ds d\sigma d\tau$$

$$= \int_{\epsilon t}^{t} \left[(I - Q)^{-1} (e^{\epsilon \cdot A(\cdot)} - 1) f \right] (\tau) d\tau$$

$$+ \int_{0}^{(1 - \epsilon) \epsilon t} \int_{\epsilon t}^{t} A(s) e^{(\tau - s) A(s)} d\tau f(s) ds$$

$$+ \int_{(1 - \epsilon) \epsilon t}^{t - \epsilon t} \int_{s/(1 - \epsilon)}^{t} A(s) e^{(\tau - s) A(s)} d\tau f(s) ds$$

$$+ \int_{\epsilon t}^{t} \int_{0}^{\tau} \int_{0}^{\sigma - \epsilon \sigma} (R(\tau, \sigma) - R(\tau, s)) A(s) e^{(\sigma - s) A(s)} f(s) ds d\sigma d\tau$$

$$+ \int_{\epsilon t}^{t} \int_{0}^{\tau - \epsilon \tau} R(\tau, s) \int_{s/(1 - \epsilon)}^{\tau} A(s) e^{(\sigma - s) A(s)} d\sigma f(s) ds d\tau$$

$$= \int_{\epsilon t}^{t} \left[(I - Q)^{-1} (e^{\epsilon \cdot A(\cdot)} - 1) f \right] (\tau) d\tau$$

$$+ \int_{0}^{(1 - \epsilon) \epsilon t} (e^{(t - s) A(s)} - e^{(\epsilon t - s) A(s)}) f(s) ds$$

$$+ \int_{(1 - \epsilon) \epsilon t}^{t - \epsilon t} (e^{(t - s) A(s)} - e^{(s/(1 - \epsilon) - s) A(s)}) f(s) ds$$

$$+ \int_{\epsilon t}^{t} \int_{0}^{\tau} \int_{0}^{\sigma - \epsilon \sigma} (R(\tau, \sigma) - R(\tau, s)) A(s) e^{(\sigma - s) A(s)} f(s) ds d\sigma d\tau$$

$$+ \int_{\epsilon t}^{t} \int_{0}^{\tau - \epsilon \tau} R(\tau, s)(e^{(\tau - s)A(s)} - e^{(s/(1-\epsilon) - s)A(s)})f(s)ds d\tau$$

$$= \int_{\epsilon t}^{t} \left[(I - Q)^{-1}(e^{\epsilon \cdot A(\cdot)}f - e^{(\epsilon \cdot /(1-\epsilon))A(\cdot)}f) \right](\tau)d\tau$$

$$+ \int_{\epsilon t}^{t} e^{(\epsilon \tau/(1-\epsilon))A(\tau)}f(\tau)d\tau + \int_{\epsilon t}^{t} \int_{0}^{\tau} R(\tau, s)e^{(\epsilon s/(1-\epsilon))A(s)}f(s)ds d\tau$$

$$- \int_{\epsilon t}^{t} \left[(I - Q)^{-1}f \right](\tau)d\tau + \int_{0}^{(1-\epsilon)\epsilon t} (e^{(t-s)A(s)} - e^{(\epsilon t - s)A(s)})f(s)ds$$

$$+ \int_{(1-\epsilon)\epsilon t}^{t-\epsilon t} e^{(t-s)A(s)}f(s)ds - \int_{(1-\epsilon)\epsilon t}^{t-\epsilon t} e^{(\epsilon s/(1-\epsilon))A(s)}f(s)ds$$

$$+ \int_{\epsilon t}^{t} \int_{0}^{\tau} \int_{0}^{\sigma - \epsilon \sigma} (R(\tau, \sigma) - R(\tau, s))A(s)e^{(\sigma - s)A(s)}f(s)ds d\sigma d\tau$$

$$+ \int_{\epsilon t}^{t} \int_{0}^{\tau - \epsilon \tau} R(\tau, s)e^{(\tau - s)A(s)}f(s)ds d\tau$$

$$- \int_{\epsilon t}^{t} \int_{0}^{\tau - \epsilon \tau} R(\tau, s)e^{(\epsilon s/(1-\epsilon))A(s)}f(s)ds d\tau$$

$$= \int_{\epsilon t}^{t} \left[(I - Q)^{-1}(e^{\epsilon \cdot A(\cdot)}f - e^{(\epsilon \cdot /(1-\epsilon))A(\cdot)}f) \right](\tau)d\tau$$

$$+ \left[\int_{\epsilon t}^{t} e^{(\epsilon s/(1-\epsilon))A(s)}f(s)ds - \int_{(1-\epsilon)\epsilon t}^{t-\epsilon t} e^{(\epsilon s/(1-\epsilon))A(s)}f(s)ds \right]$$

$$+ \int_{\epsilon t}^{t} \int_{\tau - \epsilon \tau}^{\tau} R(\tau, s)e^{(\epsilon s/(1-\epsilon))A(s)}f(s)ds d\tau - \int_{\epsilon t}^{t} \left[(I - Q)^{-1}f \right](\tau)d\tau$$

$$+ \int_{0}^{(1-\epsilon)\epsilon t} (e^{(t-s)A(s)} - e^{(\epsilon t - s)A(s)})f(s)ds + \int_{(1-\epsilon)\epsilon t}^{t-\epsilon t} e^{(t-s)A(s)}f(s)ds$$

$$+ \int_{\epsilon t}^{t} \int_{0}^{\tau} \int_{0}^{\sigma - \epsilon \sigma} (R(\tau, \sigma) - R(\tau, s))A(s)e^{(\sigma - s)A(s)}f(s)ds d\sigma d\tau$$

$$+ \int_{\epsilon t}^{t} \int_{0}^{\tau - \epsilon \tau} R(\tau, s)e^{(\tau - s)A(s)}f(s)ds d\tau. \tag{6.126}$$

Since

$$\| (e^{\epsilon \tau A(\tau)} - e^{(\epsilon \tau/(1-\epsilon))A(\tau)})f(\tau) \|$$

$$\leq C \int_{\epsilon \tau}^{\epsilon \tau/(1-\epsilon)} \frac{d\sigma}{\sigma} \|f(\tau)\| = C \log \frac{1}{1-\epsilon} \|f(\tau)\|,$$

we see that the first term of the last side of (6.126) goes to 0 as $\epsilon \to 0$. The second term also tends to 0 since it is equal to

$$\left(\int_{t-\epsilon t}^{t} - \int_{(1-\epsilon)\epsilon t}^{\epsilon t} \right) e^{(\epsilon s/(1-\epsilon))A(s)} f(s) ds.$$

It is easily seen that the third and the fifth terms also go to 0. Hence we obtain

$$\lim_{\epsilon \to 0} \int_{\epsilon t}^{t} \left[(I-Q)^{-1}(I_3 + I_8) \right] (\tau) d\tau$$

$$= - \int_0^t \left[(I-Q)^{-1} f \right] (\tau) d\tau + \int_0^t e^{(t-s)A(s)} f(s) ds$$

$$+ \int_0^t \int_0^\tau \int_0^\sigma (R(\tau, \sigma) - R(\tau, s)) A(s) e^{(\sigma-s)A(s)} f(s) ds d\sigma d\tau$$

$$+ \int_0^t \int_0^\tau R(\tau, s) e^{(\tau-s)A(s)} f(s) ds d\tau$$

$$= - \int_0^t f(s) ds - \int_0^t \int_0^\tau R(\tau, s) f(s) ds d\tau + \int_0^t e^{(t-s)A(s)} f(s) ds$$

$$+ \int_0^t \int_0^\tau \int_0^\sigma (R(\tau, \sigma) - R(\tau, s)) A(s) e^{(\sigma-s)A(s)} f(s) ds d\sigma d\tau$$

$$+ \int_0^t \int_0^\tau R(\tau, s) e^{(\tau-s)A(s)} f(s) ds d\tau$$

$$= \int_0^t e^{(t-s)A(s)} f(s) ds - \int_0^t f(s) ds$$

$$+ \int_0^t \int_s^t R(\tau, s)(e^{(\tau-s)A(s)} - 1) d\tau f(s) ds$$

$$+ \int_0^t \int_s^t \int_s^\tau (R(\tau, \sigma) - R(\tau, s)) A(s) e^{(\sigma-s)A(s)} d\sigma d\tau f(s) ds. \quad (6.127)$$

From (6.120),(6.121),(6.122),(6.124),(6.125),(6.127) and noting (6.115) we obtain (6.116).

(ii) If u is a classical solution, then it is a strict solution in $[\epsilon, T]$ for any $\epsilon \in (0, T)$. Since $A(\epsilon)u(\epsilon) + f(\epsilon) \in \overline{D(A(\epsilon))}$ in view of Lemma 6.12, we have applying (i) in $[\epsilon, T]$

$$u(t) = U(t, \epsilon)u(\epsilon) + \int_\epsilon^t U(t, s) f(s) ds, \quad t \in [\epsilon, T].$$

Letting $\epsilon \to 0$ for a fixed $t \in (0, T]$ we obtain (6.116).

6.11   Approximate Equations

We consider a sequence of equations

$$\frac{d}{dt}u_n(t) = A_n(t)u_n(t) + f(t), \quad t \in [0, T] \tag{6.128}$$

$$u_n(0) = x, \tag{6.129}$$

and $n = 1, 2, \ldots$, where

$$A_n(t) = nA(t)R(n, A(t)) = A(t)(1 - n^{-1}A(t))^{-1}$$

is the *Yosida approximation* of $A(t)$.

Lemma 6. 13 *For $\lambda \in \Sigma$ and positive integers n we have*

$$\frac{1}{|\lambda + n|} \le \min\left\{\frac{3}{(1 + |\lambda|)\sin\theta_0}, \frac{1}{n\sin\theta_0}\right\}. \tag{6.130}$$

Proof. The assertion is easily shown if we note that $\text{Re}\lambda > |\lambda|\cos\theta_0$ for $\lambda \in \Sigma$.

Lemma 6. 14 *For $\lambda \in \Sigma$, positive integers n and $0 \le s < t \le T$ we have*
(i) $\rho(A_n(t)) \supset \Sigma$ *and*

$$R(\lambda, A_n(t)) = \frac{1}{\lambda + n}(n - A(t))R\left(\frac{\lambda n}{\lambda + n}, A(t)\right)$$

$$= \left(\frac{n}{\lambda + n}\right)^2 R\left(\frac{\lambda n}{\lambda + n}, A(t)\right) + \frac{1}{\lambda + n}, \tag{6.131}$$

(ii)

$$\|R(\lambda, A_n(t))\| \le \frac{C}{1 + |\lambda|}, \tag{6.132}$$

(iii)

$$R(\lambda, A_n(t)) - R(\lambda, A(t))$$

$$= \frac{1}{\lambda + n}A(t)R\left(\frac{\lambda n}{\lambda + n}, A(t)\right)A(t)R(\lambda, A(t)), \tag{6.133}$$

(iv)

$$A_n(t)R(\lambda, A_n(t)) = \frac{n}{\lambda + n}A(t)R\left(\frac{\lambda n}{\lambda + n}, A(t)\right), \tag{6.134}$$

(v)

$$R(\lambda, A_n(t)) - R(\lambda, A_n(s)) = - \left(\frac{n}{\lambda + n} \right)^2$$
$$\times A(t)R\left(\frac{\lambda n}{\lambda + n}, A(t) \right) (A(t)^{-1} - A(s)^{-1})A(s)R\left(\frac{\lambda n}{\lambda + n}, A(s) \right).$$

$$(6.135)$$

Proof. (i) As is easily seen we have

$$\lambda - A_n(t) = (\lambda + n) \left(\frac{\lambda n}{\lambda + n} - A(t) \right) (n - A(t))^{-1}. \tag{6.136}$$

Hence $\rho(A_n(t)) \supset \Sigma$ follows if we note that $0 \neq \lambda \in \Sigma$ if and only if $\lambda^{-1} \in \Sigma$ and $\frac{\lambda n}{\lambda + n} \in \Sigma$ if and only if $\lambda^{-1} + n^{-1} \in \Sigma$. (6.131) is a simple consequence of (6.136).

(ii) This is a simple consequence of (i) and Lemma 6.13.

(iii),(iv) can be shown straightforwardly using (i), and (v) is established with the aid of (i) and (6.102).

Lemma 6. 15 *For $\tau > 0, t \in [0, T], m = 0, 1,$ and positive integers n*

(i)

$$\| A_n(t)^m e^{\tau A_n(t)} \| \le \frac{C}{\tau^m}, \tag{6.137}$$

(ii) *for $0 \le \eta \le 1$*

$$\| A_n(t)^m e^{\tau A_n(t)} - A(t)^m e^{\tau A(t)} \| \le \frac{C}{n^\eta \tau^{m+\eta}}, \tag{6.138}$$

(iii)

$$\| A_n(t)e^{\tau A_n(t)} - A_n(s)e^{\tau A_n(s)} \| \le C \sum_{i=1}^{k} \frac{(t-s)^{\alpha_i}}{\tau^{1+\beta_i}}. \tag{6.139}$$

Proof. (i) (6.137) is a simple consequence of Lemma 6.14 (ii).

(ii) With the aid of Lemma 6.14 (iii) we have

$$\| R(\lambda, A_n(t)) - R(\lambda, A(t)) \| \le \frac{C}{|\lambda + n|}. \tag{6.140}$$

The inequality (6.138) is a simple consequence of (6.140) and Lemma 6.13.

(iii) With the aid of Lemma 6.13, Lemma 6.14 (v) and the hypothesis (P4) we get

$$\| R(\lambda, A_n(t)) - R(\lambda, A_n(s)) \| \le C \sum_{i=1}^{k} (t-s)^{\alpha_i} |\lambda|^{\beta_i - 1}, \tag{6.141}$$

from which (6.139) follows easily.

Lemma 6.16 *For $\tau > 0, m = 1, 2, 0 \leq s < t \leq T$ and positive integers n*
(i)

$$\|A_n(t)^m e^{\tau A_n(t)}(A(t)^{-1} - A(s)^{-1})\| \leq C \sum_{i=1}^{k} \frac{(t-s)^{\alpha_i}}{\tau^{m+\beta_i-1}}, \qquad (6.142)$$

(ii) *for $0 \leq \eta \leq 1$*

$$\left\|(A_n(t)^m e^{\tau A_n(t)} - A(t)^m e^{\tau A(t)})(A(t)^{-1} - A(s)^{-1})\right\| \leq \frac{C}{n^\eta} \sum_{i=1}^{k} \frac{(t-s)^{\alpha_i}}{\tau^{m+\beta_i+\eta-1}}. \qquad (6.143)$$

Proof. (i) With the aid of Lemma 6.14 (iv) and the hypothesis (P4) we have

$$\|A_n(t)R(\lambda, A_n(t))(A(t)^{-1} - A(s)^{-1})\|$$
$$= \left\|\frac{n}{\lambda+n}A(t)R\left(\frac{\lambda n}{\lambda+n}, A(t)\right)(A(t)^{-1} - A(s)^{-1})\right\|$$
$$\leq C \sum_{i=1}^{k}(t-s)^{\alpha_i}\left|\frac{n}{\lambda+n}\right|^{\beta_i}|\lambda|^{\beta_i-1}. \qquad (6.144)$$

The result follows from (6.144), Lemma 6.13 and

$$A_n(t)^m e^{\tau A_n(t)}(A(t)^{-1} - A(s)^{-1})$$
$$= \frac{1}{2\pi i}\int_\Gamma \lambda^{m-1}e^{\lambda\tau}A_n(t)R(\lambda, A_n(t))(A(t)^{-1} - A(s)^{-1})d\lambda.$$

(ii) We have by Lemma 6.14 (iii) and the assumption (P4)

$$\|(R(\lambda, A_n(t)) - R(\lambda, A(t)))(A(t)^{-1} - A(s)^{-1})\|$$
$$= \left\|\frac{1}{\lambda+n}A(t)R\left(\frac{\lambda n}{\lambda+n}, A(t)\right)A(t)R(\lambda, A(t))(A(t)^{-1} - A(s)^{-1})\right\|$$
$$\leq C\left|\frac{1}{\lambda+n}\right|\sum_{i=1}^{k}(t-s)^{\alpha_i}|\lambda|^{\beta_i-1}. \qquad (6.145)$$

The inequality (6.143) follows from (6.145) and

$$(A_n(t)^m e^{\tau A_n(t)} - A(t)^m e^{\tau A(t)})(A(t)^{-1} - A(s)^{-1})$$
$$= \frac{1}{2\pi i}\int_\Gamma \lambda^m e^{\lambda\tau}(R(\lambda, A_n(t)) - R(\lambda, A(t)))(A(t)^{-1} - A(s)^{-1})d\lambda.$$

Set for $0 \leq s < t \leq T$ and $n = 1, 2, \ldots$

$$Q_n(t, s) = A_n(t)^2 e^{(t-s)A_n(t)} (A_n(t)^{-1} - A_n(s)^{-1}). \tag{6.146}$$

Since $A_n(t)^{-1} = A(t)^{-1} - 1/n$, we have

$$Q_n(t, s) = A_n(t)^2 e^{(t-s)A_n(t)} (A(t)^{-1} - A(s)^{-1}). \tag{6.147}$$

Lemma 6. 17 (i) *For $0 \leq s < t \leq T$ and positive integers n*

$$\|Q_n(t, s)\| \leq C(t - s)^{\delta - 1}. \tag{6.148}$$

(ii) *For $0 \leq s < \tau < t \leq T, \eta \in (0, \delta)$ and positive integers n*

$$\|Q_n(t, s) - Q_n(t, \tau)\| \leq C_\eta (\tau - s)^\eta (t - \tau)^{\delta - \eta - 1}. \tag{6.149}$$

Proof. (i) (6.148) is a direct consequence of (6.147) and Lemma 6.16 (i).
(ii) Following the proof of Lemma 6.9 (ii) and using Lemma 6.13 and (6.144) we obtain (6.149).

Lemma 6. 18 *For $0 \leq s < t \leq T$, positive integers n and $0 \leq \eta \leq 1$*

$$\|Q_n(t, s) - Q(t, s)\| \leq \frac{C}{n^\eta} (t - s)^{\delta - \eta - 1}. \tag{6.150}$$

Proof. This is an easy consequence of Lemma 6.16 (ii).

Let $R_n(t, s), 0 \leq s < t \leq T$, be the solution of the integral equation

$$R_n(t, s) = Q_n(t, s) + \int_s^t R_n(t, \tau) Q_n(\tau, s) d\tau$$

$$= Q_n(t, s) + \int_s^t Q_n(t, \tau) R_n(\tau, s) d\tau. \tag{6.151}$$

Lemma 6. 19 (i) *For $0 \leq s < t \leq T$ and positive integers n*

$$\|R_n(t, s)\| \leq C(t - s)^{\delta - 1}. \tag{6.152}$$

(ii) *For $0 \leq s < \tau < t \leq T, \eta \in (0, \delta)$ and positive integers n*

$$\|R_n(t, s) - R_n(t, \tau)\| \leq C_\eta (\tau - s)^\eta (t - \tau)^{\delta - \eta - 1}. \tag{6.153}$$

Proof. The results are established exactly as Lemma 6.10.

We have

$$R_n(t, s) - R(t, s)$$

$$= Q_n(t, s) - Q(t, s) + \int_s^t (Q_n(t, \tau) - Q(t, \tau)) R_n(\tau, s) d\tau$$

$$+ \int_s^t Q(t, \tau) (R_n(\tau, s) - R(\tau, s)) d\tau. \tag{6.154}$$

Hence

$$\int_s^t R(t,\sigma)\,(R_n(\sigma,s) - R(\sigma,s))\,d\sigma = \int_s^t R(t,\sigma)\,(Q_n(\sigma,s) - Q(\sigma,s))\,d\sigma$$

$$+ \int_s^t R(t,\sigma) \int_s^\sigma (Q_n(\sigma,\tau) - Q(\sigma,\tau))\,R_n(\tau,s)d\tau d\sigma$$

$$+ \int_s^t \int_\tau^t R(t,\sigma)Q(\sigma,\tau)d\sigma\,(R_n(\tau,s) - R(\tau,s))\,d\tau. \tag{6.155}$$

Using (6.96) we obtain from (6.155)

$$\int_s^t Q(t,\tau)\,(R_n(\tau,s) - R(\tau,s))\,d\tau = \int_s^t R(t,\sigma)\,(Q_n(\sigma,s) - Q(\sigma,s))\,d\sigma$$

$$+ \int_s^t R(t,\sigma) \int_s^\sigma (Q_n(\sigma,\tau) - Q(\sigma,\tau))\,R_n(\tau,s)d\tau d\sigma. \tag{6.156}$$

Lemma 6. 20 *For $0 \le s < t \le T, \eta \in (0,\delta)$ and positive integers n*

$$\|R_n(t,s) - R(t,s)\| \le \frac{C_\eta}{n^\eta}(t - s)^{\delta - \eta - 1}. \tag{6.157}$$

Proof. The inequality (6.157) is established with the aid of (6.154),(6.150), (6.152),(6.156),(6.109).

We define

$$U_n(t,s) = e^{(t-s)A_n(s)} + \int_s^t Z_n(\tau,s)d\tau, \tag{6.158}$$

$$Z_n(t,s) = A_n(t)e^{(t-s)A_n(t)} - A_n(s)e^{(t-s)A_n(s)}$$

$$+ \int_s^t R_n(t,\tau)(A_n(\tau)e^{(\tau-s)A_n(\tau)} - A_n(s)e^{(\tau-s)A_n(s)})d\tau$$

$$+ \int_s^t (R_n(t,\tau) - R_n(t,s))\,A_n(s)e^{(\tau-s)A_n(s)}d\tau$$

$$+ R_n(t,s)(e^{(t-s)A_n(s)} - 1). \tag{6.159}$$

Then $U_n(t,s)$ is the evolution operator of (6.128),(6.129). The solution of (6.128),(6.129) is given by

$$u_n(t) = U_n(t,0)x + \int_0^t U_n(t,s)f(s)ds. \tag{6.160}$$

In view of Lemma 6.15 (iii) we get

$$\|A_n(t)e^{(t-s)A_n(t)} - A_n(s)e^{(t-s)A_n(s)}\| \le C(t - s)^{\delta - 1}. \tag{6.161}$$

With the aid of (6.161), Lemmas 6.15, 6.19 we obtain

$$\|U_n(t,s)\| \leq C, \quad \|Z_n(t,s)\| \leq C(t-s)^{\delta-1}. \tag{6.162}$$

Lemma 6. 21 *As* $n \to \infty$

$$U_n(t,s) \to U(t,s), \quad 0 \leq s \leq t \leq T, \tag{6.163}$$
$$Z_n(t,s) \to Z(t,s), \quad 0 \leq s < t \leq T, \tag{6.164}$$

in the norm of $\mathcal{L}(X)$.

Proof. In view of (6.161), Corollary 6.1 and Lemma 6.15 (ii) we have

$$\|A_n(\tau)e^{(\tau-s)A_n(\tau)} - A_n(s)e^{(\tau-s)A_n(s)} - A(\tau)e^{(\tau-s)A(\tau)}$$
$$+A(s)e^{(\tau-s)A(s)}\| \leq \min\left\{C(\tau-s)^{\delta-1}, \frac{C}{n(\tau-s)^2}\right\},$$

which implies

$$\|A_n(\tau)e^{(\tau-s)A_n(\tau)} - A_n(s)e^{(\tau-s)A_n(s)} - A(\tau)e^{(\tau-s)A(\tau)}$$
$$+A(s)e^{(\tau-s)A(s)}\| \leq \frac{C}{n^\eta}(\tau-s)^{\delta-\eta-1-\delta\eta} \tag{6.165}$$

for $0 \leq \eta \leq 1$. By Lemma 6.10 (ii), Lemma 6.19 (ii) and Lemma 6.20 we have

$$\|R_n(t,\tau) - R_n(t,s) - R(t,\tau) + R(t,s)\|$$
$$\leq \min\left\{\frac{C_\eta}{n^\eta}(t-\tau)^{\delta-\eta-1}, \ C_\eta(\tau-s)^\eta(t-\tau)^{\delta-\eta-1}\right\}$$

for $\eta \in (0,\delta)$. Hence

$$\|R_n(t,\tau)-R_n(t,s)-R(t,\tau)+R(t,s)\| \leq \frac{C_\eta}{n^{\eta/2}}(t-\tau)^{\delta-\eta-1}(\tau-s)^{\eta/2}. \tag{6.166}$$

Using (6.161),(6.165),(6.166), Lemmas 6.10, 6.15, 6.20 we obtain the desired results without difficulty.

6.12 Existence of Solutions

In this section we show the existence of classical and strict solutions of (6.1),(6.2) under the assumptions (P1),(P2),(P4). The definitions of these solutions is the same as Definitions 6.1 and 6.2 of section 6.1.

Lemma 6. 22 *For $0 \le s < t \le T$*

$$\|[A(t)e^{(t-s)A(t)} - A(s)e^{(t-s)A(s)}]A(s)^{-1}\| \le C\sum_{i=1}^{k}(t-s)^{\alpha_i - \beta_i}. \qquad (6.167)$$

Proof. From (6.102) and the assumption (P4) it follows that

$$\|(R(\lambda, A(t)) - R(\lambda, A(s)))A(s)^{-1}\| \le C\sum_{i=1}^{k}(t-s)^{\alpha_i}|\lambda|^{\beta_i - 2}. \qquad (6.168)$$

The inequality (6.167) follows from (6.168).

Lemma 6. 23 *For $0 \le s < t \le T$*

$$\|Z(t,s)A(s)^{-1}\| \le C(t-s)^{\delta}. \qquad (6.169)$$

Proof. The inequality (6.169) follows from Lemmas 6.10, 6.22 and

$$\|(e^{(t-s)A(s)} - 1)A(s)^{-1}\| = \left\|\int_{0}^{t-s} e^{\tau A(s)}d\tau\right\| \le C(t-s).$$

Proposition 6. 15 *Assume $f \in C^{\alpha}([0,T]; X)$ for some $\alpha \in (0,1)$ and $x \in \overline{D(A(0))}$. Set*

$$u(t) = U(t,0)x + \int_{0}^{t} U(t,s)f(s)ds, \quad t \in (0,T]. \qquad (6.170)$$

Then $u \in C^{1}((0,T]; X)$ and $\lim_{t\to 0} u(t) = x$. If moreover $x \in D(A(0))$ and $A(0)x + f(0) \in \overline{D(A(0))}$, then

$$\lim_{t\to 0}\frac{d}{dt}u(t) = A(0)x + f(0). \qquad (6.171)$$

Proof. It is obvious that

$$\frac{d}{dt}U(t,0)x = A(0)e^{tA(0)}x + Z(t,0)x, \qquad (6.172)$$

$$\frac{d}{dt}\int_{0}^{t}\int_{s}^{t} Z(\tau,s)d\tau f(s)ds = \int_{0}^{t} Z(t,s)f(s)ds. \qquad (6.173)$$

For $0 < \epsilon < t$ we have

$$\frac{d}{dt}\int_{0}^{t-\epsilon} e^{(t-s)A(s)}f(s)ds$$

$$= e^{\epsilon A(t-\epsilon)}f(t-\epsilon) + \int_{0}^{t-\epsilon} A(s)e^{(t-s)A(s)}(f(s) - f(t))ds$$

$$+ \int_{0}^{t-\epsilon}(A(s)e^{(t-s)A(s)} - A(t)e^{(t-s)A(t)})f(t)ds$$

$$- e^{\epsilon A(t)}f(t) + e^{tA(t)}f(t).$$

With the aid of Lemma 6.8 we have

$$\|e^{\epsilon A(t-\epsilon)} - e^{\epsilon A(t)}\| \le C \sum_{i=1}^{k} \epsilon^{\alpha_i - \beta_i}.$$

Hence

$$\|e^{\epsilon A(t-\epsilon)} f(t-\epsilon) - e^{\epsilon A(t)} f(t)\|$$
$$\le \|e^{\epsilon A(t-\epsilon)}(f(t-\epsilon) - f(t))\| + \|(e^{\epsilon A(t-\epsilon)} - e^{\epsilon A(t)})f(t)\|$$
$$\le C\|f(t-\epsilon) - f(t)\| + C \sum_{i=1}^{k} \epsilon^{\alpha_i - \beta_i} \|f(t)\| \to 0$$

as $\epsilon \to 0$. Therefore we get

$$\lim_{\epsilon \to 0} \frac{d}{dt} \int_0^{t-\epsilon} e^{(t-s)A(s)} f(s) ds = \int_0^t A(s) e^{(t-s)A(s)} (f(s) - f(t)) ds$$
$$+ \int_0^t (A(s) e^{(t-s)A(s)} - A(t) e^{(t-s)A(t)}) f(t) ds + e^{tA(t)} f(t). \quad (6.174)$$

Combining (6.172),(6.173),(6.174) yields

$$\frac{d}{dt} u(t) = A(0) e^{tA(0)} x + Z(t,0) x + \int_0^t A(s) e^{(t-s)A(s)} (f(s) - f(t)) ds$$
$$+ \int_0^t (A(s) e^{(t-s)A(s)} - A(t) e^{(t-s)A(t)}) f(t) ds$$
$$+ e^{tA(t)} f(t) + \int_0^t Z(t,s) f(s) ds. \quad (6.175)$$

It is easy to see that $u(t) \to x$ as $t \to 0$. It is also easily seen that the third, fourth and last terms of the right hand side of (6.175) tend to 0 as $t \to 0$. Suppose that $x \in D(A(0))$ and $A(0)x + f(0) \in \overline{D(A(0))}$. Then by virtue of Lemma 6.23 $Z(t,0)x \to 0$ as $t \to 0$. In view of Lemma 6.8 we have

$$\cdot \|e^{tA(t)} - e^{tA(0)}\| \le C \sum_{i=1}^{k} t^{\alpha_i - \beta_i}.$$

Hence

$$A(0) e^{tA(0)} x + e^{tA(t)} f(t)$$
$$= e^{tA(0)}(A(0)x + f(0)) + (e^{tA(t)} - e^{tA(0)})f(t) + e^{tA(0)}(f(t) - f(0))$$
$$\to A(0)x + f(0)$$

as $t \to 0$. Therefore (6.171) is established.

Theorem 6. 6 *Suppose that the assumptions (P1),(P2),(P4) are satisfied. Let $f \in C^{\alpha}([0,T]; X)$ for some $\alpha \in (0,1]$ and $x \in D(A(0))$. Then a classical solution of (6.1), (6.2) exists and is unique, and is given by (6.170). If moreover $x \in D(A(0))$ and $A(0)x + f(0) \in \overline{D(A(0))}$, then the classical solution is a strict solution.*

Proof. By virtue of Theorem 6.5 and Proposition 6.15 it remains to show that the function u given by (6.170) satisfies the equation (6.1) in $(0,T]$. Let u_n be the solution of (6.128),(6.129). Then u_n is given by (6.160) and analogously to (6.175) we have

$$\frac{d}{dt}u_n(t) = A_n(0)e^{tA_n(0)}x + Z_n(t,0)x$$

$$+ \int_0^t A_n(s)e^{(t-s)A_n(s)}(f(s) - f(t))ds$$

$$+ \int_0^t (A_n(s)e^{(t-s)A_n(s)} - A_n(t)e^{(t-s)A_n(t)})f(t)ds$$

$$+ e^{tA_n(t)}f(t) + \int_0^t Z_n(t,s)f(s)ds. \tag{6.176}$$

Using Lemmas 6.15, 6.21 and (6.161),(6.162),(6.175),(6.176) it can be easily shown that $u_n(t) \to u(t)$, $du_n(t)/dt \to du(t)/dt$ for $t \in (0,T]$. Since

$$u_n(t) = A_n(t)^{-1}\left(\frac{d}{dt}u_n(t) - f(t)\right)$$

$$= \left(A(t)^{-1} - \frac{1}{n}\right)\left(\frac{d}{dt}u_n(t) - f(t)\right) \to A(t)^{-1}\left(\frac{d}{dt}u(t) - f(t)\right),$$

we have $u(t) = A(t)^{-1}(du(t)/dt - f(t))$. Therefore $u(t) \in D(A(t))$ and (6.1) holds.

Remark 6. 5 A maximal regularity result analogous to Theorem 6.4 holds. See Acquistapace and Terreni [8: Theorem 6.1].

6.13   Application

In [1] Acquistapace showed that the hypotheses (P1),(P2),(P4) are satisfied by the realizations of strongly elliptic operators in the space $C(\bar{\Omega})$, where Ω is a bounded open set of R^n, under suitable boundary conditions making use of the result of H. B. Stewart [145]. The results of [1],[6],[8] are powerful for such equations having operators which are not densely defined. However in this section we consider the same kind of problem in $L^1(\Omega)$. This is a generalization of D. G. Park [122] where the differentiability in t of the coefficients is assumed.

Let Ω be a bounded open set of R^n of class C^{2m}. Let

$$L(x, t, D_x) = \sum_{|\alpha| \le m} a_\alpha(x, t) D_x^\alpha$$

be a linear differential operator of order m with coefficients defined in $\bar{\Omega} \times [0, T]$. We assume that $L(x, t, D_x)$ is uniformly strongly elliptic in $\bar{\Omega} \times [0, T]$, i.e.

$$(-1)^{m/2} \text{Re} \sum_{|\alpha| \le m} a_\alpha(x, t) \xi^\alpha \ge c_0 |\xi|^m, \quad c_0 > 0,$$

for any $(x, t) \in \bar{\Omega} \times [0, T]$ and $\xi \in R^n$. Let

$$B_j(x, t, D_x) = \sum_{|\beta| \le m_j} b_{j\beta}(x, t) D_x^\beta, \quad j = 1, \ldots, m/2,$$

be a set of boundary operators with coefficients defined on $\partial\Omega \times [0, T]$. We assume that all the hypotheses of section 3 of Chapter 5 are satisfied by $L(x, t, D_x)$ and $\{B_j(x, t, D_x)\}_{j=1}^{m/2}$ for each fixed $t \in [0, T]$.

Let

$$L'(x, t, D_x) = \sum_{|\alpha| \le m} a'_\alpha(x, t) D_x^\alpha,$$

$$B'_j(x, t, D_x) = \sum_{|\beta| \le m'_j} b'_{j\beta}(x, t) D_x^\alpha, \quad j = 1, \ldots, m/2$$

be the adjoint system of $L(x, t, D_x), \{B_j(x, t, D_x)\}_{j=1}^{m/2}$. Let all the coefficients of $B_j(x, t, D_x), B'_j(x, t, D_x), j = 1, \ldots, m/2$, be extended to the whole of $\bar{\Omega} \times [0, T]$. We assume that

$$a_\alpha, a'_\alpha, \ |\alpha| \le m, \ D_x^\gamma b_{j\beta}, \ |\beta| \le m_j, \ |\gamma| \le m - m_j,$$
$$D_x^\gamma b'_{j\beta}, \ |\beta| \le m'_j, \ |\gamma| \le m - m'_j, \ j = 1, \ldots, m/2$$

are uniformly Hölder continuous in t of order h. We denote by $A(t)$ the realization of $L(x, t, D_x)$ in $L^1(\Omega)$ under the boundary conditions

$$B_j(x, t, D_x) u|_{\partial\Omega} = 0, \ j = 1, \ldots, m/2,$$

and those of $L(x, t, D_x)$ and $L'(x, t, D_x)$ in $L^p(\Omega), 1 < p < \infty$, under the boundary conditions

$$B_j(x, t, D_x) u|_{\partial\Omega} = 0, \ j = 1, \ldots, m/2, \ \text{and}$$
$$B'_j(x, t, D_x) u|_{\partial\Omega} = 0, \ j = 1, \ldots, m/2,$$

by $A_p(t)$ and $A'_p(t)$ respectively. We assume for the sake of simplicity that $0 \in \rho(A(t)) \cap \rho(A_p(t))$ for all $p \in (1, \infty)$. Following the argument of section 3 of Chapter 5 we establish the estimate of the kernel of $A(t)^{-1} - A(\tau)^{-1}$ and its derivatives.

As in Chapter 5 for a real vector η let $A_p^\eta(t)$ and $A'_p{}^\eta(t)$ be the realizations of $L(x, t, D_x + \eta)$ and $L'(x, t, D_x + \eta)$ in $L^p(\Omega)$ under the boundary conditions

$$B_j(x, t, D_x + \eta)u|_{\partial\Omega} = 0, \quad j = 1, \ldots, m/2, \text{ and}$$
$$B'_j(x, t, D_x + \eta)u|_{\partial\Omega} = 0, \quad j = 1, \ldots, m/2,$$

respectively. The assertions of Lemmas 5.11-5.15 hold for each $t \in [0, T]$ with constants independent of t. Hence there exists an angle $\theta_0 \in (0, \pi/2)$ such that if we set

$$\Sigma = \{\lambda; \theta_0 \leq \arg \lambda \leq 2\pi - \theta_0\} \cup \{0\}$$

the following assertion holds: for $1 < p < \infty$ there exist positive constants δ and C such that if $\lambda \in \Sigma, |\lambda| \geq C, |\eta| \leq \delta|\lambda|^{1/m}, t \in [0, T]$, then $\lambda \in \rho(A_p^\eta(t)) \cap \rho(A'_{p'}{}^\eta(t))$ and

$$\sum_{k=0}^{m} |\lambda|^{(m-k)/m}\|(A_p^\eta(t) - \lambda)^{-1}f\|_{k,p,\Omega} \leq C\|f\|_{0,p,\Omega}, \qquad (6.177)$$

$$\sum_{k=0}^{m} |\lambda|^{(m-k)/m}\|(A'_{p'}{}^\eta(t) - \lambda)^{-1}f\|_{k,p',\Omega} \leq C\|f\|_{0,p',\Omega} \qquad (6.178)$$

for $f \in L^p(\Omega)$ or $f \in L^{p'}(\Omega)$.

Lemma 6. 24 Let $u(t) = (A_p^\eta(t) - \lambda)^{-1}f$ for $1 < p < \infty, \lambda \in \Sigma$ with $|\lambda|$ sufficiently large and $|\eta| \leq \delta|\lambda|^{1/m}$. Then for $t, \tau \in [0, T]$

$$\sum_{k=0}^{m} |\lambda|^{(m-k)/m}\|u(t) - u(\tau)\|_{k,p,\Omega} \leq C|t - \tau|^h\|f\|_{0,p,\Omega}. \qquad (6.179)$$

Proof. By the definition of $u(t)$ we have

$$(L(x, t, D_x + \eta) - \lambda)u(t) = f \quad \text{in} \quad \Omega,$$
$$B_j(x, t, D_x + \eta)u(t) = 0 \quad \text{on} \quad \partial\Omega, \quad j = 1, \ldots, m/2.$$

Therefore

$$(L(x, t, D_x + \eta) - \lambda)(u(t) - u(\tau))$$
$$= (L(x, \tau, D_x + \eta) - L(x, t, D_x + \eta))u(\tau) \quad \text{in} \quad \Omega,$$
$$B_j(x, t, D_x + \eta)(u(t) - u(\tau))$$
$$= (B_j(x, \tau, D_x + \eta) - B_j(x, t, D_x + \eta))u(\tau) \quad \text{on} \quad \partial\Omega.$$

By Theorem 5.5 and (6.177)

$$\sum_{k=0}^{m} |\lambda|^{(m-k)/m} \|u(t) - u(\tau)\|_{k,p,\Omega}$$

$$\leq C\Big[\|(L(x,\tau,D_x+\eta) - L(x,t,D_x+\eta))u(\tau)\|_{0,p,\Omega}$$

$$+ \sum_{j=1}^{m/2} |\lambda|^{(m-m_j)/m} \|(B_j(x,\tau,D_x+\eta) - B_j(x,t,D_x+\eta))u(\tau)\|_{0,p,\Omega}$$

$$+ \sum_{j=1}^{m/2} \|(B_j(x,\tau,D_x+\eta) - B_j(x,t,D_x+\eta))u(\tau)\|_{m-m_j,p,\Omega}\Big]$$

$$\leq C|t-\tau|^h\Big[\sum_{k=0}^{m} |\eta|^{m-k} \|u(\tau)\|_{k,p,\Omega}$$

$$+ \sum_{j=1}^{m/2} |\lambda|^{(m-m_j)/m} \sum_{k=0}^{m_j} |\eta|^{m_j-k} \|u(\tau)\|_{k,p,\Omega}$$

$$+ \sum_{j=1}^{m/2} \sum_{k=m-m_j}^{m} |\eta|^{m-k} \|u(\tau)\|_{k,p,\Omega}\Big]$$

$$\leq C|t-\tau|^h \sum_{k=0}^{m} |\lambda|^{(m-k)/m} \|u(\tau)\|_{k,p,\Omega} \leq C|t-\tau|^h \|f\|_{0,p,\Omega}.$$

For the sake of simplicity of notations we write in what follows $A(t)$, $A^\eta(t)$, $A'(t)$, $A'^\eta(t)$ for $A_p(t)$, $A_p^\eta(t)$, $A'_p(t)$, $A'_p{}^\eta(t)$.

We choose exponents $q_1, q_2, \ldots, q_{s+1}, r_1, r_2, \ldots, r_{l-s+1}$ of section 3 of Chapter 5 so that we have $q_s > n$, $r_{l-s} > n$. Then, in view of (5.114) we have for $k = 0, \ldots, m-1$

$$\|u\|_{k,\infty,\Omega} \leq C\|u\|_{m,q_s,\Omega}^{(n+kq_s)/mq_s} \|u\|_{0,q_s,\Omega}^{(mq_s-n-kq_s)/mq_s}. \tag{6.180}$$

We denote the kernels of the operators $(A(t)-\lambda_1)^{-1}\cdots(A(t)-\lambda_l)^{-1}$ and $(A^\eta(t)-\lambda_1)^{-1}\cdots(A^\eta(t)-\lambda_l)^{-1}$ by $K_{\lambda_1,\ldots,\lambda_l}(x,y;t)$ and $K^\eta_{\lambda_1,\ldots,\lambda_l}(x,y;t)$ respectively. If we set

$$S(t) = (A^\eta(t)-\lambda_s)^{-1}\cdots(A^\eta(t)-\lambda_1)^{-1},$$
$$T(t) = (A^\eta(t)-\lambda_{s+1})^{-1}\cdots(A^\eta(t)-\lambda_l)^{-1},$$

then

$$K^\eta_{\lambda_1,\ldots,\lambda_l}(x,y;t) - K^\eta_{\lambda_1,\ldots,\lambda_l}(x,y;\tau)$$

is the kernel of

$$S(t)T(t) - S(\tau)T(\tau) = (S(t) - S(\tau))T(t) + S(\tau)(T(t) - T(\tau)). \quad (6.181)$$

From Lemma 6.24, (5.113), (6.180) it follows that

$$\left\| \left[(A^\eta(t) - \lambda_j)^{-1} - (A^\eta(\tau) - \lambda_j)^{-1} \right] f \right\|_{0, q_{j+1}, \Omega}$$
$$\leq C|t - \tau|^h |\lambda_j|^{a_j - 1} \|f\|_{0, q_j, \Omega}, \quad j = 1, \dots, s - 1, \quad (6.182)$$
$$\left\| \left[(A^\eta(t) - \lambda_s)^{-1} - (A^\eta(\tau) - \lambda_s)^{-1} \right] f \right\|_{k, \infty, \Omega}$$
$$\leq C|t - \tau|^h |\lambda_s|^{a_s - 1 + k/m} \|f\|_{0, q_s, \Omega}, \quad k = 0, \dots, m - 1. \quad (6.183)$$

Using

$$S(t) - S(\tau) = \sum_{j=1}^{s} (A^\eta(\tau) - \lambda_s)^{-1} \cdots (A^\eta(\tau) - \lambda_{j+1})^{-1}$$
$$\times \left[(A^\eta(t) - \lambda_j)^{-1} - (A^\eta(\tau) - \lambda_j)^{-1} \right] (A^\eta(t) - \lambda_{j-1})^{-1} \cdots (A^\eta(t) - \lambda_1)^{-1}$$

and (6.182),(6.183) we obtain

$$\|S(t) - S(\tau)\|_{\mathcal{L}(L^2(\Omega), W^{k, \infty}(\Omega))}$$
$$\leq C|t - \tau|^h |\lambda_s|^{k/m} \prod_{j=1}^{s} |\lambda_j|^{a_j - 1}, \quad k = 0, \dots, m - 1. \quad (6.184)$$

With the aid of Lemma 5.10, (6.184) and (5.126) with T replaced by $T(t)$ we can estimate the derivatives in x of order up to $m - 1$ of the kernel of the first term of the right hand side of (6.181). Those of the kernel of the second term can be estimated similarly, and we get

$$\left| D_x^\alpha \left(K_{\lambda_1, \dots, \lambda_l}^\eta (x, y; t) - K_{\lambda_1, \dots, \lambda_l}^\eta (x, y; \tau) \right) \right|$$
$$\leq C|t - \tau|^h |\lambda_s|^{|\alpha|/m} \prod_{j=1}^{l} |\lambda_j|^{a_j - 1}, \quad |\alpha| < m. \quad (6.185)$$

With the aid of the argument by which we deduced (5.131) from (5.129) we obtain

$$|D_x^\alpha (K_{\lambda_1, \dots, \lambda_l}(x, y; t) - K_{\lambda_1, \dots, \lambda_l}(x, y; \tau))| \leq C|t - \tau|^h |\lambda_s|^{|\alpha|/m}$$
$$\times \sum_{k=1}^{l} \exp \left(-\delta |\lambda_k|^{1/m} |x - y| \right) \prod_{j=1}^{l} |\lambda_j|^{a_j - 1}, \quad |\alpha| < m. \quad (6.186)$$

We denote the kernel of $A(t)^{-1}$ by $K(x, y; t)$. We follow the argument by which we derived (5.168) from (5.131) in Chapter 5 using the same notations there and omitting the details. The kernel of $D_x^\alpha(e^{-l\sigma A(t)} - e^{-l\sigma A(\tau)})$ is equal to

$$\left(\frac{1}{2\pi i}\right)^l \int_{\Gamma_{x,y,\sigma}} \cdots \int_{\Gamma_{x,y,\sigma}} e^{-\lambda_1\sigma - \cdots - \lambda_l\sigma}$$
$$\times D_x^\alpha \left(K_{\lambda_1,\dots,\lambda_l}(x, y; t) - K_{\lambda_1,\dots,\lambda_l}(x, y; \tau)\right) d\lambda_1 \cdots \lambda_l,$$

and in view of (6.186) it is dominated by

$$C|t - \tau|^h \sigma^{-|\alpha|/m - n/m} \exp(2l\epsilon\rho - \delta\epsilon^{1/m}\rho)$$

in absolute value. We choose ϵ so small that $2l\epsilon - \delta\epsilon^{1/m} < 0$. Since

$$D_x^\alpha(A(t)^{-1} - A(\tau)^{-1}) = l \int_0^\infty D_x^\alpha(e^{-l\sigma A(t)} - e^{-l\sigma A(\tau)}) d\sigma,$$

we obtain

$$|D_x^\alpha \left(K(x, y; t) - K(x, y; \tau)\right)|$$
$$\leq C|t - \tau|^h e^{-c|x-y|} \begin{cases} |x - y|^{m-n-|\alpha|} & m < n + |\alpha| \\ 1 & m > n + |\alpha| \\ 1 + \log^+ |x - y|^{-1} & m = n + |\alpha| \end{cases} \quad (6.187)$$

In [122] P. Grisvard's result on interpolation in $L^1(\Omega)$ with boundary conditions was used. However, the corresponding result in $L^p(\Omega)$, $1 < p < \infty$, answers the purpose there and also in what follows. Let $m_0 = \min_{j=1,\dots,m/2} m_j$. Suppose

$$h > \max\left\{\frac{m - m_0 - 1}{m}, \frac{1}{m}\right\}$$

Then there exist numbers ρ, θ satisfying

$$h > \rho > \theta > \frac{m - m_0 - 1}{m}, \quad \theta \geq \frac{1}{m}. \quad (6.188)$$

Let $(1 - \theta)m - m_0 < 1/q < 1, 1 - 1/n < 1/q < 1$. In view of P. Grisvard [71] (see also H. Triebel [154; pp.329-321]) we have

$$(D(A_q(t)), L^q(\Omega))_{\theta,q} = W^{(1-\theta)m,q}(\Omega), \quad (6.189)$$

and by Lemma 1.6

$$(D(A_q(t)), L^q(\Omega))_{\theta,q} \subset D(A_q(t)^{1-\rho}). \quad (6.190)$$

Since $\theta \geq 1/m$, we have

$$W^{m-1,q}(\Omega) \subset W^{(1-\theta)m,q}(\Omega). \tag{6.191}$$

Combining (6.189),(6.190),(6.191) yields

$$W^{m-1,q}(\Omega) \subset D(A_q(t)^{1-\rho}). \tag{6.192}$$

Using (1.21),(6.192) and that Ω is bounded we get

$$
\begin{aligned}
\|A(t)R(\lambda, A(t))&(A(t)^{-1} - A(\tau)^{-1})f\|_{0,1,\Omega} \\
&= \|A(t)^\rho R(\lambda, A(t))A(t)^{1-\rho}(A(t)^{-1} - A(\tau)^{-1})f\|_{0,1,\Omega} \\
&\leq C|\lambda|^{\rho-1}\|A(t)^{1-\rho}(A(t)^{-1} - A(\tau)^{-1})f\|_{0,1,\Omega} \\
&\leq C|\lambda|^{\rho-1}\|A(t)^{1-\rho}(A(t)^{-1} - A(\tau)^{-1})f\|_{0,q,\Omega} \\
&\leq C|\lambda|^{\rho-1}\|(A(t)^{-1} - A(\tau)^{-1})f\|_{m-1,q,\Omega}.
\end{aligned}
$$

With the aid of (6.187), $(n-1)q < n$ and Lemma 1.1 we get

$$\|(A(t)^{-1} - A(\tau)^{-1})f\|_{m-1,q,\Omega} \leq C|t - \tau|^h \|f\|_{0,1,\Omega}.$$

Hence we obtain

$$\|A(t)R(\lambda, A(t))(A(t)^{-1} - A(\tau)^{-1})\|_{\mathcal{L}(L^1(\Omega))} \leq C|t - \tau|^h |\lambda|^{\rho-1}$$

to conclude that the assumption (P4) or rather (II),(III) of A. Yagi [161] are satisfied.

Remark 6. 6 It can be shown with the aid of Theorem 2.3 of Acquistapace and Terreni [7] that a similar result holds in the space $C(\bar{\Omega})$ of continuous functions.

Chapter 7

Hyperbolic Evolution Equations

This chapter is devoted to evolution equations of hyperbolic type. The results are due to T. Kato [88], Y. Kobayashi [95], N. Okazawa [119], N. Okazawa and A. Unai [120]. See also A. Pazy [127: Chapter 5] and J. R. Dorroh [56].

7.1 Admissible Subspaces

Throughout this section X denotes a Banach space with norm $\| \cdot \|$, and $\{T(t); t \geq 0\}$ a C_0-semigroup in X with infinitesimal generator A. In view of Theorem 1.4 there exist a constant $M \geq 1$ and a real number β such that

$$\|T(t)\| \leq M e^{\beta t}, \quad t \geq 0. \tag{7.1}$$

A subspace Y of X is said to be an *invariant subspace* of the semigroup $\{T(t); t \geq 0\}$ if Y is invariant under the mapping $T(t)$ for any $t \geq 0$.

The *realization* A_Y of A in a subspace Y of X is defined by

$$D(A_Y) = \{y \in D(A) \cap Y; Ay \in Y\}, \tag{7.2}$$

$$A_Y y = Ay \quad \text{for} \quad y \in D(A_Y). \tag{7.3}$$

In what follows in this chapter we assume that Y is a subspace of X and furthermore that it is a Banach space with norm $\| \cdot \|_Y$ which is stronger than that of X. The norm of $\mathcal{L}(Y)$ is also denoted by $\| \cdot \|_Y$.

Definition 7.1 A subspace Y of X is called A-*admissible* if it is an invariant subspace of the semigroup $\{T(t); t \geq 0\}$ and the restriction of $T(t)$ to Y is a C_0-semigroup in Y.

Clearly Y is A-admissible if and only if $T(\cdot)y \in C([0,\infty); Y)$ for each $y \in Y$.

Proposition 7. 1 *A necessary and sufficient condition in order that Y is A-admissible is that the following two conditions are satisfied:*
(i) there exists a real number β such that Y is invariant under the mapping $R(\lambda, A)$ for any $\lambda > \max\{\beta, \tilde{\beta}\}$,
(ii) the realization A_Y of A in Y generates a C_0-semigroup in Y.
Moreover, if Y is A-admissible, then A_Y is the infinitesimal generator of the restriction of $\{T(t); t \geq 0\}$ to Y.

Proof. Suppose that Y is A-admissible. Then the restriction of $T(t)$ to Y is a C_0-semigroup in Y whose infinitesimal generator we denote by $\tilde{A}$. By virtue of Theorem 1.4 there exists a constant $\tilde{M} \geq 1$ and a real number $\tilde{\beta}$ such that

$$\|T(t)\|_Y \leq \tilde{M}e^{\tilde{\beta}t}, \quad t \geq 0. \tag{7.4}$$

For any $y \in Y$ and $\lambda > \max\{\beta, \tilde{\beta}\}$

$$R(\lambda, A)y = \int_0^\infty e^{-\lambda t}T(t)ydt = R(\lambda, \tilde{A})y$$

both in the strong topology of X and that of Y. Hence $R(\lambda, A)y \in Y$, i.e. Y is invariant under $R(\lambda, A)$. It follows from the definition of the infinitesimal generator that if $y \in D(\tilde{A})$ then $y \in D(A) \cap Y$ and $Ay = \tilde{A}y \in Y$. This shows that $y \in D(A_Y)$ and $A_Yy = \tilde{A}y$. Hence $\tilde{A} \subset A_Y$. On the other hand if $y \in D(A_Y)$, then $Ay = A_Yy \in Y$, and

$$T(t)y - y = \int_0^t T(s)Ayds = \int_0^t T(s)A_Yyds \tag{7.5}$$

holds in the strong topology of Y. Dividing (7.5) by t and letting $t \to 0$ we find that $y \in D(\tilde{A})$ and $\tilde{A}y = A_Yy$. Hence $A_Y \subset \tilde{A}$. Therefore we have proved $A_Y = \tilde{A}$.

Conversely suppose that (i) and (ii) are satisfied. Let $\lambda > \max\{\beta, \tilde{\beta}\}, y \in Y$, and set $x = R(\lambda, A)y$. Then by (i) $x \in Y$, and so $Ax = \lambda x - y \in Y$. Hence $x \in D(A_Y)$ and $y = (\lambda - A_Y)x$. Clearly x is uniquely determined by y, and so $\lambda \in \rho(A_Y), R(\lambda, A_Y)y = R(\lambda, A)y$. Therefore for $t > 0$ and $n/t > \max\{\beta, \tilde{\beta}\}$ we have

$$\left(1 - \frac{t}{n}A_Y\right)^{-n}y = \left(1 - \frac{t}{n}A\right)^{-n}y.$$

Letting $n \to \infty$ we get $e^{tA_Y}y = T(t)y$. This shows that Y is invariant under $T(t)$ and the restriction of $T(t)$ to Y is a C_0-semigroup.

Corollary 7. 1 *Y is A-admissible if and only if there exist a constant $\tilde{M} \geq 1$ and a real number $\tilde{\beta}$ such that*
(i) *Y is invariant under $R(\lambda, A)$ for $\mathrm{Re}\lambda > \max\{\beta, \tilde{\beta}\}$,*
(ii) *for $\mathrm{Re}\lambda > \max\{\beta, \tilde{\beta}\}, n = 1, 2, \ldots$*

$$\|R(\lambda, A)^n\|_Y \leq \tilde{M}(\mathrm{Re}\lambda - \tilde{\beta})^{-n}, \tag{7.6}$$

(iii) *for $\lambda > \max\{\beta, \tilde{\beta}\}, R(\lambda, A)Y$ is dense in Y.*

Proof. If Y is A-admissible, then it follows from Proposition 7.1 and its proof that there exist a constant $\tilde{M} \geq 1$ and a real number $\tilde{\beta}$ such that (7.4) holds, $T(t)|_Y = e^{tA_Y}$ for $t \geq 0$ and $R(\lambda, A)|_Y = R(\lambda, A_Y)$ for $\lambda > \max\{\beta, \tilde{\beta}\}$. Furthermore for $\lambda > \max\{\beta, \tilde{\beta}\}, R(\lambda, A)Y = R(\lambda, A_Y)Y = D(A_Y)$. Hence (i),(ii),(iii) hold. Conversely suppose that (i),(ii),(iii) are satisfied. Then from the proof of Proposition 7.1 $\lambda \in \rho(A_Y)$ and $R(\lambda, A)y = R(\lambda, A_Y)y$ for $\lambda > \max\{\beta, \tilde{\beta}\}$ and $y \in Y$. Hence (ii) is rewritten as

$$\|R(\lambda, A_Y)^n\|_Y \leq \tilde{M}(\lambda - \tilde{\beta})^{-n}.$$

Furthermore by (iii) $D(A_Y)$ is dense in Y. Hence in view of Theorem 1.4 A_Y generates a C_0-semigroup in T. Therefore Y is A-admissible.

Remark 7. 1 If Y is reflexive, then in Corollary 7.1 the condition (iii) follows from (i) and (ii). Indeed it follows from the resolvent equation

$$R(\lambda, A) - R(\mu, A) = (\mu - \lambda)R(\lambda, A)R(\mu, A)$$

that $D = R(\lambda, A)Y$ is independent of $\lambda \in \rho(A)$. In view of (7.6) with $n = 1$ for each $y \in Y$, $\lambda R(\lambda, A)y$ is bounded in Y as $\lambda \to \infty$. If Y is reflexive, there exists a subsequence $\lambda_n \to \infty$ such that $\lambda_n R(\lambda_n, A)y$ is weakly convergent in Y to so some element $z \in Y$. Since $\lambda R(\lambda, A)y \to y$ strongly in X as $\lambda \to \infty$, we have $y = z$. Since $\lambda_n R(\lambda_n, A)y \in D$, y is in the weak closure and hence in the strong closure of D.

Proposition 7. 2 *Let $\overline{Y}$ be the closure of Y in X. Let S be an isomorphism from Y onto $\overline{Y}$. Then Y is A-admissible if and only if $A_1 = SAS^{-1}$ is the infinitesimal generator of a C_0-semigroup $\{T_1(t); t \geq 0\}$ in $\overline{Y}$. In this case we have*

$$T_1(t) = ST(t)S^{-1}. \tag{7.7}$$

Proof. By the definition of A_1 we have

$$D(A_1) = \{x \in \overline{Y}; S^{-1}x \in D(A), AS^{-1}x \in Y\}$$
$$= \{x \in \overline{Y}; S^{-1}x \in D(A_Y)\} = SD(A_Y).$$

Hence $D(A_1)$ is dense in $\overline{Y}$ if and only if $D(A_Y)$ is dense in Y. For $x \in D(A_1)$ we have

$$(\lambda - A_1)x = (\lambda - SAS^{-1})x = S(\lambda - A)S^{-1}x = S(\lambda - A_Y)S^{-1}x. \quad (7.8)$$

Therefore

$$\{(\lambda - A_1)x; x \in D(A_1)\} = S\{(\lambda - A_Y)y; y \in D(A_Y)\},$$
$$\{x; (\lambda - A_1)x = 0\} = \{0\} \quad \text{if and only if} \quad \{y; (\lambda - A_Y)y = 0\} = \{0\}.$$

These imply that $\rho(A_1) = \rho(A_Y)$ and

$$R(\lambda, A_1) = SR(\lambda, A)S^{-1} = SR(\lambda, A_Y)S^{-1} \quad \text{for} \quad \lambda \in \rho(A_1) = \rho(A_Y). \tag{7.9}$$

Next we claim that for $\lambda \in \rho(A)$, $\lambda \in \rho(A_1)$ if and only if Y is invariant under $R(\lambda, A)$. Suppose first that $\lambda \in \rho(A_1)$. Then for any $y \in Y$ there exists an element $x \in D(A_1)$ such that

$$Sy = (\lambda - A_1)x = S(\lambda - A)S^{-1}x,$$

which implies $y = (\lambda - A)S^{-1}x$, and hence $R(\lambda, A)y = S^{-1}x \in Y$. Consequently $R(\lambda, A)Y \subset Y$. Conversely suppose that $R(\lambda, A)Y \subset Y$. Let $x \in \overline{Y}$. Set $y = R(\lambda, A)S^{-1}x$. Then $y \in Y \cap D(A)$ and $Ay = \lambda y - S^{-1}x \in Y$. This implies $y \in D(A_Y)$, and hence $Sy \in D(A_1)$. Therefore with the aid of (7.8)

$$x = S(\lambda - A)y = S(\lambda - A)S^{-1}Sy = (\lambda - A_1)Sy \in R(\lambda - A_1).$$

With the aid of (7.8) we see that $(\lambda - A_1)x = 0$ implies $x = 0$. Thus we conclude $\lambda \in \rho(A_1)$.

Suppose that A_1 generates a C_0-semigroup in $\overline{Y}$. Then $D(A_1)$ is dense in $\overline{Y}$, and there exist a constant $M_1 \geq 1$ and a real number β_1 such that $\rho(A_1) \supset \{\lambda; \text{Re}\lambda > \beta_1\}$ and for $\text{Re}\lambda > \beta_1$

$$\|R(\lambda, A_1)^n\|_{\overline{Y}} \leq M_1(\text{Re}\lambda - \beta_1)^{-n}, \quad n = 1, 2, \ldots. \tag{7.10}$$

Since $D(A_1)$ is dense in $\overline{Y}$, $D(A_Y)$ is dense in Y. If $\text{Re}\lambda > \beta_1$, then in view of (7.9),(7.10) $\lambda \in \rho(A_1) = \rho(A_Y)$ and

$$\|R(\lambda, A_Y)^n\|_Y = \|S^{-1}R(\lambda, A_1)^n S\|_Y$$
$$\leq \|S^{-1}\|_{\mathcal{L}(\overline{Y},Y)}\|S\|_{\mathcal{L}(Y,\overline{Y})}M_1(\text{Re}\lambda - \beta_1)^{-n}, \quad n = 1, 2, \ldots.$$

This implies that A_Y generates a C_0-semigroup. By the previous step for $\lambda > \max\{\beta, \beta_1\}$, Y is invariant under $R(\lambda, A)$. Therefore by Proposition 7.1 Y is A-admissible.

Conversely suppose that Y is A-admissible. Then by Corollary 7.1 there exist a constant $\tilde{M} \geq 1$ and a real number $\tilde{\beta}$ such that (i),(ii),(iii) in the corollary hold. Since $D(A_Y)$ is dense in Y, $D(A_1)$ is dense in $\overline{Y}$. If $\text{Re}\lambda > \tilde{\beta}$, then $\lambda \in \rho(A_Y) = \rho(A_1)$ and by virtue of (7.9),(7.6)

$$\|R(\lambda, A_1)^n\|_{\overline{Y}} = \|SR(\lambda, A)^n S^{-1}\|_{\overline{Y}}$$
$$\leq \|S\|_{\mathcal{L}(Y, \overline{Y})} \|S^{-1}\|_{\mathcal{L}(\overline{Y}, Y)} \tilde{M}(\text{Re}\lambda - \tilde{\beta})^{-n}, \quad n = 1, 2, \ldots.$$

Thus A_1 generates a C_0-semigroup in $\overline{Y}$. Finally (7.7) follows from (7.9) and Theorem 1.5.

7.2   Stable Families of Infinitesimal Generators

In this section we consider a family $\{A(t); t \in [0, T]\}$ of infinitesimal generators of C_0-semigroups in a Banach space X.

Definition 7.2 A family $\{A(t); t \in [0, T]\}$ of infinitesimal generators of C_0-semigroups in X is called *stable* with *stability constants* $M \geq 1, \beta$ if

$$\rho(A(t)) \supset (\beta, \infty), \quad t \in [0, T] \tag{7.11}$$

and for every finite sequence $0 \leq t_1 \leq t_2 \leq \cdots \leq t_k \leq T$, $k = 1, 2, \ldots$, and $\lambda > \beta$

$$\left\| \prod_{j=1}^{k} R(\lambda, A(t_j)) \right\| \leq M(\lambda - \beta)^{-k}. \tag{7.12}$$

In (7.12) and in what follows products containing $\{t_j\}$ are always "time ordered", i.e. a factor with a larger t_j stands to the left of ones with smaller t_j.

If $A(t) \in G(X, 1, \beta)$ for any $t \in [0, T]$, then the family $\{A(t); t \in [0, T]\}$ is clearly stable with stability constants $1, \beta$.

Proposition 7.3 *A family* $\{A(t); t \in [0, T]\}$ *of infinitesimal generators of* C_0-semigroups in X is stable if and only if there exist constants $M \geq 1$ and β such that (7.11) holds and either one of the following conditions is satisfied: for any finite sequence $0 \leq t_1 \leq t_2 \leq \cdots \leq t_k \leq T, k = 1, 2, \ldots$, and $s_j \geq 0, j = 1, \ldots, k$

$$\left\| \prod_{j=1}^{k} e^{s_j A(t_j)} \right\| \leq M \exp\left(\beta \sum_{j=1}^{k} s_j \right) \tag{7.13}$$

or for any finite sequence $0 \leq t_1 \leq t_2 \cdots \leq t_k \leq T, k = 1, 2, \ldots,$ *and* $\lambda_j > \beta, j = 1, \ldots, k,$

$$\left\| \prod_{j=1}^{k} R(\lambda_j, A(t_j)) \right\| \leq M \prod_{j=1}^{k} (\lambda_j - \beta)^{-1}. \qquad (7.14)$$

Proof. Suppose that $\{A(t); t \geq 0\}$ is stable with stability constants M, β. Let $0 \leq t_1 \leq t_2 \leq \cdots \leq t_k \leq T$ and $s_j, j = 1, \ldots, k$, be positive rational numbers. We write $s_j = m_j/N$ with a common denominator N. Set $m = \sum_{j=1}^{k} m_j = N \sum_{j=1}^{k} s_j$. Then in view of (7.12)

$$\left\| \prod_{j=1}^{k} \left(1 - \frac{s_j}{m_j} A(t_j) \right)^{-m_j} \right\| = \left\| \prod_{j=1}^{k} \left[\frac{m_j}{s_j} R \left(\frac{m_j}{s_j}, A(t_j) \right) \right]^{m_j} \right\|$$

$$= \left\| \prod_{j=1}^{k} [NR(N, A(t_j))]^{m_j} \right\| = N^m \left\| \prod_{j=1}^{k} R(N, A(t_j))^{m_j} \right\|$$

$$\leq N^m M(N - \beta)^{-m} = M \left(1 - \frac{\beta}{N} \right)^{-N \sum_{j=1}^{k} s_j}.$$

Letting $N \to \infty$ and using Theorem 1.5 we obtain (7.13). Since C_0-semigroups are strongly continuous, we conclude that (7.13) holds also for positive real numbers s_j.

Suppose that (7.13) holds. Let $\lambda_j > \beta, j = 1, \ldots, k$. With the aid of (1.4) we have

$$\prod_{j=1}^{k} R(\lambda_j, A(t_j)) = \prod_{j=1}^{k} \int_0^\infty e^{-\lambda_j s} e^{sA(t_j)} ds$$

$$= \int_0^\infty \cdots \int_0^\infty \exp \left(-\sum_{j=1}^{k} \lambda_j s_j \right) \prod_{j=1}^{k} e^{s_j A(t_j)} ds_1 \cdots ds_k.$$

Making use of (7.13)

$$\left\| \prod_{j=1}^{k} R(\lambda_j, A(t_j)) \right\| \leq M \prod_{j=1}^{k} \int_0^\infty e^{-(\lambda_j - \beta)s_j} ds_j = M \prod_{j=1}^{k} (\lambda_j - \beta)^{-1}.$$

Thus (7.14) holds. It is obvious that (7.14) implies (7.12).

A useful criterion of the stability is given in the following proposition.

Proposition 7.4 *Let $\{A(t); t \in [0,T]\}$ be a stable family of infinitesimal generators of C_0-semigroups in X with stability constants M and β. Let $B(t), t \in [0,T]$, be bounded linear operators in X. If $\|B(t)\| \leq K$ for any $t \in [0,T]$, then $\{A(t) + B(t); t \in [0,T]\}$ is a stable family of infinitesimal generators with stability constants M and $\beta + KM$.*

Proof. If $\lambda > \beta + KM$, then $\lambda \in \rho(A(t) + B(t))$ and

$$R(\lambda, A(t) + B(t)) = \sum_{n=0}^{\infty} R(\lambda, A(t))[B(t)R(\lambda, A(t))]^n.$$

Consequently for $0 \leq t_1 \leq t_2 \leq \cdots \leq t_k \leq T$

$$\prod_{j=1}^{k} R(\lambda, A(t_j) + B(t_j)) = \prod_{j=1}^{k} \left\{ \sum_{n=0}^{\infty} R(\lambda, A(t_j)) [B(t_j)R(\lambda, A(t_j))]^n \right\}. \tag{7.15}$$

If we expand the right hand side of (7.15), a general term is of the form

$$R(\lambda, A(t_k)) [B(t_k)R(\lambda, A(t_k))]^{n_k} \cdots R(\lambda, A(t_1)) [B(t_1)R(\lambda, A(t_1))]^{n_1},$$

the norm of which does not exceed $M^{n+1}K^n(\lambda - \beta)^{-n-k}$ in view of the stability condition (7.12), where $n = \sum_{j=1}^{k} n_j$. The number of the terms in which $\sum_{j=1}^{k} n_j = n$ is equal to the coefficient of a^n in the expansion of $\left(\sum_{m=0}^{\infty} a^m\right)^k$. Since

$$\left(\sum_{m=0}^{\infty} a^m \right)^k = (1-a)^{-k} = \sum_{n=0}^{\infty} \binom{k+n-1}{n} a^n,$$

this coefficient is equal to $\binom{k+n-1}{n}$. Therefore

$$\left\| \prod_{j=1}^{k} R(\lambda, A(t_j) + B(t_j)) \right\| \leq \sum_{n=0}^{\infty} \binom{k+n-1}{n} M^{n+1}K^n(\lambda - \beta)^{-n-k}$$

$$= M(\lambda - \beta)^{-k} \sum_{n=0}^{\infty} \binom{k+n-1}{n} [MK(\lambda - \beta)^{-1}]^n$$

$$= M(\lambda - \beta)^{-k} \left[1 - MK(\lambda - \beta)^{-1}\right]^{-k} = M(\lambda - \beta - MK)^{-k},$$

and the proof is complete.

Proposition 7. 5 *Let Y be a Banach space which is densely and continuously embbeded in X, and $\{S(t); t \in [0, T]\}$ a family of isomorphisms from Y onto X satisfying the following conditions:*
(i) $\|S(t)\|_{\mathcal{L}(Y,X)}$ and $\|S(t)^{-1}\|_{\mathcal{L}(X,Y)}$ are uniformly bounded,
(ii) $S(t)$ is of bounded variation in $[0, T]$ in the norm of $\mathcal{L}(Y, X)$.
Let $\{A(t); t \in [0, T]\}$ be a family of infinitesimal generators in X and set $A_1(t) = S(t)A(t)S(t)^{-1}$. If $\{A_1(t); t \in [0, T]\}$ is a stable family of infinitesimal generators in X, then Y is $A(t)$-admissible for each $t \in [0, T]$ and the family $\{A_Y(t); t \in [0, T]\}$ of the realizations of $A(t)$ in Y is a stable family of infinitesimal generators in Y.

Proof. From Proposition 7.2 it follows that Y is $A(t)$-admissible for each $t \in [0, T]$, and in view of Proposition 7.1 $A_Y(t)$ is the infinitesimal generator of a C_0-semigroup in Y. As in the proof of Proposition 7.2 $D(A_1(t)) = S(t)D(A_Y(t))$ and $R(\lambda, A_Y(t)) = S(t)^{-1}R(\lambda, A_1(t))S(t)$ for large enough real numbers λ. Therefore

$$\prod_{j=1}^{k} R(\lambda, A_Y(t_j)) = \prod_{j=1}^{k} S(t_j)^{-1}R(\lambda, A(t_j))S(t_j). \tag{7.16}$$

If we set $P_j = (S(t_j) - S(t_{j-1}))S(t_{j-1})^{-1}$, then right hand side of (7.16) is equal to

$$S(t_k)^{-1}R(\lambda, A_1(t_k))(1 + P_k)\cdots(1 + P_2)R(\lambda, A_1(t_1))S(t_1). \tag{7.17}$$

Let M_1 and β_1 be the stability constants of $\{A_1(t); t \in [0, T]\}$. When we estimate the norm of a term containing m of the P_j in the expansion of (7.17), we need only $m + 1$ of M_1. Hence the norm of (7.17) does not exceed

$$\|S(t_k)^{-1}\|_{\mathcal{L}(X,Y)}M_1(\lambda - \beta_1)^{-k} \prod_{j=2}^{k} (1 + M_1\|P_j\|) \|S(t_1)\|_{\mathcal{L}(Y,X)}.$$

If we put

$$C = \max \left\{ \sup_{t \in [0,T]} \|S(t)\|_{\mathcal{L}(Y,X)}, \sup_{t \in [0,T]} \|S(t)^{-1}\|_{\mathcal{L}(X,Y)} \right\},$$

we obtain

$$\left\| \prod_{j=1}^{k} R(\lambda, A_Y(t_j)) \right\|_Y \leq C^2 M_1(\lambda - \beta_1)^{-k} \exp\left(M_1 \sum_{j=2}^{k} \|P_j\| \right)$$

$$\leq C^2 M_1(\lambda - \beta_1)^{-k} \exp\left[M_1 C \sum_{j=2}^{k} \|S(t_j) - S(t_{j-1})\|_{\mathcal{L}(Y,X)} \right]$$

$$\leq C^2 M_1 e^{M_1 C V}(\lambda - \beta_1)^{-k},$$

where V is the total variation of $S(t)$. Hence $\{A_Y(t)\}$ is stable.

7.3 Construction of Evolution Operator (1)

In this section we construct an evolution operator for the equation of hyperbolic type

$$\frac{d}{dt}u(t) = A(t)u(t) + f(t), \quad t \in [0,T], \tag{7.18}$$

$$u(0) = x. \tag{7.19}$$

Let X and Y be Banach spaces with norm $\|\cdot\|$ and $\|\cdot\|_Y$ respectively. We assume that Y is a dense subspace of X and the norm of Y is stronger that that of X.

We state the basic assumptions of this section.

(H1) $\{A(t); t \in [0,T]\}$ is a stable family of infinitesimal generators of C_0-semigroups in X.

(H2) Y is $A(t)$-admissible for any $t \in [0,T]$, and the family $\{A_Y(t); t \in [0,T]\}$ of the realizations $A_Y(t)$ of $A(t)$ in Y is stable in Y.

(H3) For $t \in [0,T], D(A(t)) \supset Y$, and $A(t)$ is a bounded linear operator from Y into X, and $A(t)$ is continuous in $[0,T]$ in the norm topology of $\mathcal{L}(Y,X)$.

We denote the stability constants of $\{A(t); t \in [0,T]\}$ by M, β, and those of $\{A_Y(t); t \in [0,T]\}$ by $\tilde{M}, \tilde{\beta}$.

Theorem 7.1 *Under the assumptions* (H1),(H2),(H3) *there exists an evolution operator* $U(t,s), 0 \le s \le t \le T$, *of* (7.18), (7.19) *which is a strongly continuous family of bounded linear operators in* X *satisfying*

$$U(t,r)U(r,s) = U(t,s), \quad 0 \le s \le r \le t \le T, \tag{7.20}$$

$$U(s,s) = I, \quad s \in [0,T], \tag{7.21}$$

$$\|U(t,s)\| \le Me^{\beta(t-s)}, \quad 0 \le s \le t \le T, \tag{7.22}$$

$$\left(\frac{\partial}{\partial t}\right)^+ U(t,s)y\bigg|_{t=s} = A(s)y, \quad 0 \le s < T, \ y \in Y, \tag{7.23}$$

$$\frac{\partial}{\partial s}U(t,s)y = -U(t,s)A(s)y, \quad 0 \le s \le t \le T, \ y \in Y, \tag{7.24}$$

where the right derivative in (7.23) *and the derivative in* (7.24) *are in the strong topology of* X. *Moreover, the evolution operator having the above properties is unique.*

Proof. For a positive integer n let $t_k^n = (k/n)T, k = 0, 1, \ldots, n$. Set

$$
A_n(t) = \begin{cases} A(t_k^n) & t_k^n \le t < t_{k+1}^n, \quad k = 0, 1, \ldots, n-1, \\ A(T) & t = T. \end{cases} \tag{7.25}
$$

Since $A(\cdot)$ is continuous in the norm of $\mathcal{L}(Y, X)$, we have

$$
\|A(t) - A_n(t)\|_{\mathcal{L}(Y,X)} \to 0 \quad \text{as} \quad n \to \infty \tag{7.26}
$$

uniformly in $t \in [0, T]$. Clearly $\{A_n(t); t \in [0, T]\}$ is stable in X with stability constants M, β, and $\{A_{n,Y}(t); t \in [0, T]\}$ which is a family of the realizations of $A_n(t)$ in Y is stable with stability constants $\tilde{M}, \tilde{\beta}$.

We define an operator valued function $U_n(t, s), 0 \le s \le t \le T$, by

$$
U_n(t, s) = \begin{cases} e^{(t - t_l^n)A(t_l^n)} \left[\displaystyle\prod_{j=k+1}^{l-1} e^{(T/n)A(t_j^n)} \right] e^{(t_{k+1}^n - s)A(t_k^n)}, \\ \qquad\qquad t_l^n \le t \le t_{l+1}^n,\ t_k^n \le s \le t_{k+1}^n, \quad \text{if } l > k, \\ e^{(t-s)A(t_k^n)}, \qquad\qquad t_k^n \le s \le t \le t_{k+1}^n. \end{cases} \tag{7.27}
$$

It is easy to show that

$$
U_n(t, s) = U_n(t, r)U_n(r, s), \quad 0 \le s \le r \le t \le T, \tag{7.28}
$$
$$
U_n(s, s) = I, \tag{7.29}
$$

and $U_n(t, s)$ is strongly continuous in s, t in $0 \le s \le t \le T$. From (H1) and Proposition 7.3 it follows that

$$
\|U_n(t, s)\| \le M e^{\beta(t-s)}, \quad 0 \le s \le t \le T, \tag{7.30}
$$

and from (H2) and Proposition 7.3

$$
U_n(t, s)Y \subset Y \quad \text{and} \quad \|U_n(t, s)\|_Y \le \tilde{M} e^{\tilde{\beta}(t-s)}, \quad 0 \le s \le t \le T. \tag{7.31}
$$

Since $D(A(t)) \supset Y$ we have for $y \in Y$

$$
\frac{\partial}{\partial t} U_n(t, s)y = A_n(t)U_n(t, s)y, \quad t \ne t_k^n, \quad k = 0, 1, \ldots, n, \tag{7.32}
$$
$$
\frac{\partial}{\partial s} U_n(t, s)y = -U_n(t, s)A_n(s)y, \quad s \ne t_k^n, \quad k = 0, 1, \ldots, n. \tag{7.33}
$$

With the aid of (7.32),(7.33) we have for $y \in Y, 0 \le s < t \le T$

$$
U_n(t, s)y - U_m(t, s)y = -\int_s^t \frac{\partial}{\partial r} U_n(t, r)U_m(r, s)y\, dr
$$
$$
= \int_s^t U_n(t, r)(A_n(r) - A_m(r))U_m(r, s)y\, dr. \tag{7.34}
$$

Setting $\gamma = \max\{\beta, \tilde{\beta}\}$ we get from (7.34)

$$\|U_n(t,s)y - U_m(t,s)y\| \leq M\tilde{M}e^{\gamma(t-s)}\|y\|_Y \int_s^t \|A_n(r) - A_m(r)\|_{\mathcal{L}(Y,X)} dr.$$
(7.35)

From (7.26),(7.35) it follows that $U_n(t,s)y$ converges uniformly in $0 \leq s \leq t \leq T$ in the strong topology of X. Since Y is dense in X and $\{U_n(t,s)\}$ is uniformly bounded in $\mathcal{L}(X)$ in view of (7.30), $U_n(t,s)x$ converges uniformly in $0 \leq s \leq t \leq T$ in the strong topology of X also for $x \in X$. Let

$$U(t,s)x = \lim_{n\to\infty} U_n(t,s)x, \quad 0 \leq s \leq t \leq T, \ x \in X. \tag{7.36}$$

Clearly $U(t,s)x$ is continuous in $0 \leq s \leq t \leq T$ in the strong topology of X, and it follows from (7.28) and (7.29) that (7.20) and (7.21) hold. Next we show (7.23),(7.24). For $y \in Y, 0 \leq s < t \leq T, 0 \leq \tau \leq T$ we have

$$U_n(t,s)y - e^{(t-s)A(\tau)}y = -\int_s^t \frac{\partial}{\partial r}U_n(t,r)e^{(r-s)A(\tau)}y\,dr$$

$$= \int_s^t U_n(t,r)(A_n(r) - A(\tau))e^{(r-s)A(\tau)}y\,dr. \tag{7.37}$$

Hence

$$\|U_n(t,s)y - e^{(t-s)A(\tau)}y\| \leq M\tilde{M}e^{\gamma(t-s)}\|y\|_Y \int_s^t \|A_n(r) - A(\tau)\|_{\mathcal{L}(Y,X)} dr.$$
(7.38)

Letting $n \to \infty$ and dividing by $t - s$ we get

$$\left\|\frac{U(t,s)y - y}{t-s} - \frac{e^{(t-s)A(\tau)}y - y}{t-s}\right\|$$

$$\leq M\tilde{M}e^{\gamma(t-s)}\|y\|_Y \frac{1}{t-s}\int_s^t \|A(r) - A(\tau)\|_{\mathcal{L}(Y,X)} dr. \tag{7.39}$$

Choosing $\tau = s$ in (7.39) and letting $t \to s$ we obtain (7.23). If we choose $\tau = t$ in (7.38) and let $s \to t$ we obtain

$$\left(\frac{\partial}{\partial s}\right)^- U(t,s)y\bigg|_{s=t} = \lim_{s\uparrow t} \frac{U(t,t)y - U(t,s)y}{t-s}$$

$$= \lim_{s\uparrow t} \frac{y - U(t,s)y}{t-s} = -A(t)y. \tag{7.40}$$

For $0 \leq s < t \leq T$ we get with the aid of (7.23)

$$\left(\frac{\partial}{\partial s}\right)^+ U(t,s)y = \lim_{h\downarrow 0} \frac{1}{h}[U(t,s+h)y - U(t,s)y]$$

$$= \lim_{h \downarrow 0} U(t, s+h) \frac{y - U(s+h, s)y}{h}$$

$$= -U(t, s) \left(\frac{\partial}{\partial t} \right)^{+} U(t, s)y \bigg|_{t=s} = -U(t, s)A(s)y, \qquad (7.41)$$

and for $0 \leq s \leq t \leq T$ by (7.40)

$$\left(\frac{\partial}{\partial s} \right)^{-} U(t, s)y = \lim_{h \downarrow 0} \frac{1}{h} [U(t, s)y - U(t, s - h)y]$$

$$= \lim_{h \downarrow 0} U(t, s) \frac{y - U(s, s - h)y}{h} = -U(t, s)A(s)y. \qquad (7.42)$$

Combining (7.41) and (7.42) we obtain (7.24).

Finally we show the uniqueness of an evolution operator. Suppose that $V(t, s)$ is an evolution operator satisfying (7.20),(7.21),(7.22),(7.23),(7.24). Then using (7.24) we get for $0 \leq s < t \leq T, y \in Y$

$$V(t, s)y - U_n(t, s)y = - \int_s^t \frac{\partial}{\partial r} V(t, r)U_n(r, s)y dr$$

$$= \int_s^t V(t, r)(A(r) - A_n(r))U_n(r, s)y dr.$$

Therefore

$$\|V(t, s)y - U_n(t, s)y\| \leq M \tilde{M} e^{\gamma(t-s)} \|y\|_Y \int_s^t \|A(r) - A_n(r)\|_{\mathcal{L}(Y,X)} dr. \qquad (7.43)$$

Letting $n \to \infty$ in (7.43) and using (7.26) we obtain $V(t, s)y = U(t, s)y$. Since Y is dense in X we conclude $V(t, s) = U(t, s)$.

7.4   Uniqueness of Regular Solutions

In this section we consider the initial value problem (7.18),(7.19) assuming that $f \in C([0, T]; X)$ and $x \in Y$. Following A. Pazy [127] we call a function u a *Y-valued solution* of (7.18),(7.19) if $u \in C^1([0, T]; X) \cap C([0, T]; Y)$ and (7.18),(7.19) hold. A *Y*-valued solution is a more restrictive solution than a strict one since it is required to satisfy $u(t) \in Y \subset D(A(t))$.

Theorem 7.2 *Suppose that* (H1),(H2),(H3) *are satisfied. Let* $x \in Y$ *and* $f \in C([0, T]; X)$. *If* u *is a Y-valued solution of* (7.18),(7.19), *then it is represented by*

$$u(t) = U(t, 0)x + \int_0^t U(t, s)f(s) ds, \qquad (7.44)$$

where $U(t, s)$ is the evolution operator constructed in Theorem 1. Hence a Y-valued solution is unique.

Proof. Let $U_n(t, s)$ be the operator valued function defined by (7.27). Then $U_n(t, r)u(r)$ is a continuously differentiable function of r except at a finite number of points and

$$\frac{\partial}{\partial r}U_n(t, r)u(r) = -U_n(t, r)(A_n(r) - A(r))u(r) + U_n(t, r)f(r). \qquad (7.45)$$

Integrating (7.45) from 0 to t we get

$$u(t) = U_n(t, 0)x - \int_0^t U_n(t, r)(A_n(r) - A(r))u(r)dr + \int_0^t U_n(t, r)f(r)dr. \qquad (7.46)$$

Since

$$\left\| \int_0^t U_n(t, r)(A_n(r) - A(r))u(r)dr \right\|$$

$$\leq Me^{\beta t} \sup_{r \in [0,T]} \|u(r)\|_Y \int_0^t \|A_n(r) - A(r)\|_{\mathcal{L}(Y,X)}dr \to 0$$

as $n \to \infty$, we obtain (7.44) from (7.46).

7.5 Construction of Evolution Operator (2)

In [88] T. Kato proved replacing (H2) by the following more restrictive assumption

(H2') There is a family $\{S(t); t \in [0, T]\}$ of isomorphisms of Y onto X such that $S(t)$ is strongly continuously differentiable in $[0, T]$ and

$$S(t)A(t)S(t)^{-1} = A(t) + B(t), \quad t \in [0, T], \qquad (7.47)$$

where $B(t)$ is strongly continuous in $[0, T]$ to $\mathcal{L}(X)$

that the evolution operator constructed in section 7.3 has stronger regularity properties so that it provides with a Y-valued solution. In this section following K. Kobayashi [95] we prove a similar result replacing the norm continuity of $A(t)$ in t by the strong continuity:

(H3') $D(A(t)) \supset Y$ for each $t \in [0, T]$, and $A(t)$ is strongly continuous in $[0, T]$ to $\mathcal{L}(Y, X)$.

Theorem 7. 3 *Under the assumptions* (H1), (H2'), (H3') *there exists a unique evolution operator* $U(t, s), 0 \leq s \leq t \leq T$, *of* (7.18), (7.19) *which is a strongly continuous family of bounded linear operators in* X *satisfying* (7.20), (7.21), (7.22), (7.24) *and*

$$U(t, s)Y \subset Y, \text{ and } U(t, s) \text{ is strongly continuous}$$
$$\text{in } 0 \leq s \leq t \leq T \text{ to } \mathcal{L}(Y), \tag{7.48}$$

$$\frac{\partial}{\partial t}U(t, s)y = A(t)U(t, s)y, \quad y \in Y, \quad 0 \leq s \leq t \leq T, \tag{7.49}$$

where the derivatives in (7.24), (7.49) *are in the strong topology of* X.

Proof. We divide the proof into several steps.

Step 1. We first note that $\|A(t)\|_{\mathcal{L}(Y,X)}, \|S(t)\|_{\mathcal{L}(Y,X)}, \|S(t)^{-1}\|_{\mathcal{L}(X,Y)}, \|B(t)\|$ are bounded in $t \in [0, T]$. By vitue of Proposition 7.4 and the assumption (H2') $S(t)A(t)S(t)^{-1}$ is stable in X, and so applying Proposition 7.5 we see that Y is $A(t)$-admissible for each $t \in [0, T]$ and $A_Y(t)$ is stable in Y. We denote the stability constants of $\{A_Y(t)\}$ by $\tilde{M}, \tilde{\beta}$.

Let $P = \{t_k; k = 0, 1, 2, \ldots\}$ be a sequence such that $0 \leq t_0 < t_1 < \cdots < t_k < \cdots \leq T$ and $t_\infty = \lim_{k \to \infty} t_k$. For this sequence P we define the operator $U(t, s; P), t_0 \leq s \leq t < t_\infty$, by

$$U(t, s; P) = \begin{cases} e^{(t-t_l)A(t_l)} \displaystyle\prod_{j=k+1}^{l-1} e^{(t_{j+1}-t_j)A(t_j)} \cdot e^{(t_{k+1}-s)A(t_k)} \\ \qquad \text{if } t_k \leq s < t_{k+1}, t_l \leq t < t_{l+1}, k < l, \\ e^{(t-s)A(t_k)} \quad \text{if } t_k \leq s \leq t < t_{k+1}. \end{cases} \tag{7.50}$$

For an operator valued function $F(t), t \in [0, T]$, let $F(t; P)$ be the step function defined by

$$F(t; P) = F(t_k), \quad t \in [t_k, t_{k+1}), \quad k = 0, 1, 2, \ldots.$$

By the above remark $U(t, s; P)$ leaves Y invariant and

$$\|U(t, s; P)\| \leq M e^{\beta(t-s)}, \quad \|U(t, s; P)\|_Y \leq \tilde{M} e^{\tilde{\beta}(t-s)}. \tag{7.51}$$

Step 2. *Suppose that* $P = \{t_k; k = 0, 1, 2, \ldots\}$ *an infinite sequence. Let* $t_k \leq t_k'' < t_{k+1}, k = 0, 1, 2, \ldots$. *Then we have*

$$\lim_{k \to \infty} U(t_k'', t_0; P)x \quad \text{exists in} \quad X \quad \text{for any} \quad x \in X, \tag{7.52}$$

$$\lim_{k \to \infty} U(t_k'', t_0; P)y \quad \text{exists in} \quad Y \quad \text{for any} \quad y \in Y. \tag{7.53}$$

Proof. If $x \in Y$, then by (7.51)

$$\|(d/dt)U(t, t_0; P)x\| = \|A(t; P)U(t, t_0; P)x\|$$
$$\leq \|A(t; P)\|_{\mathcal{L}(Y, X)}\|U(t, t_0; P)\|_Y\|x\|_Y \leq C\|x\|_Y.$$

Hence $\lim_{k \to \infty} U(t_k'', t_0; P)x$ exists in X. The same result remains valid for $x \in X$ since $\|U(t, s; P)\|$ is uniformly bounded by (7.51).

In order to prove (7.53) we first show that

$$\|S(t_k'')U(t_k'', t_i; P)S(t_i)^{-1}x - U(t_k'', t_i; P)x\| \leq C(t_k'' - t_i)\|x\| \qquad (7.54)$$

for $x \in X$ and $0 \leq i \leq k$. It suffices to show (7.54) for $x \in Y$ since Y is dense in X. Using (7.47) we get for $t_j \leq \sigma < t_{j+1}, j = i, \ldots, k - 1$,

$$\frac{\partial}{\partial \sigma}\left[S(t_k'')U(t_k'', \sigma; P)S(\sigma; P)^{-1}U(\sigma, t_i; P)x\right]$$
$$= -S(t_k'')U(t_k'', \sigma; P)A(t_j)S(t_j)^{-1}U(\sigma, t_i; P)x$$
$$+S(t_k'')U(t_k'', \sigma; P)S(t_j)^{-1}A(t_j)U(\sigma, t_i; P)x$$
$$= -S(t_k'')U(t_k'', \sigma; P)S(t_j)^{-1}B(t_j)U(\sigma, t_i; P)x.$$

Hence we have

$$\frac{\partial}{\partial \sigma}\left[S(t_k'')U(t_k'', \sigma; P)S(\sigma; P)^{-1}U(\sigma, t_i; P)x\right]$$
$$= -S(t_k'')U(t_k'', \sigma; P)S(\sigma; P)^{-1}B(\sigma; P)U(\sigma, t_i; P)x \qquad (7.55)$$

for $t_i \leq \sigma < t_k$. The same equality also holds for $t_k \leq \sigma \leq t_k''$. Integrating (7.55) from t_j to t_{j+1} and rewriting the obtained equality appropriately we get

$$S(t_k'')U(t_k'', t_{j+1}; P)S(t_{j+1})^{-1}U(t_{j+1}, t_i; P)x$$
$$+S(t_k'')U(t_k'', t_{j+1}; P)(S(t_j)^{-1} - S(t_{j+1})^{-1})U(t_{j+1}, t_i; P)x$$
$$-S(t_k'')U(t_k'', t_j; P)S(t_j)^{-1}U(t_j, t_i; P)x$$
$$= -\int_{t_j}^{t_{j+1}} S(t_k'')U(t_k'', \sigma; P)S(\sigma; P)^{-1}B(\sigma; P)U(\sigma, t_i; P)x d\sigma. \qquad (7.56)$$

Adding (7.56) for $j = i, \ldots, k - 1$ yields

$$S(t_k'')U(t_k'', t_k; P)S(t_k)^{-1}U(t_k, t_i; P)x - S(t_k'')U(t_k'', t_i; P)S(t_i)^{-1}x$$
$$+\sum_{j=i}^{k-1} S(t_k'')U(t_k'', t_{j+1}; P)(S(t_j)^{-1} - S(t_{j+1})^{-1})U(t_{j+1}, t_i; P)x$$
$$= -\int_{t_i}^{t_k} S(t_k'')U(t_k'', \sigma; P)S(\sigma; P)^{-1}B(\sigma; P)U(\sigma, t_i; P)x d\sigma. \qquad (7.57)$$

Integrating (7.55) from t_k to t_k''

$$S(t_k'')S(t_k)^{-1}U(t_k'', t_i; P)x - S(t_k'')U(t_k'', t_k; P)S(t_k)^{-1}U(t_k, t_i; P)x$$

$$= -\int_{t_k}^{t_k''} S(t_k'')U(t_k'', \sigma; P)S(\sigma; P)^{-1}B(\sigma; P)U(\sigma, t_i; P)x\,d\sigma. \quad (7.58)$$

From (7.57) and (7.58) we obtain

$$S(t_k'')U(t_k'', t_i; P)S(t_i)^{-1}x - U(t_k'', t_i; P)x$$
$$= S(t_k'')(S(t_k)^{-1} - S(t_k'')^{-1})U(t_k'', t_i; P)x$$
$$+ \sum_{j=i}^{k-1} S(t_k'')U(t_k'', t_{j+1}; P)(S(t_j)^{-1} - S(t_{j+1})^{-1})U(t_{j+1}, t_i; P)x$$
$$+ \int_{t_i}^{t_k''} S(t_k'')U(t_k'', \sigma; P)S(\sigma; P)^{-1}B(\sigma; P)U(\sigma, t_i; P)x\,d\sigma. \quad (7.59)$$

Since $S(t)^{-1}$ is Lipschitz continuous in the norm of $\mathcal{L}(X, Y)$, we readily obtain (7.54) from (7.59) using (7.51).

For $x \in X$ set $w_i = S(t_i)U(t_i, t_0; P)S(t_0)^{-1}x$, and

$$W(t, s; P) = S(t)U(t, s; P)S(s)^{-1} - U(t, s; P).$$

Then by (7.51) and (7.54) we have

$$\|W(t_k'', t_i; P)w_i\| = \|(S(t_k'')U(t_k'', t_i; P)S(t_i)^{-1} - U(t_k'', t_i; P))w_i\|$$
$$\leq C(t_k'' - t_i)\|w_i\| \leq C(t_k'' - t_i)\|x\|. \quad (7.60)$$

Since

$$S(t_k'')U(t_k'', t_0; P)S(t_0)^{-1}x$$
$$= S(t_k'')U(t_k'', t_i; P)S(t_i)^{-1}w_i = W(t_k'', t_i; P)w_i + U(t_k'', t_i; P)w_i,$$

it follows from (7.60) that

$$a_{k,j} \equiv \|S(t_j'')U(t_j'', t_0; P)S(t_0)^{-1}x - S(t_k'')U(t_k'', t_0; P)S(t_0)^{-1}x\|$$
$$\leq \|W(t_j'', t_i; P)w_i\| + \|W(t_k'', t_i; P)w_i\|$$
$$+ \|U(t_j'', t_i; P)w_i - U(t_k'', t_i; P)w_i\|$$
$$\leq C(t_j'' - t_i + t_k'' - t_i)\|x\| + \|U(t_j'', t_i; P)w_i - U(t_k'', t_i; P)w_i\|.$$

Since $\lim_{k\to\infty} U(t_k'', t_i; P)w_i$ exists in X in view of (7.52) (from the proof it is obvious that the same result remains valid with t_0 replaced by any other value in (t_0, t_∞) in (7.52)), we obtain

$$\limsup_{k,j\to\infty} a_{k,j} \leq C(t_\infty - t_i)\|x\|$$

for each i. Hence we get $\limsup_{k,j \to \infty} a_{k,j} = 0$, and

$$\lim_{k \to \infty} S(t_k'')U(t_k'', t_0; P)S(t_0)^{-1}x$$

exists in X. Therefore for any $y \in Y$

$$U(t_k'', t_0; P)y = S(t_k'')^{-1} \cdot S(t_k'')U(t_k'', t_0; P)S(t_0)^{-1} \cdot S(t_0)y$$

converges in Y for any $y \in Y$.

Step 3. *For each $\epsilon > 0, y \in Y$ and $s \in [0, T)$ there exists a finite partition $P(\epsilon, s, y) : s = t_0 < t_1 < \cdots < t_N = T$ of the interval $[s, T]$ such that*

$$t_{k+1} - t_k \le \epsilon, \quad k = 0, 1, \ldots, N-1, \tag{7.61}$$
$$\|(A(t') - A(t))U(t, s; P)y\| \le \epsilon \quad \text{for} \quad t, t' \in [t_k, t_{k+1}],$$
$$k = 0, 1, \ldots, N-1. \tag{7.62}$$

Proof. Set $t_0 = s$ and define t_{k+1} inductively in the following manner. If $t_k = T$, then set $t_{k+1} = t_k$; and if $t_k < T$, then set $t_{k+1} = t_k + h_k$ where h_k is the largest number such that the following conditions hold:

$$0 < h_k \le \epsilon, \quad t_k + h_k \le T, \tag{7.63}$$
$$\|(A(t') - A(t))u_k(t - t_k)\| \le \epsilon \text{ for } t, t' \in [t_k, t_k + h_k], \tag{7.64}$$

where

$$u_k(t) = e^{tA(t_k)} \prod_{j=0}^{k-1} e^{(t_{j+1} - t_j)A(t_j)}y.$$

Since $(A(t') - A(t))u_k(t - t_k)$ is strongly continuous in X in t, t' by virtue of the assumption (H3'), we see that $h_k > 0$. If we verify that $t_N = T$ for some N, then the proof will be complete. On the contrary suppose that $t_k < T$ for all k. Clearly $h_k \to 0$ as $k \to \infty$. Set $P' = \{t_k; k = 0, 1, 2, \ldots\}$. Since $(A(t') - A(t))u_k(t - t_k)$ is strongly continuous in t, t' in X, we can show without difficulty that if k is so large that $h_k < \epsilon$, then there exist $t_k', t_k'' \in [t_k, t_{k+1})$ such that

$$\|(A(t_k') - A(t_k''))u_k(t_k'' - t_k)\| \ge \epsilon/2. \tag{7.65}$$

According to (7.53) $\lim_{k \to \infty} U(t_k'', t_0; P')y$ exists in Y. Hence by (H2') we have

$$(A(t_k') - A(t_k'')) u_k(t_k'' - t_k)$$
$$= (A(t_k') - A(t_k'')) e^{(t_k'' - t_k)A(t_k)} \prod_{j=0}^{k-1} e^{(t_{j+1} - t_j)A(t_j)}y$$
$$= (A(t_k') - A(t_k'')) U(t_k'', t_0; P')y \to 0$$

in X as $k \to \infty$. Therefore letting $k \to \infty$ in (7.65) we obtain $0 \geq \epsilon/2$, which is obviously a contradiction.

Step 4. Let $\epsilon_i > 0$, $s_i \in [0, T)$ and $y_i \in Y$, $i = 1, 2$, and let $P_i = P(\epsilon_i, s_i, y_i)$ be a partition of $[s_i, T]$ satisfying (7.61),(7.62) with ϵ_i, s_i, y_i in place of ϵ, s, y. Let $\tilde{P}_i$ be any partition of $[s_i, T]$ which is a refinement of P_i. Then we have

$$\|U(t_1, s_1; \tilde{P}_1)y_1 - U(t_2, s_2; \tilde{P}_2)y_2\|$$
$$\leq C\left[\|y_1 - y\| + \|y_2 - y\| + \epsilon_1 + \epsilon_2 + (|t_1 - t_2| + |s_1 - s_2|)\|y\|_Y\right]$$

$$(7.66)$$

for all $t_i \in [s_i, T]$, $i = 1, 2$, and $y \in Y$.

Proof. Differentiating $U(t_i, \sigma; \tilde{P}_i)U(\sigma, s_i; P_i)y_i$ in σ and integrating over $[s_i, t_i]$ we obtain

$$U(t_i, s_i; P_i)y_i - U(t_i, s_i; \tilde{P}_i)y_i$$
$$= \int_{s_i}^{t_i} U(t_i, \sigma; \tilde{P}_i)(A(\sigma; P_i) - A(\sigma; \tilde{P}_i))U(\sigma, s_i; P_i)y_i d\sigma. \quad (7.67)$$

Since $\tilde{P}_i$ is a refinement of P_i we have in view of (7.62)

$$\|(A(\sigma; P_i) - A(\sigma, \tilde{P}_i))U(\sigma, s_i; P_i)y_i\|$$
$$\leq \|(A(\sigma; P_i) - A(\sigma))U(\sigma, s_i; P_i)y_i\| + \|(A(\sigma, \tilde{P}_i) - A(\sigma))U(\sigma, s_i; P_i)y_i\|$$
$$\leq 2\epsilon_i$$

for $\sigma \in [s_i, T]$. Hence from (7.67) we get

$$\|U(t_i, s_i; P_i)y_i - U(t_i, s_i; \tilde{P}_i)y_i\| \leq C\epsilon_i, \quad i = 1, 2. \quad (7.68)$$

Therefore

$$\|U(t_1, s_1; \tilde{P}_1)y_1 - U(t_2, s_2; \tilde{P}_2)y_2\|$$
$$\leq C(\epsilon_1 + \epsilon_2) + \|U(t_1, s_1; P_1)y_1 - U(t_2, s_2; P_2)y_2\|$$
$$\leq C(\epsilon_1 + \epsilon_2) + I_1 + I_2 + I_3, \quad (7.69)$$

where

$$I_1 = \|U(t_1, s_1; P_1)y_1 - U(t_1, s_1; P_3)y_1\|,$$
$$I_2 = \|U(t_1, s_1; P_3)y_1 - U(t_2, s_2; P_3)y_2\|,$$
$$I_3 = \|U(t_2, s_2; P_3)y_2 - U(t_2, s_2; P_2)y_2\|,$$

and P_3 is the superposition of P_1 and P_2. By virtue of (7.68) we have

$$I_1 \leq C\epsilon_1, \quad I_3 \leq C\epsilon_2. \quad (7.70)$$

Therefore it remains to estimate I_2. We may assume $s_2 \leq s_1$. By (7.51) we have

$$I_2 \leq Me^{\beta T}(\|y_1 - y\| + \|y_2 - y\|) + \|U(t_1, s_1; P_3)y - U(t_2, s_2; P_3)y\|. \quad (7.71)$$

Noting $\|(d/dt)U(t, s_2; P_3)y\| \leq C\|y\|_Y$, we obtain

$$\begin{aligned}
\|U&(t_1, s_1; P_3)y - U(t_2, s_2; P_3)y\| \\
&\leq \|U(t_1, s_1; P_3)y - U(t_1, s_2; P_3)y\| + \|U(t_1, s_2; P_3)y - U(t_2, s_2; P_3)y\| \\
&\leq \|U(t_1, s_1; P_3)(1 - U(s_1, s_2; P_3))y\| + C|t_1 - t_2|\|y\|_Y \\
&\leq C\|(1 - U(s_1, s_2; P_3))y\| + C|t_1 - t_2|\|y\|_Y \\
&\leq C(|s_1 - s_2| + |t_1 - t_2|)\|y\|_Y. \quad (7.72)
\end{aligned}$$

Combining (7.69), (7.70), (7.71), (7.72) we conclude (7.66).

Step 5. Let $x \in X$ and $0 \leq s \leq t \leq T$ be fixed. Let $\{s_n\}, \{t_n\}, \{y_n\}$ be sequences such that $0 \leq s_n \leq t_n \leq T, y_n \in Y, s_n \to s, t_n \to t, y_n \to x$ in X. For each n let P_n be a partition of $[s_n, T]$ satisfying (7.61), (7.62) with ϵ, s, y replaced by $1/n, s_n, y_n$ respectively. Then in view of (7.66) we have for each $y \in Y$

$$\begin{aligned}
\limsup_{n,m \to \infty} &\|U(t_n, s_n; P_n)y_n - U(t_m, s_m; P_m)y_m\| \\
&\leq \lim_{n,m \to \infty} C[\|y_n - y\| + \|y_m - y\| + 1/n + 1/m \\
&\quad + (|t_n - t_m| + |s_n - s_m|)\|y\|_Y] = 2C\|x - y\|.
\end{aligned}$$

Since Y is dense in X, this implies

$$U(t, s)x = \lim_{n \to \infty} U(t_n, s_n; P_n)y_n \quad (7.73)$$

exists in X, and furthermore using (7.66) we see that the limit $U(t, s)x$ does not depend on the choice of the sequences $\{s_n\}, \{t_n\}, \{y_n\}, \{P_n\}$. We are going to show that

For each s, t, $U(t, s)$ is a bounded linear operator in X satisfying

$$\|U(t, s)\| \leq Me^{\beta(t-s)}, \quad (7.74)$$

and is strongly continuous in $0 \leq s \leq t \leq T$ to $\mathcal{L}(X)$. Furthermore it satisfies (7.20), (7.21).

From Step 4 or (7.68) in its proof it follows that

$$U(t, s)x = \lim_{n \to \infty} U(t_n, s_n; \tilde{P}_n)y_n$$

for any refinement $\tilde{P}_n$ of P_n. Hence it is easily seen that $U(t, s)$ is a linear operator in X. The inequality (7.74) is a direct consequence of (7.51). Let $s \leq r \leq t$ and r_n be a point of P_n such that $r_n \to r$ as $n \to \infty$. If we set $z_n = U(r_n, s_n; P_n)y_n$, then $z_n \in Y$ and $z_n \to U(r, s)x$. Let $P'_n = P_n \cap [r_n, T]$. Then P'_n satisfies (7.61), (7.62) with ϵ, s, y replaced by $1/n, r_n, z_n$, since

$$U(t, r_n; P'_n)z_n = U(t, r_n; P'_n)U(r_n, s_n; P_n)y_n = U(t, s_n; P_n)y_n. \qquad (7.75)$$

Therefore letting $n \to \infty$ in (7.75) with $t = t_n$ we obtain

$$U(t, r)U(r, s)x = U(t, s)x.$$

With the aid of Step 4 we can show without difficulty that for $0 \leq s \leq t \leq T, 0 \leq s' \leq t' \leq T$ and $x \in X, y \in Y$

$$\|U(t, s)x - U(t', s')x\| \leq C\left[2\|x - y\| + (|t - t'| + |s - s'|)\|y\|_Y\right].$$

Hence $U(t, s)x$ is continuous in $0 \leq s \leq t \leq T$ in the strong topology of X.

Step 6. We prove (7.24) in this step. The statement is established with the aid of an argument similar to that used in the proof of the same statement in Theorem 7.1 via (7.39), (7.40), (7.41) if we have the following inequality: for $0 \leq s \leq t \leq T, r \in [0, T]$ and $y \in Y$

$$\|U(t, s)y - e^{(t-s)A(r)}y\| \leq C \int_s^t \|(A(\sigma) - A(r))e^{(\sigma-s)A(r)}y\|d\sigma. \qquad (7.76)$$

Proof of (7.76). Let $P_n = P(1/n, s, y)$ be a partition of $[s, T]$ as in Step 3. Differentiating $U(t, \sigma; P_n)e^{(\sigma-s)A(r)}y$ in σ and integrating over $[s, t]$ yields

$$U(t, s; P_n)y - e^{(t-s)A(r)}y$$
$$= \int_s^t U(t, \sigma; P_n)(A(\sigma; P_n) - A(r))e^{(\sigma-s)A(r)}yd\sigma.$$

Hence with the aid of (7.51) we obtain

$$\|U(t, s; P_n)y - e^{(t-s)A(r)}y\|$$
$$\leq Me^{\beta T}\int_s^t \|(A(\sigma; P_n) - A(r))e^{(\sigma-s)A(r)}y\|d\sigma. \qquad (7.77)$$

The inequality (7.76) is obtained by letting $n \to \infty$ in (7.77).

We also note that if we have (7.76) we can show (7.23) since

$$\left\|\frac{U(t, s)y - y}{t - s} - \frac{e^{(t-s)A(s)}y - y}{t - s}\right\| \leq \frac{1}{t - s}\|U(t, s)y - e^{(t-s)A(s)}y\|$$

$$\leq \frac{C}{t-s} \int_s^t \|(A(\sigma) - A(s))e^{(\sigma-s)A(s)}y\| d\sigma,$$

$$\|(A(\sigma) - A(s))e^{(\sigma-s)A(s)}y\|$$
$$\leq \|(A(\sigma) - A(s))(e^{(\sigma-s)A(s)}y - y)\| + \|(A(\sigma) - A(s))y\|$$
$$\leq C\|e^{(\sigma-s)A(s)}y - y\|_Y + \|(A(\sigma) - A(s))y\| \to 0$$

as $\sigma \to s$.

Step 7. We complete the proof of the theorem in this step. We begin with the proof of (7.48). Let $s \in [0, T)$ and $y \in Y$. Let $P_n = P(1/n, s, y)$ be a partition of $[s, T]$ as in Step 3. Set $V(t, r) = U(t, r)S(r)^{-1}$ for $s \leq r \leq t \leq T$. Then using(7.24)

$$\frac{\partial}{\partial r} V(t,r)U(r,s; P_n)y = -U(t,r)A(r)S(r)^{-1}U(r,s; P_n)y$$

$$+U(t,r)\frac{d}{dr}S(r)^{-1} \cdot U(r,s; P_n)y + U(t,r)S(r)^{-1}A(r; P_n)U(r,s; P_n)y$$

$$= -U(t,r)A(r)S(r)^{-1}U(r,s; P_n)y + U(t,r)\frac{d}{dr}S(r)^{-1} \cdot U(r,s; P_n)y$$

$$+U(t,r)S(r)^{-1}(A(r; P_n) - A(r))U(r,s; P_n)y$$
$$+U(t,r)S(r)^{-1}A(r)U(r,s; P_n)y$$

$$= -U(t,r)S(r)^{-1}B(r)U(r,s; P_n)y + U(t,r)\frac{d}{dr}S(r)^{-1} \cdot U(r,s; P_n)y$$

$$+U(t,r)S(r)^{-1}(A(r; P_n) - A(r))U(r,s; P_n)y$$
$$= V(t,r)C(r)U(r,s; P_n)y$$
$$+V(t,r)(A(r; P_n) - A(r))U(r,s; P_n)y, \tag{7.78}$$

where

$$C(r) = S(r)\frac{d}{dr}S(r)^{-1} - B(r) = \dot{S}(r)S(r)^{-1} - B(r),$$

which is strongly continuous in $[0, T]$ to $\mathcal{L}(X)$. Integration of (7.78) from s to t yields

$$S(t)^{-1}U(t,s; P_n)y - V(t,s)y = \int_s^t V(t,r)C(r)U(r,s; P_n)ydr$$

$$+ \int_s^t V(t,r)(A(r; P_n) - A(r))U(r,s; P_n)ydr. \tag{7.79}$$

Since

$$\|(A(r; P_n) - A(r))U(r,s; P_n)y\| \leq 1/n$$

in view of (7.62) we see that the second term of the right hand side of (7.79) tends to 0 as $n \to \infty$. Hence we obtain

$$S(t)^{-1}U(t, s)y - V(t, s) = \int_s^t V(t, r)C(r)U(r, s)ydr. \tag{7.80}$$

Since Y is dense in X, (7.80) implies

$$V(t, s) = S(t)^{-1}U(t, s) - \int_s^t V(t, r)C(r)U(r, s)dr. \tag{7.81}$$

Let $W(t, s)$ be the solution of the integral equation

$$W(t, s) = U(t, s) - \int_s^t W(t, r)C(r)U(r, s)dr. \tag{7.82}$$

This equation can be solved by successive approximation and $W(t, s)$ is continuous in the strong topology of $\mathcal{L}(X)$. Since $S(t)^{-1}W(t, s)$ also satisfies the integral equation (7.81) whose solution is unique, we get

$$S(t)^{-1}W(t, s) = V(t, s) = U(t, s)S(s)^{-1}.$$

Hence we obtain

$$U(t, s)|Y = S(t)^{-1}W(t, s)S(s),$$

from which (7.48) follows.

From (7.23) which was proved in the last step and (7.48) it easily follows that

$$\left(\frac{\partial}{\partial t}\right)^+ U(t, s)y = A(t)U(t, s)y \tag{7.83}$$

for $y \in Y$. Since the right hand side of (7.83) is continuous in $0 \le s \le t \le T$ in the strong topology of X, we conclude (7.49) with the aid of Lemma 1.3. Thus the proof of Theorem 7.3 is complete.

Remark 7. 2 The partition $P(\epsilon, s, y)$ of Step 3 depends on ϵ, s, y. In [160] A. Yagi obtained a partition P of the whole inteval $[0, T]$ independent of s such that we have

$$\sup_{0 \le s \le t \le T} \|(A(t) - A(t; P))U(t, s; P)y\| \le \epsilon.$$

This enables us to construct the evolution operator also by using the Yosida approximation of $A(t)$. But the proof is rather delicate and laborious, and we prefered the simpler proof of Kobayashi [95].

Theorem 7.4 *Suppose that $\{A(t); t \in [0, T]\}$ is a stable family of infinitesimal generators of C_0-semigroups in X. If $D(A(t)) = D$ is independent of t and for every $y \in D$, $A(t)y$ is strongly continuously differentiable in X, then there exists a unique evolution operator $U(t, s)$ satisfying (7.20),(7.21),(7.22),(7.24),(7.48),(7.49), where Y is the space D equipped with the graph norm of $A(0)$:*

$$\|y\|_Y = \|y\| + \|A(0)y\|.$$

Proof. Let M, β be the stability constants of $\{A(t); t \in [0, T]\}$. If we set $S(t) = \lambda_0 - A(t)$ for some $\lambda_0 > \beta$, then $S(t)$ is an isomorphism of Y onto X and the assumption (H2') is satisfied with $B(t) = 0$. Clearly the assumption (H3), and hence (H3') is satisfied. Therefore the assertion of the theorem follows from Theorem 7.3.

7.6 Existence of Regular Solutions

This section is concerned with the initial value problem (7.18), (7.19).

Theorem 7.5 *Suppose that the assumptions of Theorem 7.3 are satisfied. If $x \in Y$, $f \in C([0, T]; Y)$, then the function u given by*

$$u(t) = U(t, 0)x + \int_0^t U(t, s)f(s)ds \tag{7.84}$$

is the unique Y-valued solution of (7.18), (7.19).

Proof. In view of (7.21), (7.49) $U(t, 0)x$ is a Y-valued solution of (7.18), (7.19) with $f(t) \equiv 0$. By virtue of (7.48), $U(t, s)f(s)$ is continuous in $0 \le s \le t \le T$ to Y and with the aid of (7.49) we see that

$$w(t) = \int_0^t U(t, s)f(s)ds$$

is continuously differentiable in X and

$$dw(t)/dt = A(t)w(t) + f(t).$$

Hence $u(t)$ given by (7.84) is a desired Y-valued solution. The uniqueness is proved just as Theorem 7.2.

Theorem 7.6 *Suppose $\{A(t); t \in [0, T]\}$ is a stable family of infinitesimal generators of C_0-semigroups in X such that $D(A(t)) = D$ is independent of t and for each $y \in D$, $A(t)y$ is continuously differentiable in X. If $x \in X$, $f \in C^1([0, T]; X)$, then the function given by (7.84) is the unique strict solution of (7.18), (7.19).*

Proof. We use the notation of the proof of Theorem 7.4. By assumption

$$(d/ds)(S(s)^{-1}f(s)) = -S(s)^{-1}\dot{S}(s)S(s)^{-1}f(s) + S(s)^{-1}f'(s)$$

is a continuous function taking values in Y. By virtue of (7.24) we have

$$U(t,s) = \partial U(t,s)/\partial s \cdot S(s)^{-1} + \lambda_0 U(t,s)S(s)^{-1}.$$

Using this and integrating by parts we get

$$\int_0^t U(t,s)f(s)ds$$

$$= \int_0^t \frac{\partial}{\partial s}U(t,s)\cdot S(s)^{-1}f(s)ds + \lambda_0\int_0^t U(t,s)S(s)^{-1}f(s)ds$$

$$= S(t)^{-1}f(t) - U(t,0)S(0)^{-1}f(0) - \int_0^t U(t,s)\frac{d}{ds}(S(s)^{-1}f(s))ds$$

$$+\lambda_0\int_0^t U(t,s)S(s)^{-1}f(s)ds. \tag{7.85}$$

From (7.85) it follows that

$$\frac{d}{dt}\int_0^t U(t,s)f(s)ds = -A(t)U(t,0)S(0)^{-1}f(0)$$

$$-\int_0^t A(t)U(t,s)\frac{d}{ds}(S(s)^{-1}f(s))ds + \lambda_0 S(t)^{-1}f(t)$$

$$+\lambda_0\int_0^t A(t)U(t,s)S(s)^{-1}f(s)ds,$$

$$A(t)\int_0^t U(t,s)f(s)ds = A(t)S(t)^{-1}f(t) - A(t)U(t,0)S(0)^{-1}f(0)$$

$$-\int_0^t A(t)U(t,s)\frac{d}{ds}(S(s)^{-1}f(s))ds + \lambda_0\int_0^t A(t)U(t,s)S(s)^{-1}f(s)ds.$$

Therefore we obtain

$$\frac{d}{dt}\int_0^t U(t,s)f(s)ds - A(t)\int_0^t U(t,s)f(s)ds$$
$$= \lambda_0 S(t)^{-1}f(t) - A(t)S(t)^{-1}f(t) = f(t).$$

Since the first term of the right hand side of (7.84) is a strict solution with $f(t) \equiv 0$, we complete the proof.

7.7   Equations in Hilbert Spaces

In this section following N. Okazawa and A. Unai [120] we state how to solve hyperbolic equations in Hilbert spaces.

Let X be a Hilbert space with norm $\| \cdot \|$ and innerproduct $(\cdot, \cdot)$, and $\{A(t); t \in [0, T]\}$ be a family of closed linear operators in X. Let S be a nonnegative selfadjoint operator in X, and set

$$Y = D(S^{1/2}) \quad \text{and} \quad \|y\|_Y = (\|y\|^2 + \|S^{1/2}y\|^2)^{1/2}. \qquad (7.86)$$

Then Y is a Hilbert space embedded densely and continuously in X. We make the following assumptions:

(H4) There exists a constant $\alpha \geq 0$ such that

$$|\mathrm{Re}(A(t)y, y)| \leq \alpha\|y\|^2, \quad y \in D(A(t)), \quad t \in [0, T].$$

(H5) $Y \subset D(A(t)), \quad t \in [0, T]$.

(H6) There exists a constant $\beta \geq \alpha$ such that

$$|\mathrm{Re}(A(t)x, Sx)| \leq \beta\|S^{1/2}x\|^2, \quad x \in D(S), \quad t \in [0, T].$$

(H7) $A(t)$ is continuous in $[0, T]$ in the norm of $\mathcal{L}(Y, X)$.

Proposition 7. 6 *Let A and S be densely defined closed linear operators in a Banach space X such that $D(S) \subset D(A)$ and $D(S^*)$ is dense in X^*. Suppose that*
(i) there exists a complex number ξ such that for sufficiently large positive integer n

$$R\left(\frac{1}{n}S + A + \xi\right) = X,$$

and hence for any $y \in X$ there exists a sequence $\{x_n\} \subset D(S)$ such that

$$\frac{1}{n}Sx_n + Ax_n + \xi x_n = y, \qquad (7.87)$$

(ii) for every $y \in X$ there exists a sequence $\{x_n\} \subset D(S)$ satisfying (7.87) such that $\{x_n\}, \{Ax_n\}$ are bounded.
Then $(A + \xi)D(S)$ and hence $R(A + \xi)$ is dense in X.

Proof. Let $f \in X^*$ be such that for any $x \in D(S)$, $((A + \xi)x, f) = 0$. Let y be an arbitrary element of X and a sequence $\{x_n\}$ be as above. Then

$$(y, f) = \left(\frac{1}{n} Sx_n + Ax_n + \xi x_n, f \right) = \frac{1}{n}(Sx_n, f).$$

If $f \in D(S^*)$, then $\frac{1}{n}(Sx_n, f) = \frac{1}{n}(x_n, S^* f) \to 0$ as $n \to \infty$. Since $D(S^*)$ is dense in X^* and $\left\{ \frac{1}{n} Sx_n \right\}$ is bounded, $\frac{1}{n}(Sx_n, f) \to 0$ for any $f \in X^*$. This implies $(y, f) = 0$, and hence $f = 0$.

Definition 7. 3 A linear operator A in a Hilbert space X is called *accretive* if

$$\mathrm{Re}(Ax, x) \geq 0 \tag{7.88}$$

for any $x \in D(A)$. An accretive operator A is called *m-accretive* if $R(\mu + A) = X$ for some $\mu > 0$. If $-A$ is accretive, then A is called *dissipative*, and if $-A$ is m-accretive, then A is called *m-dissipative*.

From (7.88) it follows that

$$\|(A + \lambda)x\| \geq \mathrm{Re}\lambda \|x\|.$$

Hence if A is accretive and $\mathrm{Re}\lambda > 0$, then $A + \lambda$ has a continuous inverse. Therefore an m-accretive operator is closed. If A is accretive and $R(A+\mu) = X, \mu > 0$, then $-\mu \in \rho(A)$, and $y = (\lambda + A)x$ is equivalent to

$$x = (\mu + A)^{-1}y - (\lambda - \mu)(\mu + A)^{-1}x.$$

Consequently with the aid of a fixed point theorem we can easily show that an accretive operator A is m-accretive if and only if $R(\lambda + A) = X$ for any λ such that $\mathrm{Re}\lambda > 0$. If A is m-accretive, then for any $x \in X$ and a positive integer n we have $\|n(n + A)^{-1}x\| \leq \|x\|$. Hence some subsequence of $\{n(n + A)^{-1}x\}$ converges weakly to some element $z \in X$. Since

$$A(n + A)^{-1}x = x - n(n + A)^{-1}x,$$

$(n + A)^{-1}x \to 0$ and A is closed, it follows that $x = z \in \overline{D(A)}$. Therefore $D(A)$ is dense in X. Thus we have shown that A is m-accretive if and only if $-A$ generates a contraction C_0-semigroup: $\|e^{-tA}\| \leq 1$. It is easily seen that a bounded accrretive operator is m-accretive.

A subset Z of X is called a *core* for an operator A if $Z \subset D(A)$ and A coincides with the smallest closed extension of $A|_Z$, i.e. for any $x \in D(A)$ there exists a sequence $\{x_n\} \subset Z$ such that $x_n \to x$ and $Ax_n \to Ax$.

Propostion 7. 7 *In Proposition 7.6 if A is accretive and $\xi > 0$, then A is m-accretive and $D(S)$ is a core for A.*

Proof. By Proposition 7.6 $R(A + \xi)$ is dense. Since $A + \xi$ has a continuous inverse and A is closed, $R(A + \xi)$ is closed. Therefore $R(A + \xi) = X$, and hence A is m-accretive. Let $x \in D(A)$. Then there exists a sequence $\{x_n\} \subset D(S)$ such that $(A + \xi)x_n \to (A + \xi)x$. Since $A + \xi$ has a bounded inverse, $x_n \to x$ and hence $Ax_n \to Ax$. This means that $D(S)$ is a core for A.

Lemma 7. 1 *Let A be a linear closed operator in X such that $A + \alpha$ is accretive for some $\alpha \geq 0$. Let S be a nonnegative self-adjoint operator in X such that $D(S) \subset D(A)$. Suppose that there exist nonnegative constants β and γ such that*

$$\mathrm{Re}(Ax, Sx) \geq -\gamma\|x\|^2 - \beta\|x\|\|Sx\|$$

for any $x \in D(S)$. Then $A + \alpha$ is m-accretive and $D(S)$ is a core for A.

Proof. Let $\xi > \beta$. Set $\epsilon = \xi^{-1}$. Then $(A + \alpha)\left(1 + \dfrac{\epsilon}{n}S\right)^{-1}$ is a bounded linear operator for any positive integer n, and for any $x \in X$ we have

$$\mathrm{Re}\left(\left((A + \alpha)\left(1 + \frac{\epsilon}{n}S\right)^{-1} + \xi\right)x, x\right)$$

$$= \mathrm{Re}\left((A + \alpha)\left(1 + \frac{\epsilon}{n}S\right)^{-1}x, \left(1 + \frac{\epsilon}{n}S\right)^{-1}x\right)$$

$$+ \frac{\epsilon}{n}\mathrm{Re}\left(A\left(1 + \frac{\epsilon}{n}S\right)^{-1}x, S\left(1 + \frac{\epsilon}{n}S\right)^{-1}x\right)$$

$$+ \frac{\epsilon}{n}\alpha\left(\left(1 + \frac{\epsilon}{n}S\right)^{-1}x, S\left(1 + \frac{\epsilon}{n}S\right)^{-1}x\right) + \xi\|x\|^2$$

$$\geq -\frac{\epsilon}{n}\left[\gamma\left\|\left(1 + \frac{\epsilon}{n}S\right)^{-1}x\right\|^2 + \beta\left\|\left(1 + \frac{\epsilon}{n}S\right)^{-1}x\right\|\left\|S\left(1 + \frac{\epsilon}{n}S\right)^{-1}x\right\|\right]$$

$$+ \xi\|x\|^2 \geq \left(\xi - \beta - \frac{\epsilon}{n}\gamma\right)\|x\|^2.$$

Therefore if n is so large that $\xi - \beta - \dfrac{\epsilon}{n}\gamma > 0$, then $(A + \alpha)\left(1 + \dfrac{\epsilon}{n}S\right)^{-1} + \xi$ is m-accretive. Consequently

$$\frac{1}{n}S + A + \alpha + \xi = \left[(A + \alpha)\left(1 + \frac{\epsilon}{n}S\right)^{-1} + \xi\right]\left(1 + \frac{\epsilon}{n}S\right)$$

is surjective, and

$$\left\| \left(\frac{1}{n} S + A + \alpha + \xi \right)^{-1} \right\| \leq \frac{1}{\xi - \beta - \epsilon\gamma/n}.$$

Hence for any $y \in X$ there exists a sequence $\{x_n\}$ such that

$$\left(\frac{1}{n} S + A + \alpha + \xi \right) x_n = y$$

and $\{x_n\}$ is bounded. Since

$$\left\| \frac{1}{n} S x_n \right\|^2 = \left(y - A x_n - (\alpha + \xi) x_n, \frac{1}{n} S x_n \right)$$

$$= \mathrm{Re} \left(y, \frac{1}{n} S x_n \right) - \frac{1}{n} \mathrm{Re}(A x_n, S x_n) - \frac{\alpha + \xi}{n}(x_n, S x_n)$$

$$\leq \|y\| \left\| \frac{1}{n} S x_n \right\| + \frac{\gamma}{n} \|x_n\|^2 + \beta \|x_n\| \left\| \frac{1}{n} S x_n \right\|,$$

$\left\{ \frac{1}{n} S x_n \right\}$ is bounded. Therefore in view of Proposition 7.7 $A + \alpha$ is m-accretive and $D(S)$ is a core for A.

Since $A + \alpha$ is m-accretive under the assumptions of Lemma 7.1, we have $\rho(A) \supset \{\lambda; \mathrm{Re}\lambda < -\alpha\}$, and

$$\|(A + \lambda)^{-1}\| \leq \frac{1}{\mathrm{Re}\lambda - \alpha} \quad \text{for} \quad \mathrm{Re}\lambda > \alpha. \tag{7.89}$$

Furthermore $-A$ generates a C_0-semigroup satisfying

$$\|e^{-tA}\| \leq e^{\alpha t}, \quad t \geq 0. \tag{7.90}$$

Let $S_\epsilon = S(1 + \epsilon S)^{-1} = \epsilon^{-1}(1 - (1 + \epsilon S)^{-1}), \epsilon > 0$, be the Yosida approximation of S.

Lemma 7. 2 *Let A and S be as in Lemma 7.1. Suppose that there exists a nonnegative constant β such that $\beta \geq \alpha$ and*

$$\mathrm{Re}(Ax, Sx) \geq -\beta(x, Sx), \quad x \in D(S). \tag{7.91}$$

Then

$$\mathrm{Re}(Ax, S_\epsilon x) \geq -\beta(x, S_\epsilon x), \quad x \in D(A). \tag{7.92}$$

Proof. For $x \in D(A)$

$$(Ax, S_\epsilon x) = (Ax - A(1 + \epsilon S)^{-1}x, S_\epsilon x) + (A(1 + \epsilon S)^{-1}x, S_\epsilon x)$$
$$= \epsilon(AS_\epsilon x, S_\epsilon x) + (A(1 + \epsilon S)^{-1}x, S(1 + \epsilon S)^{-1}x).$$

Hence

$$\mathrm{Re}(Ax, S_\epsilon x) \geq -\alpha\epsilon\|S_\epsilon x\|^2 - \beta((1 + \epsilon S)^{-1}x, S(1 + \epsilon S)^{-1}x)$$
$$= -\alpha\epsilon\|S_\epsilon x\|^2 - \beta((1 + \epsilon S)^{-1}x, S_\epsilon x)$$
$$= (\beta - \alpha)\epsilon\|S_\epsilon x\|^2 - \beta(x, S_\epsilon x) \geq -\beta(x, S_\epsilon x).$$

In what follows S will always denote a nonnegative selfadjoint operator and Y the space defined by (7.86).

Lemma 7.3 *Let A and S be as in Lemma 7.1. Suppose that (7.91) is satisfied. Then Y is $-A$-admissible. Furthermore for any $y \in Y$ we have*

$$\|S^{1/2}(1 + \lambda A)^{-1}y\| \leq (1 - \lambda\beta)^{-1}\|S^{1/2}y\|, \quad 0 < \lambda < \beta^{-1}, \quad (7.93)$$
$$\|S^{1/2}e^{-tA}y\| \leq e^{\beta t}\|S^{1/2}y\|, \quad t \geq 0. \quad (7.94)$$

Proof. Let $y \in D(S^{1/2})$ and $y(\lambda)$ be the solution of

$$y(\lambda) + \lambda Ay(\lambda) = y, \quad 0 < \lambda < \alpha^{-1}. \quad (7.95)$$

Let $S_\epsilon, \epsilon > 0$, be the Yosida approximation of S. Then with the aid of (7.92) we get

$$\|S_\epsilon^{1/2}y(\lambda)\|^2 = (y(\lambda), S_\epsilon y(\lambda)) = (y - \lambda Ay(\lambda), S_\epsilon y(\lambda))$$
$$\leq \mathrm{Re}(S_\epsilon^{1/2}y, S_\epsilon^{1/2}y(\lambda)) + \beta\lambda(y(\lambda), S_\epsilon y(\lambda))$$
$$\leq \|S_\epsilon^{1/2}y\|\|S_\epsilon^{1/2}y(\lambda)\| + \beta\lambda\|S_\epsilon^{1/2}y(\lambda)\|^2.$$

This implies

$$\|S_\epsilon^{1/2}y(\lambda)\| \leq (1 - \beta\lambda)^{-1}\|S^{1/2}y\| \quad (7.96)$$

for $0 < \lambda < \beta^{-1} \leq \alpha^{-1}$. Hence $\{S_\epsilon^{1/2}y(\lambda)\}$ is bounded as $\epsilon \to 0$, and so it contains a weakly convergent subsequence. Since $(1 + \epsilon S)^{-1/2}y(\lambda) \to y(\lambda)$ as $\epsilon \to 0$ and $S_\epsilon^{1/2} = S^{1/2}(1 + \epsilon S)^{-1/2}$, it follows that $y(\lambda) \in D(S^{1/2})$ and

$$S^{1/2}y(\lambda) = \text{w-}\lim_{\epsilon \to 0} S_\epsilon^{1/2}y(\lambda).$$

From (7.95) and (7.96) it follows that Y is invariant under $(1 + \lambda A)^{-1}$ and (7.93) holds. A repeating application of (7.93) yields

$$\|S^{1/2}(1 + \lambda A)^{-n}y\| \leq (1 - \beta\lambda)^{-n}\|S^{1/2}y\| \quad (7.97)$$

for any positive integer n. Let $t > 0$. If n is so large that $n > \beta t$, then we get from (7.97)

$$\left\| S^{1/2} \left(1 + \frac{t}{n} A \right)^{-n} y \right\| \le \left(1 - \frac{\beta}{n} t \right)^{-n} \| S^{1/2} y \|.$$

Letting $n \to \infty$ we conclude that Y is invariant under e^{-tA} and (7.94) holds.

In order to prove that Y is $-A$-admissible it remains to verify that for $y \in Y$

$$\| S^{1/2} e^{tA} y - S^{1/2} y \| \to 0 \tag{7.98}$$

as $t \to 0$. With the aid of (7.94) we can easily show that

$$S^{1/2} y = \text{w-} \lim_{t \to 0} S^{1/2} e^{-tA} y. \tag{7.99}$$

Hence

$$\| S^{1/2} y \| \le \liminf_{t \to 0} \| S^{1/2} e^{-tA} y \|.$$

On the other hand we see from (7.94)

$$\limsup_{t \to 0} \| S^{1/2} e^{-tA} y \| \le \| S^{1/2} y \|.$$

Therefore

$$\lim_{t \to 0} \| S^{1/2} e^{-tA} y \| = \| S^{1/2} y \|. \tag{7.100}$$

We conclude (7.98) from (7.99) and (7.100).

From (7.93),(7.94) and $\beta \ge \alpha$ we obtain

$$\| (1 + \lambda A)^{-1} \|_Y \le (1 - \beta \lambda)^{-1}, \quad 0 < \lambda < \beta^{-1}, \tag{7.101}$$
$$\| e^{-tA} \|_Y \le e^{\beta t}, \quad t \ge 0. \tag{7.102}$$

We are going to prove the following theorem.

Theorem 7.7 *Under the assumptions* (H4),(H5),(H6),(H7) *there exists a unique evolution operator* $U(t, s)$, $(s, t) \in [0, T] \times [0, T]$, *for the initial value problem* (7.18),(7.19) *satisfying* (7.20),(7.21),(7.22),(7.24),(7.48),(7,49) *for* $(s, t) \in [0, T] \times [0, T]$. *For each* $x \in Y, f \in C([0, T]; Y), s \in [0, T]$

$$u(t) = U(t, s)x + \int_s^t U(t, \tau) f(\tau) d\tau \tag{7.103}$$

is a unique Y-*valued solution of* (7.18) *in* $[0, T]$ *satisfying the initial condition* $u(s) = x$.

In view of Lemmas 7.1, 7.3 and (7.101),(7.102), $\{A(t); t \in [0,T]\}$ is stable in X with stability constants 1, α, Y is $A(t)$-admissible for each $t \in [0,T]$ and $\{A_Y(t); t \in [0,T]\}$ is stable with stability constants 1, β. Hence, we can apply Theorem 7.1 to find that there exists a unique evolution operator $U(t,s), 0 \leq s \leq t \leq T$, satisfying (7.20),(7.21),(7.22),(7.23),(7.24) with 1, α in place of M, β. Let $\{U_n(t,s)\}$ be the approximating sequence defind by (7.27). Then for each $x \in X$, $\{U_n(t,s)x\}$ converges to $U(t,s)x$ in X uniformly in $0 \leq s \leq t \leq T$, and in view of (7.30),(7.31),(7.94) we have

$$\|U_n(t,s)\| \leq e^{\alpha(t-s)}, \quad \|U_n(t,s)\|_Y \leq e^{\beta(t-s)}, \tag{7.104}$$
$$\|S^{1/2}U_n(t,s)y\| \leq e^{\beta(t-s)}\|S^{1/2}y\| \quad for \quad y \in Y. \tag{7.105}$$

Lemma 7.4 *Let $y \in Y$. Then*
(i) $U(t,s)Y \subset Y$ and

$$\|U(t,s)\|_Y \leq e^{\beta(t-s)}, \tag{7.106}$$
$$\|S^{1/2}U(t,s)y\| \leq e^{\beta(t-s)}\|S^{1/2}y\|, \tag{7.107}$$

(ii) $S^{1/2}U(t,s)y$ is weakly continuous in $0 \leq s \leq t \leq T$,
(iii) for $t_0 \in [0,T]$

$$S^{1/2}U(t,s)y \to S^{1/2}y \quad as \quad (t,s) \to (t_0,t_0),$$

(iv) for $t \in (0,T], U(t,\cdot)y \in C([0,t];Y)$,
(v) for $s \in [0,T), S^{1/2}U(\cdot,s)y$ is right continuous in $[s,T)$.

Proof. (i) and (ii) can be established by means of the standard argument with the aid of (7.104),(7.105).
(iii) In view of (ii) it suffices to show that

$$\|S^{1/2}U(t,s)y\| \to \|S^{1/2}y\| \tag{7.108}$$

as $(t,s) \to (t_0,t_0)$. Since $S^{1/2}y =$w-$\lim_{(s,t)\to(t_0,t_0)} S^{1/2}U(t,s)y$ we have

$$\|S^{1/2}y\| \leq \liminf_{(s,t)\to(t_0,t_0)} \|S^{1/2}U(t,s)y\|.$$

On the other hand in view of (i)

$$\limsup_{(s,t)\to(t_0,t_0)} \|S^{1/2}U(t,s)y\| \leq \|S^{1/2}y\|.$$

Hence (7.108) follows.
(iv) Let $y \in Y, 0 \leq s \leq s' \leq t \leq T$. Then as $s' \to s$ or $s \to s'$

$$\|S^{1/2}U(t,s')y - S^{1/2}U(t,s)y\|$$
$$= \|S^{1/2}U(t,s')(y - U(s',s)y)\| \leq e^{\beta(t-s')}\|S^{1/2}(y - U(s',s)y)\| \to 0$$

by virtue of (7.107) and (iii).

(v) Let $t_0 \in [s, T]$. Then it follows from (iii) that as $t \downarrow t_0$

$$\|S^{1/2}U(t, s)y - S^{1/2}U(t_0, s)y\| = \|S^{1/2}(U(t, t_0) - 1)U(t_0, s)y\| \to 0.$$

Lemma 7.5 *Let $y \in Y$. Then*

(i) *we have*

$$A(t)U(t, s)y = \text{w-} \lim_{n \to \infty} A_n(t)U_n(t, s)y, \quad 0 \leq s \leq t \leq T, \qquad (7.109)$$

where $\{A_n(t)\}$ is the approximating sequence defined by (7.25), and

$$\|A(t)U(t, s)y\| \leq M e^{\beta(t-s)} \|y\|_Y, \qquad (7.110)$$

where $M = \sup\{\|A(t)\|_{\mathcal{L}(Y,X)}; t \in [0, T]\}$,

(ii) *$A(t)U(t, s)y$ is weakly continuous in $0 \leq s \leq t \leq T$,*

(iii) *for each $s \in [0, T)$, $A(\cdot)U(\cdot, s)y$ is right continuous in $[s, T]$.*

Proof. (i) In view of (7.26), (7.104)

$$\|A_n(t)U_n(t, s)y - A(t)U_n(t, s)y\|$$
$$\leq \|A_n(t) - A(t)\|_{\mathcal{L}(Y,X)} \|U_n(t, s)y\|_Y \to 0.$$

Hence in order to prove (7.109) it suffices to show that $A(t)U_n(t, s)y \to A(t)U(t, s)y$ weakly in X. This can be shown with the aid of

$$\|A(t)U_n(t, s)y\| \leq M\|U_n(t, s)y\|_Y \leq M e^{\beta T} \|y\|_Y$$

and the usual argument. (7.110) follows from (7.109) and

$$\|A_n(t)U_n(t, s)y\| \leq M e^{\beta(t-s)} \|y\|_Y. \qquad (7.111)$$

(ii) Let $y \in Y$ and $0 \leq s_0 \leq t_0 \leq T$. Then

$$\|A(t)U(t, s)y - A(t_0)U(t, s)y\| \leq \|A(t) - A(t_0)\|_{\mathcal{L}(Y,X)} \|U(t, s)y\|_Y \to 0 \qquad (7.112)$$

as $t \to t_0$. Hence it suffices to show

$$A(t_0)U(t, s)y \to A(t_0)U(t_0, s_0)y$$

weakly in X as $(s, t) \to (t_0, s_0)$. But this follows from Lemma 7.4 (ii) since $A(t_0)(1 + S)^{-1/2} \in \mathcal{L}(X)$.

(iii) Let $t_0 \in [s, T)$. Then in view of (7.112) it suffices to show that

$$A(t_0)U(t_0, s)y = \lim_{t \downarrow t_0} A(t_0)U(t, s)y. \qquad (7.113)$$

Since

$$\|A(t_0)(U(t, s)y - U(t_0, s)y)\| \leq M\|U(t, s)y - U(t_0, s)y\|_Y,$$

(7.113) follows from Lemma 7.4 (v).

We followed T. Kato [88; section 5] in the proof of the above two lemmas.

Lemma 7.6 *Let $y \in Y$. Then*
(i) for each $s \in [0, T)$ and $z \in X$, $(U(\cdot, s)y, z) \in C^1([s, T])$ and

$$\frac{\partial}{\partial t}(U(t, s)y, z) = (A(t)U(t, s)y, z),$$

(ii) for each $s \in [0, T)$, $A(\cdot)U(\cdot, s)y$ is Bochner integrable in $[s, T]$, and

$$U(t, s)y = y - \int_s^t A(r)U(r, s)y\,dr, \quad t \in [s, T], \tag{7.114}$$

(iii) for each $s \in [0, T)$, $U(\cdot, s)y$ is right differentiable in $[s, T)$, and

$$\left(\frac{\partial}{\partial t}\right)^+ U(t, s)y = A(t)U(t, s)y, \quad t \in [s, T),$$

(iv)

$$\frac{\partial}{\partial t}U(t, s)y = A(t)U(t, s)y \quad a.e. \quad t \in (s, T).$$

Proof. (i) Letting $n \to \infty$ in

$$(U_n(t, s)y, z) = (y, z) + \int_s^t (A_n(r)U_n(r, s)y, z)dr,$$

we obtain with the aid of Lemma 7.5 (i) and (7.111)

$$(U(t, s)y, z) = (y, z) + \int_s^t (A(r)U(r, s)y, z)dr. \tag{7.115}$$

Hence the assertion follows from Lemma 7.5 (ii).
(ii) Since $A(\cdot)U(\cdot, s)y$ is weakly continuous in $[s, T]$, it is Bochner integrable there. Therefore (7.114) follows from (7.115) (see K. Yosida [166], p.133).
(iii) is a consequence of (7.114) and Lemma 7.5 (iii).
(iv) is a consequence of (7.114), since the assertion of Theorem 1.2 remains valid for functions with values in a Banach space.

With the aid of Lemma 7.2 we obtain

$$|\text{Re}(A(t)x, S_\epsilon x)| \le \beta \|S^{1/2}x\|^2, \ x \in D(A(t)), \ t \in [0, T], \ \epsilon > 0. \tag{7.116}$$

Lemma 7.7 *We have*
(i) for $s \in [0, T)$ and $y \in Y$, $U(\cdot, s)y \in C([s, T]; Y)$,
(ii) $U(t, s)$ is strongly continuous in $0 \le s \le t \le T$ to $\mathcal{L}(Y)$,
(iii) for $s \in [0, T)$ and $y \in Y$, $U(\cdot, s)y \in C^1([s, T]; X)$ and

$$\frac{\partial}{\partial t}U(t, s)y = A(t)U(t, s)y.$$

Proof. (i) Since $U(\cdot, s)y \in C([s, T]; X)$ it suffices to show that $S^{1/2}U(\cdot, s)y \in C([s, T]; X)$. Let $t_0 \in [s, T]$. Since $S^{1/2}U(\cdot, s)y$ is weakly continuous in $[s, T]$ by virtue of Lemma 7.4 (ii), it remains to show that

$$\|S^{1/2}U(t, s)y\| \to \|S^{1/2}U(t_0, s)y\| \quad \text{as} \quad t \to t_0. \tag{7.117}$$

In view of Lemma 7.6 (iv) we have for $\epsilon > 0$

$$\frac{\partial}{\partial r}\|S_\epsilon^{1/2}U(r, s)y\|^2 = 2\text{Re}(A(r)U(r, s)y, S_\epsilon U(r, s)y) \quad a.e. \quad (s, t).$$

Integrating both sides from t_0 to t and using (7.116),(7.107)

$$\left| \|S_\epsilon^{1/2}U(t, s)y\|^2 - \|S_\epsilon^{1/2}U(t_0, s)y\|^2 \right|$$

$$= 2\left| \int_{t_0}^{t} \text{Re}(A(r)U(r, s)y, S_\epsilon U(r, s)y)dr \right|$$

$$\leq 2\beta \left| \int_{t_0}^{t} \|S^{1/2}U(r, s)y\|^2 dr \right| \leq 2\beta e^{2\beta T}\|S^{1/2}y\|^2|t - t_0|.$$

Letting $\epsilon \to 0$ we obtain

$$\left| \|S^{1/2}U(t, s)y\|^2 - \|S^{1/2}U(t_0, s)y\|^2 \right| \leq 2\beta e^{2\beta T}\|S^{1/2}y\|^2|t - t_0|,$$

from which (7.117) follows.

(ii) Let $0 \leq s_0 < t_0 \leq T$. Choosing $a \in (s_0, t_0)$ we write for (s, t) sufficiently close to (s_0, t_0)

$$U(t, s)y - U(t_0, s_0)y$$
$$= U(t, a)(U(a, s)y - U(a, s_0)y) + (U(t, a) - U(t_0, a))U(a, s_0)y.$$

Hence the assertion follows from (i) and Lemma 7.4 (i),(iv).

(iii) is a direct consequence of (ii) and Lemma 7.6.

So far we have proved that there exists a unique evolution operator satisfying (7.20),(7.21),(7.22),(7.24),(7.48),(7.49) in $0 \leq s \leq t \leq T$. We can directly construct the desired evolution operator $U(t, s)$ for $(s, t) \in [0, T] \times [0, T]$ by defining the approximating sequence $\{U_n(t, s)\}$ for $(s, t) \in [0, T] \times [0, T]$. However, we can also define $U(t, s)$ for $0 \leq t \leq s \leq T$ in the following manner. Since $\{-A(T - t); t \in [0, T]\}$ satisfies (H4),(H5),(H6), (H7), we can construct an evolution operator $V(t, s), 0 \leq s \leq t \leq T$, for the equation

$$du(t)/dt = -A(T - t)u(t)$$

as above. Set

$$U(t, s) = V(T - t, T - s)$$

for $0 \le t \le s \le T$. Then for $y \in Y$

$$\frac{\partial}{\partial t}U(t,s)y = A(t)U(t,s)y, \quad 0 \le t \le s \le T.$$

Hence for $0 < s < t$ and $y \in Y$ we have

$$\frac{\partial}{\partial s}U(t,s)U(s,t)y = 0.$$

Therefore we obtain

$$U(t,s)U(s,t) = I. \tag{7.118}$$

With the aid of (7.118) we can easily show that $U(t,s)$, $(s,t) \in [0,T] \times [0,T]$ satisfies the assertions of Theorem 7.7.

Remark 7.3 In an unpublished paper [119] N. Okazawa showed that the conclusion of Theorem 7.4 remains valid replacing the norm continuity of $t \mapsto A(t)$ by its strong continuity. The method is to construct the evolution operator $U(t,s)$ as the limit of the sequence of the evolution operators $U_n(t,s)$ for the equations with $A(t)$ replaced by the Yosida approximations $A_n(t) = A(t)(1 - n^{-1}A(t))^{-1}$. In the proof of the convergence of $\{U_n(t,s)\}$ Okazawa uses the method of Y. Komura [97] and T. Kato [87]. Namely, if we set

$$u_{nm}(r,s) = U_n(r,s)y - U_m(r,s)y,$$

then

$$\frac{1}{2}\frac{\partial}{\partial r}\|u_{nm}(r,s)\|^2$$
$$= \text{Re}(A_n(r)U_n(r,s)y - A_m(r)U_m(r,s)y, u_{nm}(r,s))$$
$$= \text{Re}(A_n(r)U_n(r,s)y - A_m(r)U_m(r,s)y, u_{nm}(r,s) - w_{nm}(r,s))$$
$$+\text{Re}(A_n(r)U_n(r,s)y - A_m(r)U_m(r,s)y, w_{nm}(r,s)) \equiv I_1 + I_2,$$

where

$$w_{nm}(r,s) = J_n(r)U_n(r,s)y - J_m(r)U_m(r,s)y,$$
$$J_n(r) = (1 - n^{-1}A(r))^{-1}.$$

We have

$$I_1 = \text{Re}\bigg(A_n(r)U_n(r,s)y - A_m(r)U_m(r,s)y,$$
$$-\frac{1}{n}A_n(r)U_n(r,s)y + \frac{1}{m}A_m(r)U_m(r,s)y\bigg)$$

$$= \left(\frac{1}{n} + \frac{1}{m}\right) \mathrm{Re}(A_n(r)U_n(r, s)y, A_m(r)U_m(r, s)y)$$

$$- \frac{1}{n}\|A_n(r)U_n(r, s)y\|^2 - \frac{1}{m}\|A_m(r)U_m(r, s)y\|^2$$

$$\leq \frac{1}{2}\left|\frac{1}{n} - \frac{1}{m}\right| \max\left\{\|A_nU_n(r, s)y\|^2, \|A_m(r)U_m(r, s)y\|^2\right\},$$

and in view of the accretivity of $\alpha - A(r)$

$$I_2 \leq \alpha\|w_{nm}(r, s)\|^2.$$

Combining this with

$$\|S^{1/2}U_n(t, s)y\| \leq \exp\{\beta(1 - n^{-1}\beta)^{-2}(t - s)\}\|S^{1/2}y\|,$$

the convergence of $\{U_n(t, s)y\}$ is obtained.

7.8   Remark on Applications

F. J. Massey III [110] applied the results of T. Kato [88] to the mixed problem for the following symmetric hyperbolic system of partial differential equations:

$$\frac{\partial u}{\partial t} + \sum_{j=1}^{m} a_j(x, t)\frac{\partial u}{\partial x_j} + b(x, t)u = f(x, t), \ x \in \Omega, \ t \in [0, T], \qquad (7.119)$$

$$u(x, 0) = \phi(x), \quad x \in \Omega, \qquad (7.120)$$

$$u(x, t) \in P(x, t), \quad x \in \partial\Omega, \quad t \in [0, T]. \qquad (7.121)$$

The unknown $u = (u_1, \ldots, u_N)$ is a real vector-valued function, the coefficients, a_j and b, are real $N \times N$ matrix-valued functions, and the a_j are symmetric. It is assumed that $a_j \in C^2(\bar{\Omega} \times [0, T])$ and $b \in C^1(\bar{\Omega} \times [0, T])$. Ω is a bounded open subset of R^m of class C^3.

It is also assumed that the boundary matrix

$$a_n(x, t) = \sum_{j=1}^{m} n_j(x)a_j(x, t), \quad x \in \partial\Omega, \quad t \in [0, T],$$

is nonsingular on $\partial\Omega \times [0, T]$, where $n = (n_1, \ldots, n_m)$ is the exterior unit normal to $\partial\Omega$.

The boundary subspace $P(x, t)$ is a linear subspace of R^N which varies in a C^3 manner with $(x, t) \in \partial\Omega \times [0, T]$. $P(x, t)$ is *maximal nonnegative* for each x, t, i.e.

$$(a_n(x, t)u, u) \geq 0, \quad u \in P(x, t),$$

and $P(x,t)$ is not a proper subset of any other subspace of R^N having this property.

Let $A(t)$ be the smallest closed extension of the operator $A_0(t)$ defined by

$$A_0(t) = \sum_{j=1}^{N} a_j(x,t) \frac{\partial u}{\partial x_j} + b(x,t)u$$

with

$$D(A_0(t)) = H_{P_t}^1(\Omega) = \{u \in H^1(\Omega)^N; u(x) \in P(x,t) \quad a.e. \quad \partial\Omega\}.$$

Massey proved the following theorems.

Theorem M1 *There exists an isomorphism $S(t)$ from $H_{P_t}^1(\Omega)$ onto $L^2(\Omega)^N$ such that*

$$S(t)A(t)S(t)^{-1} = A(t) + B(t),$$

where $B(t)$ is a bounded linear operator from $L^2(\Omega)^N$ to itself.

Theorem M2 *If $P(x,t) = P(x)$ is independent of t, then $S(t)$ in Theorem M1 may be chosen so that $S(\cdot) \in C^1([0,T]; \mathcal{L}(H_P^1(\Omega), L^2(\Omega)^N))$ and $B(\cdot) \in C([0,T]; \mathcal{L}(L^2(\Omega)^N))$.*

Theorem M3 *Suppose that $\phi \in H_{P_0}^1(\Omega)$ and $f \in C([0,T]; H^1(\Omega))$ so that $f(t) \in H_{P_t}^1(\Omega)$ for $t \in [0,T]$. Then (7.119),(7.120),(7.121) has a unique solution u such that $u \in C^1([0,T]; L^2(\Omega)^N)$ and $u(t) \in H_{P_t}^1(\Omega)$ for $t \in [0,T]$.*

If $P(x,t)$ is independent of t so that the conclusion of Theorem M2 is true, then Theorem 7.3 (or Theorem 6.1 of T. Kato [88]) may be applied to the present family $\{A(t)\}$ taking $Y = H_P^1(\Omega)$. Moreover Massey shows that the general case where $P(x,t)$ varies with t may be reduced to the case $P(x,t) = P(x,0)$ by an orthogonal transformation of the dependent variables.

There are also a number of applications to nonlinear hyperbolic equations, T. Kato [90],[91],[92], K. Kobayashi and N. Sanekata [96], N. Okazawa and A. Unai [121], N. Sanekata [131] and also see the Reference of [92].

Chapter 8

Retarded Functional Differential Equations

This chapter is concerned with the following retarded functional differential equations

$$\frac{d}{dt}u(t) = A_0 u(t) + A_1 u(t-h) + \int_{-h}^{0} a(s) A_2 u(t+s) ds + f(t),$$

$$t \in [0, T], \qquad (8.1)$$

$$u(0) = g^0, \quad u(s) = g^1(s), \quad s \in [-h, 0) \qquad (8.2)$$

in a Hilbert space. First following G. Di Blasio, K. Kunisch and E. Sinestrari [51] the solvability of (8.1),(8.2) is described. After that we state the control theory by C. Bernier and A. Manitius [20], M. C. Delfour and A. Manitius [44],[45], A. Manitius [107],[108],[109] etc. for equations in finite dimensional spaces and by S. Nakagiri [112], J.-M. Jeong, S. Nakagiri and H. Tanabe [82], S. Nakagiri and H. Tanabe [114], etc. for equations in infinite dimensional spaces.

8.1 Maximal Regularity Result

We begin with a maximal regularity result concerning equations in a Hilbert space. Let A be a densely defined closed linear operator which generates an analytic semigroup in a Hilbert space H. Hence there exist positive constants M, C_0 and an angle $\theta \in (\pi/2, \pi]$ such that for $\lambda \in \Sigma = \{\lambda; |\arg \lambda| < \theta, |\lambda| > C_0\}$

$$\|\lambda R(\lambda, A)\| \leq M, \qquad (8.3)$$

where $R(\lambda, A) = (\lambda - A)^{-1}$. In this section the semigroup generated by A is denoted by $U(t)$: $U(t) = e^{tA}$, and we use the notation

$$(U * f)(t) = \int_0^t U(t - s)f(s)ds.$$

We always endow $D(A)$ with the graph norm of A denoted by $\|\cdot\|_{D(A)}$ so that $D(A)$ is a Hilbert space. The following proposition is due to J. L. Lions and E. Magenes [99].

Proposition 8. 1 *If $u = U * f$, $f \in L^2(0, T; H)$, then*

$$u \in L^2(0, T; D(A)) \cap W^{1,2}(0, T; H), \tag{8.4}$$

and hence

$$u \in C([0, T]; (D(A), H)_{1/2,2}). \tag{8.5}$$

There exists a constant c_1 such that

$$\|u\|_{L^2(0,T;D(A)) \cap W^{1,2}(0,T;H)} \leq c_1 \|f\|_{L^2(0,T;H)}. \tag{8.6}$$

We have

$$du(t)/dt = Au(t) + f(t) \quad a.e. \quad (0, T), \tag{8.7}$$

$$u(0) = 0. \tag{8.8}$$

Furthermore u is a unique function satisfying (8.4), (8.7), (8.9).

Proof. First we note that (8.5) follows from (8.4) in view of (1.18). We begin with the proof of the uniqueness. Let u satisfy (8.4),(8.7),(8.8) and $0 < t < T$. If we integrate the equality

$$(\partial/\partial s)(U(t - s)u(s)) = U(t - s)(u'(s) - Au(s)) = U(t - s)f(s)$$

from 0 to t, we obtain $u = U * f$. However, if we want to avoid verifying the absolute continuity of $U(t - s)u(s)$ in $(0, t)$, we use the mollifier $\rho_n(t) = n\rho(nt)$, where $0 \leq \rho \in C^\infty(-\infty, \infty)$, $\int_{-\infty}^\infty \rho(t)dt = 1$, $\rho(t) = 0$ for $|t| \geq 1$. Let

$$u_n(t) = \int_0^T \rho_n(t - s)u(s)ds, \quad f_n(t) = \int_0^T \rho_n(t - s)f(s)ds.$$

Then by a standard argument we see that $u_n \to u$ in $L^2(0, T; D(A))$ and $f_n \to f$ in $L^2(0, T; H)$. Since

$$u_n'(t) = -\rho_n(t - T)u(T) + \int_0^T \rho_n(t - s)u'(s)ds,$$

we have $u'_n \to u'$ in $L^2(0, T-\epsilon; H)$ for any $\epsilon > 0$. Let $0 < t < T$. Integrating both sides of

$$(\partial/\partial s)(U(t - s)u_n(s)) = U(t - s)(u'_n(s) - Au_n(s))$$

from 0 to t we get

$$u_n(t) - U(t)u_n(0) = \int_0^t U(t - s)(u'_n(s) - Au_n(s))ds.$$

Letting $n \to \infty$ we obtain

$$u(t) = \int_0^t U(t - s)(u'(s) - Au(s))ds = \int_0^t U(t - s)f(s)ds.$$

Clearly this equality also holds for $t = T$. Thus the proof of the uniqueness is complete.

We denote the Laplace transform of a function f defined in $(-\infty, \infty)$ and taking values in H by $\hat{f}$:

$$\hat{f}(\lambda) = \frac{1}{\sqrt{2\pi}} \int_{-\infty}^{\infty} e^{-t\lambda} f(t)dt.$$

Since for real variables ξ, η $\hat{f}(\xi + i\eta)$ is the Fourier transform of $e^{-t\xi}f(t)$, we have in view of Plancherel's theorem

$$\int_{-\infty}^{\infty} \|\hat{f}(\xi + i\eta)\|^2 d\eta = \int_{-\infty}^{\infty} e^{-2t\xi} \|f(t)\|^2 dt.$$

Let $f \in L^2(0, T; H)$. We extend f to $(-\infty, \infty)$ putting $f(t) = 0$ outside $[0, T]$. Let $w(t)$ be the function such that

$$\hat{w}(\lambda) = R(\lambda, A)\hat{f}(\lambda), \quad \operatorname{Re}\lambda > C_0.$$

Then for $\xi > C_0$

$$e^{-t\xi}w(t) = \frac{1}{\sqrt{2\pi}} \int_{-\infty}^{\infty} e^{it\eta}\hat{w}(\xi + i\eta)d\eta$$

$$= \frac{1}{\sqrt{2\pi}} \int_{-\infty}^{\infty} e^{it\eta}R(\xi + i\eta, A)\hat{f}(\xi + i\eta)d\eta.$$

Hence using (8.3)

$$\int_{-\infty}^{\infty} \|e^{-t\xi}w(t)\|^2_{D(A)}dt = \int_{-\infty}^{\infty} \|R(\xi + i\eta, A)\hat{f}(\xi + i\eta)\|^2_{D(A)}d\eta$$

$$\leq \left(\frac{M}{C_0}\right)^2 \int_{-\infty}^{\infty} \|\hat{f}(\xi + i\eta)\|^2 d\eta = \left(\frac{M}{C_0}\right)^2 \int_{-\infty}^{\infty} e^{-2t\xi} \|f(t)\|^2 dt$$

$$= \left(\frac{M}{C_0}\right)^2 \int_0^T e^{-2t\xi} \|f(t)\|^2 dt.$$

Since the rightmost side goes to 0 as $\xi \to +\infty$, we see that $w(t) = 0$ for $t < 0$. Also we have

$$w \in L^2_{loc}(-\infty, \infty; D(A)). \tag{8.9}$$

Differentiating both sides of

$$w(t) = \frac{1}{\sqrt{2\pi}} \int_{-\infty}^{\infty} e^{t(\xi+i\eta)} R(\xi + i\eta, A) \hat{f}(\xi + i\eta) d\eta$$

we get

$$\frac{d}{dt} w(t) = \frac{1}{\sqrt{2\pi}} \int_{-\infty}^{\infty} e^{t(\xi+i\eta)} (\xi + i\eta) R(\xi + i\eta, A) \hat{f}(\xi + i\eta) d\eta. \tag{8.10}$$

Hence

$$\int_{-\infty}^{\infty} \left\| e^{-t\xi} \frac{d}{dt} w(t) \right\|^2 dt = \int_{-\infty}^{\infty} \|(\xi + i\eta) R(\xi + i\eta, A) \hat{f}(\xi + i\eta)\|^2 d\eta$$

$$\leq M^2 \int_{\infty}^{\infty} \|\hat{f}(\xi + i\eta)\|^2 d\eta = M^2 \int_0^T e^{-2t\xi} \|f(t)\|^2 dt.$$

Consequently $w \in W^{1,2}_{loc}(-\infty, \infty; H)$. Combining this with (8.9) we get

$$w \in C(-\infty, \infty; (D(A), H)_{1/2,2}).$$

From (8.10) and

$$Aw(t) = \frac{1}{\sqrt{2\pi}} \int_{-\infty}^{\infty} e^{t(\xi+i\eta)} AR(\xi + i\eta, A) \hat{f}(\xi + i\eta) d\eta,$$

it follows that

$$\frac{d}{dt} w(t) - Aw(t) = \frac{1}{\sqrt{2\pi}} \int_{-\infty}^{\infty} e^{t(\xi+i\eta)} \hat{f}(\xi + i\eta) d\eta = f(t).$$

Therefore if we define u as the restriction of w to $[0, T]$, then u satisfies (8.4),(8.7),(8.8).

Proposition 8. 2 *If $u = U(\cdot)x$, $x \in (D(A), H)_{1/2,2}$, then*

$$u \in L^2(0, T; D(A)) \cap W^{1,2}(0, T; H),$$

and there exists a constant c_2 such that

$$\|u\|_{L^2(0,T;D(A))\cap W^{1,2}(0,T;H)} \leq c_2 \|x\|_{(D(A),H)_{1/2,2}}. \tag{8.11}$$

Proof. This is a direct consequence of the definition of $(D(A), H)_{1/2,2}$ and Theorem 1.10.

Combining Propositions 8.1 and 8.2 we obtain

Proposition 8.3 *If* $f \in L^2(0,T;H)$ *and* $x \in (D(A), H)_{1/2,2}$, *then the function* u *defined by*

$$u = U(\cdot)x + U * f$$

belongs to $L^2(0,T;D(A)) \cap W^{1,2}(0,T;H)$ *and satisfies*

$$du(t)/dt = Au(t) + f(t) \quad a.e. \quad (0,T), \tag{8.12}$$
$$u(0) = x, \tag{8.13}$$
$$\|u\|_{L^2(0,T;D(A)) \cap W^{1,2}(0,T;H)}$$
$$\leq c_1 \|f\|_{L^2(0,T;H)} + c_2 \|x\|_{(D(A),H)_{1/2,2}}. \tag{8.14}$$

We shall need the following regularity result when the data are more regular.

Proposition 8.4 (i) *If* $f \in W^{1,2}(0,T;H)$, $x \in D(A)$ *and* $Ax + f(0) \in (D(A), H)_{1/2,2}$, *then the solution* u *of* (8.12),(8.13) *satisfies*

$$u \in W^{1,2}(0,T;D(A)) \cap W^{2,2}(0,T;H) \subset C^1([0,T]; (D(A),H)_{1/2,2}),$$
$$\|u'\|_{L^2(0,T;D(A)) \cap W^{1,2}(0,T;H)}$$
$$\leq c_1 \|f'\|_{L^2(0,T;H)} + c_2 \|Ax + f(0)\|_{(D(A),H)_{1/2,2}} \tag{8.15}$$

and we have

$$du(t)/dt = U(\cdot)(Ax + f(0)) + U * f'. \tag{8.16}$$

(ii) *If* $f \in L^2(0,T;D(A))$, $x \in D(A)$ *and* $Ax \in (D(A), H)_{1/2,2}$, *then the solution* u *of* (8.12),(8.13) *satisfies*

$$u \in L^2(0,T;D(A^2)) \cap W^{1,2}(0,T;D(A)),$$
$$u, Au \in C([0,T]; (D(A),H)_{1/2,2}),$$
$$\|Au\|_{L^2(0,T;D(A)) \cap W^{1,2}(0,T;H)}$$
$$\leq c_1 \|Af\|_{L^2(0,T;H)} + c_2 \|Ax\|_{(D(A),H)_{1/2,2}}, \tag{8.17}$$

and we have

$$Au(\cdot) = U(\cdot)Ax + U * Af(\cdot). \tag{8.18}$$

Proof. (i) Let $\{f_n\}$ be a sequence in $C^1([0,T];H)$ such that $f_n \to f$ in $W^{1,2}(0,T;H)$. Set

$$u_n = U(\cdot)x + U * f_n.$$

Differentiating both sides of

$$u_n(t) = U(t)x + \int_0^t U(s)f_n(t-s)ds$$

we get

$$u_n'(t) = U(t)(Ax + f_n(0)) + \int_0^t U(t-s)f_n'(s)ds.$$

As $n \to \infty$

$$u_n \to u, \quad u_n' \to U(\cdot)(Ax + f(0)) + U * f'$$

in $L^2(0, T; H)$. Consequently $u \in W^{1,2}(0, T; H)$ and (8.16) holds. Hence the remaining part of the assertion is an easy consequence of Proposition 8.3.

(ii)   It is easy to verify (8.18). Hence we complete the proof applying Proposition 8.3.

8.2   Assumptions and Notations

Let H be a Hilbert space with norm $\| \cdot \|$. We state the assumptions.

(R1) A_0 is a densely defined closed linear operator and generates an analytic semigroup in H. Hence there exist positive constants M, C_0 and an angle $\theta \in (\pi/2, \pi]$ such that $\rho(A_0) \supset \Sigma \equiv \{\lambda; |\arg \lambda| < \theta, |\lambda| > C_0\}$ and for $\lambda \in \Sigma$ the following inequality holds

$$\|\lambda R(\lambda, A_0)\| \le M. \tag{8.19}$$

(R2) A_1, A_2 are bounded linear operators from $D(A_0)$ to H.
(R3) $a(\cdot)$ is a real valued function belonging to $L^2(-h, 0)$, where h is a positive number.

Let T be a fixed positive number. For a function u defined in $[-h, T]$ and taking values in H we set

$$u_t(s) = u(t+s) \quad s \in [-h, 0], \quad t \in [0, T].$$

Hence for each $t \in [0, T]$ u_t is a function with values in H defined in $[-h, 0]$. For $u \in L^2(-h, 0; D(A_0))$ set

$$Lu = \int_{-h}^0 a(s)A_2u(s)ds. \tag{8.20}$$

Then $L \in \mathcal{L}(L^2(-h, 0; D(A_0)), H)$ and

$$\|L\| \le \|a\|_{L^2(-h,0)}\|A_2\|_{\mathcal{L}(D(A_0),H)}. \tag{8.21}$$

Since

$$Lu_t = \int_{-h}^{0} a(s)A_2 u_t(s)ds = \int_{-h}^{0} a(s)A_2 u(t+s)ds,$$

the problem (8.1),(8.2) is rewritten as

$$\frac{d}{dt}u(t) = A_0 u(t) + A_1 u(t-h) + Lu_t + f(t), \quad t \in [0,T], \quad (8.22)$$

$$u(0) = g^0, \quad u(s) = g^1(s) \quad s \in [-h, 0). \quad (8.23)$$

We are going to find a solution of (8.22),(8.23) in the space

$$L^2(-h, T; D(A_0)) \cap W^{1,2}(0, T; H) \subset C([0,T]; (D(A_0), H)_{1/2,2}).$$

Hence we assume that

$$g^0 \in (D(A_0), H)_{1/2,2}, \ g^1 \in L^2(-h, 0; D(A_0)), \ f \in L^2(0, T; H).$$

8.3 Solvability and Regularity

In this section we prove the existence and uniqueness of solutions of (8.22), (8.23) together with some regularity property.

Theorem 8.1 *Suppose that the assumptions* (R1),(R2),(R3) *are satisfied. Then for any* $g^0 \in (D(A_0), H)_{1/2,2}$, $g^1 \in L^2(-h, 0; D(A_0))$, $f \in L^2(0, T; H)$ *there exists a unique solution* u *of* (8.1),(8.2) *belonging to* $L^2(-h, T; D(A_0)) \cap W^{1,2}(0, T; H)$. *Moreover there exists a constant* c_3 *such that the following inequality holds:*

$$\|u\|_{L^2(0,T;D(A_0)) \cap W^{1,2}(0,T;H)}$$
$$\leq c_3 \left(\|g^0\|_{(D(A_0),H)_{1/2,2}} + \|g^1\|_{L^2(-h,0;D(A_0))} + \|f\|_{L^2(0,T;H)} \right). \quad (8.24)$$

Proof. By Proposition 8.1 for $u(t) = \int_0^t e^{(t-s)A_0} f(s)ds$ we have

$$\|u\|_{L^2(0,T;D(A_0)) \cap W^{1,2}(0,T;H)} \leq c_1 \|f\|_{L^2(0,T;H)}, \quad (8.25)$$

and by Proposition 8.2 for $x = e^{tA_0}x$ we have

$$\|u\|_{L^2(0,T;D(A_0)) \cap W^{1,2}(0,T;H)} \leq c_2 \|x\|_{(D(A_0),H)_{1/2,2}}. \quad (8.26)$$

We note that the constants c_1, c_2 are taken independently of T in a bounded interval. For the sake of simplicity we denote the operator norm all by $\| \cdot \|$.

First we consider the case $0 < T \leq h$. Then

$$\|A_1 g^1(\cdot - h)\|^2_{L^2(0,T;H)} = \int_0^T \|A_1 g^1(t-h)\|^2 dt$$

$$= \int_{-h}^{T-h} \|A_1 g^1(t)\|^2 dt \leq \int_{-h}^0 \|A_1 g^1(t)\|^2 dt$$

$$\leq \|A_1\|^2 \int_{-h}^0 \|g^1(t)\|^2_{D(A_0)} dt = \|A_1\|^2 \|g^1\|^2_{L^2(-h,0;D(A_0))}.$$

Hence
$$\|A_1 g^1(\cdot - h)\|_{L^2(0,T;H)} \leq \|A_1\| \|g^1\|_{L^2(-h,0;D(A_0))}. \qquad (8.27)$$

Similarly

$$\|Lu.\|^2_{L^2(0,T;H)} = \int_0^T \|Lu_t\|^2 dt \leq \|L\|^2 \int_0^T \|u_t\|^2_{L^2(-h,0;D(A_0))} dt$$

$$= \|L\|^2 \int_0^T \int_{-h}^0 \|u(t+s)\|^2_{D(A_0)} ds dt = \|L\|^2 \int_0^T \int_{t-h}^t \|u(s)\|^2_{D(A_0)} ds dt$$

$$\leq \|L\|^2 \int_0^T \int_{-h}^T \|u(s)\|^2_{D(A_0)} ds dt = \|L\|^2 T \|u\|^2_{L^2(-h,T;D(A_0))}.$$

Hence
$$\|Lu.\|_{L^2(0,T;H)} \leq \|L\|\sqrt{T}\|u\|_{L^2(-h,T;D(A_0))}. \qquad (8.28)$$

For a given $\bar{u} \in L^2(0,T; D(A_0))$ define

$$u(t) = \begin{cases} g^1(t) & t \in [-h,0) \\ \bar{u}(t) & t \in [0,T] \end{cases}. \qquad (8.29)$$

Then $u \in L^2(-h,T; D(A_0))$. Noting $u(t-h) = g^1(t-h)$ for $t \in [0,T]$ since we are assuming $T \leq h$, we set for $0 \leq t \leq T$

$$(S\bar{u})(t) = e^{tA_0} g^0 + \int_0^t e^{(t-s)A_0} \left(A_1 g^1(s-h) + Lu_s + f(s)\right) ds. \qquad (8.30)$$

By Proposition 8.3 and (8.27),(8.28) S is a mapping from $L^2(0,T; D(A_0))$ into itself. We are going to apply a fixed point theorem to S. Let $\bar{u}_1, \bar{u}_2$ be two elements of $L^2(0,T; D(A_0))$ and set

$$u_i(t) = \begin{cases} g^1(t) & t \in [-h,0) \\ \bar{u}_i(t) & t \in [0,T] \end{cases}, \quad i = 1,2. \qquad (8.31)$$

Then

$$(S\bar{u}_1)(t) - (S\bar{u}_2)(t) = \int_0^t e^{(t-s)A_0} L(u_{1s} - u_{2s}) ds. \qquad (8.32)$$

In view of (8.25),(8.28)

$$
\begin{aligned}
\|S\bar{u}_1 - S\bar{u}_2\|_{L^2(0,T;D(A_0))} &\leq c_1 \|L(u_{1\cdot} - u_{2\cdot})\|_{L^2(0,T;H)} \\
&\leq c_1 \|L\| \sqrt{T} \|u_{1\cdot} - u_{2\cdot}\|_{L^2(-h,T;D(A_0))} \\
&= c_1 \|L\| \sqrt{T} \|\bar{u}_1 - \bar{u}_2\|_{L^2(0,T;D(A_0))}. \qquad (8.33)
\end{aligned}
$$

Hence if $c_1\|L\|\sqrt{T} < 1$, S is a strict contraction and there exists a unique fixed point $\bar{u}$. The function defined as (8.29) is a unique solution of (8.1),(8.2). The magnitude of T depends only on c_1 and $\|L\|$, and is independent of the initial values. Since in $(0,T)$

$$u(t) = e^{tA_0} g^0 + \int_0^t e^{(t-s)A_0} \left(A_1 g^1(s-h) + Lu_s + f(s) \right) ds,$$

we have by (8.25),(8.26),(8.27),(8.28)

$$
\begin{aligned}
&\|u\|_{L^2(0,T;D(A_0)) \cap W^{1,2}(0,T;H)} \\
&\quad \leq c_2 \|g^0\|_{(D(A_0),H)_{1/2,2}} + c_1 \|A_1 g^1(\cdot - h) + Lu_\cdot + f\|_{L^2(0,T;H)} \\
&\quad \leq c_2 \|g^0\|_{(D(A_0),H)_{1/2,2}} + c_1 \|A_1\| \|g^1\|_{L^2(-h,0;D(A_0))} \\
&\qquad + c_1 \|L\| \sqrt{T} \|u\|_{L^2(-h,T;D(A_0))} + c_1 \|f\|_{L^2(0,T;H)} \\
&\quad \leq c_2 \|g^0\|_{(D(A_0),H)_{1/2,2}} + c_1 \|A_1\| \|g^1\|_{L^2(-h,0;D(A_0))} \\
&\qquad + c_1 \|L\| \sqrt{T} \|g^1\|_{L^2(-h,0;D(A_0))} + c_1 \|L\| \sqrt{T} \|u\|_{L^2(0,T;D(A_0))} \\
&\qquad + c_1 \|f\|_{L^2(0,T;H)}.
\end{aligned}
$$

Hence

$$
\begin{aligned}
&\|u\|_{L^2(0,T;D(A_0)) \cap W^{1,2}(0,T;H)} \\
&\quad \leq \frac{1}{1 - c_1 \|L\| \sqrt{T}} \Big[c_2 \|g^0\|_{(D(A_0),H)_{1/2,2}} \\
&\qquad + c_1 (\|A_1\| + \|L\| \sqrt{T}) \|g_1\|_{L^2(-h,0;D(A_0))} + c_1 \|f\|_{L^2(0,T;H)} \Big]. \qquad (8.34)
\end{aligned}
$$

Next we continue the solution to an arbitrary time interval. Suppose that $t_1 > 0, t_2 > 0$ and u is a solution in $[-h, t_1 + t_2]$. If $t \in [t_1, t_1 + t_2]$

$$u(t) = e^{tA_0} g^0 + \left(\int_0^{t_1} + \int_{t_1}^t \right) e^{(t-s)A_0} \left(A_1 u(s-h) + Lu_s + f(s) \right) ds$$

$$= e^{(t-t_1)A_0} \left[e^{t_1 A_0} g^0 + \int_0^{t_1} e^{(t_1-s)A_0} \left(A_1 u(s-h) + Lu_s + f(s) \right) ds \right]$$

$$+ \int_{t_1}^t e^{(t-s)A_0} \left(A_1 u(s-h) + Lu_s + f(s) \right) ds$$

$$= e^{(t-t_1)A_0} u(t_1)$$

$$+ \int_0^{t-t_1} e^{(t-t_1-s)A_0} \left(A_1 u(t_1+s-h) + Lu_{t_1+s} + f(t_1+s) \right) ds.$$

Hence for $t \in [0, t_2]$

$$u(t_1+t) = e^{tA_0} u(t_1) + \int_0^t e^{(t-s)A_0} \left(A_1 u(t_1+s-h) + Lu_{t_1+s} + f(t_1+s) \right) ds.$$

This shows that $u(t_1+t)$ is a solution in $[-h, t_2]$ with initial values $u(t_1), u_{t_1}$ and the inhomogeneous term $f(t_1 + t)$.

Conversely suppose that v is a solution in $[-h, t_1]$ and w is a solution in $[-h, t_2]$ with initial values $v(t_1), v_{t_1}$ and the inhomogeneous term $f(t_1 + \cdot)$. Set

$$u(t) = \begin{cases} v(t) & -h \leq t \leq t_1 \\ w(t-t_1) & t_1 \leq t \leq t_1 + t_2 \end{cases} . \tag{8.35}$$

We are going to verify that u is a solution in $[-h, t_1 + t_2]$. If $-h \leq t < 0$, $w(t) = v_{t_1}(t) = v(t_1 + t) = u(t_1 + t)$, and if $0 \leq t \leq t_2$, $w(t) = u(t_1 + t)$. Hence

$$w(t) = u(t_1 + t) \quad \text{for} \quad -h \leq t \leq t_2. \tag{8.36}$$

If $t_1 < t \leq t_1 + t_2$, using (8.36) we have

$$u(t) = w(t - t_1) = e^{(t-t_1)A_0} v(t_1)$$

$$+ \int_0^{t-t_1} e^{(t-t_1-s)A_0} \left(A_1 w(s-h) + Lw_s + f(t_1+s) \right) ds$$

$$= e^{(t-t_1)A_0} \left[e^{t_1 A_0} g^0 + \int_0^{t_1} e^{(t_1-s)A_0} \left(A_1 v(s-h) + Lv_s + f(s) \right) ds \right]$$

$$+ \int_0^{t-t_1} e^{(t-t_1-s)A_0} \left(A_1 u(t_1+s-h) + Lu_{t_1+s} + f(t_1+s) \right) ds$$

$$= e^{tA_0} g^0 + \int_0^{t_1} e^{(t-s)A_0} \left(A_1 u(s-h) + Lu_s + f(s) \right) ds$$

$$+ \int_{t_1}^t e^{(t-s)A_0} \left(A_1 u(s-h) + Lu_s + f(s) \right) ds$$

$$= e^{tA_0} g^0 + \int_0^t e^{(t-s)A_0} \left(A_1 u(s-h) + Lu_s + f(s) \right) ds.$$

This implies that u is a solution in $[-h, t_1 + t_2]$.

Therefore if we have a solution u in $[-h, T]$ where $0 < T < (c_1\|L\|)^{-2}$, then choosing $t_1 = t_2 = T$ in the above we obtain a solution u^* in $[-h, T]$ with the initial values $u(T), u_T$. If we set $u(t) = u^*(t-T)$ in $[T, 2T]$, then u is a solution in $[-h, 2T]$. Continuing this process we obtain a global solution.

In the next section we shall need the following regularity result.

Theorem 8. 2 *If $g^0 \in D(A_0)$, $g^1 \in W^{1,2}(-h, 0; D(A_0))$, $g^1(0) = g^0$, $f \in W^{1,2}(0, T; H)$ and*

$$A_0 g^0 + A_1 g^1(-h) + Lg^1 + f(0) \in (D(A_0), H)_{1/2,2},$$

then the solution of (8,1),(8.2) obtained in Theorem 8.1 belongs to

$$W^{1,2}(-h, T; D(A_0)) \cap W^{2,2}(0, T; H) \subset C^1([0, T]; (D(A_0), H)_{1/2,2}).$$

Proof. First consider the case $T \le h, c_1\|L\|\sqrt{T} < 1$. Set

$$K = \{\bar{u} \in W^{1,2}(0, T; D(A_0)); \bar{u}(0) = g^0\}.$$

We are going to show that the mapping S defined by (8.30) is a strict contraction in K. Let $\bar{u} \in K$ and define u by (8.29). Then from the assumptions it follows that $u \in W^{1,2}(-h, T; D(A_0))$. Set

$$\tilde{f}(s) = A_1 g^1(s - h) + Lu_s + f(s).$$

Then $\tilde{f} \in W^{1,2}(0, T; H)$ and using (8.27), (8.28) we get

$$\|\tilde{f}\|_{W^{1,2}(0,T;H)} \le \|A_1\|\|g^1\|_{W^{1,2}(-h,0;D(A_0))}$$
$$+ \|L\|\sqrt{T}\|u\|_{W^{1,2}(-h,T;D(A_0))} + \|f\|_{W^{1,2}(0,T;H)}.$$

Since

$$A_0 g^0 + \tilde{f}(0) = A_0 g^0 + A_1 g^1(-h) + Lg^1 + f(0) \in (D(A_0), H)_{1/2,2},$$

we have in view of Proposition 8.4

$$S\bar{u} \in W^{1,2}(0, T; D(A_0)) \cap W^{2,2}(0, T; H).$$

It is clear that $(S\bar{u})(0) = g^0$. Hence $S\bar{u} \in K$ and S maps K into itself. Next, let $\bar{u}_1, \bar{u}_2 \in K$ and define u_1, u_2 by (8.31). Since $L(u_{10} - u_{20}) = L(g^1 - g^1) = 0$, we get differentiating both sides of (8.32)

$$\frac{d}{dt}((S\bar{u}_1)(t) - (S\bar{u}_2)(t))$$

$$= \int_0^t e^{(t-s)A_0} \frac{d}{ds} L(u_{1s} - u_{2s}) ds = \int_0^t e^{(t-s)A_0} L(u'_{1s} - u'_{2s}) ds.$$

Therefore applying (8.33) we get

$$\|S\bar{u}_1 - S\bar{u}_2\|^2_{W^{1,2}(0,T;D(A_0))}$$
$$= \|S\bar{u}_1 - S\bar{u}_2\|^2_{L^2(0,T;D(A_0))} + \|(S\bar{u}_1 - S\bar{u}_2)'\|^2_{L^2(0,T;D(A_0))}$$
$$\leq c_1^2\|L\|^2 T\left(\|u_1 - u_2\|^2_{L^2(-h,T;D(A_0))} + \|u_1' - u_2'\|^2_{L^2(-h,T;D(A_0))}\right)$$
$$= c_1^2\|L\|^2 T\|\bar{u}_1 - \bar{u}_2\|^2_{W^{1,2}(0,T;D(A_0))}.$$

Consequently S is a strict contraction in K, and the solution u belongs to $W^{1,2}(0,T;D(A_0))$. Using the argument by which we deduced (8.34) we obtain

$$\|u'\|_{L^2(0,T;D(A_0))\cap W^{1,2}(0,T;H)}$$
$$\leq \frac{1}{1 - c_1\|L\|\sqrt{T}}\Big[c_2\|A_0 g^0 + A_1 g^1(-h) + L g^1 + f(0)\|_{(D(A_0),H)_{1/2,2}}$$
$$+ c_1(\|A_1\| + \|L\|\sqrt{T})\|g^{1'}\|_{L^2(-h,0;D(A_0))} + c_1\|f'\|_{L^2(0,T;H)}\Big].$$

Therefore we can proceed as in Theorem 8.1 to obtain the conclusion for an arbitrary $T > 0$.

8.4   Solution Semigroup

In this section we consider the problem (8.1),(8.2) with $f(t) \equiv 0$:

$$\frac{d}{dt}u(t) = A_0 u(t) + A_1 u(t - h) + \int_{-h}^{0} a(s)A_2 u(t + s)ds, \quad t \in [0, \infty), \quad (8.37)$$

$$u(0) = g^0, \quad u(s) = g^1(s) \quad s \in [-h, 0). \quad (8.38)$$

Set

$$Z = (D(A_0), H)_{1/2,2} \times L^2(-h, 0; D(A_0)) \quad (8.39)$$

and for $g = (g^0, g^1) \in Z$

$$\|g\|^2_Z = \left(\|g^0\|^2_{(D(A_0),H)_{1/2,2}} + \|g^1\|^2_{L^2(-h,0;D(A_0))}\right)^{1/2}. \quad (8.40)$$

Then Z is Hilbert space with norm (8.40). By virtue of Theorem 8.1 for $g = (g^0, g^1) \in Z$ there exists a unique solution u of (8.37),(8.38) in $[-h, \infty)$, and using (1.17) we have for each $T > 0$

$$C_T^{-1} \max_{t \in [0,T]} \|u(t)\|_{(D(A_0),H)_{1/2,2}}$$
$$\leq \|u\|_{L^2(-h,T;D(A_0))\cap W^{1,2}(0,T;H)} \leq C_T\|g\|_Z, \quad (8.41)$$

where C_T is a positive constant depending on T.

We denote the solution of (8.37),(8.38) by $u(t; g) = u(t; g^0, g^1)$. Set

$$S(t)g = (u(t; g), u_t(\cdot; g)). \qquad (8.42)$$

Theorem 8.3 $\{S(t); t \geq 0\}$ *is a C_0-semigroup in Z.*

Proof. In view of (8.41) for $g = (g^0, g^1) \in Z$

$$\|S(t)g\|_Z^2 = \|u(t; g)\|_{(D(A_0),H)_{1/2,2}}^2 + \|u_t(\cdot, g)\|_{L^2(-h,0;D(A_0))}^2$$
$$= \|u(t; g)\|_{(D(A_0),H)_{1/2,2}}^2 + \|u(\cdot; g)\|_{L^2(t-h,t;D(A_0))}^2$$
$$\leq \|u(t; g)\|_{(D(A_0),H)_{1/2,2}}^2 + \|u(\cdot; g)\|_{L^2(-h,t;D(A_0))}^2 \leq C_t^2(C_t^2 + 1)\|g\|_Z^2.$$

Hence $S(t) \in \mathcal{L}(Z, Z)$. For $t, \tau \in [0, \infty)$

$$\|S(t)g - S(\tau)g\|_Z^2$$
$$= \|u(t; g) - u(\tau; g)\|_{(D(A_0),H)_{1/2,2}}^2 + \|u_t(\cdot; g) - u_\tau(\cdot; g)\|_{L^2(-h,0;D(A_0))}^2$$
$$= \|u(t; g) - u(\tau; g)\|_{(D(A_0),H)_{1/2,2}}^2$$
$$+ \int_{-h}^0 \|u(t+s; g) - u(\tau + s; g)\|_{D(A_0)}^2 ds \to 0 \quad \text{as} \quad \tau \to t.$$

Consequently $S(t)g$ is strongly continuous in $[0, \infty)$. Choosing $t = 0$ in the above we obtain that $S(\tau)g \to g$ as $\tau \to 0$.

Let $t_1 > 0, t_2 > 0$. Let v be the solution of (8.37),(8.38) in $[-h, t_1]$ with the initial values $g = (g^0, g^1)$, and w be the solution in $[-h, t_2]$ with the initial values $(v(t_1), v_{t_1})$. Then as was shown in the proof of Theorem 8.1 the function defined by (8.35) is the solution in $[-h, t_1 + t_2]$ with the initial values g^0, g^1, i.e.

$$v(t) = u(t; g) \quad -h \leq t \leq t_1, \quad w(t) = u(t; v(t_1), v_{t_1}) \quad -h \leq t \leq t_2,$$
$$u(t) = u(t; g) \quad -h \leq t \leq t_1 + t_2.$$

Hence

$$u(t_1 + t_2; g) = u(t_1 + t_2) = w(t_2) = u(t_2; v(t_1), v_{t_1})$$

and by (8.36) for $s \in [-h, 0]$

$$u(t_1 + t_2 + s; g) = w(t_2 + s) = u(t_2 + s; v(t_1), v_{t_1}) = u_{t_2}(s; v(t_1), v_{t_1}).$$

Therefore

$$S(t_1 + t_2)g = (u(t_1 + t_2; g), u_{t_1+t_2}(\cdot; g))$$
$$= (u(t_2; v(t_1), v_{t_1}), u_{t_2}(\cdot; v(t_1), v_{t_1})) = S(t_2)(v(t_1), v_{t_1})$$
$$= S(t_2)(u(t_1; g), u_{t_1}(\cdot; g)) = S(t_2)S(t_1)g.$$

Thus we have shown $S(t_1 + t_2) = S(t_2)S(t_1)$ and the proof is complete.

Definition 8. 1 The semigrooup $S(t)$ in Theorem 8.3 is called the *solution semigroup* of (8.1),(8.2).

Theorem 8. 4 *The infinitesimal generator A of the semigroup $S(t)$ is given by*

$$D(A) = \{g = (g^0, g^1) \in Z; g^1 \in W^{1,2}(-h, 0; D(A_0)),$$
$$g^0 = g^1(0), A_0 g^0 + A_1 g^1(-h) + Lg^1 \in (D(A_0), H)_{1/2,2}\}, \quad (8.43)$$
$$Ag = (A_0 g^0 + A_1 g^1(-h) + Lg^1, g^{1'}). \quad (8.44)$$

Proof. Let A be the operator defined by (8.43),(8.44), and $\tilde{A}$ the infinitesimal generator of $S(t)$. We first show that

(i) $S(t)D(A) \subset D(A)$ for any $t \geq 0$,
(ii) $D(A)$ is dense in Z,
(iii) $A \subset \tilde{A}$,
(iv) A is closed in Z.

Proof of (i). Let $g \in D(A)$. In view of Theorem 8.2 we have for any $T > 0$

$$u(\cdot; g) \in W^{1,2}(-h, T; D(A_0)) \cap W^{2,2}(0, T; H)$$
$$\subset C^1([0, T]; (D(A_0), H)_{1/2,2}). \quad (8.45)$$

Hence for any $t \geq 0$ $u(t; g) \in D(A_0)$ and $u_t(\cdot; g) \in W^{1,2}(-h, 0; D(A_0))$. Clearly $u_t(0; g) = u(t; g)$. By the equation (8.37) satisfied by $u(\cdot; g)$ and (8.45) we have

$$A_0 u(t; g) + A_1 u_t(-h; g) + Lu_t(\cdot; g)$$
$$= A_0 u(t; g) + A_1 u(t - h; g) + Lu_t(\cdot; g) = u'(t; g) \in (D(A_0), H)_{1/2,2}.$$

Therefore $S(t)g = (u(t; g), u_t(\cdot; g)) \in D(A)$.

Proof of (ii). Let $g \in Z$. Since $\epsilon^{-1} \int_0^\epsilon S(t)gdt \to g$ in Z as $\epsilon \to 0$, it suffices to show that $\int_0^\epsilon S(t)gdt \in D(A)$ for $\epsilon > 0$. Note that

$$\int_0^\epsilon S(t)gdt = \left(\int_0^\epsilon u(t; g)dt, \int_0^\epsilon u_t(\cdot; g)dt \right).$$

For $-h \leq s < 0$

$$\frac{d}{ds} \left(\int_0^\epsilon u_t(\cdot, g)dt \right)(s) = \frac{d}{ds} \int_0^\epsilon u(t + s; g)dt$$
$$= \int_0^\epsilon u'(t + s; g)dt = u(\epsilon + s; g) - u(s; g) = u(\epsilon + s; g) - g^1(s).$$

Hence

$$\int_{-h}^0 \left\| A_0 \frac{d}{ds} \left(\int_0^\epsilon u_t(\cdot; g) dt \right)(s) \right\|^2 ds$$

$$= \int_{-h}^0 \| A_0(u(\epsilon + s; g) - g^1(s)) \|^2 ds$$

$$\leq 2 \left(\int_{-h}^0 \| A_0 u(\epsilon + s; g) \|^2 ds + \int_{-h}^0 \| A_0 g^1(s) \|^2 ds \right) < \infty.$$

Therefore $\int_0^\epsilon u_t(\cdot; g) dt \in W^{1,2}(-h, 0; D(A_0))$. Clearly

$$\left(\int_0^\epsilon u_t(\cdot; g) dt \right)(0) = \int_0^\epsilon u(t; g) dt.$$

By the equation (8.37)

$$A_0 \int_0^\epsilon u(t; g) dt + A_1 \left(\int_0^\epsilon u_t(\cdot; g) dt \right)(-h) + L \int_0^\epsilon u_t(\cdot; g) dt$$

$$= A_0 \int_0^\epsilon u(t; g) dt + A_1 \int_0^\epsilon u(t - h; g) dt + L \int_0^\epsilon u_t(\cdot; g) dt$$

$$= \int_0^\epsilon \left(A_0 u(t; g) + A_1 u(t - h; g) + L u_t(\cdot; g) \right) dt$$

$$= \int_0^\epsilon u'(t; g) dt = u(\epsilon; g) - u(0; g) = u(\epsilon; g) - g^0 \in (D(A_0), H)_{1/2,2}.$$

Consequently $\int_0^\epsilon S(t) g dt \in D(A)$.

Proof of (iii). Suppose that $g \in D(A)$. By virtue of Theorem 8.2 (8.45) holds for any $T > 0$. Therefore as $t \to 0$

$$\left\| \frac{1}{t}(u(t; g) - g^0) - u'(0; g) \right\|_{(D(A_0), H)_{1/2,2}}$$

$$= \left\| \frac{1}{t}(u(t; g) - u(0; g)) - u'(0; g) \right\|_{(D(A_0), H)_{1/2,2}} \to 0. \qquad (8.46)$$

Since

$$\frac{1}{t} \left(u_t(\cdot; g) - g^1 \right)(s) = \frac{1}{t}(u(t + s; g) - u(s; g))$$

$$= \frac{1}{t} \int_0^1 \frac{d}{d\theta} u(\theta t + s; g) d\theta = \int_0^1 u'(\theta t + s; g) d\theta,$$

we have

$$\int_{-h}^{0} \left\| \frac{1}{t} \left(u_t(\cdot; g) - g^1 \right)(s) - u'(s; g) \right\|_{D(A_0)}^2 ds$$

$$= \int_{-h}^{0} \left\| \int_{0}^{1} \left(u'(\theta t + s; g) - u'(s; g) \right) d\theta \right\|_{D(A_0)}^2 ds$$

$$\leq \int_{-h}^{0} \int_{0}^{1} \| u'(\theta t + s; g) - u'(s; g) \|_{D(A_0)}^2 d\theta ds$$

$$= \int_{0}^{1} \int_{-h}^{0} \| u'(\theta t + s; g) - u'(s; g) \|_{D(A_0)}^2 ds d\theta.$$

For each fixed $\theta \in (0, 1)$

$$\mathbf{I} \equiv \int_{-h}^{0} \| u'(\theta t + s; g) - u'(s; g) \|_{D(A_0)}^2 ds \to 0$$

as $t \to 0$. Furthermore

$$\mathbf{I} \leq 2 \int_{-h}^{0} \| u'(\theta t + s; g) \|_{D(A_0)}^2 ds + 2 \int_{-h}^{0} \| u'(s; g) \|_{D(A_0)}^2 ds$$

$$\leq 2 \int_{-h}^{t} \| u'(s; g) \|_{D(A_0)}^2 ds + 2 \int_{-h}^{0} \| g^1(s) \|_{D(A_0)}^2 ds$$

is bounded as $t \to 0$. Therefore

$$\int_{-h}^{0} \left\| \frac{1}{t} \left(u_t(\cdot; g) - g^1 \right)(s) - u'(s; g) \right\|_{D(A_0)}^2 ds \to 0. \qquad (8.47)$$

Combining (8.46),(8.47) we obtain

$$\frac{1}{t}(S(t)g - g) = \left(\frac{1}{t} \left(u(t; g) - g^0 \right), \frac{1}{t} \left(u_t(\cdot; g) - g^1 \right) \right)$$

$$\to (u'(0; g), u'(\cdot; g)) = \left(A_0 g^0 + A_1 g^1(-h) + L g^1, g^{1'} \right) = Ag$$

in Z as $t \to 0$. Therefore $g \in D(\tilde{A})$, $\tilde{A}g = Ag$.

Proof of (iv). Let $g_n = (g_n^0, g_n^1)$ be a sequence of elements of $D(A)$ such that $g_n \to g = (g^0, g^1)$, $Ag_n \to w = (w^0, w^1)$ in Z. By the definition and assumptions we have

$$g_n^1 \in W^{1,2}(-h, 0; D(A_0)), \quad g_n^1(0) = g_n^0,$$

$$A_0 g_n^0 + A_1 g_n^1(-h) + L g_n^1 \in (D(A_0), H)_{1/2,2}$$

$$g_n^0 \to g^0 \text{ in } (D(A_0), H)_{1/2,2}, \ g_n^1 \to g^1 \text{ in } L^2(-h, 0; D(A_0)), \quad (8.48)$$

$$A_0 g_n^0 + A_1 g_n^1(-h) + L g_n^1 \to w^0 \text{ in } (D(A_0), H)_{1/2,2}, \quad (8.49)$$

$$g_n^{1'} \to w^1 \text{ in } L^2(-h, 0; D(A_0)). \quad (8.50)$$

From (8.48),(8.50) it follows that $g^1 \in W^{1,2}(-h, 0; D(A_0)), g_n^1 \to g^1$ in $W^{1,2}(-h, 0; D(A_0))$ and

$$g^{1'} = w^1. \quad (8.51)$$

Therefore $g_n^0 = g_n^1(0) \to g^1(0), g_n^1(-h) \to g^1(-h)$ in $D(A_0)$. Combining this with (8.48) we get $g^0 = g^1(0)$, and $g_n^0 \to g^0$ in $D(A_0)$. Therefore

$$A_0 g_n^0 + A_1 g_n^1(-h) + L g_n^1 \to A_0 g^0 + A_1 g^1(-h) + L g^1$$

in H. This and (8.49) implies

$$A_0 g^0 + A_1 g^1(-h) + L g^1 = w^0 \in (D(A_0), H)_{1/2,2}. \quad (8.52)$$

Therefore $g \in D(A)$, and by virtue of (8.51),(8.52) $Ag = w$. Thus the proof of (i),(ii),(iii),(iv) is complete.

Let $g \in D(A)$. Then in view of (i) $S(t)g \in D(A)$. By (iii)

$$AS(t)g = \tilde{A}S(t)g = S(t)\tilde{A}g = S(t)Ag,$$

which implies that $AS(t)g$ is continuous in $t \in [0, \infty)$, and $AS(t)g = dS(t)g/dt$. Combining this with (iv) we obtain

$$A \int_0^t S(\tau)g d\tau = \int_0^t AS(\tau)g d\tau = \int_0^t \frac{d}{d\tau} S(\tau)g d\tau = S(\tau)g - g.$$

Let g be an arbitrary element of Z. Then by (ii) there exists a sequence $\{g_n\} \subset D(A)$ such that $g_n \to g$ in Z. Then

$$A \int_0^t S(\tau)g_n d\tau = S(t)g_n - g_n \to S(t)g - g, \quad \int_0^t S(\tau)g_n d\tau \to \int_0^t S(\tau)g d\tau.$$

Hence, in view of (iv)

$$\int_0^t S(\tau)g d\tau \in D(A) \quad \text{and} \quad A \int_0^t S(\tau)g d\tau = S(t)g - g.$$

Therefore if $g \in D(\tilde{A})$,

$$A \frac{1}{t} \int_0^t S(\tau)g d\tau = \frac{S(t)g - g}{t} \to \tilde{A}g, \quad \frac{1}{t} \int_0^t S(\tau)g d\tau \to g$$

as $t \to 0$. This implies that $g \in D(A)$ and $Ag = \tilde{A}g$. Consequently we conclude $\tilde{A} = A$.

8.5   Mild Solutions

Suppose that $g = (g^0, g^1) \in D(A)$ and $f \in W^{1,2}(0, T; (D(A_0), H)_{1/2,2})$. Then

$$g^1 \in W^{1,2}(-h, 0; D(A_0)), \quad g^0 = g^1(0),$$
$$A_0 g^0 + A_1 g^1(-h) + Lg^1 \in (D(A_0), H)_{1/2,2}, \quad f(0) \in (D(A_0), H)_{1/2,2}.$$

Hence the assumptions of Theorem 8.2 are satisfied. Therefore the solution u of (8.1),(8.2) satisfies

$$u \in W^{1,2}(-h, T; D(A_0)) \cap W^{2,2}(0, T; H)$$
$$\subset C^1([0, T]; (D(A_0), H)_{1/2,2}). \tag{8.53}$$

Set $x(t) = (u(t), u_t)$ for $t \geq 0$. In view of (8.53) $u(t) \in D(A_0), u_t \in W^{1,2}(-h, 0; D(A_0))$. Clearly $u_t(0) = u(t)$. By (8.1),(8.53)

$$A_0 u(t) + A_1 u_t(-h) + Lu_t = u'(t) - f(t) \in (D(A_0), H)_{1/2,2}.$$

Hence $x(t) \in D(A)$ and

$$Ax(t) = (A_0 u(t) + A_1 u(t - h) + Lu_t, (u_t)') = (u'(t) - f(t), (u_t)'), \tag{8.54}$$

where $(u_t)'$ is the function

$$\frac{d}{ds}(u_t)(s) = \frac{d}{ds} u(t + s) = u'(t + s) = (u')_t(s). \tag{8.55}$$

Lemma 8.1   For $u \in W^{1,2}(-h, T; H)$ we have

$$du_t/dt = (u')_t, \quad t \in [0, T].$$

Proof. For $0 \leq t \leq T, -h \leq s \leq 0$

$$u(t + s) - u(s) = \int_0^t u'(\tau + s)d\tau = \int_0^t (u')_\tau(s)d\tau = \left(\int_0^t (u')_\tau d\tau \right)(s).$$

Hence

$$u_t - u_0 = \int_0^t (u')_\tau d\tau$$

from which the conclusion follows.

By Lemma 8.1 and (8.54),(8.55) we get

$$Ax(t) = (u'(t) - f(t), du_t/dt) = x'(t) - (f(t), 0).$$

Therefore if we put $F(t) = (f(t), 0)$, $x(t)$ satisfies

$$dx(t)/dt = Ax(t) + F(t), \quad 0 \le t \le T, \tag{8.56}$$
$$x(0) = g. \tag{8.57}$$

It is easy to see that x', Ax are continuous functions with values in Z in $[0, T]$. Namely x is a strict solution of (8.56),(8.57).

Conversely suppose that $x(t) = (x^0(t), x^1(t))$ is a strict solution of (8.56), (8.57). Since $x(t) \in D(A)$ we have

$$x^1(t) \in W^{1,2}(-h, 0; D(A_0)), \quad x^0(t) = x^1(t)(0), \tag{8.58}$$
$$A_0 x^0(t) + A_1 x^1(t)(-h) + L x^1(t) \in (D(A_0), H)_{1/2,2}. \tag{8.59}$$

Furthermore

$$\frac{d}{dt} x^0(t) = A_0 x^0(t) + A_1 x^1(t)(-h) + L x^1(t) + f(t), \tag{8.60}$$
$$\left(\frac{d}{dt} x^1(t) \right)(s) = \frac{d}{ds} x^1(t)(s). \tag{8.61}$$

By (8.61) $x^1(t)(s)$ is a function of $t + s$, and we write $x^1(t + s)$ instead of $x^1(t)(s)$. By virtue of the second part of (8.58) we have $x^0(t) = x^1(t)$ for $t \ge 0$ in the new notation for x^1. Therefore if we define the function u by

$$u(t) = \begin{cases} x^0(t) & t \ge 0, \\ g^1(t) & -h \le t < 0, \end{cases}$$

then we have $u(t) = x^1(t)$ for $t \ge -h$. Clearly (8.60) implies (8.1), and (8.57) implies (8.2). Since $x(t) = (u(t), u_t)$ is a strict solution of (8.56),(8.57),

$$\|(u')_t\|_{L^2(-h,0;D(A_0))}^2 = \int_{-h}^0 \|(u')_t(s)\|_{D(A_0)}^2 ds$$
$$= \int_{-h}^0 \|u'(t+s)\|_{D(A_0)}^2 ds = \int_{t-h}^t \|u'(s)\|_{D(A_0)}^2 ds$$

is bounded in $[0, T]$. Therefore $u \in W^{1,2}(-h, T; D(A_0))$. From the equation (8.1) it also follows that $u \in W^{2,2}(0, T; H)$. Hence u is a solution of (8.1),(8.2) satisfying (8.53).

The strict solution of (8.56),(8.57) is represented by

$$u(t) = S(t)g + \int_0^t S(t - \tau)F(\tau)d\tau. \tag{8.62}$$

For arbitrary $g \in Z$ and $F = (f, 0)$, $f \in L^2(-h, 0; (D(A_0), H)_{1/2,2})$ the right hand side of (8.62) is meaningful, and is called a *mild solution* of (8.56),(8.57).

8.6   Regularly Accretive Operators

If we take as A_0 in (8.1) a regularly accretive operator whose definition will be given below, we can develop a semigroup treatment for (8.1),(8.2) in a more easily handled space. Such a treatment is convenient in applications to control problems. As a preparation we consider operators associated with sesquilinear forms on Hilbert spaces.

Let H, V be two Hilbert spaces such that V is a dense subspace of H and the imbedding $V \subset H$ is continuous. We denote the innerproduct and norm of H by $(\cdot, \cdot)$ and $|\cdot|$ respectively. The norm of V is denoted by $\|\cdot\|$. Then by assumption there exists a positive constant C_0 such that

$$|u| \le C_0 \|u\|, \quad u \in V. \tag{8.63}$$

In this and the following sections H^*, V^* stand for the sets of all conjugate linear continuous functionals defined in H, V respectively. By assumption the mapping which maps elements of H^* to their restrictions to V is an injection from H^* to V^*. Therefore identifying an element of H^* and its restriction to V we may consider $H^* \subset V^*$. Using Riesz's theorem we identify H and H^*. Hence we may consider

$$V \subset H \subset V^*. \tag{8.64}$$

The pairing between V and V^* is also denoted by $(\cdot, \cdot)$: for $l \in V^*, u \in V$, (l, v) is the value of l at v. In particular if $l \in H$, then (l, v) is the innerproduct of l and v considered as elements of H. The norm of V^* is denoted by $\|\cdot\|_*$:

$$\|l\|_* = \sup_{0 \ne v \in V} |(l, v)| / \|v\|. \tag{8.65}$$

As is easily seen $\|f\|_* \le C_0 |f|$ for any $f \in H$. Hence the imbedding $H \subset V^*$ is also continuous.

In what follows in this chapter we denote the complex number field by $\mathbf{C}$.

Let $a(u, v)$ be a sesquilinear form defined on $V \times V$: for $u, v \in V$ $a(u, v) \in \mathbf{C}$ and

$$a(u_1 + u_2, v) = a(u_1, v) + a(u_2, v),$$
$$a(u, v_1 + v_2) = a(u, v_1) + a(u, v_2)$$
$$a(\lambda u, v) = \lambda a(u, v), \quad a(u, \lambda v) = \bar{\lambda} a(u, v).$$

We assume that there exist positive constants C_1, c_0 and a real number k such that

$$|a(u, v)| \le C_1 \|u\| \|v\|, \quad u, v \in V, \tag{8.66}$$
$$\operatorname{Re} a(u, u) \ge c_0 \|u\|^2 - k|u|^2, \quad u \in V. \tag{8.67}$$

The inequality (8.67) is called Gårding's inequality. In view of (8.66) for each fixed $u \in V$ the mapping $V \ni v \mapsto a(u,v) \in \mathbf{C}$ is continuous and conjugate linear. Therefore it defines an element $\tilde{\Lambda} u \in V^*$ such that

$$(\tilde{\Lambda} u, v) = a(u, v), \quad v \in V. \tag{8.68}$$

Since

$$|(\tilde{\Lambda} u, v)| = |a(u, v)| \leq C_1 \|u\| \|v\|,$$

we have $\|\tilde{\Lambda} u\|_* \leq C_1 \|u\|$. Therefore $\tilde{\Lambda} \in \mathcal{L}(V, V^*)$ and $\|\tilde{\Lambda}\| \leq C_1$. Let Λ be the realization of $\tilde{\Lambda}$ in H:

$$D(\Lambda) = \{u \in V; \tilde{\Lambda} u \in H\}, \quad \Lambda u = \tilde{\Lambda} u \quad \text{for} \quad u \in D(\Lambda). \tag{8.69}$$

Then it is well known that $-\Lambda$ generates an analytic C_0-semigroup in H. Such an operator is called *regularly accretive* (T. Kato [85]). It is shown in [149] that $-\tilde{\Lambda}$ generates an analytic C_0-semigroup in V^*. For the sake of convenience we reproduce the proof here.

Lemma 8. 2 (Lax-Milgram's theorem) *Let X be a Hilbert space with in-nerproduct $(\cdot, \cdot)$ and norm $\|\cdot\|$. Let $B[u, v]$ be a sesquilinear form on $X \times X$, and suppose that there exist positive constants C, c such that*

$$|B[u, v]| \leq C \|u\| \|v\|, \quad u, v \in X, \tag{8.70}$$
$$|B[u, u]| \geq c \|u\|^2, \quad u \in X. \tag{8.71}$$

Then for any $F \in X^*$ there exists an element u of X such that for any $v \in X$ we have $F(v) = B[u, v]$.

Proof. Let $u \in X$ be fixed. Then the functional $X \ni v \mapsto B[u, v] \in \mathbf{C}$ is conjugate linear and continuous. Therefore there exists an element Su of X such that $B[u, v] = (Su, v)$ for any $v \in X$. S is a linear mapping from X to X and by (8.70)

$$|(Su, v)| = |B[u, v]| \leq C \|u\| \|v\|$$

which implies $\|S\| \leq C$. In view of (8.71)

$$c \|u\|^2 \leq |B[u, u]| = |(Su, u)| \leq \|Su\| \|u\|,$$

from which it follows that $c \|u\| \leq \|Su\|$, and hence S has a continuous inverse. Therefore $R(S)$ is closed. If v is orthogonal to $R(S)$, then for any $u \in X$ we have $(Su, v) = 0$. Choosing $u = v$ we have

$$c \|v\|^2 \leq |B[v, v]| = |(Sv, v)| = 0,$$

which implies $v = 0$. Consequently we have shown that S is an isomorphism from X onto X. Suppose $F \in X^*$. Then, in view of Riesz's theorem there exists an element w of X such that $F(v) = (w, v)$ for any $v \in X$. If u is an element of X such that $w = Su$, then

$$F(v) = (Su, v) = B[u, v], \quad v \in X.$$

If there exists another element $u' \in X$ such that $F(v) = B[u', v]$ for any $v \in X$, then $B[u - u', v] = 0$ for any $v \in X$. Hence

$$c\|u - u'\|^2 \leq B[u - u', u - u'] = 0,$$

which implies $u = u'$.

Lemma 8.3 V *is dense in* V^*.

Proof. It suffices to show that if $l \in V^{**}$ and $l(v) = 0$ for any $v \in V$, then $l = 0$. Since V is reflexive, this reduces to proving that if $u \in V$ and $(u, v) = 0$ for any $v \in V$, then $u = 0$. This follows from the denseness of V in H.

Theorem 8.5 *If* $\mathrm{Re}\lambda \leq -k$, *then* $\tilde{\Lambda} - \lambda$ *is an isomorphism from* V *to* V^*, *and for any* $f \in V^*$ *or* $\in H$ *we have*

$$|(\tilde{\Lambda} - \lambda)^{-1}f| \leq C_2|\lambda|^{-1}|f|, \tag{8.72}$$

$$|(\tilde{\Lambda} - \lambda)^{-1}f| \leq C_3|\lambda|^{-1/2}\|f\|_*, \tag{8.73}$$

$$\|(\tilde{\Lambda} - \lambda)^{-1}f\| \leq C_3|\lambda|^{-1/2}|f|, \tag{8.74}$$

$$\|(\tilde{\Lambda} - \lambda)^{-1}f\| \leq c_0^{-1}\|f\|_*, \tag{8.75}$$

$$\|(\tilde{\Lambda} - \lambda)^{-1}f\|_* \leq C_2|\lambda|^{-1}\|f\|_*, \tag{8.76}$$

where $C_2 = 1 + C_1/c_0, C_3 = \sqrt{C_1 + c_0}/c_0$. *Especially* $-\Lambda, -\tilde{\Lambda}$ *generate analytic* C_0-*semigroups in* H, V^* *respectively.*

Proof. Let $\mathrm{Re}\lambda \leq -k$. For $u, v \in V$ set

$$B[u, v] = ((\tilde{\Lambda} - \lambda)u, v) = a(u, v) - \lambda(u, v).$$

Then

$$|B[u, v]| \leq C_1\|u\|\|v\| + |\lambda||u||v| \leq \left(C_1 + |\lambda|C_0^2\right)\|u\|\|v\|,$$
$$\mathrm{Re}B[u, u] = \mathrm{Re}\,a(u, u) - \mathrm{Re}\lambda|u|^2$$
$$\geq c_0\|u\|^2 - (k + \mathrm{Re}\lambda)|u|^2 \geq c_0\|u\|^2. \tag{8.77}$$

Hence we can apply Lemma 8.2 to conclude that $\tilde{\Lambda} - \lambda$ is an isomorphism from V onto V^*.

Let $(\tilde{\Lambda} - \lambda)u = f$ for $f \in V^*$ or $\in H$. Then

$$a(u, v) - \lambda(u, v) = (f, v), \quad v \in V. \tag{8.78}$$

Letting $v = u$ in (8.78) we get

$$a(u, u) - \lambda|u|^2 = (f, u). \tag{8.79}$$

Therefore

$$\operatorname{Re} a(u, u) - \operatorname{Re}\lambda|u|^2 = \operatorname{Re}(f, u). \tag{8.80}$$

Combining (8.77) and (8.80)

$$c_0\|u\|^2 \leq \operatorname{Re}(f, u) \leq \|f\|_*\|u\|, \tag{8.81}$$

which implies

$$c_0\|u\| \leq \|f\|_*, \tag{8.82}$$

or (8.75). From (8.79) and (8.82) it follows that

$$|\lambda||u|^2 = |a(u, u) - (f, u)| \leq C_1\|u\|^2 + \|f\|_*\|u\|$$
$$\leq C_1\|f\|_*^2/c_0^2 + \|f\|_*^2/c_0 = C_3^2\|f\|_*^2,$$

which implies (8.73). Again from (8.77) and (8.80)

$$c_0\|u\|^2 \leq \operatorname{Re}(f, u) \leq |f||u|. \tag{8.83}$$

Combining (8.79) and (8.83) we get

$$|\lambda||u|^2 = |a(u, u) - (f, u)| \leq C_1\|u\|^2 + |f||u| \leq C_1|f||u|/c_0 + |f||u| = C_2|f||u|,$$

which implies

$$|\lambda||u| \leq C_2|f|, \tag{8.84}$$

or (8.72). From (8.83) and (8.84) it follows that

$$\|u\|^2 \leq |f||u|/c_0 \leq C_2|\lambda|^{-1}|f|^2/c_0 = C_3^2|\lambda|^{-1}|f|^2,$$

which implies

$$\|u\| \leq C_3|\lambda|^{-1/2}|f|$$

or (8.74). From (8.78) and (8.82)

$$|\lambda||(u, v)| \leq |a(u, v)| + |(f, v)| \leq C_1\|u\|\|v\| + \|f\|_*\|v\|$$
$$\leq C_1\|f\|_*\|v\|/c_0 + \|f\|_*\|v\| = C_2\|f\|_*\|v\|$$

for any $v \in V$. Therefore we conclude

$$|\lambda|\|u\|_* \leq C_2\|f\|_*$$

or (8.76).

8.7 Semigroup Approach Revised and Control Problem

We study a semigroup theoretical treatment of (8.1),(8.2) on different spaces from those in the preceding sections. These spaces are more useful in considering the following control problem for the equation (8.1):

$$\frac{d}{dt}u(t) = A_0 u(t) + A_1 u(t-h) + \int_{-h}^{0} a(s)A_2 u(t+s)ds + \Phi_0 w(t),$$

$$t \in [0,T], \quad (8.85)$$

$$u(0) = g^0, \quad u(s) = g^1(s), \quad s \in [-h,0), \quad (8.86)$$

where Φ_0 is a controller. Following C. Bernier and A. Manitius [20], M. C. Delfour and A. Manitius [44],[45], M. C. Manitius [107], S. Nakagiri [112], etc. and using the result of G. Di Blasio, K. Kunisch and E. Sinestrari [51],[52] we rewrite the problem (8.55),(8.56) as an equation in the space $Z = (D(A_0), H)_{1/2,2} \times L^2(-h,0;D(A_0))$:

$$\frac{d}{dt}x(t) = Ax(t) + \Phi w(t), \quad 0 \le t \le T, \quad (8.87)$$

$$x(0) = g, \quad (8.88)$$

where A is the infinitesimal generator of the solution semigroup associated with the problem (8.1),(8.2) and $\Phi w = (\Phi_0 w, 0)$. When we discuss the controllability, observability, etc. we consider also the adjoint equation

$$dz(t)/dt = A^* z(t), \quad z(0) = \psi.$$

This is an equation in $Z^* = (D(A_0), H)_{1/2,2}^* \times L^2(-h,0;D(A_0)^*)$. However, $D(A_0)^*$ is not a space easily dealt with, namely $D(A_0)$ is too small a space to consider its adjoint space. Therefore in what follows we consider the equation (8.85) in the space V^* in case $-A_0$ is a regularly accretive operator associated with a sesquilinear form satisfying (8.66),(8.67). Then $Z = H \times L^2(-h,0;V)$, since the domain of A_0 in this situation coincides with V and $(V,V^*)_{1/2,2} = H$ by virtue of Proposition 8.5 below, and hence $Z^* = H \times L^2(-h,0;V^*)$.

We state our fundamental hypothesis.

Let Hilbert spaces H, V and a sesquilinear form $a(\cdot,\cdot)$ be as in section 8.6. We denote the operators $\Lambda, \tilde{\Lambda}$ of section 8.6 both by the same notation $-A_0$, and $D(A_0)$ denotes the domain of its realization in H:

$$(-A_0 u, v) = a(u,v) \quad (8.89)$$

for $u \in D(A_0)$ or $\in V$ and $v \in V$, and $D(A_0) = \{u \in V; A_0 u \in H\}$. In this notation A_0 generates an analytic semigroup both in H and V^*. We assume that

(R2') A_1, A_2 are bounded linear operators from V to V^*, and and their restrictions to $D(A_0)$ are bounded linear operators from $D(A_0)$ to H.
As for $a(\cdot)$ we assume (R3) of section 8.6. Φ_0 is a bounded linear operator from some Banach space U to H.

Let $A_i^*, i = 0, 1, 2$, be the adjoint operators of A_i:

$$(A_i^* u, v) = (u, A_i v), \quad i = 0, 1, 2, \tag{8.90}$$

where the right hand side is the pairing between $u \in V$ and $A_i v \in V^*$. It is well known that A_0^* is the operator associated with the sesquilinear form $a^*(u, v) = \overline{a(v, u)}$ adjoint to $a(u, v)$, which also satisfies (8.66), (8.67). The domain $D(A_0^*)$ of the realization of A_0^* in H coincides with the domain of the adjoint operator of the realization of A_0 in H. We assume

(R4) The restrictions of A_1^*, A_2^* to $D(A_0^*)$ are bounded linear operators from $D(A_0^*)$ to H.

Lemma 8. 4 *If $u \in L^2(0, T; V)$, $u' \in L^2(0, T; V^*)$, $T > 0$, then $u \in C([0, T]; H)$. If v is another function having the same property, then $(u(t), v(t))$ is absolutely continuous in $[0, T]$ and*

$$d(u(t), v(t))/dt = (u'(t), v(t)) + (u(t), v'(t)). \tag{8.91}$$

Proof. If we set $u(t) = u(-t)$ for $-a < t < 0$, $u(t) = u(2T - t)$ for $T < t < T + a$, where $0 < a < T$, then $u \in L^2(-a, T + a; V)$ and $u' \in L^2(-a, T+a; V^*)$. Let ψ be a continuously differentiable real valued function such that $\psi(t) = 1$ for $t \in [0, T]$ and $\psi(t) = 0$ for t in some neighborhoods of $-a$ and $T + a$, and set $w(t) = \psi(t)u(t)$. Let ρ_n be a mollifier, and set

$$w_n(t) = \int_{-a}^{T+a} \rho_n(t - s)w(s)ds.$$

Then as $n \to \infty$ $w_n \to w$ in $L^2(-a, T + a; V)$ and $w_n' \to w'$ in $L^2(-a, T + a; V^*)$. Noting that $w_n \in C^1([-a, T + a]; V)$, we obtain

$$|w_n(t) - w_m(t)|^2 = \int_{-a}^{t} \frac{d}{ds}|w_n(s) - w_m(s)|^2 ds$$

$$= \int_{-a}^{t} 2\text{Re}\,(w_n'(s) - w_m'(s), w_n(s) - w_m(s))\,ds$$

$$\leq \int_{-a}^{T+a} \|w_n'(s) - w_m'(s)\|_*^2 ds + \int_{-a}^{T+a} \|w_n(s) - w_m(s)\|^2 ds.$$

Hence $\{w_n\}$ is a Cauchy sequence in $C([-a, T + a]; H)$. Therefore if we modify the values of u in some null set, we have $|w_n(t) - u(t)| \to 0$ uniformly in $[0, T]$. Thus we conclude $u \in C([0, T]; H)$. Next we prove (8.91). If $\phi \in C_0^\infty(0, T)$, then

$$\int_0^T (w_n(t), v(t))\phi'(t)dt$$
$$= \int_0^T \left(\frac{d}{dt}(\phi(t)w_n(t)), v(t)\right) dt - \int_0^T \phi(t)\, (w_n'(t), v(t))\, dt$$
$$= -\int_0^T \phi(t)\, (w_n(t), v'(t))\, dt - \int_0^T \phi(t)\, (w_n'(t), v(t))\, dt.$$

Letting $n \to \infty$

$$\int_0^T (u(t), v(t))\phi'(t)dt = -\int_0^T \phi(t)(u(t), v'(t))dt - \int_0^T \phi(t)(u'(t), v(t))dt.$$

This means that the distribution derivative of $(u(t), v(t))$ coincides with the right hand side of (8.91). Since this is integrable in $[0, T]$, the conclusion of the lemma follows.

Proposition 8.5 $(V, V^*)_{1/2,2} = H$.

Proof. Suppose $x \in H$. If we put $u(t) = e^{tA_0}x$, then

$$\frac{1}{2}\frac{d}{dt}|u(t)|^2 = \mathrm{Re}(u'(t), u(t))$$
$$= \mathrm{Re}(A_0 u(t), u(t)) = -\mathrm{Re}\, a(u(t), u(t)) \leq -c_0\|u\|^2 + k|u(t)|^2.$$

Integrating this differential inequality we obtain

$$\frac{1}{2}|u(t)|^2 + c_0 \int_0^t e^{2k(t-s)}\|u(s)\|^2 ds \leq \frac{1}{2}e^{2kt}|x|^2.$$

Therefore for any $T > 0$ $u \in L^2(0, T; V)$, $u' = A_0 u \in L^2(0, T; V^*)$. Hence $x = u(0) \in (V, V^*)_{1/2,2}$. Conversely suppose $x \in (V, V^*)_{1/2,2}$. If we set $u(t) = e^{tA_0}x$, then in view of Theorem 1.10 $u \in L^2(0, T; V)$ and $u' \in L^2(0, T; V^*)$. By Lemma 8.4 $u \in C([0, T]; H)$. Hence $x = u(0) \in H$.

As was stated in the beginning of this section we consider the problem (8.85), (8.86) in V^*. In view of Proposition 8.5 the associated solution semigroup $S(t)$ is a C_0-semigroup in $H \times L^2(-h, 0; V)$. In what follows in this chapter we denote this space by Z:

$$Z = H \times L^2(-h, 0; V). \tag{8.92}$$

The problem (8.85),(8.86) is rewritten as

$$\frac{d}{dt}x(t) = Ax(t) + \Phi w(t), \quad 0 \le t \le T, \tag{8.93}$$

$$x(0) = g, \tag{8.94}$$

where $x(t) = (u(t), u_t)$,

$$\Phi w = (\Phi_0 w, 0) \tag{8.95}$$

and $g = (g^0, g^1) \in Z$. The adjoint problem of (8.93),(8.94) is

$$\frac{d}{dt}z(t) = A^* z(t), \tag{8.96}$$

$$z(0) = \psi \in Z^*. \tag{8.97}$$

Since Z is reflexive, $S^*(t)$ is a C_0-semigroup in

$$Z^* = H \times L^2(-h, 0; V^*), \tag{8.98}$$

and the solution of (8.96),(8.97) is represented by

$$z(t) = S^*(t)\psi. \tag{8.99}$$

We consider also the following problem in addition to (8.85),(8.86) and (8.96),(8.97):

$$\frac{d}{dt}v(t) = A_0^* v(t) + A_1^* v(t-h) + \int_{-h}^{0} a(s) A_2^* v(t+s) ds, \quad t \in [0, T], \tag{8.100}$$

$$v(0) = \phi^0, \quad v(s) = \phi^1(s) \quad s \in [-h, 0). \tag{8.101}$$

The associated solution semigroup is denoted by $S_T(t)$, and its infinitesimal generator by A_T. In view of Theorem 8.4 the infinitesimal generator A of $S(t)$ is characterized by

$$D(A) = \{g = (g^0, g^1) \in Z; g^1 \in W^{1,2}(-h, 0; V),$$
$$g^0 = g^1(0), A_0 g^0 + A_1 g^1(-h) + L g^1 \in H\}, \tag{8.102}$$

$$Ag = \left(A_0 g^0 + A_1 g^1(-h) + L g^1, g^{1'}\right), \tag{8.103}$$

and A_T by

$$D(A_T) = \{g = (g^0, g^1) \in Z; g^1 \in W^{1,2}(-h, 0; V),$$
$$g^0 = g^1(0), A_0^* g^0 + A_1^* g^1(-h) + L^* g^1 \in H\}, \tag{8.104}$$

$$A_T g = \left(A_0^* g^0 + A_1^* g^1(-h) + L^* g^1, g^{1'}\right), \tag{8.105}$$

where

$$Lg^1 = \int_{-h}^0 a(s)A_2 g^1(s)ds, \quad L^*g^1 = \int_{-h}^0 a(s)A_2^* g^1(s)ds.$$

For $\lambda \in \mathbf{C}$ set

$$\Delta(\lambda) = \lambda - A_0 - e^{-\lambda h}A_1 - \int_{-h}^0 e^{\lambda s}a(s)A_2 ds, \qquad (8.106)$$

$$\Delta_T(\lambda) = \lambda - A_0^* - e^{-\lambda h}A_1^* - \int_{-h}^0 e^{\lambda h}a(s)A_2^* ds. \qquad (8.107)$$

Then $\Delta(\lambda), \Delta_T(\lambda)$ are bounded linear operators from V to V^*, and $\Delta(\lambda)^* = \Delta(\bar{\lambda})$. As is easily seen if $\mathrm{Re}\lambda$ is sufficiently large, then $\Delta(\lambda), \Delta_T(\lambda)$ have bounded inverses $\Delta(\lambda)^{-1}, \Delta_T(\lambda)^{-1}$, which are called *retarded resolvents* of A and A_T respectively.

Lemma 8.5 *Suppose* $g = (g^0, g^1)$, $f = (f^0, f^1) \in Z$. *Then*

$$g \in D(A) \quad and \quad (\lambda - A)g = f \qquad (8.108)$$

if and only if

$$g^0 \in V, \qquad (8.109)$$

$$\Delta(\lambda)g^0 = f^0 + \int_{-h}^0 e^{-\lambda(h+\tau)}A_1 f^1(\tau)d\tau$$

$$+ \int_{-h}^0 a(s)\int_s^0 e^{\lambda(s-\tau)}A_2 f^1(\tau)d\tau ds, \qquad (8.110)$$

$$g^1(s) = e^{\lambda s}g^0 + \int_s^0 e^{\lambda(s-\tau)}f^1(\tau)d\tau. \qquad (8.111)$$

Consequently if $\mathrm{Re}\lambda$ *is sufficiently large, then* $\lambda \in \rho(A) \cap \rho(A_T)$.

Proof. First assume that (8.108) holds. Then in view of (8.102),(8.103)

$$\lambda g^0 - A_0 g^0 - A_1 g^1(-h) - Lg^1 = f^0, \qquad (8.112)$$

$$\lambda g^1 - g^{1'} = f^1, \quad g^1(0) = g^0 \in V. \qquad (8.113)$$

Solving the initial value problem (8.113) we obtain (8.111). Substituting this in (8.112) we conclude (8.110).

Conversely suppose that (8.109),(8.110),(8.111) hold. From (8.111) we get

$$g^{1'} = \lambda g^1 - f^1 \in L^2(-h, 0; V), \quad g^1(0) = g^0 \in V. \qquad (8.114)$$

It follows from (8.110),(8.111) and the definition of $\Delta(\lambda)$

$$A_0 g^0 + A_1 g^1(-h) + \int_{-h}^0 a(s) A_2 g^1(s) ds$$

$$= A_0 g^0 + A_1 \left(e^{-\lambda h} g^0 + \int_{-h}^0 e^{-\lambda(h+\tau)} f^1(\tau) d\tau \right)$$

$$+ \int_{-h}^0 a(s) A_2 \left(e^{\lambda s} g^0 + \int_s^0 e^{\lambda(s-\tau)} f^1(\tau) d\tau \right) ds$$

$$= A_0 g^0 + e^{-\lambda h} A_1 g^0 + \int_{-h}^0 e^{-\lambda(h+\tau)} A_1 f^1(\tau) d\tau$$

$$+ \int_{-h}^0 e^{\lambda s} a(s) A_2 g^0 ds + \int_{-h}^0 a(s) \int_s^0 e^{\lambda(s-\tau)} A_2 f^1(\tau) d\tau ds$$

$$= A_0 g^0 + e^{-\lambda h} A_1 g^0 + \int_{-h}^0 e^{\lambda s} a(s) A_2 g^0 ds + \Delta(\lambda) g^0 - f^0$$

$$= \lambda g^0 - f^0 \in H.$$

This and (8.114) imply that $g \in D(A)$ and

$$(\lambda - A)g = \lambda g - (\lambda g^0 - f^0, \lambda g^1 - f^1) = f.$$

8.8 Structural Operator

Following M. C. Delfour and A. Manitius [44],[45] we define the *structural operator* associated with the problem (8.85),(8.86) as follows: for $f = (f^0, f^1) \in Z$

$$Ff = \left(f^0, A_1 f^1(-h - \cdot) + \int_{-h}^{\cdot} a(\tau) A_2 f^1(\tau - \cdot) d\tau \right). \tag{8.115}$$

We denote the first and second components of Ff by $[Ff]^0$ and $[Ff]^1$ respectively:

$$[Ff]^0 = f^0, \tag{8.116}$$

$$[Ff]^1(s) = A_1 f^1(-h - s) + \int_{-h}^s a(\tau) A_2 f^1(\tau - s) d\tau. \tag{8.117}$$

It is clear that $F \in \mathcal{L}(Z, Z^*)$. It is easy to see that the adjoint operator F^* of F is given by

$$F^* f = \left(f^0, A_1^* f^1(-h - \cdot) + \int_{-h}^{\cdot} a(\tau) A_2^* f^1(\tau - \cdot) d\tau \right), \tag{8.118}$$

namely F^* is the operator with A_1^*, A_2^* in place of A_1, A_2 in the definition of F. Clearly $F^* \in \mathcal{L}(Z, Z^*)$. We find it to convenient to define the following operators introduced by M. C. Delfour and A. Manitius [44],[45]: for $\lambda \in \mathbb{C}$

$$E_\lambda \in \mathcal{L}(V, Z), \quad [E_\lambda u]^0 = u, \quad [E_\lambda u]^1(s) = e^{\lambda s} u \text{ for } u \in V, \quad (8.119)$$

$$T_\lambda \in \mathcal{L}(Z, Z), \quad [T_\lambda g]^0 = 0, \quad [T_\lambda g]^1(s) = \int_s^0 e^{\lambda(s-\tau)} g^1(\tau) d\tau,$$

$$\text{for } g = (g^0, g^1) \in Z, \quad (8.120)$$

$$H_\lambda \in \mathcal{L}(Z^*, V^*), \quad H_\lambda f = f^0 + \int_{-h}^0 e^{\lambda s} f^1(s) ds$$

$$\text{for } f = (f^0, f^1) \in Z^*, \quad (8.121)$$

$$K_\lambda \in \mathcal{L}(Z^*, Z^*), \quad [K_\lambda f]^0 = 0, \quad [K_\lambda f]^1(s) = \int_{-h}^s e^{\lambda(\tau-s)} f^1(\tau) d\tau$$

$$\text{for } f = (f^0, f^1) \in Z^*. \quad (8.122)$$

Lemma 8. 6 *For $\lambda \in \mathbb{C}$*
(i) $FT_\lambda = K_\lambda F$, (ii) $F^* T_\lambda = K_\lambda F^*$, (iii) $E_\lambda^* = H_{\bar{\lambda}}$, (iv) $H_\lambda^* = E_{\bar{\lambda}}$,
(v) $T_\lambda^* = K_{\bar{\lambda}}$, (vi) $K_\lambda^* = T_{\bar{\lambda}}$.

Proof. (i) and (ii) We have for $g \in Z$

$$[FT_\lambda g]^0 = [T_\lambda g]^0 = 0, \quad [K_\lambda F g]^0 = 0,$$

$$[FT_\lambda g]^1(s) = A_1 [T_\lambda g]^1(-h-s) + \int_{-h}^s a(\tau) A_2 [T_\lambda g]^1(\tau - s) d\tau$$

$$= A_1 \int_{-h-s}^0 e^{\lambda(-h-s-\tau)} g^1(\tau) d\tau + \int_{-h}^s a(\tau) A_2 \int_{\tau-s}^0 e^{\lambda(\tau-s-\sigma)} g^1(\sigma) d\sigma d\tau,$$

$$[K_\lambda F g]^1(s) = \int_{-h}^s e^{\lambda(\tau-s)} [Fg]^1(\tau) d\tau$$

$$= \int_{-h}^s e^{\lambda(\tau-s)} \left[A_1 g^1(-h-\tau) + \int_{-h}^\tau a(\sigma) A_2 g^1(\sigma - \tau) d\sigma \right] d\tau$$

$$= A_1 \int_{-h}^s e^{\lambda(\tau-s)} g^1(-h-\tau) d\tau + \int_{-h}^s e^{\lambda(\tau-s)} \int_{-h}^\tau a(\sigma) A_2 g^1(\sigma - \tau) d\sigma d\tau$$

$$= A_1 \int_{-h-s}^0 e^{\lambda(-h-\tau-s)} g^1(\tau) d\tau + \int_{-h}^s a(\sigma) \int_\sigma^s e^{\lambda(\tau-s)} A_2 g^1(\sigma - \tau) d\tau d\sigma$$

$$= A_1 \int_{-h-s}^0 e^{\lambda(-h-\tau-s)} g^1(\tau) d\tau + \int_{-h}^s a(\sigma) A_2 \int_{\sigma-s}^0 e^{\lambda(\sigma-\tau-s)} g^1(\tau) d\tau d\sigma$$

$$= [FT_\lambda g]^1(s).$$

Thus (i) is established. Since F^* has the same property as F, and T_λ, K_λ are independent of A_i, $i = 0, 1, 2$, (ii) is a consequence of (i).

(iii) and (iv) We have for $f \in Z^*$ and $u \in V$

$$(E_\lambda^* f, u) = (f, E_\lambda u) = (f^0, u) + \int_{-h}^0 (f^1(s), e^{\lambda s} u) ds$$

$$= \left(f^0 + \int_{-h}^0 e^{\bar{\lambda} s} f^1(s) ds, u \right) = (H_{\bar{\lambda}} f, u),$$

which implies (iii). Taking adjoints we obtain (iv).
(v) and (vi) We have for $f \in Z^*$ and $g \in Z$

$$(T_\lambda^* f, g) = (f, T_\lambda g) = \int_{-h}^0 \left(f^1(s), \int_s^0 e^{\lambda(s-\tau)} g^1(\tau) d\tau \right) ds$$

$$= \int_{-h}^0 \left(\int_{-h}^\tau e^{\bar{\lambda}(s-\tau)} f^1(s) ds, g^1(\tau) \right) d\tau$$

$$= \int_{-h}^0 \left([K_{\bar{\lambda}} f]^1(\tau), g^1(\tau) \right) d\tau = (K_{\bar{\lambda}} f, g),$$

which shows that (v) holds. Taking adjoints we get (vi).

The following theorem was proved by M. C. Delfour and A. Manitius [44] for finite dimensional equations, and by S. Nakagiri [112] and J.-M. Jeong, S. Nakagiri and H. Tanabe [82] for equations in infinite dimensional spaces.

Theorem 8. 6 (i) *For $t \geq 0$*

$$FS(t) = S_T(t)^* F, \quad S(t)^* F^* = F^* S_T(t). \tag{8.123}$$

(ii) *If $\mathrm{Re}\lambda$ is sufficiently large,*

$$F(\lambda - A)^{-1} = (\lambda - A_T^*)^{-1} F, \quad (\lambda - A^*)^{-1} F^* = F^* (\lambda - A_T)^{-1}. \tag{8.124}$$

Remark 8. 1 The second statement of (i) is the relation which connects two kinds of adjoints of $S(t)$.

Proof of Theorem 8.6. Let $f = (f^0, f^1)$ be an arbitrary elements of Z. Let $\mathrm{Re}\lambda$ be so large that both $\Delta(\lambda)$, $\Delta_T(\bar{\lambda})$ have bounded inverses. Then in view of Lemma 8.5 there exists an element $g = (g^0, g^1)$ of $D(A)$ such that $(\lambda - A)g = f$. By virtue of Lemma 8.5

$$H_\lambda F f = f^0 + \int_{-h}^0 e^{\lambda s} \left[A_1 f^1(-h-s) + \int_{-h}^s a(\tau) A_2 f^1(\tau-s) d\tau \right] ds$$

$$= f^0 + \int_{-h}^0 e^{\lambda s} A_1 f^1(-h-s) ds + \int_{-h}^0 e^{\lambda s} \int_{-h}^s a(\tau) A_2 f^1(\tau-s) d\tau ds$$

$$= f^0 + \int_{-h}^0 e^{-\lambda(h+\tau)} A_1 f^1(\tau) d\tau + \int_{-h}^0 a(\tau) \int_\tau^0 e^{\lambda s} A_2 f^1(\tau - s) ds d\tau$$

$$= f^0 + \int_{-h}^0 e^{-\lambda(h+\tau)} A_1 f^1(\tau) d\tau + \int_{-h}^0 a(\tau) \int_\tau^0 e^{\lambda(\tau-s)} A_2 f^1(s) ds d\tau$$

$$= \Delta(\lambda) g^0.$$

Consequently we have
$$\Delta(\lambda)^{-1} H_\lambda F f = g^0.$$

Again by Lemma 8.5

$$E_\lambda \Delta(\lambda)^{-1} H_\lambda F f + T_\lambda f = (g^0, e^{\lambda \cdot} g^0) + \left(0, \int_\cdot^0 e^{\lambda(\cdot - \tau)} f^1(\tau) d\tau \right)$$

$$= \left(g^0, e^{\lambda \cdot} g^0 + \int_\cdot^0 e^{\lambda(\cdot - \tau)} f^1(\tau) d\tau \right) = (g^0, g^1) = g,$$

and so
$$(\lambda - A)^{-1} = E_\lambda \Delta(\lambda)^{-1} H_\lambda F + T_\lambda. \tag{8.125}$$

Analogously
$$(\bar\lambda - A_T)^{-1} = E_{\bar\lambda} \Delta_T(\bar\lambda)^{-1} H_{\bar\lambda} F^* + T_{\bar\lambda}. \tag{8.126}$$

Taking adjoint of (8.126) and using Lemma 8.6
$$(\lambda - A_T^*)^{-1} = F E_\lambda \Delta(\lambda)^{-1} H_\lambda + K_\lambda. \tag{8.127}$$

From (8.125),(8.127) and Lemma 8.6 it follows that

$$F(\lambda - A)^{-1} = F E_\lambda \Delta(\lambda)^{-1} H_\lambda F + F T_\lambda$$
$$= F E_\lambda \Delta(\lambda)^{-1} H_\lambda F + K_\lambda F = (\lambda - A_T^*)^{-1} F. \tag{8.128}$$

Hence the first equality of (8.124) is established. Taking the adjoint and replacing $\bar\lambda$ by λ we obtain the second one. From (8.128) it follows that if $f \in D(A)$, then

$$Ff = F(\lambda - A)^{-1}(\lambda - A)f = (\lambda - A_T^*)^{-1} F(\lambda - A)f \in D(A_T^*)$$

and
$$(\lambda - A_T^*)Ff = F(\lambda - A)f,$$

and so
$$A_T^* F f = F A f. \tag{8.129}$$

If we put $u(t) = F S(t) f$ for $f \in D(A)$, then applying (8.129) to $S(t) f \in D(A)$ we get

$$u'(t) = F A S(t) f = A_T^* F S(t) f = A_T^* u(t), \quad u(0) = F f,$$

which implies
$$u(t) = S_T^*(t)Ff.$$

Hence we see that
$$FS(t)f = S_T^*(t)Ff \tag{8.130}$$

holds if $f \in D(A)$. Since $FS(t)$ and $S_T^*(t)F$ are both bounded linear operators from Z to Z^* we see that (8.130) holds for any $f \in Z$. Thus the first part of (8.123) is established. The second part follows by taking adjoint.

From the theorem or its proof (cf. (8.129)) we get

Corollary 8. 1 *If $f \in D(A)$, then $Ff \in D(A_T^*)$ and $A_T^*Ff = FAf$, and if $f \in D(A_T)$, then $F^*f \in D(A^*)$ and $A^*F^*f = F^*A_Tf$.*

Definition 8. 2 $\rho(\Delta) = \{\lambda; \Delta(\lambda)$ is an isomorphism from V onto $V^*\}$, $\sigma(\Delta) = \mathbf{C} \setminus \rho(\Delta)$. Replacing $\Delta(\lambda)$ by $\Delta_T(\lambda)$ the sets $\rho(\Delta_T)$ and $\sigma(\Delta_T)$ are defined.

The following corollary was first proved by M. C. Delfour and A. Manitius [45: Proposition 4.2 (ii)]

Corollary 8. 2 (i) *If $\lambda \in \rho(\Delta)$, then $\lambda \in \rho(A)$ and*
$$(\lambda - A)^{-1} = E_\lambda \Delta(\lambda)^{-1}H_\lambda F + T_\lambda.$$

(ii) *If $\lambda \in \rho(\Delta_T)$, then $\lambda \in \rho(A_T)$ and*
$$(\lambda - A_T)^{-1} = E_\lambda \Delta_T(\lambda)^{-1}H_\lambda F^* + T_\lambda.$$

Proof. See (8.125) and (8.126) in the proof of Theorem 8.6.

8.9 Controllability and Observability

In this section we study the controllability problem for (8.93),(8.94). For the sake of simplicity we consider the case $g = 0$:

$$(S) \quad \begin{cases} \dfrac{d}{dt}x(t) = Ax(t) + \Phi w(t) & t \geq 0, \\ x(0) = 0. \end{cases}$$

Definition 8. 3 The system (S) is said to be *approximately controllable* if
$$Cl\left\{ \int_0^t S(t - \tau)\Phi w(\tau)d\tau; \; w \in L^2(0, t; U), \; t > 0 \right\} = Z,$$

where Cl means the closure in Z.

The observability of the system

$$(S^T) \quad \begin{cases} \dfrac{d}{dt}y(t) = A_T y(t) & t \geq 0, \\ y(0) = \phi \in Z \end{cases}$$

is defined as follows.

Definition 8. 4 The system (S^T) is said to be *observable* if $\Phi_0^*[S_T(t)\phi]^0 \equiv 0$ implies $\phi = 0$.

The observability means that $\Phi_0^*[S_T(t)\phi_1]^0 \equiv \Phi_0^*[S_T(t)\phi_2]^0$ implies $\phi_1 = \phi_2$, i.e. the initial value ϕ is observable through the measured values $\Phi_0^* v(t)$, where $v(t)$ is the solution of (8.100),(8.101).

Theorem 8. 7 *Suppose F is an isomorphism from Z to Z^*. Then (S) is approximately controllable if and only if (S^T) is observable.*

Proof. (S) is approximately controllable if and only if

$$f \in Z^*, \ \left(f, \int_0^t S(t-\tau)\Phi w(\tau)d\tau \right) = 0 \text{ for any } w \in L^2(0,t;U), \ t > 0$$

implies $f = 0$, or

$$\Phi^* S(t)^* f \equiv 0 \quad \text{implies} \quad f = 0. \tag{8.131}$$

Since F^* is an isomorphism from Z to Z^* under our assumption, (8.131) is equivalent to saying that

$$\phi \in Z, \quad \Phi^* S(t)^* F^* \phi \equiv 0 \quad \text{implies} \quad \phi = 0. \tag{8.132}$$

By Theroem 8.6 this holds if and only if

$$\phi \in Z, \quad \Phi^* F^* S_T(t)\phi \equiv 0 \quad \text{implies} \quad \phi = 0. \tag{8.133}$$

As is easily seen

$$\Phi^* F^* S_T(t)\phi = \Phi_0^* [F^* S_T(t)\phi]^0 = \Phi_0^*[S_T(t)\phi]^0.$$

Hence (8.133) is equivalent to the observability of (S^T).

A sufficient condition in order that the assumption of Theorem 8.7 is satisfied is given by the following proposition.

Proposition 8. 6 *If A_1 is an isomorphism from V to V^*, then F is an isomorphism from Z to Z^*.*

Proof. The relation $Fg = f$ is rewritten as

$$g^0 = f^0,$$

$$A_1 g^1(-h - s) + \int_{-h}^{s} a(\tau) A_2 g^1(\tau - s) d\tau = f^1(s). \qquad (8.134)$$

We want to solve (8.134) for a given $f^1 \in L^2(-h, 0; V^*)$. Letting A_1^{-1} operate on both sides and making a suitable change of variable we see that (8.134) is equivalent to

$$g^1(-s) + \int_0^s b(s - \tau) A_1^{-1} A_2 g^1(-\tau) d\tau = A_1^{-1} f^1(s - h), \quad s \in [0, h],$$

where $b(s) = a(s - h), s \in [0, h]$. Hence the problem is reduced to solving the following integral equation

$$\phi(s) + \int_0^s b(s - \tau) A_1^{-1} A_2 \phi(\tau) d\tau = \psi(s), \quad s \in [0, h], \qquad (8.135)$$

where $\psi \in L^2(0, h; V)$. Let $a * b$ denote the convolution

$$(a * b)(s) = \int_0^s a(s - \sigma) b(\sigma) d\sigma.$$

The resolvent kernel of (8.135) is the function R with values in $\mathcal{L}(V, V)$ satisfying

$$R + b A_1^{-1} A_2 + R * b A_1^{-1} A_2 = R + b A_1^{-1} A_2 + b A_1^{-1} A_2 * R = 0,$$

and is constructed by successive approximation:

$$R = \sum_{n=1}^{\infty} (-1)^n b_n \left(A_1^{-1} A_2 \right)^n,$$

where $b_n = b * \cdots * b$ is the convolution of n of b. Since $b \in L^2(0, h)$, b_2 is bounded: ess $\sup |b_2(s)| = \|b_2\|_\infty < \infty$. By induction we can show that

$$|b_{2n}(s)| \leq \|b_2\|_\infty^n \frac{s^{n-1}}{(n-1)!},$$

$$|b_{2n+1}(s)| \leq \|b_2\|_\infty^n \int_0^s \frac{(s - \sigma)^{n-1}}{(n-1)!} |b(\sigma)| d\sigma,$$

for $n = 1, 2, \ldots$. Therefore $R \in L^2(0, h; \mathcal{L}(V, V))$. The unique solution of (8.135) is given by $\phi = \psi + R * \psi$.

The results of this section are due to A. Manitius [109] and S. Nakagiri and M. Yamamoto [115].

8.10 Characterization of $\text{Ker}(\lambda - A)^l$

Supposing that $\lambda \in \sigma_p(A) = $ the set of point spectra of A, we characterize

$$\text{Ker}(\lambda - A)^l = \{\phi \in Z; (\lambda - A)^l \phi = 0\}.$$

Under this hypothesis $\Delta(\lambda)$ is not invertible by Lemma 8.5.

Theorem 8.8 *For* $l = 1, 2, \cdots$

$$\text{Ker}(\lambda - A)^l = \left\{ \left(\phi_0^0, e^{\lambda \cdot} \sum_{i=0}^{l-1} \frac{(-\cdot)^i}{i!} \phi_i^0 \right); \right.$$

$$\left. \sum_{i=j-1}^{l-1} \frac{(-1)^{i-j}}{(i-j+1)!} \Delta^{(i-j+1)}(\lambda) \phi_i^0 = 0, j = 1, \ldots, l \right\}, \quad (8.136)$$

where $\Delta^{(i-j+1)}(\lambda) = (d/d\lambda)^{(i-j+1)} \Delta(\lambda)$. *The same result holds with* A *and* Δ *replaced by* A_T *and* Δ_T *respectively.*

Proof. In case $l = 1$ the assertion is

$$\text{Ker}(\lambda - A) = \left\{ (\phi^0, e^{\lambda \cdot} \phi^0); \Delta(\lambda) \phi^0 = 0 \right\}. \quad (8.137)$$

This is a direct consequence of Lemma 8.5.

We denote the set of the right hand side of (8.136) by K_l. We first show that $K_l \subset \text{Ker}(\lambda - A)^l$. Let

$$\phi = \left(\phi_0^0, e^{\lambda \cdot} \sum_{i=0}^{l-1} \frac{(-\cdot)^i}{i!} \phi_i^0 \right) \in K_l,$$

$$\sum_{i=j-1}^{l-1} \frac{(-1)^{i-j}}{(i-j+1)!} \Delta^{(i-j+1)}(\lambda) \phi_i^0 = 0, \quad j = 1, \ldots, l. \quad (8.138)$$

Then

$$A\phi = \left(\psi^0, \lambda e^{\lambda \cdot} \sum_{i=0}^{l-1} \frac{(-\cdot)^i}{i!} \phi_i^0 - e^{\lambda \cdot} \sum_{i=1}^{l-1} \frac{(-\cdot)^{i-1}}{(i-1)!} \phi_i^0 \right),$$

where

$$\psi^0 = A_0 \phi_0^0 + A_1 e^{-\lambda h} \sum_{i=0}^{l-1} \frac{h^i \phi_i^0}{i!} + \int_{-h}^{0} a(s) e^{\lambda s} \sum_{i=0}^{l-1} \frac{(-s)^i}{i!} A_2 \phi_i^0 ds$$

$$= A_0 \phi_0^0 + \sum_{i=0}^{l-1} \frac{1}{i!} \left(h^i e^{-\lambda h} A_1 + \int_{-h}^{0} (-s)^i e^{\lambda s} a(s) ds A_2 \right) \phi_i^0$$

$$= A_0\phi_0^0 + \left(e^{-\lambda h}A_1 + \int_{-h}^0 e^{\lambda s}a(s)ds A_2\right)\phi_0^0$$

$$+ \sum_{i=1}^{l-1}\frac{1}{i!}\left(-\delta_{i,1} - (-1)^i\Delta^{(i)}(\lambda)\right)\phi_i^0$$

$$= (\lambda - \Delta(\lambda))\phi_0^0 - \phi_1^0 - \sum_{i=1}^{l-1}\frac{(-1)^i}{i!}\Delta^{(i)}(\lambda)\phi_i^0, \tag{8.139}$$

where we used that for $i = 1, 2, \ldots$

$$(-1)^i\Delta^{(i)}(\lambda) = -\delta_{i,1} - h^i e^{-\lambda h}A_1 - \int_{-h}^0 (-s)^i e^{\lambda s}a(s)ds A_2. \tag{8.140}$$

Using (8.138) for $j = 1$ we get from (8.139)

$$\psi^0 = \lambda\phi_0^0 - \phi_1^0.$$

Hence

$$(\lambda - A)\phi = \left(\phi_1^0, e^{\lambda \cdot}\sum_{i=0}^{l-2}\frac{(-1)^i}{i!}\phi_{i+1}^0\right).$$

The condition (8.138) for $j = 2, \ldots, l$ is rewritten as

$$\sum_{i=j-1}^{l-2}\frac{(-1)^{i-j}}{(i-j+1)!}\Delta^{(i-j+1)}(\lambda)\phi_{i+1}^0 = 0, \quad j = 1, \ldots, l-1.$$

Consequently we see that $(\lambda - A)\phi \in K_{l-1}$. Repeating this we obtain $(\lambda - A)^{l-1}\phi \in K_1 = \mathrm{Ker}(\lambda - A)$, or $\phi \in \mathrm{Ker}(\lambda - A)^l$.

Next supposing that $\mathrm{Ker}(\lambda - A)^l \subset K_l$ holds, we are going to show that the same inclusion relation holds for $l+1$ in place of l. Let $\phi \in \mathrm{Ker}(\lambda-A)^{l+1}$. Then $(\lambda - A)\phi \in \mathrm{Ker}(\lambda - A)^l$. Hence

$$(\lambda - A)\phi = \left(\psi_0^0, e^{\lambda \cdot}\sum_{i=0}^{l-1}\frac{(-\cdot)^i}{i!}\psi_i^0\right), \tag{8.141}$$

$$\sum_{i=j-1}^{l-1}\frac{(-1)^{i-j}}{(i-j+1)!}\Delta^{(i-j+1)}(\lambda)\psi_i^0 = 0, \quad j = 1, \cdots, l. \tag{8.142}$$

In view of Lemma 8.5

$$\Delta(\lambda)\phi^0 = \psi_0^0 + \int_{-h}^0 e^{-\lambda(h+\tau)}A_1 e^{\lambda\tau}\sum_{i=0}^{l-1}\frac{(-\tau)^i}{i!}\psi_i^0 d\tau$$

$$+ \int_{-h}^{0} a(s) \int_{s}^{0} e^{\lambda(s-\tau)} A_2 e^{\lambda\tau} \sum_{i=0}^{l-1} \frac{(-\tau)^i}{i!} \psi_i^0 d\tau ds, \quad (8.143)$$

$$\phi^1(s) = e^{\lambda s} \phi^0 + \int_{s}^{0} e^{\lambda(s-\tau)} e^{\lambda\tau} \sum_{i=0}^{l-1} \frac{(-\tau)^i}{i!} \psi_i^0 d\tau. \quad (8.144)$$

From (8.143) it follows that

$$\Delta(\lambda)\phi^0 = \psi_0^0 + e^{-\lambda h} \sum_{i=0}^{l-1} \frac{h^{i+1}}{(i+1)!} A_1 \psi_i^0 + \sum_{i=0}^{l-1} \int_{-h}^{0} e^{\lambda s} a(s) \frac{(-s)^{i+1}}{(i+1)!} A_2 \psi_i^0 ds$$

$$= \psi_0^0 + \sum_{i=0}^{l-1} \frac{1}{(i+1)!} \left[h^{i+1} e^{-\lambda h} A_1 + \int_{-h}^{0} (-s)^{i+1} e^{\lambda s} a(s) ds A_2 \right] \psi_i^0$$

$$= \psi_0^0 + \sum_{i=0}^{l-1} \frac{1}{(i+1)!} \left[-\delta_{i+1,1} - (-1)^{i+1} \Delta^{(i+1)}(\lambda) \right] \psi_i^0$$

$$= -\sum_{i=0}^{l-1} \frac{(-1)^{i+1}}{(i+1)!} \Delta^{(i+1)}(\lambda) \psi_i^0, \quad (8.145)$$

where we again used (8.140), and from (8.144)

$$\phi^1(s) = e^{\lambda s} \phi^0 + e^{\lambda s} \sum_{i=0}^{l-1} \frac{(-s)^{i+1}}{(i+1)!} \psi_i^0. \quad (8.146)$$

If we set $\phi_0^0 = \phi^0, \phi_1^0 = \psi_0^0, \ldots, \phi_l^0 = \psi_{l-1}^0$, then (8.145),(8.146) imply

$$\sum_{i=0}^{l} \frac{(-1)^i}{i!} \Delta^{(i)}(\lambda) \phi_i^0 = 0, \quad (8.147)$$

$$\phi^1(s) = e^{\lambda s} \sum_{i=0}^{l} \frac{(-s)^i}{i!} \phi_i^0. \quad (8.148)$$

By (8.142) we have

$$\sum_{i=j-1}^{l} \frac{(-1)^{i-j}}{(i-j+1)!} \Delta^{(i-j+1)}(\lambda) \phi_i^0 = 0, \quad j = 2, \ldots, l+1. \quad (8.149)$$

Combining (8.147),(8.148),(8.149) we conclude $\phi \in K_{l+1}$ to complete the proof of the theorem.

8.11   A Special Case

We consider the special case where

$$A_1 = \gamma A_0, \quad A_2 = A_0, \tag{8.150}$$

γ being a real number. In this case

$$\Delta(\lambda) = \lambda - m(\lambda)A_0, \tag{8.151}$$

$$m(\lambda) = 1 + \gamma e^{-\lambda h} + \int_{-h}^{0} e^{\lambda s} a(s) ds. \tag{8.152}$$

As is easily seen $m(\lambda)$ is an entire function of λ and

$$m(\lambda) \to 1 \quad \text{as} \quad \text{Re}\lambda \to \infty. \tag{8.153}$$

In G. Di Blasio, K. Kunisch and E. Sinestrari [52] and S. Nakagiri and H. Tanabe [114] the relations between the spectrum of A and that of $\Delta(\cdot)$ are discussed in detail.

We prepare some results from the spectral theory of closed linear operators in a Banach space.

Definition 8. 5 Let T be a closed linear operator in some Banach space, and λ be an isolated point of the spectra $\sigma(T)$ of T. Then the *spectral projection* of T at λ is the operator P defined by

$$P = \frac{1}{2\pi i} \oint_{|\mu - \lambda| = \epsilon} (\mu - T)^{-1} d\mu,$$

where ϵ is a positive number such that λ is the only spectrum of T in the closed disk $\{\mu; |\mu - \lambda| \leq \epsilon\}$.

In what follows in this chapter we denote by $\text{Im}T$ the image (range) of the operator T.

The following lemma is rather well known (cf. T. Kato [86: p.180]).

Lemma 8. 7 *Let T be a closed linear operator in a Banach space X, and λ an isolated point of $\sigma(T)$. Then $(\mu - T)^{-1}$ has the Laurent expansion near λ:*

$$(\mu - T)^{-1} = \sum_{n=0}^{\infty} A_n (\mu - \lambda)^n + \sum_{n=1}^{\infty} B_n (\mu - \lambda)^{-n}. \tag{8.154}$$

A_n and B_n are bounded linear operators and

$$B_n = (\lambda - T)^{n-1} P, \quad n = 1, 2, \ldots, \tag{8.155}$$

where P is the spectral projection of T at λ. If moreover λ is a pole of $(\mu - T)^{-1}$ of order ν, then λ is an eigenvalue of T and

$$\mathrm{Ker}((\lambda - T)^{n-1}) \neq \mathrm{Ker}((\lambda - T)^n), \quad n = 1, 2, \ldots, \nu,$$
$$\mathrm{Ker}((\lambda - T)^n) = \mathrm{Ker}((\lambda - T)^\nu) = \mathrm{Im}P, \quad n = \nu + 1, \nu + 2, \ldots,$$
$$\mathrm{Im}((\lambda - T)^{n-1}) \neq \mathrm{Im}((\lambda - T)^n), \quad n = 1, 2, \ldots, \nu,$$
$$\mathrm{Im}((\lambda - T)^n) = \mathrm{Im}((\lambda - T)^\nu) = \mathrm{Im}(I - P), \quad n = \nu + 1, \nu + 2, \ldots.$$

Hence

$$X = \mathrm{Ker}((\lambda - T)^\nu) \oplus \mathrm{Im}((\lambda - T)^\nu). \qquad (8.156)$$

The first part of the lemma is proved by a direct calculation, and the second part is proved by using that λ is the only spectrum of T in the space $\mathrm{Im}P$ and the spectra of T in the space $\mathrm{Im}(I - P)$ coincides with $\sigma(T) \setminus \{\lambda\}$.

If the imbedding $V \subset H$ is compact, then by the Riesz-Schauder theory the spectra of A_0 consists only of discrete eigenvalues:

$$\sigma(A_0) = \{\mu_j; j = 1, 2, \ldots\}, \quad \mu_j \to \infty \quad \text{as} \quad j \to \infty. \qquad (8.157)$$

It is easily seen that the spectrum of A_0 considered as an operator in V^* coincides with that of its realization in H.

Theorem 8. 9 (J.-M. Jeong [80]) *Suppose that* (8.150) *holds for some real constant* γ, *and that the imbedding* $V \subset H$ *is compact. Then we have*

$$\sigma(A) = \sigma_e(A) \cup \sigma_p(A), \qquad (8.158)$$

where $\sigma_e(A) = \{\lambda; m(\lambda) = 0\}$ *and* $\sigma_p(A) = \{\lambda; m(\lambda) \neq 0, \lambda/m(\lambda) \in \sigma(A_0)\}$. *Each nonzero point of* $\sigma_e(A)$ *is not an eigenvalue of* A *and is a limit point of* $\sigma(A)$. $\sigma_p(A)$ *consists only of discrete eigenvalues.*

Suppose $m(0) = 0$. *Then* 0 *is an eigenvalue of* A *of infinite multiplicity. The point* 0 *is an isolate point of* $\sigma(A)$ *if it is a simple zero of* $m(\cdot)$, *and is a limit point of* $\sigma(A)$ *if it is a multiple zero of* $m(\cdot)$.

Proof. If $\lambda \in \rho(A)$, then in view of Lemma 8.5 the equation $(\lambda - m(\lambda)A_0)g^0 = f^0$ has a unique solution $g^0 \in V$ for any $f^0 \in H$. Hence $m(\lambda) \neq 0$ and $\lambda/m(\lambda) \in \rho(A_0)$. Conversely if λ satisfies this condition, then the equation $(\lambda - A)g = f$ has a unique solution g for any $f \in Z$ in view of Lemma 8.5. Therefore we have

$$\rho(A) = \{\lambda; m(\lambda) \neq 0, \lambda/m(\lambda) \in \rho(A_0)\}.$$

Suppose $\lambda_0 \neq 0$ is a zero of $m(\lambda)$ of order k. Then there exists a function $h(\lambda)$ which is holomorphic and does not vanish in some neighborhood of λ_0

such that $m(\lambda)/\lambda = (\lambda - \lambda_0)^k h(\lambda)^k$. Applying the inverse function theorem to $(\lambda - \lambda_0)h(\lambda)$ and noting $\mu_j \to \infty$ we see that for sufficiently large j there exists a complex number λ_j such that $(\lambda_j - \lambda_0)h(\lambda_j) = \mu_j^{-1/k}$ and $\lambda_j \to \lambda_0$. Then $\lambda_j/m(\lambda_j) = \mu_j \in \sigma(A_0)$. Hence λ_0 is a limit point of $\{\lambda_j\} \subset \sigma(A)$.

Next suppose that $\lambda_0 \in \sigma_p(A)$, i.e. $m(\lambda_0) \neq 0, \lambda_0/m(\lambda_0) \in \sigma(A_0)$. If there exists a sequence $\{\lambda_j\}$ such that $\lambda_0 \neq \lambda_j \in \sigma(A), m(\lambda_j) \neq 0$ and $\lambda_j \to \lambda_0$. Then $\lambda_j/m(\lambda_j) \to \lambda_0/m(\lambda_0), \lambda_j/m(\lambda_j) \in \sigma(A_0)$. Since $\sigma(A_0)$ consists only of isolated points we have $\lambda_j/m(\lambda_j) = \lambda_0/m(\lambda_0)$ for sufficiently large j. By virtue of the theorem of identity we have $\lambda/m(\lambda) \equiv \lambda_0/m(\lambda_0)$, which contradicts (8.153). Next we show that λ_0 is a pole of $(\lambda - A)^{-1}$. From the theorem of identity and (8.153) it follows that if $\lambda \neq \lambda_0$ lies in a sufficiently small neighborhood of λ_0, then $\lambda/m(\lambda) \in \rho(A_0)$, and hence $\lambda \in \rho(\Delta)$. By the Riesz-Schauder theory $\lambda_0/m(\lambda_0)$ is a pole of the resolvent of A_0 whose order we denote by ν. Therefore if we denote by k the order of λ_0 as a zero of $\lambda/m(\lambda) - \lambda_0/m(\lambda_0)$, then λ_0 is a pole of

$$\Delta(\lambda)^{-1} = \frac{1}{m(\lambda)}\left(\frac{\lambda}{m(\lambda)} - A_0\right)^{-1}$$

of order $k\nu$. Hence the result follows from Corollary 8.2 or (8.125) since all other functions in the right hand sides of (8.125) are entire functions of λ. It is easily seen that the elements $\phi_0^0, \phi_1^0, \cdots, \phi_{l-1}^0$ of the right hand side of (8.136) satisfy

$$\Delta(\lambda)\phi_{l-1}^0 = \Delta(\lambda)^2\phi_{l-2}^0 = \cdots = \Delta(\lambda)^l\phi_0^0 = 0.$$

Moreover $\mathrm{Ker}\,\Delta(\lambda_0)^\nu$ is finite dimensional. Therefore $\mathrm{Ker}(\lambda_0 - A)^{k\nu}$ is finite dimensional.

If $m(0) = 0$, then for any $g^0 \in V$ $g = (g^0, g^0)$ satisfies $Ag = 0$ in view of Lemma 8.5. Hence 0 is an eigenvalue of A of infinite multiplicity. Suppose 0 is a simple zero of $m(\lambda)$. Then there exists a function $h(\lambda)$ which is holomorphic and does not vanish in some neighborhood of 0 such that $m(\lambda) = \lambda h(\lambda)$. If there exists a sequence $0 \neq \lambda_j \in \sigma(A)$ such that $\lambda_j \to 0$, then for large j $m(\lambda_j) \neq 0, \lambda_j/m(\lambda_j) \in \sigma(A_0)$ and $\lambda_j/m(\lambda_j) = h(\lambda_j)^{-1} \to h(0)^{-1}$. Since $\sigma(A_0)$ consists only of isolated points we have $\lambda_j/m(\lambda_j) = 1/h(0)$ for large j. Hence $m(\lambda) \equiv \lambda h(0)$, which contradicts (8.153). Next suppose that 0 is a zero of $m(\lambda)$ of order $k > 1$. Then there exists a function $h(\lambda)$ with the same property as above such that $m(\lambda)/\lambda = \lambda^{k-1}h(\lambda)^{k-1}$. Applying the inverse function theorem to $\lambda h(\lambda)$ we find that for large j there exists a point λ_j such that $\lambda_j h(\lambda_j) = \mu_j^{-1/(k-1)}, \lambda_j \to 0$. Then $\lambda_j/m(\lambda_j) = \mu_j \in \sigma(A_0)$. Hencce $0 \neq \lambda_j \in \sigma(A)$. This means that 0 is a limit point of $\sigma(A)$.

8.12   Eigenmanifold Decomposition

The adjoint operator A^* of A is characterized by the following theorem, the proof of which is due to S. Nakagiri [112].

Theorem 8.10 *We have for the operator A^**

$$D(A^*) = \{f = (f^0, f^1); f^0 \in V, f^1 \in W^{1,2}(-h, 0; V^*),$$
$$A_0^* f^0 + f^1(0) \in H, f^1(-h) = A_1^* f^0\}, \qquad (8.159)$$
$$A^* f = (A_0^* f^0 + f^1(0), a(\cdot)A_2^* f^0 - f^{1'}) \text{ for } f \in D(A^*). \quad (8.160)$$

Proof. Suppose $f \in D(A^*)$ and $A^* f = \psi$. Then for any $g \in D(A)$ we have $(Ag, f) = (g, \psi)$, or

$$\left(A_0 g^0 + A_1 g^1(-h) + \int_{-h}^0 a(s) A_2 g^1(s) ds, f^0 \right)$$
$$+ \int_{-h}^0 (g^{1'}(s), f^1(s)) ds = (g^0, \psi^0) + \int_{-h}^0 (g^1(s), \psi^1(s)) ds. \quad (8.161)$$

Let ϕ be a real valued function such that $\phi \in W^{1,2}(-h, 0), \phi(-h) = 0, \phi(0) = 1, \int_{-h}^0 a(s)\phi(s)ds = 0$. Then for any $g^0 \in D(A_0)$ $(g^0, \phi(\cdot)g^0) \in D(A)$. Substituting this in (8.161) we get

$$(A_0 g^0, f^0) + \int_{-h}^0 \phi'(s)(g^0, f^1(s)) ds = (g^0, \psi^0) + \int_{-h}^0 \phi(s)(g^0, \psi^1(s)) ds,$$

or

$$(A_0 g^0, f^0) = \left(g^0, \psi^0 + \int_{-h}^0 \phi(s)\psi^1(s) ds - \int_{-h}^0 \phi'(s) f^1(s) ds \right).$$

Since the right hand side is continuous in g^0 in the topology of V, we find that $f^0 \in V$. Set

$$M(s) = \int_{-h}^s \psi^1(\tau) d\tau, \quad N(s) = A_1^* f^0 + \int_{-h}^s a(\tau) A_2^* f^0 d\tau.$$

Then $M, N \in W^{1,2}(-h, 0; V^*)$. Let $y \in W^{1,2}(-h, 0; D(A_0))$. Then

$$\int_{-h}^0 (y'(s), M(s)) ds = (y(0), M(0)) - \int_{-h}^0 (y(s), \psi^1(s)) ds, \qquad (8.162)$$
$$\int_{-h}^0 (y'(s), N(s)) ds$$

$$= (y(0), N(0)) - (y(-h), N(-h)) - \int_{-h}^{0} (y(s), a(s)A_2^* f^0) ds$$

$$= (y(0), N(0)) - (y(-h), A_1^* f^0)$$

$$- \left(\int_{-h}^{0} a(s)A_2 y(s) ds, f^0 \right). \tag{8.163}$$

Substituting $(y(0), y) \in D(A)$ in (8.161) we get

$$\left(A_0 y(0) + A_1 y(-h) + \int_{-h}^{0} a(s)A_2 y(s) ds, f^0 \right)$$

$$+ \int_{-h}^{0} (y'(s), f^1(s)) ds = (y(0), \psi^0) + \int_{-h}^{0} (y(s), \psi^1(s)) ds. \tag{8.164}$$

Combining (8.162),(8.163),(8.164) we obtain

$$(A_0 y(0) + A_1 y(-h), f^0) + (y(0), N(0)) - (y(-h), A_1^* f^0)$$

$$- \int_{-h}^{0} (y'(s), N(s)) ds + \int_{-h}^{0} (y'(s), f^1(s)) ds$$

$$= (y(0), \psi^0) + (y(0), M(0)) - \int_{-h}^{0} (y'(s), M(s)) ds,$$

and hence using $(A_1 y(-h), f^0) = (y(-h), A_1^* f^0)$

$$\int_{-h}^{0} (y'(s), f^1(s) - N(s) + M(s)) ds$$

$$= (y(0), \psi^0 - N(0) + M(0)) - (A_0 y(0), f^0). \tag{8.165}$$

From this it follows that

$$f^1(s) \equiv N(s) - M(s). \tag{8.166}$$

Therefore $f^1 \in W^{1,2}(-h, 0; V^*)$, $f^1(-h) = N(-h) - M(-h) = A_1^* f^0$ and $\psi^1 = a(\cdot)A_2^* f^0 - f^{1'}$. Letting $y(s) \equiv g^0 \in D(A_0)$ we get from (8.165)

$$(g^0, \psi^0 - N(0) + M(0)) = (A_0 g^0, f^0)$$

which implies $A_0^* f^0 = \psi^0 - N(0) + M(0)$. From this and (8.166) we conclude

$$A_0^* f^0 + f^1(0) = A_0^* f^0 + N(0) - M(0) = \psi^0 \in H.$$

Conversely if f belongs to the right hand side of (8.159) and ψ is equal to the right hand side of (8.160), then by a direct calculation we can easily verify that $(f, Ag) = (\psi, g)$ for any $g \in D(A)$.

Lemma 8.8 *Let $f, g \in Z^*$. Then*

$$f \in D(A^*) \quad and \quad (A^* - \lambda)f = g \tag{8.167}$$

if and only if

$$f^0 \in V \ and \ \Delta_T(\lambda)f^0 = -H_\lambda g, \ f = F^* E_\lambda f^0 - K_\lambda g. \tag{8.168}$$

The same result remains valid replacing $A^*, \Delta_T(\lambda), F^*$ by $A_T^*, \Delta(\lambda), F$ respectively.

Proof. Suppose (8.167) holds. Then by Theorem 8.10 $f^0 \in V$, $f^1 \in W^{1,2}(-h, 0; V^*)$ and

$$A_0^* f^0 + f^1(0) = \lambda f^0 + g^0, \tag{8.169}$$

$$a(s)A_2^* f^0 - f^{1\prime}(s) = \lambda f^1(s) + g^1(s), \tag{8.170}$$

$$f^1(-h) = A_1^* f^0. \tag{8.171}$$

We find from (8.170) that

$$e^{\lambda s} f^1(s) = e^{-\lambda h} f^1(-h) + \int_{-h}^s e^{\lambda \tau} a(\tau) A_2^* f^0 d\tau - \int_{-h}^s e^{\lambda \tau} g^1(\tau) d\tau. \tag{8.172}$$

This equality with $s = 0$ and (8.169),(8.171) imply

$$\int_{-h}^0 e^{\lambda \tau} a(\tau) A_2^* f^0 d\tau = \lambda f^0 + g^0 - A_0^* f^0 - e^{-\lambda h} A_1^* f^0 + \int_{-h}^0 e^{\lambda \tau} g^1(\tau) d\tau.$$

Substituting this in the right hand side of

$$\Delta_T(\lambda)f^0 = \lambda f^0 - A_0^* f^0 - e^{-\lambda h} A_1^* f^0 - \int_{-h}^0 e^{\lambda s} a(s) A_2^* f^0 ds$$

we obtain

$$\Delta_T(\lambda)f^0 = -g^0 - \int_{-h}^0 e^{\lambda \tau} g^1(\tau) d\tau = -H_\lambda g.$$

From (8.171),(8.172) we get

$$f^1(s) = e^{-\lambda(h+s)} A_1^* f^0 + \int_{-h}^s e^{\lambda(\tau-s)} a(\tau) A_2^* f^0 d\tau$$

$$- \int_{-h}^s e^{\lambda(\tau-s)} g^1(\tau) d\tau = [F^* E_\lambda f^0]^1(s) - [K_\lambda g]^1(s). \tag{8.173}$$

Clearly $[F^* E_\lambda f^0 - K_\lambda g]^0 = f^0$. This and (8.173) imply (8.168). The converse statement can be proved analogously.

We use the following notations:

$$\sigma_p(\Delta) = \{\lambda; \Delta(\lambda) \text{ is not invertible}\},$$
$$\sigma_p(\Delta_T) = \{\lambda; \Delta_T(\lambda) \text{ is not invertible}\}.$$

Corollary 8.3 (i) $\sigma_p(\Delta) = \sigma_p(A_T^*) = \sigma_p(A)$, (ii) $\sigma_p(\Delta_T) = \sigma_p(A^*) = \sigma_p(A_T)$.

Proof. The result follows from Lemmas 8.5 and 8.8.

In addition to Lemma 8.6 we need the following lemma (S. Nakagiri [112: Proposition 6.1]).

Lemma 8.9 *For* $\lambda \in \mathbf{C}$ *and* $k = 1, 2, \ldots$

$$FT_\lambda^k = K_\lambda^k F, \quad F^*T_\lambda^k = K_\lambda^k F^*, \tag{8.174}$$

$$T_\lambda^k E_\lambda = \frac{(-1)^k}{k!} E_\lambda^{(k)}, \tag{8.175}$$

$$H_\lambda FT_\lambda^k E_\lambda = \frac{(-1)^k}{(k+1)!} \Delta^{(k+1)}(\lambda), \tag{8.176}$$

$$H_\lambda F^*T_\lambda^k E_\lambda = \frac{(-1)^k}{(k+1)!} \Delta_T^{(k+1)}(\lambda). \tag{8.177}$$

Proof. The equalities (8.174) can be shown with the aid of Lemma 8.6 (i),(ii) inductively on k. (8.175) is established by a straightforward calculation. We show (8.176). Let $u \in V$. In view of (8.140)

$$\frac{(-1)^k}{(k+1)!} \Delta^{(k+1)}(\lambda)u = \delta_{k,0}u + \frac{h^{k+1}}{(k+1)!} e^{-\lambda h} A_1 u$$
$$+ \frac{1}{(k+1)!} \int_{-h}^0 (-s)^{k+1} e^{\lambda s} a(s) A_2 u\, ds. \tag{8.178}$$

On the other hand

$$H_\lambda FT_\lambda^k E_\lambda u = [FT_\lambda^k E_\lambda u]^0 + \int_{-h}^0 e^{\lambda s} [FT_\lambda^k E_\lambda u]^1(s)ds. \tag{8.179}$$

In view of (8.175)

$$[FT_\lambda^k E_\lambda u]^0 = [T_\lambda^k E_\lambda u]^0 = \frac{(-1)^k}{k!} [E_\lambda^{(k)} u]^0 = \delta_{k,0}u, \tag{8.180}$$

$$[FT_\lambda^k E_\lambda u]^1(s) = A_1 [T_\lambda^k E_\lambda u]^1(-h-s) + \int_{-h}^s a(\tau) A_2 [T_\lambda^k E_\lambda u]^1(\tau - s)d\tau$$

$$= \frac{(-1)^k}{k!} A_1 [E_\lambda^{(k)} u]^1(-h-s) + \frac{(-1)^k}{k!} \int_{-h}^s a(\tau) A_2 [E_\lambda^{(k)} u]^1(\tau - s)d\tau$$

$$= \frac{1}{k!}(h+s)^k e^{-\lambda(h+s)} A_1 u + \frac{1}{k!} \int_{-h}^s a(\tau)(s-\tau)^k e^{\lambda(\tau - s)} A_2 u d\tau.$$

Hence

$$\int_{-h}^{0} e^{\lambda s}[FT_{\lambda}^{k}E_{\lambda}u]^{1}(s)ds = \frac{1}{k!}e^{-\lambda h}\int_{-h}^{0}(h+s)^{k}ds A_{1}u$$

$$+\frac{1}{k!}\int_{-h}^{0}\int_{-h}^{s}e^{\lambda \tau}a(\tau)(s-\tau)^{k}A_{2}ud\tau ds$$

$$=\frac{h^{k+1}}{(k+1)!}e^{-\lambda h}A_{1}u + \frac{(-1)^{k+1}}{(k+1)!}\int_{-h}^{0}\tau^{k+1}e^{\lambda \tau}a(\tau)A_{2}ud\tau.$$

Combining this with (8.178),(8.179),(8.180) we complete the proof.

Proposition 8.7 (i) *For $\lambda \in \sigma_{p}(A)$ and $l = 1, 2, \ldots$, F is injective on* $\text{Ker}(\lambda - A)^{l}$.
(ii) *For $\lambda \in \sigma_{p}(A_{T})$ and $l = 1, 2, \ldots$, F^{*} is injective on* $\text{Ker}(\lambda - A_{T})^{l}$.

Proof. Suppose $\phi \in \text{Ker}(\lambda - A)$ and $F\phi = 0$. By Theorem 8.8 $\phi = (\phi^{0}, e^{\lambda \cdot}\phi^{0})$. Then $\phi^{0} = [F\phi]^{0} = 0$, and hence $\phi = 0$. If $\phi \in \text{Ker}(\lambda - A)^{l}, l > 1$, and $F\phi = 0$, then $(\lambda - A)^{l-1}\phi \in \text{Ker}(\lambda - A)$. By virtue of Corollary 8.1 $F\phi \in \cap_{k=1}^{l-1}D((A_{T}^{*})^{k})$ and

$$F(\lambda - A)^{l-1}\phi = (\lambda - A_{T}^{*})^{l-1}F\phi = 0.$$

From the result of the case $l = 1$ it follows that $(\lambda - A)^{l-1}\phi = 0$. Continuing this process we conclude $\phi = 0$. The following theorem was first established by M. C. Delfour and A. Manitius [45] for equations in finite dimensional equations, and by S. Nakagiri [112] and S. Nakagiri and H. Tanabe [114] for equations in infinite dimensional spaces.

Theorem 8.11 (i) *If $\lambda \in \sigma_{p}(\Delta) = \sigma_{p}(A_{T}^{*}) = \sigma_{p}(A)$, then for $l = 1, 2, \ldots$*

$$\text{Ker}(\lambda - A_{T}^{*})^{l} = F\text{Ker}(\lambda - A)^{l}, \qquad (8.181)$$

$$\dim \text{Ker}(\lambda - A_{T}^{*})^{l} = \dim \text{Ker}(\lambda - A)^{l}. \qquad (8.182)$$

(ii) *If $\lambda \in \sigma_{p}(\Delta_{T}) = \sigma_{p}(A^{*}) = \sigma_{p}(A_{T})$, then for $l = 1, 2, \ldots$*

$$\text{Ker}(\lambda - A^{*})^{l} = F^{*}\text{Ker}(\lambda - A_{T})^{l}, \qquad (8.183)$$

$$\dim \text{Ker}(\lambda - A^{*})^{l} = \dim \text{Ker}(\lambda - A_{T})^{l}. \qquad (8.184)$$

Proof. It suffices to show only (ii). With the aid of Corollary 8.1 we can inductively show that

$$(\lambda - A^{*})^{l}F^{*} = F^{*}(\lambda - A_{T})^{l} \quad \text{on} \quad D(A_{T}^{l}).$$

Therefore we have

$$\text{Ker}(\lambda - A^*)^l \supset F^*\text{Ker}(\lambda - A_T)^l.$$

To prove the converse inclusion relation let $\psi \in \text{Ker}(\lambda - A^*)^l$. Set $\phi_j = (\lambda - A^*)^j\psi, j = 0, \ldots, l-1$. Then $A^*\phi_{j-1} = \lambda\phi_{j-1} - \phi_j, j = 1, \ldots, l$, where $\phi_l = 0$. By Lemma 8.8

$$\Delta_T(\lambda)\phi_{j-1}^0 = H_\lambda\phi_j, \tag{8.185}$$

$$\phi_{j-1} = F^*E_\lambda\phi_{j-1}^0 + K_\lambda\phi_j, \tag{8.186}$$

for $j = 1, \ldots, l$. Hence $\phi_{l-1} = F^*E_\lambda\phi_{l-1}^0$. Using (8.186) we can show inductively that

$$\phi_{j-1} = \sum_{i=j-1}^{l-1} K_\lambda^{i-j+1} F^* E_\lambda \phi_i^0$$

$$= \sum_{i=j-1}^{l-1} F^* T_\lambda^{i-j+1} E_\lambda \phi_i^0, \quad j = 1, \ldots, l, \tag{8.187}$$

where we used (8.174). Substituting this in the right hand side of (8.185) and using (8.177) we get

$$\Delta_T(\lambda)\phi_{j-1}^0 = H_\lambda \sum_{i=j}^{l-1} F^* T_\lambda^{i-j} E_\lambda \phi_i^0 = \sum_{i=j}^{l-1} \frac{(-1)^{i-j}}{(i-j+1)!} \Delta_T^{(i-j+1)}(\lambda)\phi_i^0,$$

which is rewritten as

$$\sum_{i=j-1}^{l-1} \frac{(-1)^{i-j}}{(i-j+1)!} \Delta_T^{(i-j+1)}(\lambda)\phi_i^0 = 0. \tag{8.188}$$

On the other hand

$$\psi = \phi_0 = \sum_{i=0}^{l-1} F^* T_\lambda^i E_\lambda \phi_i^0$$

$$= F^* \sum_{i=0}^{l-1} \frac{(-1)^i}{i!} E_\lambda^{(i)} \phi_i^0 = F^* \left(\phi_0^0, e^\lambda \sum_{i=0}^{l-1} \frac{(-\cdot)^i}{i!} \phi_i^0 \right). \tag{8.189}$$

Theorem 8.8 and (8.188),(8.189) imply

$$\psi \in F^*\text{Ker}(\lambda - A_T)^l.$$

The assertion (8.184) follows from (8.183) and Proposition 8.7.

Let λ be an isolate point of $\sigma(A)$. Then $\bar{\lambda}$ is an isolate point of $\sigma(A^*)$. We denote the corresponding spectral projections by P_λ and $P_{\bar{\lambda}}^*$:

$$P_\lambda = \frac{1}{2\pi i} \oint_{|\mu - \lambda| = \epsilon} (\mu - A)^{-1} d\mu, \quad P_{\bar{\lambda}}^* = \frac{1}{2\pi i} \oint_{|\mu - \bar{\lambda}| = \epsilon} (\mu - A^*)^{-1} d\mu.$$

Analogously if λ is an isolate point of $\sigma(A_T)$, then $\bar{\lambda}$ is an isolate point of $\sigma(A_T^*)$. The corresponding spectral projections are denoted by P_λ^T and $P_{\bar{\lambda}}^{T*}$. As is easily seen

$$P_{\bar{\lambda}}^* = (P_\lambda)^*, \quad P_{\bar{\lambda}}^{T*} = \left(P_\lambda^T\right)^*. \tag{8.190}$$

We set

$$\mathcal{M}_\lambda = \operatorname{Im} P_\lambda, \ \mathcal{M}_{\bar{\lambda}}^* = \operatorname{Im} P_{\bar{\lambda}}^*, \ \mathcal{M}_\lambda^T = \operatorname{Im} P_\lambda^T, \ \mathcal{M}_{\bar{\lambda}}^{T*} = \operatorname{Im} P_{\bar{\lambda}}^{T*}. \tag{8.191}$$

Theorem 8. 12 (i) *If λ is an isolate point of $\sigma(A)$ and $\sigma(A_T^*)$, then*

$$FP_\lambda = P_{\bar{\lambda}}^{T*} F. \tag{8.192}$$

(ii) *If λ is an isolate point of $\sigma(A_T)$ and $\sigma(A^*)$, then*

$$F^* P_\lambda^T = P_{\bar{\lambda}}^* F^*. \tag{8.193}$$

Proof. The assertions are simple consequences of Theorem 8.6.

The following proposition follows from Lemma 8.7 and (8.191).

Proposition 8. 8 (i) *Let λ be a pole of $(\mu - A)^{-1}$ of order k_λ. Then $\bar{\lambda}$ is a pole of $(\mu - A^*)^{-1}$ of order k_λ. We have*

$$\mathcal{M}_\lambda = \operatorname{Ker}(\lambda - A)^{k_\lambda}, \quad \mathcal{M}_{\bar{\lambda}}^* = \operatorname{Ker}(\bar{\lambda} - A^*)^{k_\lambda},$$
$$\dim \mathcal{M}_\lambda = \dim \mathcal{M}_{\bar{\lambda}}^* \leq \infty.$$

(ii) *Let λ be a pole of $(\mu - A_T)^{-1}$ of order $k_{\bar{\lambda}}^T$. Then $\bar{\lambda}$ is a pole of $(\mu - A_T^*)^{-1}$ of order $k_{\bar{\lambda}}^T$. We have*

$$\mathcal{M}_\lambda^T = \operatorname{Ker}(\lambda - A_T)^{k_{\bar{\lambda}}^T}, \quad \mathcal{M}_{\bar{\lambda}}^{T*} = \operatorname{Ker}(\bar{\lambda} - A_T^*)^{k_{\bar{\lambda}}^T},$$
$$\dim \mathcal{M}_\lambda^T = \dim \mathcal{M}_{\bar{\lambda}}^{T*} \leq \infty.$$

Remark 8. 2 In the special case of section 8.11 we have $\sigma(A) = \overline{\sigma(A_T)} = \overline{\sigma(A^*)} = \sigma(A_T^*)$ in view of Theorem 8.9 (note that γ is real and $a(\cdot)$ is real valued), and (8.124) holds in $\rho(A)$ or $\overline{\rho(A)}$. If moreover $0 \in \rho(A_0)$ and $\gamma \neq 0$, then by virtue of Proposition 8.6 F is an isomorphism from Z onto Z^*. Therefore by (8.124) we have $k_{\bar{\lambda}}^T = k_\lambda$. Hence by Theorem 8.11 and Proposition 8.8 we have

$$F\mathcal{M}_\lambda = \mathcal{M}_{\bar{\lambda}}^{T*}, \quad F^* \mathcal{M}_{\bar{\lambda}}^T = \mathcal{M}_{\bar{\lambda}}^*. \tag{8.194}$$

If λ is an isolate point of $\sigma(A)$ and $\dim \operatorname{Im} P_\lambda < \infty$, then λ is a discrete spectrum of A. We denote by $\sigma_d(A)$ the set of discrete spectra of A. If $\lambda \in \sigma_d(A)$, then $\bar{\lambda} \in \sigma_d(A^*)$. If moreover $\bar{\lambda} \in \sigma_d(A_T)$, then $\lambda \in \sigma_d(A_T^*)$, and by Lemma 8.7 and Theorem 8.11 we have

$$d_\lambda \equiv \dim \mathcal{M}_\lambda = \dim \mathcal{M}_{\bar{\lambda}}^{T*} = \dim \mathcal{M}_{\bar{\lambda}}^* = \dim \mathcal{M}_{\bar{\lambda}}^T < \infty.$$

Let $\{\phi_1, \ldots, \phi_{d_\lambda}\}$ and $\{\psi_1, \ldots, \psi_{d_\lambda}\}$ be the bases of $\mathcal{M}_\lambda$ and $\mathcal{M}_{\bar{\lambda}}^T$ respectively. Since $\mathcal{M}_{\bar{\lambda}}^* = F^* \mathcal{M}_{\bar{\lambda}}^T$, the $d_\lambda \times d_\lambda$ matrix $\{(\phi_i, F^*\psi_j)\}$ is nonsingular. Hence we may suppose $(\phi_i, F^*\psi_j) = \delta_{i,j}$. If we set for $g \in Z$

$$\tilde{P}_\lambda g = \sum_{i=1}^{d_\lambda} (g, F^*\psi_i)\phi_i,$$

then as is easily seen $\operatorname{Im}\tilde{P}_\lambda = \mathcal{M}_\lambda$ and $\operatorname{Ker}\tilde{P}_\lambda = \operatorname{Im}(\lambda - A)^{k_\lambda}$. Hence we have $\tilde{P}_\lambda = P_\lambda$:

$$P_\lambda g = \sum_{i=1}^{d_\lambda} (g, F^*\psi_i)\phi_i. \tag{8.195}$$

8.13 Second Structural Operator

Let $g^0 \in H$. We represent the solution of (8.1),(8.2) with $g^1(s) \equiv 0$, $f(t) \equiv 0$ by $u(t) = W(t)g^0$. Then by Theorem 8.1 the mapping $g^0 \mapsto W(\cdot)g^0$ is a bounded linear operator from H to

$$L^2(-h, T; V) \cap W^{1,2}(0, T; V^*) \subset C([0, T]; H)$$

for any $T > 0$. $W(t)$ is continuous in $t \in [0, \infty)$ in the strong topology of H and

$$W(0) = I, \quad W(s) = 0 \quad s \in [-h, 0). \tag{8.196}$$

$W(\cdot)$ is called the *fundamental solution* of (8.1),(8.2).

Set $\tilde{V} = (D(A_0), H)_{1/2,2}$. In many cases $\tilde{V} = V$. In view of Theorem 8.1 if $g = (g^0, g^1) \in \tilde{Z} \equiv \tilde{V} \times L^2(-h, 0; D(A_0))$, then the solution of (8.1),(8.2) satisfies

$$\|u\|_{L^2(-h,T;D(A_0)) \cap W^{1,2}(0,T;H)} \leq C_T \left(\|g\|_{\tilde{Z}} + \|f\|_{L^2(0,T;H)} \right). \tag{8.197}$$

Hence if $g^0 \in \tilde{V}$, then

$$W(\cdot)g^0 \in L^2(0, T; D(A_0)) \cap W^{1,2}(0, T; H) \subset C([0, T]; \tilde{V}).$$

Analogously the fundamental solution $W_T(\cdot)$ of the transposed equation (8.100),

(8.101) is defined.

Let u be the solution of (8.1),(8.2) with $g = (g^0, g^1) \in Z, f \in L^2(0, T; V^*)$ and let $\phi \in H$. Then $u, W_T(\cdot)\phi \in L^2(-h, T; V) \cap W^{1,2}(0, T; V^*)$. Hence in view of Lemma 8.4 $(u(\tau), W_T(t - \tau)\phi)$ is absolutely continuous in $\tau \in [0, t]$ for any $t > 0$, and

$$\frac{d}{d\tau}(u(\tau), W_T(t - \tau)\phi) = (u'(\tau), W_T(t - \tau)\phi) - (u(\tau), W_T'(t - \tau)\phi)$$

$$= \left(A_0 u(\tau) + A_1 u(\tau - h) + \int_{-h}^0 a(s) A_2 u(\tau + s) ds + f(\tau), \right.$$

$$\left. W_T(t - \tau)\phi \right) - \left(u(\tau), A_0^* W_T(t - \tau)\phi + A_1^* W_T(t - \tau - h)\phi \right.$$

$$\left. + \int_{-h}^0 a(s) A_2^* W_T(t - \tau + s)\phi ds \right)$$

$$= \left(A_1 u(\tau - h) + \int_{-h}^0 a(s) A_2 u(\tau + s) ds + f(\tau), W_T(t - \tau)\phi \right)$$

$$- \left(u(\tau), A_1^* W_T(t - \tau - h)\phi + \int_{-h}^0 a(s) A_2^* W_T(t - \tau + s)\phi ds \right).$$

Integrating this equality from 0 to t we get

$$(u(t), \phi) - (g^0, W_T(t)\phi) = \int_0^t (A_1 u(\tau - h), W_T(t - \tau)\phi) d\tau$$

$$+ \int_0^t \left(\int_{-h}^0 a(s) A_2 u(\tau + s) ds, W_T(t - \tau)\phi \right) d\tau$$

$$+ \int_0^t (f(\tau), W_T(t - \tau)\phi) d\tau - \int_0^t (u(\tau), A_1^* W_T(t - \tau - h)\phi) d\tau$$

$$- \int_0^t \left(u(\tau), \int_{-h}^0 a(s) A_2^* W_T(t - \tau + s)\phi ds \right) d\tau = \sum_{i=1}^5 I_i. \quad (8.198)$$

In view of the initial conditions (8.1),(8.196)

$$I_1 + I_4 = \int_{-h}^{t-h} (A_1 u(\tau), W_T(t - \tau - h)\phi) d\tau$$

$$- \int_0^t (A_1 u(\tau), W_T(t - \tau - h)\phi) d\tau$$

$$= \int_{-h}^0 \left(A_1 g^1(\tau), W_T(t - \tau - h)\phi \right) d\tau. \quad (8.199)$$

Analogously

$$
\begin{aligned}
I_2 + I_5 &= \int_{-h}^{0} \int_{0}^{t} (a(s)A_2 u(\tau + s), W_T(t - \tau)\phi) d\tau ds \\
&\quad - \int_{-h}^{0} \int_{0}^{t} (u(\tau), a(s)A_2^* W_T(t - \tau + s)\phi) \, d\tau ds \\
&= \int_{-h}^{0} \int_{s}^{t+s} (a(s)A_2 u(\tau), W_T(t - \tau + s)\phi) d\tau ds \\
&\quad - \int_{-h}^{0} \int_{0}^{t} (a(s)A_2 u(\tau), W_T(t - \tau + s)\phi) d\tau ds \\
&= \int_{-h}^{0} \int_{s}^{0} \left(a(s)A_2 g^1(\tau), W_T(t - \tau + s)\phi \right) d\tau ds. \quad (8.200)
\end{aligned}
$$

From (8.198),(8.199),(8.200) it follows that

$$
\begin{aligned}
(u(t), \phi) &= (g^0, W_T(t)\phi) + \int_{-h}^{0} \left(A_1 g^1(\tau), W_T(t - \tau - h)\phi \right) d\tau \\
&\quad + \int_{-h}^{0} \int_{-h}^{\tau} \left(a(s)A_2 g^1(\tau), W_T(t - \tau + s)\phi \right) ds d\tau \\
&\quad + \int_{0}^{t} (f(\tau), W_T(t - \tau)\phi) d\tau. \quad (8.201)
\end{aligned}
$$

In particular if $g^1 = 0, f = 0$, we have

$$
(W(t)g^0, \phi) = (g^0, W_T(t)\phi)
$$

for any $g^0, \phi \in H$. Therefore we obtain

$$
W(t)^* = W_T(t). \quad (8.202)
$$

If $g \in \tilde{Z}$ and $f \in L^2(0, T; H)$, then we have

$$
u \in L^2(-h, T; D(A_0)) \cap W^{1,2}(0, T; H) \subset C([0, T]; \tilde{V}).
$$

Hence by (8.201),(8.202)

$$
\begin{aligned}
(u(t), \phi) &= (W(t)g^0, \phi) + \int_{-h}^{0} (W(t - \tau - h)A_1 g^1(\tau), \phi) d\tau \\
&\quad + \int_{-h}^{0} \int_{-h}^{\tau} \left(a(s)W(t - \tau + s)A_2 g^1(\tau), \phi \right) ds d\tau + \int_{0}^{t} (W(t - \tau)f(\tau), \phi) d\tau.
\end{aligned}
$$

Consequently we have

$$u(t) = W(t)g^0 + \int_{-h}^0 U_t(\tau)g^1(\tau)d\tau + \int_0^t W(t-\tau)f(\tau)d\tau, \qquad (8.203)$$

where

$$U_t(\tau) = W(t - \tau - h)A_1 + \int_{-h}^\tau a(s)W(t - \tau + s)ds A_2. \qquad (8.204)$$

By Theorem 8.1 we have

$$\|u\|_{L^2(-h,T;V) \cap W^{1,2}(0,T;V^*)} \lesssim C_T \left(\|g\|_Z + \|f\|_{L^2(0,T;V^*)} \right). \qquad (8.205)$$

Letting $g = 0$ we find that the mapping

$$f \mapsto W * f = \int_0^\cdot W(\cdot - \tau)f(\tau)d\tau,$$

which is a bounded linear operator from $L^2(0, T; H)$ to $L^2(-h, T; D(A_0))$ $\cap W^{1,2}(0, T; H)$ can be extended to a bounded linear operator from $L^2(0, T; V^*)$ to $L^2(-h, T; V) \cap W^{1,2}(0, T; V^*)$, where we set $(W * f)(t) = 0$ for $t < 0$ and T is an arbitrary positive number. This extended operator is also denoted by $f \mapsto W * f$.

Let $g^0 \in \tilde{V}$ and u be the solution of (8.1),(8.2) with $g^1 = 0$, $f = 0$. Let $t_1 > 0, t_2 > 0$. Then $u(t + t_2)$ is the solution of (8.1),(8.2) with the initial value $(u(t_2), u_{t_2}) \in \tilde{Z}$ and the inhomogeneous term 0. Hence in view of (8.203)

$$u(t + t_2) = W(t)u(t_2) + \int_{-h}^0 U_t(\tau)u(t_2 + \tau)d\tau.$$

Since $u(t) = W(t)g^0$, we have for $g^0 \in \tilde{V}$

$$W(t_1 + t_2)g^0 = W(t_1)W(t_2)g^0 + \int_{-h}^0 U_{t_1}(\tau)W(t_2 + \tau)g^0 d\tau. \qquad (8.206)$$

Let $f^0 \in \tilde{V}, f^1 \in L^2(-h, 0; H), t > 0$. Suggested by A. Manitius [107: p.7] we set

$$G_t f = ([G_t f]^0, [G_t f]^1(\cdot)), \qquad (8.207)$$

$$[G_t f]^1(s) = W(t + s)f^0 + \int_{-h}^0 W(t + s + \tau)f^1(\tau)d\tau, \quad s \in [-h, 0], \qquad (8.208)$$

$$[G_t f]^0 = [G_t f]^1(0) = W(t)f^0 + \int_{-h}^0 W(t + \tau)f^1(\tau)d\tau. \qquad (8.209)$$

Since $W(\cdot)f^0 \in L^2(-h, T; D(A_0)) \cap W^{1,2}(0, T; H)$ for any $T > 0$, we have

$$W(t + \cdot)f^0 \in L^2(-h, 0; D(A_0)) \cap W^{1,2}((-h) \vee (-t), 0; H),$$

and furthermore

$$\|W(t + \cdot)f^0\|_{L^2(-h,0;V) \cap W^{1,2}((-h) \vee (-t),0;V^*)}$$
$$= \|W(\cdot)f^0\|_{L^2(t-h,t;V) \cap W^{1,2}((t-h) \vee 0,t;V^*)} \leq C_T |f^0| \quad (8.210)$$

for $0 \leq t \leq T$. If we extend $f^1(s)$ to $(-\infty, 0]$ putting $f^1(s) = 0$ for $s < -h$, then

$$\int_{-h}^0 W(t + s + \tau)f^1(\tau)d\tau = \int_{-t-s}^0 W(t + s + \tau)f^1(\tau)d\tau$$
$$= \int_0^{t+s} W(t + s - \tau)f^1(-\tau)d\tau = (W * \check{f}^1)(t + s), \quad (8.211)$$

where $\check{f}^1(\tau) = f^1(-\tau)$. Since $W * \check{f}^1 \in L^2(-h, T; D(A_0)) \cap W^{1,2}(0, T; H)$ for any $T > 0$ we have

$$\int_{-h}^0 W(t + \cdot + \tau)f^1(\tau)d\tau \in L^2(-h, 0; D(A_0)) \cap W^{1,2}((-h) \vee (-t), 0; H).$$

Furthermore

$$\|(W * \check{f}^1)(t + \cdot)\|_{L^2(-h,0;V) \cap W^{1,2}((-h) \vee (-t),0;V^*)}$$
$$= \|W * \check{f}^1\|_{L^2(t-h,t;V) \cap W^{1,2}((t-h) \vee 0,t;V^*)}$$
$$\leq C_T \|\check{f}^1\|_{L^2(0,T;V^*)} \leq C_T \|f^1\|_{L^2(-h,0;V^*)}. \quad (8.212)$$

In view of (8.208),(8.210),(8.211),(8.212) we obtain

$$\|[G_t f]^1\|_{L^2(-h,0;V) \cap W^{1,2}((-h) \vee (-t),0;V^*)} \leq C_T \|f\|_{Z^*} \quad (8.213)$$
$$|[G_t f]^0| = |[G_t f]^1(0)| \leq C_T \|f\|_{Z^*}. \quad (8.214)$$

Consequently

$$\|G_t f\|_Z \leq C_T \|f\|_{Z^*}, \quad 0 \leq t \leq T. \quad (8.215)$$

Therefore G_t which is a bounded linear operator from $\tilde{V} \times L^2(-h, 0; H)$ to $\tilde{V} \times L^2(-h, 0; D(A_0))$ can be extended to a bounded linear operator from Z^* to Z. This extended operator is again denoted by G_t.

Definition 8. 6 $G = G_h$ is called the *second structural operator*.

Analogously starting from the operator G_t^+ defined by

$$G_t^+ g = ([G_t^+ g]^0, [G_t^+ g]^1(\cdot)),$$

$$[G_t^+ g]^1(s) = W_T(t+s)g^0 + \int_{-h}^0 W_T(t+s+\tau)g^1(\tau)d\tau, \ s \in [-h, 0],$$

$$[G_t^+ g]^0 = [G_t^+ g]^1(0)$$

for $g = (g^0, g^1) \in \tilde{V} \times L^2(-h, 0; H)$, we obtain an operator $G_t^+ \in \mathcal{L}(Z^*, Z)$. Since as is easily seen for $f, g \in \tilde{V} \times L^2(-h, 0; H)$ we have

$$(G_t f, g) = (f, G_t^+ g),$$

we find that

$$G_t^+ = G_t^*. \tag{8.216}$$

Therefore if we define $G^+ = G_h^+$, then

$$G^+ = G^*. \tag{8.217}$$

If we set $u(h+s) = [Gf]^1(s), s \in [-h, 0]$, for $f \in \tilde{V} \times L^2(-h, 0; H)$, then

$$u(t) = W(t)f^0 + \int_0^t W(t-\tau)f^1(-\tau)d\tau, \ t \in [0, h].$$

Therefore

$$\frac{d}{dt}u(t) = A_0 u(t) + A_1 u(t-h) + \int_{-h}^0 a(\tau)A_2 u(t+\tau)d\tau + f^1(-t),$$

$$t \in [0, h], \quad (8.218)$$

$$u(0) = f^0, \quad u(s) = 0 \quad s \in [-h, 0). \tag{8.219}$$

By the initial condition (8.219) we get from (8.218)

$$\frac{d}{dt}u(t) = A_0 u(t) + \int_{-t}^0 a(\tau)A_2 u(t+\tau)d\tau + f^1(-t), \quad t \in [0, h]. \tag{8.220}$$

The solution u of (8.218),(8.219) exists also for $f \in Z^*$ and we have $Gf = (u(h), u(h+\cdot))$. Since $u \in L^2(0, h; V) \cap W^{1,2}(0, h; V^*)$, we have

$$\text{Im}G \subset \mathcal{W}(-h, 0)$$
$$= \{\phi \in Z; \phi^1 \in L^2(-h, 0; V) \cap W^{1,2}(-h, 0; V^*), \phi^0 = \phi^1(0) \in H\}.$$

Conversely for $\phi \in \mathcal{W}(-h, 0)$ we can find $f \in Z^*$ satisfying $Gf = \phi$, since if we set $u(t) = \phi^1(t-h), t \in [0, h]$, then $f = (f^0, f^1)$ with $f^0 = u(0)$ and

f^1 defined by the equation (8.220) is a desired one. Moreover it is easily seen that this f is a unique element satisfying this condition. Therefore the inverse G^{-1} exists and is given by

$$[G^{-1}\phi]^0 = \phi^1(-h), \tag{8.221}$$

$$[G^{-1}\phi]^1(s) = \phi^{1'}(-s-h) - A_0\phi^1(-s-h)$$

$$- \int_s^0 a(\tau)A_2\phi^1(-s+\tau-h)d\tau. \tag{8.222}$$

Thus we have established the following proposition.

Proposition 8.9 G *is an isomorphism from* Z^* *to* $\mathcal{W}(-h, 0)$ *and the inverse* G^{-1} *is given by* (8.221),(8.222). *A similar result remains valid for* G^* *with* A_0, A_2 *replaced by* A_0^*, A_2^*.

From the definition of $S(t)$ and (8.203) it follows that for $f \in \tilde{Z}$

$$[S(t)f]^1(s) = W(t+s)f^0 + \int_{-h}^0 U_{t+s}(\tau)f^1(\tau)d\tau. \tag{8.223}$$

Theorem 8.13 $S(t)G = GS_T(t)^*$, $\quad G^*S(t)^* = S_T(t)G^*$.

Proof. Let $f \in \tilde{V} \times L^2(-h, 0; \tilde{V}) \subset \tilde{V} \times L^2(-h, 0; H)$. Then $Gf \in \tilde{Z} = \tilde{V} \times L^2(-h, 0; D(A_0))$, and with the aid of (8.223) and (8.208),(8.209) with $t = h$

$$[S(t)Gf]^1(s) = W(t+s)[Gf]^0 + \int_{-h}^0 U_{t+s}(\sigma)[Gf]^1(\sigma)d\sigma$$

$$= W(t+s)\left(W(h)f^0 + \int_{-h}^0 W(h+\tau)f^1(\tau)d\tau\right)$$

$$+ \int_{-h}^0 U_{t+s}(\sigma)\left(W(h+\sigma)f^0 + \int_{-h}^0 W(h+\sigma+\tau)f^1(\tau)d\tau\right)d\sigma$$

$$= W(t+s)W(h)f^0 + \int_{-h}^0 U_{t+s}(\sigma)W(h+\sigma)f^0 d\sigma$$

$$+ \int_{-h}^0 W(t+s)W(h+\tau)f^1(\tau)d\tau$$

$$+ \int_{-h}^0 U_{t+s}(\sigma)\int_{-h}^0 W(h+\sigma+\tau)f^1(\tau)d\tau d\sigma = \sum_{i=1}^4 I_i.$$

Using (8.206) we get

$$I_1 + I_2 = W(t+s+h)f^0.$$

Since
$$|U_{t+s}(\sigma)W(h+\sigma+\tau)f^1(\tau)| \le C\|W(h+\sigma+\tau)f^1(\tau)\|_{D(A_0)},$$
and by (8.197)
$$\int_{-h}^0 \|W(h+\sigma+\tau)f^1(\tau)\|_{D(A_0)}^2 d\sigma$$
$$= \|W(\cdot)f^1(\tau)\|_{L^2(\tau,h+\tau;D(A_0))}^2 \le C\|f^1(\tau)\|_{\tilde{V}}^2,$$
we can apply Fubini's theorem to I_4 to obtain
$$I_3 + I_4 = \int_{-h}^0 \left[W(t+s)W(h+\tau)f^1(\tau) \right.$$
$$\left. + \int_{-h}^0 U_{t+s}(\sigma)W(h+\sigma+\tau)f^1(\tau)d\sigma \right] d\tau$$
$$= \int_{-h}^0 W(t+s+h+\tau)f^1(\tau)d\tau,$$
where we again used (8.206). Therefore
$$[S(t)Gf]^1(s)$$
$$= W(t+s+h)f^0 + \int_{-h}^0 W(t+s+h+\tau)f^1(\tau)d\tau = [G_{t+s}f]^1(s),$$
which implies
$$S(t)Gf = G_{t+h}f, \tag{8.224}$$
holds for any $f \in \tilde{V} \times L^2(-h,0;\tilde{V})$. We can easily show that this equality holds also for $f \in Z^*$ by approximating f by a sequence in $\tilde{V} \times L^2(-h,0;\tilde{V})$. Analogously we can show
$$S_T(t)G^* = G_{t+h}^*. \tag{8.225}$$
The assertion of the theorem follows from (8.224) and (8.225).

We denote by $\sigma_o(A)$ the set of the poles of the resolvent of A.

Theorem 8. 14 (i) *Suppose that* $\sigma_o(A^*) = \sigma_o(A_T)$. *Then*
$$\sigma_d(A^*) = \sigma_d(A_T), \tag{8.226}$$
and for each $\lambda \in \sigma_d(A^*) = \sigma_d(A_T)$ *and* $l = 1,2,\dots$
$$G^*\mathrm{Ker}(\lambda - A^*)^l = \mathrm{Ker}(\lambda - A_T)^l, \tag{8.227}$$
$$\dim \mathrm{Ker}(\lambda - A^*)^l = \dim \mathrm{Ker}(\lambda - A_T)^l < \infty. \tag{8.228}$$

In particular

$$G^* \mathcal{M}_\lambda^* = \mathcal{M}_\lambda^T. \tag{8.229}$$

(ii) *Suppose* $\sigma_o(A_T^*) = \sigma_o(A)$. *Then*

$$\sigma_d(A_T^*) = \sigma_d(A), \tag{8.230}$$

and for each $\lambda \in \sigma_d(A_T^*) = \sigma_d(A)$ *and* $l = 1, 2, \dots$

$$G\mathrm{Ker}(\lambda - A_T^*)^l = \mathrm{Ker}(\lambda - A)^l, \tag{8.231}$$
$$\dim \mathrm{Ker}(\lambda - A_T^*)^l = \dim \mathrm{Ker}(\lambda - A)^l < \infty. \tag{8.232}$$

In particular

$$G\mathcal{M}_\lambda^{T*} = \mathcal{M}_\lambda. \tag{8.233}$$

Proof. It suffices to prove only (i). Suppose $\lambda \in \sigma_d(A^*)$. Then $\lambda \in \sigma_o(A^*) = \sigma_o(A_T)$. Hence $\lambda \in \sigma_p(A_T)$ and by Theorem 8.11 (8.228) holds. Therefore the order of λ as a pole of $(\lambda - A^*)^{-1}$ coincides with that as a pole of $(\lambda - A_T)^{-1}$, and $\lambda \in \sigma_d(A_T)$. Thus we proved $\sigma_d(A^*) \subset \sigma_d(A_T)$. The converse inclusion relation is proved similarly. From Theorem 8.13 it follows that if $\psi \in D(A^*)$ then $G^*\psi \in D(A_T)$ and $A_T G^*\psi = G^* A^* \psi$. Hence

$$G^* \mathrm{Ker}(\lambda - A^*)^l \subset \mathrm{Ker}(\lambda - A_T)^l, \quad l = 1, 2, \dots. \tag{8.234}$$

By Proposition 8.9 G^* is injective. Therefore we conclude (8.227) from (8.228),(8.234).

The results of this section were first established by A. Manitius [107] for equations in finite dimensional spaces, and extended to equations in infinite dimensional spaces by S. Nakagiri [112], J.-M. Jeong, S. Nakagiri and H. Tanabe [82] and S. Nakagiri and H. Tanabe [114].

8.14 A Special Case (Continued)

In the situation of section 8.11 we first consider in this section the problem of the completeness of generalized eigenvectors of the infinitesimal generator of the associated solution semigroup.

Definition 8.7 Let A be a closed linear operator in a Banach space X. The system of generalized eigenvectors of A is said to be *complete* if they span a dense subspace of X.

We use the result of A. Manitius [107] for equations in a finite dimensional space.

Let A_0, A_1, A_2 be $n \times n$ matrices and consider the problem (8.1),(8.2) (or equations with more complicated delay terms). The associated solution semigroup $S(t) = e^{tA}$ is a C_0 semigroup in $M_2 = R^n \times L^2(-h, 0; R^n)$. $\Delta(\lambda)$ defined by (8.106) is an $n \times n$ matrix valued entire function and in view of Corollary 8.2 $\sigma(A)$ is the set of λ at which $\Delta(\lambda)$ is singular. Therefore $\sigma(A)$ consists only of poles of the resolvent of A. The structural operator F is a bounded linear operator from M_2 to itself.

Theorem of A. Manitius *The system of generalized eigenvectors of A is complete if and only if F^* is injective.*

Suppose that A_0 is nonsingular, and $A_1 = \gamma A_0, \gamma$ is a real number $\neq 0, A_2 = A_0$, and that $m(0) \neq 0$, where $m(\lambda)$ is the function defined by (8.152). If $m(\lambda) = 0$, then $\Delta(\lambda) = \lambda$ is nonsingular, since $\lambda \neq 0$ then by assumption. Therefore $\Delta(\lambda)$ is singular if and only if $m(\lambda) \neq 0$ and $\lambda/m(\lambda) \in \sigma(A_0)$, and hence

$$\sigma(A) = \{\lambda; m(\lambda) \neq 0, \lambda/m(\lambda) \in \sigma(A_0)\}.$$

Now we return to our situation of section 8.11.

Theorem 8. 15 (J.-M. Jeong [80]) *Suppose $0 \in \rho(A_0)$, $\gamma \neq 0$, $m(0) \neq 0$, and that the immedding $V \subset H$ is compact. If the system of generalized eigenvectors of A_0 is complete in H, then the system of generalized eigenvectors of A is complete in Z.*

Proof. By virtue of Proposition 8.6 F^* is an isomorphism from V to V^*. Let $\sigma(A_0) = \{\mu_n; n = 1, 2, \ldots\}$ and P_n the spectral projection of A_0 at μ_n. Set $H_n = P_n H$ and $A_{0n} = A_0|_{H_n}$. Then $P_n V = H_n$. Set $Z_n = H_n \times L^2(-h, 0; H_n)$. We denote the solution semigroup of

$$\frac{d}{dt}u(t) = A_{0n}u(t) + \gamma A_{0n}u(t - h) + \int_{-h}^{0} a(s)A_{0n}u(t + s)ds,$$
$$u(0) = g^0, \quad u(s) = g^1(s) \quad s \in [-h, 0),$$

where $g = (g^0, g^1) \in Z_n$, and its infinitesimal generator by $S_n(t)$ and A_n respectively. Then by the commutativity of A_0 and P_n we can easily verify that $S_n(t) = S(t)|_{Z_n}$, $D(A_n) = D(A) \cap Z_n$, $A_n = A|_{D(A_n)}$. By virtue of Manitius' theorem the system of generalized eigenvectors of A_n is complete in Z_n. Since $\sigma(A_{0n}) = \{\mu_n\}$, we have

$$\sigma(A_n) = \{\lambda; m(\lambda) \neq 0, \lambda/m(\lambda) = \mu_n\}. \tag{8.235}$$

We write for $\phi \in Z, \psi \in Z^*$ $P_n\phi = (P_n\phi^0, P_n\phi^1(\cdot))$, $P_n^*\psi = (P_n^*\psi^0, P_n^*\psi^1(\cdot))$. Suppose that $\psi \in Z^*$ is orthogonal to all generalized eigenvectors of A.

Let $\phi \in Z_n$ be a generalized eigenvector of A_n. Then ϕ is a generalized eigenvector of A and $\phi = P_n \phi$. Therefore

$$(P_n^* \psi, \phi) = (\psi, P_n \phi) = (\psi, \phi) = 0,$$

from which $P_n^* \psi = 0$ follows for any n. This means that $\psi^0, \psi^1(s)$ are orthogonal to all generalized eigenvectors of A_0, and hence $\psi = 0$.

Lemma 8.10 *Suppose that the assumptions of Theorem 8.15 are satisfied and the system of generalized eigenvectors of A_0 is complete in H. Then if $P_\lambda^T f = 0$ for any $\lambda \in \sigma_p(A_T)$, then $f = 0$.*

Proof. By Theorem 8.9 P_λ^T is expressed as (8.195):

$$P_\lambda^T f = \sum_{i=1}^{d_\lambda} (f, F\psi_{\bar{\lambda}i})\phi_{\lambda i},$$

where $\phi_{\lambda i}, i = 1, \ldots, d_\lambda$, is a basis of $\mathcal{M}_\lambda^T$, $\psi_{\bar{\lambda}i}, i = 1, \ldots, d_\lambda$, is a basis of $\mathcal{M}_{\bar{\lambda}}$, and $d_\lambda = \dim \mathcal{M}_\lambda^T = \dim \mathcal{M}_{\bar{\lambda}}$. Therefore, if $P_\lambda^T f = 0$ for any $\lambda \in \sigma_p(A_T)$, then $(F^* f, \psi_{\bar{\lambda}i}) = 0$ for any $i = 1, \ldots, d_\lambda$ and $\bar{\lambda} \in \sigma(A)$. By Theorem 8.15 we conclude $F^* f = 0$, and hence $f = 0$ since F^* is injective under the present hypothesis.

Theorem 8.16 *Suppose that the assumptions of Theorem 8.15 are satisfied and that the system of generalized eigenvectors of A_0 is complete in H. Then the problem (S^T) is observable if and only if for any $\lambda \in \sigma_p(A_T)$*

$$\mathrm{Ker}\Phi_0^* \cap \mathrm{Ker}\Delta_T(\lambda) = \{0\}. \tag{8.236}$$

Proof. Suppose that (8.236) holds and $\Phi_0^*[S_T(t)\phi]^0 \equiv 0$. Since

$$(\mu - A_T)^{-1} = \int_0^\infty e^{-\mu t} S_T(t) dt$$

if $\mathrm{Re}\mu$ is sufficiently large, we have

$$\Phi_0^*[(\mu - A_T)^{-1}\phi]^0 = \int_0^\infty e^{-\mu t}\Phi_0^*[S_T(t)\phi]^0 dt = 0 \tag{8.237}$$

for these values of μ. By virtue of Theorem 8.9 we see that (8.237) holds for any $\mu \in \rho(A_T)$. Let $\lambda \in \sigma_p(A_T)$ and set $Q_\lambda^T = (A_T - \lambda)P_\lambda^T$. Then

$$(Q_\lambda^T)^j = (A_T - \lambda)^j P_\lambda^T, \quad (Q_\lambda^T)^{k_\lambda} = 0,$$

where k_λ is the order of λ as a pole of $(\mu - A_T)^{-1}$. In view of (8.237) we obtain

$$\Phi_0^*[(Q_\lambda^T)^j \phi]^0 = \frac{1}{2\pi i} \int_{|\mu - \lambda| = \epsilon} (\mu - \lambda)^j \Phi_0^*[(\mu - A_T)^{-1}\phi]^0 d\mu = 0 \tag{8.238}$$

for $j = 0, \ldots, k_\lambda - 1$. Set $\phi_1 = (Q_\lambda^T)^{k_\lambda - 1}\phi$. Then

$$(A_T - \lambda)\phi_1 = (A_T - \lambda)(Q_\lambda^T)^{k_\lambda - 1}\phi = (A_T - \lambda)^{k_\lambda} P_\lambda^T \phi = (Q_\lambda^T)^{k_\lambda} \phi = 0.$$

Therefore by Theorem 8.8

$$\phi_1 = (\phi_1^0, e^{\lambda \cdot}\phi_1^0), \quad \Delta_T(\lambda)\phi_1^0 = 0. \tag{8.239}$$

We have by (8.238)

$$\Phi_0^* \phi_1^0 = \Phi_0^*[(Q_\lambda^T)^{k_\lambda - 1}\phi]^0 = 0. \tag{8.240}$$

Hence from (8.236),(8.239),(8.240) it follows that $\phi_1^0 = 0$, and hence $\phi_1 = 0$. Next set $\phi_2 = (Q_\lambda^T)^{k_\lambda - 2}\phi = 0$. Then $\phi_2 \in \mathrm{Ker}(\lambda - A_T)$. By the same argument as above, we find that $\phi_2 = 0$. Continuing this procedure we obtain $P_\lambda^T \phi = 0$ for any $\lambda \in \sigma_p(A_T)$. Therefore by Lemma 8.10 we conclude $\phi = 0$.

Conversely suppose (S^T) is observable and $\phi^0 \in \mathrm{Ker}\Phi_0^* \cap \mathrm{Ker}\Delta_T(\lambda)$ for some $\lambda \in \sigma_p(A_T)$. Then $\phi = (\phi^0, e^{\lambda \cdot}\phi^0) \in \mathrm{Ker}(\lambda - A_T)$. Therefore

$$S_T(t)\phi = e^{\lambda t}\phi,$$

and hence

$$\Phi_0^*[S_T(t)\phi]^0 = \Phi_0^*(e^{\lambda t}\phi^0) = e^{\lambda t}\Phi_0^*(\phi^0) = 0.$$

From the hypothesis we conclude $\phi = 0$, and hence $\phi^0 = 0$.

We owe the above proof of the sufficiency part to S. Nakagiri and M. Yamamoto [115] and T. Suzuki and M. Yamamoto [146].

Finally in this section we consider the case where the control space U is a finite dimensional space $\mathbf{C}^N$. Then the controller Φ_0 is represented by

$$\Phi_0 w = \sum_{i=1}^{N} w_i b_i^0,$$

where $w = (w_1, \ldots, w_N) \in \mathbf{C}^N$ and $b_i^0 \in H, i = 1, \ldots, N$. The adjoint operator $\Phi_0^* : H \to \mathbf{C}^N$ is given by

$$\Phi_0^* \phi^0 = ((\phi^0, b_1^0), \ldots, (\phi^0, b_N^0)).$$

We suppose that the basis $\phi_{\lambda i}, i = 1, \ldots, d_\lambda$, of $\mathcal{M}_\lambda^T$ are arranged so that $\phi_{\lambda i}, i = 1, \ldots, n_\lambda, n_\lambda = \dim \mathrm{Ker}(\lambda - A_T)$, is the basis of $\mathrm{Ker}(\lambda - A_T)$. Then for $i = 1, \ldots, n_\lambda$ $\phi_{\lambda i} = (\phi_{\lambda i}^0, e^{\lambda \cdot}\phi_{\lambda i}^0)$, and $\phi_{\lambda i}^0, i = 1, \ldots, n_\lambda$, is a basis of $\mathrm{Ker}\Delta_T(\lambda)$.

We refer to the Rank Condition introduced by H. O. Fattorini [58]:
RANK CONDITION: For any $\lambda \in \sigma_p(A_T)$

$$\text{rank} \begin{pmatrix} (\phi^0_{\lambda 1}, b^0_1) & \cdots & (\phi^0_{\lambda 1}, b^0_N) \\ \cdots\cdots\cdots\cdots\cdots\cdots\cdots \\ (\phi^0_{\lambda n_\lambda}, b^0_1) & \cdots & (\phi^0_{\lambda n_\lambda}, b^0_N) \end{pmatrix} = n_\lambda.$$

Theorem 8.17 *Suppose that the hypotheses of Theorem 8.16 are satisfied. Then the problem (S^T) is observable if and only if the Rank Condition is satisfied.*

Proof. Let $\phi^0 \in \text{Ker}\Delta_T(\lambda)$ for some $\lambda \in \sigma_p(A_T)$. Then $\phi^0 = \sum_{i=1}^{n_\lambda} c_i \phi^0_{\lambda i}$ for some $c_i \in \mathbb{C}, i = 1, \cdots, n_\lambda$, and we have

$$\Phi^*_0 \phi^0 = ((\phi^0, b^0_1), \cdots, (\phi^0, b^0_N))$$

$$= \left(\sum_{i=1}^{n_\lambda} c_i(\phi^0_{\lambda i}, b^0_1), \cdots, \sum_{i=1}^{n_\lambda} c_i(\phi^0_{\lambda i}, b^0_N) \right)$$

$$= (c_1, \cdots, c_{n_\lambda}) \begin{pmatrix} (\phi^0_{\lambda 1}, b^0_1) & \cdots & (\phi^0_{\lambda 1}, b^0_N) \\ \cdots\cdots\cdots\cdots\cdots\cdots\cdots \\ (\phi^0_{\lambda n_\lambda}, b^0_1) & \cdots & (\phi^0_{\lambda n_\lambda}, b^0_N) \end{pmatrix}.$$

Therefore if the Rank Condition is satisfied, $\Phi^*_0 \phi^0 = 0$ implies $c_1 = \cdots = c_{n_\lambda} = 0$, or $\phi^0 = 0$. Hence (S^T) is observable in view of Theorem 8.16. The proof of the converse is similar.

Remark 8.3 An example of an operator for which the Rank Condition is satisfied is the realization of Δ in a disk in R^2 under the Dirichlet boundary condition (see p.517 of S. Nakagiri [111]). In this case $n_\lambda = 2$ for any eigenvalue λ by G. N. Watson [158: 15.28].

8.15 *F*-completeness of Generalized Eigenfunctions

Throughout this section we assume

$$\sigma_o(\Delta) = \sigma_o(A^*_T) = \sigma_o(A), \qquad (8.241)$$

$$\sigma_o(\Delta_T) = \sigma_o(A^*) = \sigma_o(A_T), \qquad (8.242)$$

where $\sigma_o(\Delta), \sigma_o(A)$, etc. are the sets of poles of $\Delta(\cdot)^{-1}$, the resolvent of A, etc. Since $\lambda \in \sigma_o(A)$ if and only if $\bar{\lambda} \in \sigma_o(A^*)$, we have

$$\overline{\sigma_o(\Delta)} = \overline{\sigma_o(A^*_T)} = \overline{\sigma_o(A)} = \sigma_o(\Delta_T) = \sigma_o(A^*) = \sigma_o(A_T). \qquad (8.243)$$

Lemma 8. 11   (i) *If λ is an isolate point of $\sigma(A)$, then*

$$\mathrm{Ker} P_\lambda = \{g \in Z; \Delta(\mu)^{-1} H_\mu F g \text{ is holomorphic at } \mu = \lambda\}.$$

(ii) *If λ is an isolate point of $\sigma(A_T)$, then*

$$\mathrm{Ker} P_\lambda^T = \{g \in Z; \Delta_T(\mu)^{-1} H_\mu F^* g \text{ is holomorphic at } \mu = \lambda\}.$$

(iii) *If λ is an isolate point of $\sigma(A^*)$, then*

$$\mathrm{Ker} P_\lambda^* = \{f \in Z^*; \Delta_T(\mu)^{-1} H_\mu f \text{ is holomorphic at } \mu = \lambda\}.$$

(iv) *If λ is an isolate point of $\sigma(A_T^*)$, then*

$$\mathrm{Ker} P_\lambda^{T^*} = \{f \in Z^*; \Delta(\mu)^{-1} H_\mu f \text{ is holomorphic at } \mu = \lambda\}.$$

Proof. (i) In view of Lemma 8.7 $g \in \mathrm{Ker} P_\lambda$ if and only if $(\mu - A)^{-1} g$ is holomorphic at $\mu = \lambda$. By (8.125) this is equivalent to the statement that $E_\mu \Delta(\mu)^{-1} H_\mu F g$ is holomorphic at $\mu = \lambda$. By the definition (8.119)

$$E_\mu \Delta(\mu)^{-1} H_\mu F g = \left(\Delta(\mu)^{-1} H_\mu F g, e^{\mu \cdot} \Delta(\mu)^{-1} H_\mu F g \right).$$

Thus the assertion follows.

(iii) $f \in \mathrm{Ker} P_\lambda^*$ if and only if $(\mu - A^*)^{-1} f$ is holomorphic at $\mu = \lambda$. Analogously to (8.127) we have

$$(\mu - A^*)^{-1} = F^* E_\mu \Delta_T(\mu)^{-1} H_\mu + K_\mu.$$

Hence $(\mu - A^*)^{-1} f$ is holomorphic at $\mu = \lambda$ if and only if the same holds for $F^* E_\mu \Delta_T(\mu)^{-1} H_\mu f$. In view of (8.118),(8.119)

$$
\begin{aligned}
F^* E_\mu &\Delta_T(\mu)^{-1} H_\mu f \\
&= \left([E_\mu \Delta_T(\mu)^{-1} H_\mu f]^0, A_1^* [E_\mu \Delta_T(\mu)^{-1} H_\mu f]^1(-h - \cdot) \right. \\
&\quad + \left. \int_{-h}^{\cdot} a(\tau) A_2^* [E_\mu \Delta_T(\mu)^{-1} H_\mu f]^1(\tau - \cdot) d\tau \right) \\
&= \left(\Delta_T(\mu)^{-1} H_\mu f, e^{\mu(-h - \cdot)} A_1^* \Delta_T(\mu)^{-1} H_\mu f \right. \\
&\quad + \left. \int_{-h}^{\cdot} a(\tau) e^{\mu(\tau - \cdot)} A_2^* \Delta_T(\mu)^{-1} H_\mu f d\tau \right).
\end{aligned}
$$

Thus the conclusion follows.

Assume that $\sigma_p(A) = \sigma_o(\Delta)$. Then by (8.241)

$$\sigma_p(A) = \sigma_o(A). \tag{8.244}$$

It follows from Corollary 8.3 (i), (8.241),(8.244) that

$$\sigma_p(A_T^*) = \sigma_o(A) = \sigma_o(A_T^*). \tag{8.245}$$

Therefore we can define

$$\mathcal{M} = \cup_{\lambda \in \sigma_p(A)}\mathcal{M}_\lambda, \quad \mathcal{M}^{T*} = \cup_{\lambda \in \sigma_p(A_T^*)}\mathcal{M}_\lambda^{T*}. \tag{8.246}$$

If $\lambda \in \sigma_p(A)$, then by (8.244) λ is a pole of the resolvent of A, and hence $\bar{\lambda}$ is a pole of the resolvent of A^*. Therefore $\mathcal{M}_{\bar{\lambda}}^*$ can be defined. Similarly in view of (8.245) we can define $\mathcal{M}_{\bar{\lambda}}^T$.

Analogously, if $\sigma_p(A_T) = \sigma_o(\Delta_T)$, then

$$\sigma_p(A_T) = \sigma_o(A_T) \quad \sigma_p(A^*) = \sigma_o(A_T) = \sigma_o(A^*),$$

and we can define

$$\mathcal{M}^* = \cup_{\lambda \in \sigma_p(A^*)}\mathcal{M}_\lambda^*, \quad \mathcal{M}^T = \cup_{\lambda \in \sigma_p(A_T)}\mathcal{M}_\lambda^T. \tag{8.247}$$

Moreover for $\lambda \in \sigma_p(A_T) = \sigma_p(A^*)$ we can define $\mathcal{M}_{\bar{\lambda}}^{T*}, \mathcal{M}_{\bar{\lambda}}$.

Proposition 8. 10 (i) *If $\sigma_p(A) = \sigma_o(\Delta)$, then*

$$\mathcal{M}^\perp = \{f \in Z^*; \Delta_T(\mu)^{-1}H_\mu f \text{ is holomorphic in } \rho(\Delta_T) \cup \sigma_o(\Delta_T)\},$$
$$(\mathcal{M}^{T*})^\perp = \{g \in Z; \Delta_T(\mu)^{-1}H_\mu F^* g \text{ is holomorphic in } \rho(\Delta_T) \cup \sigma_o(\Delta_T)\}.$$

(ii) *If $\sigma_p(A_T) = \sigma_o(\Delta_T)$, then*

$$(\mathcal{M}^T)^\perp = \{f \in Z^*; \Delta(\mu)^{-1}H_\mu f \text{ is holomorphic in } \rho(\Delta) \cup \sigma_o(\Delta)\},$$
$$(\mathcal{M}^*)^\perp = \{g \in Z; \Delta(\mu)^{-1}H_\mu Fg \text{ is holomorphic in } \rho(\Delta) \cup \sigma_o(\Delta)\}.$$

Proof. (i) As is easily seen $f \in \mathcal{M}_\lambda^\perp$ if and only if $f \in \text{Ker}P_{\bar{\lambda}}^*$. Hence $f \in \mathcal{M}^\perp$ if and only if $f \in \text{Ker}P_{\bar{\lambda}}^*$ for any $\lambda \in \sigma_p(A)$. By Lemma 8.11 (iii) $f \in \text{Ker}P_{\bar{\lambda}}^*$ if and only if $\Delta_T(\mu)^{-1}H_\mu f$ is holomorphic at $\mu = \bar{\lambda}$. By assumption $\lambda \in \sigma_p(A)$ if and only if $\lambda \in \sigma_o(\Delta)$, and in view of (8.243) this is equivalent to $\bar{\lambda} \in \sigma_o(\Delta_T)$. Thus the first assertion follows. Similarly $g \in (\mathcal{M}_\lambda^{T*})^\perp$ if and only if $g \in \text{Ker}P_{\bar{\lambda}}^T$. Therefore $g \in (\mathcal{M}^{T*})^\perp$ if and only if $g \in \text{Ker}P_{\bar{\lambda}}^T$ for any $\lambda \in \sigma_p(A_T^*)$. By Lemma 8.11 (ii) $g \in \text{Ker}P_{\bar{\lambda}}^T$ if and only if $\Delta_T(\mu)^{-1}H_\mu F^* g$ is holomorphic at $\mu = \bar{\lambda}$. By Corollary 8.3 (i), the assumption $\sigma_p(A) = \sigma_o(\Delta)$, and (8.243), we have $\sigma_p(A_T^*) = \overline{\sigma_o(\Delta_T)}$. Hence $\lambda \in \sigma_p(A_T^*)$ if and only if $\bar{\lambda} \in \sigma_o(\Delta_T)$. Thus the second assertion follows. The proof of (ii) is similar.

Propostion 8. 11   (i) *If $\sigma_p(A) = \sigma_o(\Delta)$, then*

$$(\mathcal{M}^{T*})^\perp = (F\mathcal{M})^\perp. \tag{8.248}$$

(ii) *If $\sigma_p(A_T) = \sigma_o(\Delta_T)$, then*

$$(\mathcal{M}^*)^\perp = (F^*\mathcal{M}^T)^\perp. \tag{8.249}$$

Proof. As is easily seen $g \in (F\mathcal{M})^\perp$ if and only if $F^*g \in \mathcal{M}^\perp$, which by Proposition 8.10 (i) is equivalent to $g \in (\mathcal{M}^{T*})^\perp$. The proof of (ii) is similar.

Delfour and Manitius introduced the following concept of F-completeness and explained its importance in their papers [45] and [107].

Definition 8. 8 The system of generalized eigenfunctions of A (resp. A_T) is said to be *F-complete* (resp. *F*-complete*) if $F\mathcal{M}$ (resp. $F^*\mathcal{M}^T$) spans a dense subspace of $\mathrm{Im}F$ (resp. $\mathrm{Im}F^*$).

Remark 8. 4 The system of generalized eigenfunctions of A (resp. A_T) is F-complete (resp. F^*-complete) if and only if $(F\mathcal{M})^\perp = (\mathrm{Im}F)^\perp$ (resp. $(F^*\mathcal{M}^T)^\perp = (\mathrm{Im}F^*)^\perp$.

Definition 8. 9 The system of generalized eigenfunctions of A^* (resp. A_T^*) is said to be complete in $Cl(\mathrm{Im}F^*)$ (resp. $Cl(\mathrm{Im}F)$) if $\mathcal{M}^*$ (resp. $\mathcal{M}^{T*}$) spans a dense subspace of $\mathrm{Im}F^*$ (resp. $\mathrm{Im}F$).

Remark 8. 5 By Theorem 8.11 or Remark 8.2 we have $\mathcal{M}^* \subset \mathrm{Im}F^*$ and $\mathcal{M}^{T*} \subset \mathrm{Im}F$. The system of generalized eigenfunctions of A^* (resp. A_T^*) is complete in $\mathrm{Im}F^*$ (resp. $\mathrm{Im}F$) if and only if $(\mathcal{M}^*)^\perp = (\mathrm{Im}F^*)^\perp$ (resp. $(\mathcal{M}^{T*})^\perp = (\mathrm{Im}F)^\perp$.

Theorem 8. 18 (i) *If $\sigma_p(A) = \sigma_o(\Delta)$, then the system of generalized eigenfunctions of A is F-complete if and only if the system of generalized eigenfunctions of A_T^* is complete in $Cl(\mathrm{Im}F)$.*
(ii) *If $\sigma_p(A_T) = \sigma_o(\Delta_T)$, then the system of generalized eigenfunctions of A_T is F^*-complete if and only if the system of generalized eigenfunctions of A^* is complete in $Cl(\mathrm{Im}F^*)$.*

Proof. The assertions follow from Propositions 8.10 and 8.11.

Thorem 8.18 is due to M. C. Delfour and A. Manitius [45] in finite dimensional case and S. Nakagiri in infinite dimensional case (unpublished).

8.16 Example and Remarks

Let Ω be a bounded domain in R^n with smooth boundary. We set $H = L^2(\Omega)$ and $V = H_0^1(\Omega) = W_0^{1,2}(\Omega)$. Let $a(\cdot, \cdot)$ be the sesquilinear form in $H_0^1(\Omega) \times H_0^1(\Omega)$ defined by

$$a(u, v) = \int_\Omega \left[\sum_{i,j=1}^n a_{ij}(x) \frac{\partial u}{\partial x_i} \frac{\overline{\partial v}}{\partial x_j} + \sum_{i=1}^n b_i(x) \frac{\partial u}{\partial x_i} \overline{v} + c(x) u \overline{v} \right] dx. \quad (8.250)$$

We assume that the coefficients a_{ij}, b_i, c are real valued and satisfy

$$a_{ij} \in C^1(\bar{\Omega}), \quad b_i \in C^1(\bar{\Omega}), \quad c \in L^\infty(\Omega),$$

$a_{ij} = a_{ji}, i, j = 1, \ldots, n$, and a uniform ellipticity

$$\sum_{i,j=1}^n a_{ij}(x)\xi_i\xi_j \geq c_0|\xi|^2, \quad \xi = (\xi_1, \ldots, \xi_n) \in R^n$$

for some positive constant c_0. As is well known this sequilinear form satisfies (8.66),(8.67). The associated operator $A_0 \in \mathcal{L}(H_0^1(\Omega), H^{-1}(\Omega))$, where $H^{-1}(\Omega) = H_0^1(\Omega)^*$, has the following realization in $L^2(\Omega)$. Let

$$\mathcal{A}_0 = -\sum_{i,j=1}^n \frac{\partial}{\partial x_j}\left(a_{ij}(x)\frac{\partial}{\partial x_i}\right) + \sum_{i=1}^n b_i(x)\frac{\partial}{\partial x_i} + c(x).$$

be the associated differential operator. Then the realization of $\mathcal{A}_0$ in $L^2(\Omega)$ under the Dirichlet boundary condition is the restriction of $\mathcal{A}_0$ to $D(A_0) = H^2(\Omega) \cap H_0^1(\Omega)$, where $H^2(\Omega) = W^{2,2}(\Omega)$.

Next, let $A_\iota, \iota = 1, 2$, be the restriction to $H_0^1(\Omega)$ of the second order linear differential operator $\mathcal{A}_\iota, \iota = 1, 2$, given by

$$\mathcal{A}_\iota = -\sum_{i,j=1}^n \frac{\partial}{\partial x_j}\left(a_{ij}^\iota(x)\frac{\partial}{\partial x_i}\right) + \sum_{i=1}^n b_i^\iota(x)\frac{\partial}{\partial x_i} + c^\iota(x),$$

where

$$a_{ij}^\iota = a_{ji}^\iota \in C^1(\bar{\Omega}), \quad b_i^\iota \in C^1(\bar{\Omega}), \quad c^\iota \in L^\infty(\Omega).$$

Clearly $A_\iota \in \mathcal{L}(H_0^1(\Omega), H^{-1}(\Omega)) \cap \mathcal{L}(D(A_0), L^2(\Omega))$ for $\iota = 1, 2$.

We consider the following parabolic partial functional differential equation

$$\frac{\partial u(t, x)}{\partial t} = \mathcal{A}_0 u(t, x) + \mathcal{A}_1 u(t - h, x)$$

$$+ \int_{-h}^0 a(s)\mathcal{A}_2 u(t + s, x)ds + f(t, x), \; t \in (0, T), x \in \Omega, \quad (8.251)$$

$$u(t, x) = 0, \quad t \in (0, T), \; x \in \partial\Omega, \quad (8.252)$$

$$u(0, x) = g^0(x), \quad u(s, x) = g^1(s, x) \; s \in [-h, 0), \; x \in \Omega. \quad (8.253)$$

Here we assume that

$$f \in L^2_{loc}([0,\infty); H^{-1}(\Omega)), \tag{8.254}$$
$$g^0 \in L^2(\Omega), \quad g^1 \in L^2(-h, 0; H^1_0(\Omega)). \tag{8.255}$$

Under these assumptions the problem (8.251)-(8.253) is written in the form of the abstract equation (8.1),(8.2) in the space $H^{-1}(\Omega)$, and the hypotheses of the preceding sections are satisfied.

Next we consider the transposed system. The adjoint operator $A^*_0 \in \mathcal{L}(H^1_0(\Omega), H^{-1}(\Omega))$ of A_0 is defined by

$$(u, A^*_0 v) = -a(u, v), \quad u, v \in H^1_0(\Omega).$$

The realization of A^*_0 in $L^2(\Omega)$ is characterized as the restiction of the formal adjoint of $\mathcal{A}$

$$\mathcal{A}^* = -\sum_{i,j=1}^{n} \frac{\partial}{\partial x_j}\left(a_{ij}(x)\frac{\partial}{\partial x_i}\right) + \sum_{i=1}^{n}\frac{\partial}{\partial x_i}(b_i(x)\cdot) + c(x)$$

to $H^2(\Omega)\cap H^1_0(\Omega)$. The adjoint operator A^*_i of $A_i, i = 1, 2$, is the restriction of the formal adjoint $\mathcal{A}^*_i$ of $\mathcal{A}_i$ to $H^1_0(\Omega)$. The transposed system of (8.251)-(8.253) is given by

$$\frac{\partial v(t, x)}{\partial t} = \mathcal{A}^*_0 v(t, x) + \mathcal{A}^*_1 v(t - h, x)$$
$$+ \int_{-h}^{0} a(s)\mathcal{A}^*_2 v(t + s, x)ds + \psi(t, x), \quad t \in (0, T), x \in \Omega, \tag{8.256}$$
$$v(t, x) = 0, \quad t \in (0, T), \ x \in \partial\Omega, \tag{8.257}$$
$$v(0, x) = \phi^0(x), \quad v(s, x) = \phi^1(s, x) \ s \in [-h, 0), \ x \in \Omega. \tag{8.258}$$

This equation is also formulated as an abstract equation in the space $H^{-1}(\Omega)$.

In order to treat equations with an L^1-valued controller $\Phi_0 \in \mathcal{L}(U, L^1(\Omega))$, J.-M. Jeong [81] investigated the problem (8.251)-(8.253) in the space $W^{-1,p}(\Omega) = W^{1,p'}_0(\Omega)^*$, $1 < p < n/(n-1)$, noting that we may consider $L^1(\Omega) \subset W^{-1,p}(\Omega)$ by identifying $f \in L^1(\Omega)$ with the continuous conjugate linear functional

$$\left(C_0(\bar{\Omega}) \supset\right) W^{1,p'}_0(\Omega) \ni v \mapsto \int_{\Omega} f\bar{v}dx \in \mathbf{C}. \tag{8.259}$$

He first proved that the operator $A_0 \in \mathcal{L}(W^{1,p}_0(\Omega), W^{-1,p}(\Omega))$ defined by

$$a(u, v) = -(A_0 u, v), \quad u \in W^{1,p}_0(\Omega), \quad v \in W^{1,p'}_0(\Omega),$$

where $a(\cdot,\cdot)$ is the same sesquilinear form as (8.250), generates an analytic semigroup in $W^{-1,p}(\Omega)$ using the results on complex interpolation of A. P. Calderón [26] and T. Seeley [136]. Next with the aid of the result of T. Seeley [135] (See also J. Prüss and H. Sohr [130]) he proved that

$$\|(-A_0)^{i\theta}\|_{\mathcal{L}(W^{-1,p}(\Omega))} \le Ce^{\gamma|s|}, \quad s \in R$$

for some constant $\gamma \in (0, \pi/2)$. Hence applying the maximal regularity result of G. Dore and A. Venni [55] and following the method of G. Di Blasio, K. Kunisch and E. Sinestrari [51] he constructed the solution semigroup associated with the problem (8.251)-(8.253), which is a C_0-semigroup in the space

$$Z_{p,q} = H_{p,q} \times L^q(-h, 0; W_0^{1,p}(\Omega))$$

where $H_{p,q} = (W_0^{1,p}(\Omega), W^{-1,p}(\Omega))_{1/q,q}, q \in (1, \infty)$. The enlarged system of the transposed equation (8.100) is an equation in the space

$$Z_{p',q'} = H_{p',q'} \times L^{q'}(-h, 0; W^{1,p'}(\Omega)).$$

The space $H_{p,q}$ was characterized by J.-Y. Park and J.-M. Jeong [126] except some critical cases as follows: if $2/q - 2 + 1/p \ne 0$

$$H_{p,q} = \begin{cases} \text{the closure of } C_0^1(\Omega) \text{ in } B_{p,q}^{1-2/q}(\Omega) & 2/q < 1 - 1/p, \\ B_{p,q}^{1-2/q}(\Omega) & 2/q > 1 - 1/p. \end{cases} \quad (8.260)$$

This is a special case of a more general result on the characterization of the interpolation space $(W_0^{1,p}(\Omega), W^{-1,p}(\Omega))_{\theta,q}$ as a Besov space, which is considered to be a result of independent interest and is proved with the aid of the reiteration theorem by J. L. Lions and J. Peetre [100] and the result of P. Grisvard [71: Theorem 7.3] (see also H. Triebel [154: Theorem 4.3.3]) after showing that $L^p(\Omega)$ is of class $\mathcal{K}_{1/2}(W_0^{1,p}(\Omega), W^{-1,p}(\Omega))$ in the sense of Lions and Peetre [100]. If $1 < p < n/(n-1)$, then $p' > n$, and hence we can choose q so that $1 - 2/q' > n/p'$. Then by (8.260) and Sobolev's imbedding theorem we have $H_{p',q'} \subset C_0(\bar{\Omega})$. Therefore we may consider $L^1(\Omega) \subset H_{p',q'}^* = H_{p,q}$ by a similar identification to (8.259), and hence $\Phi_0^* \in \mathcal{L}(H_{p',q'}, U^*)$. The solution of the transposed equation (8.100) is a continuous function with values in $H_{p',q'}$, and so it is possible to establish analogous results in the preceding sections for equations with an L^1-valued controller.

Bibliographical Remarks

Chapter 1. For the theory of semigroups refer to E. Hille and R. S. Phillips [74], J. Goldstein [70], K. Yosida [166], S. D. Zaidman [167]. Goldstein's book [70] contains abundant applications including those to mathematical physics. S. Agmon and L. Nirenberg [12] is a far-reaching study of solutions of ordinary differential equations in Banach spaces with applications to partial differential equations. Zaidman's book [167] inclides the same kind of topics. The Hille-Yosida theory is extended to the case of infinitesimal generators with nondense domain by G. Da Prato and E. Sinestrari [43], Sinestrari [138],[139] and to the case in which infinitesimal generators are multivalued linear operators by A. Yagi [164].

Chapter 2. Singular integrals of several variables are due to A. P. Calderón and A. Zygmund [27]. We owe the exposition of Hilbert transforms to E. C. Titchmarsh [153] and A. Zygmund [168] in addition to [27]. Hilbert transforms for functions with values in a Banach space play an important role. A Banach space with the property that the Hilbert transform for functions with values in it is bounded is characterized as ζ-convex. See D. L. Burkholder [22] and also H. Amann [16: Chapter III] for this topic. Using this result G. Dore and A. Venni [54],[55] established a very useful maximal regularity result for evolution equations. See also G. Dore [53], M. Giga, Y. Giga and H. Sohr [69] and papers listed in its References. In the proof of these results the boundedness of pure imaginary powers of operators is important. For this topic see R. Seeley [135], J. Prüss and H. Sohr [129],[130] and A. Venni [155].

Chapter 3. The proof of the Gagliardo-Nirenberg inequality is due to L. Nirenberg [118]. See also A. Friedman [65: Part 1]. The proofs of the interpolation inequalities are also found in S. Agmon [10: section 3], A. Friedman [65: Part 1] and Nirenberg [116]. A very useful inequality of Sobolev type [140],[141] is established in Nirenberg [117].

Chapter 4. This chapter consists of a slightly simplified exposition of the L^p-estimates for general elliptic boundary value problems by S. Agmon, A.

Douglis and L. Nirenberg [11]. In [11] the Schauder estimates are also established.

Chapter 5. Section 5.1 is due to M. Schechter [132],[133],[134]. Adjoint boundary conditions were introduced and discussed by N. Aronszajn and A. N. Milgram [17]. The proof of Theorem 5.2 is due to Schechter [133]. In Schechter [134] the regularity of weak solutions and a theorem on the Fredholm alternative are established for general elliptic boundary value problems. Theorem 2.1 of S. Agmon [9] which is Theorem 5.3 in this book is very famous. There is also a related result in S. Agmon and L. Nirenberg [12]. N. Ikebe [76] proved that (5.47) holds under the assumption that the coefficients are Hölder continuous in the case of boundary value problems in contiguous two domains. The paper [21] of F. E. Browder is very useful in the study of elliptic boundary value problems in unbounded domains. In the estimate of the kernel of the semigroup $exp(-tA_2)$ in section 5.3 we followed R. Beals [19] who deduced an asymptotic behavior of the resolvent kernel of a selfadjoint realization A of an elliptic operator from that of the kernel of $(A^l - \lambda)^{-1} = (A - \lambda_1)^{-1} \cdots (A - \lambda_l)^{-1}$ where $\lambda_j, j = 1, \ldots, l$, are the l th roots of λ and l is an integer such that $[l/2] > n/2m$. The coefficients of A are not assumed to be so smooth that A^l is a differential operator in the ordinary sense. The use of the operators $L(\cdot, D + \eta), B_j(\cdot, D + \eta)$ was suggested by L. Hörmander [75] who established a sharp estimate for the resolvent kernel of a selfadjoint realization of an elliptic operator. Since the paper [75] is concerned with interior estimates, the boundary operators $B_j(\cdot, D + \eta)$ do not appear. The results of this section were announced in [148]. This method was used in the study of parabolic equations in L^1 space in D. G. Park [122],[123] and D. G. Park and H. Tanabe [125]. G. Di Blasio [46] proved the regularity of solutions in the maximal sense in interpolation spaces in an L^p setting for parabolic initial boundary value problems with time independent coefficients. In [48] Di Blasio proved that $(L^1(\Omega), D(A_1))_{\theta,p} \cong (L^1(\Omega), W^{2,1}(\Omega) \cap W_0^{1,1}(\Omega))_{\theta,p}$, $0 < \theta < 1, 1 \leq p < \infty$, for the realization A_1 of a second order elliptic operator in $L^1(\Omega)$ under the Dirichlet boundary condition, and characterized the interpolation spaces $D_{A_1}(\theta, 1)$ in terms of Besov spaces. The example in the last section of this chapter was investigated in connection with parabolic equations of higher order in the time variable by A. Favini and H. Tanabe [62]. The generation of analytic semigroups in various function spaces is established by S. Campanato [28],[29], P. Cannarsa, B. Terreni and V. Vespri [31], P. Cannarsa and V. Vespri [32],[33],[34], F. Colombo and V. Vespri [36]. In D. G. Park and S. Y. Kim [124] an example such that $\int_0^1 \|Ae^{tA}x\| dt = \infty$ for some analytic semigroup e^{tA} is constructed.

Chapter 6. The results of this chapter are due to P. Acquistapace and B.

Terreni [1],[3],[6], [8] except for the last section. In [1],[8] various function spaces are introduced and solutions are estimated in these spaces. Related results are in A. Yagi [159] and [161]. In [161] under the assumptions (P1),(P2) and (P4) with $k = 1$ Yagi constructed the fundamental solution using the fractional power of $A(t)$ with a simpler proof. In this connection we refer also to Yagi [162],[163].

Classical results on parabolic evolution equations are found in the books R. W. Carroll [35], S. G. Krein [98] and H. Tanabe [149].

A. Buttu [23] constructed the evolution operator assuming that the mapping $t \mapsto A(t)$ is merely continuous in case where $D(A(t))$ is not necessarily dense and an interpolation space $D_{A(t)}(\theta + 1, \infty)$ between $D(A(t))$ and $D(A(t)^2)$ is independent of t. In [24] replacing $D_{A(0)}(\theta, \infty), D_{A(0)}(\theta+1, \infty)$ by the continuous interpolation spaces $D_{A(0)}(\theta), D_{A(0)}(\theta + 1)$ a further regularity result is obtained. The result is applied to the initial boundary value problems for parabolic differential equations in noncylindrical domains.

H. Amann [14] constructed the fundamental solution under the assumption that $D(A(t)^\beta)$ is constant for some $\beta \in (0, 1)$ and $t \mapsto A(t)$ is Hölder continuous. The method is to extend $A(t)$ to an operator $\tilde{A}(t)$ in an extended space called an extrapolation space so that $D(\tilde{A}(t))$ is constant. Owing to the weakness of the smoothness hypothesis for $A(t)$ in t this result can be applied to quasilinear equations. He continued this study in [15]. See also Amann [16: Chapter V]. Also by considering the equation in an extrapolation space G. Da Prato [37] and G. Da Prato and P. Grisvard [40] showed the existence of a unique solution with the maximal regularity property for a merely continuous inhomogeneous term under the assumptions that $D(A(t))$ is constant, $t \mapsto A(t)$ is only continuous and $A(t)^{-1}A(s)$ has a bounded extension. G. Di Blasio [49] introduced the space $D(0, p)$ and considering the equation in it established the existence and uniqueness of a solution with maximal regularity in the L^p sense.

A. Lunardi and V. Vespri [106] established the optimal Hölder regularity of solutions of variational problems for second order parabolic equations using the method of analytic semigroups. Lunardi [102] constructed an evolution operator using new maximal regularity results in case $D(A(t))$ is constant but not necessarily dense and $t \mapsto A(t)$ is Hölder continuous. See also her book [104]. Based on the estimates of this evolution operator various regularity results are established. In the subsequent paper [103] she established further regularity results in case $A(t)$ is of class $C^{1+\alpha}$ in t with an application to nonhomogeneous initial boundary value problems for parabolic differential equations.

G. Da Prato and E. Sinestrari [42] proved maximal Hölder regularity results in case $D(A(t))$ is constant without using the fundamental solution but using some sharp regularity results in the autonomous case. Sinestrari [138] is a systematic study of autonomous parabolic equations with infinitesimal generator not necessarily densely defined. He proved the unique solvability with maximal regularity, and applied it to initial boundary value problems for parabolic differential equations in the space of continuous functions.

The result of A. Yagi [159] is the most general one in the case where domains are densely defined but totally variable. His main assumption is that for some $-1 \leq \alpha_i < \beta_i \leq 1$

$$\|A(t)(\lambda - A(t))^{-1} dA(t)^{-1}/dt - A(s)(\lambda - A(s))^{-1} dA(s)^{-1}/ds\|$$
$$\leq C \sum_{i=1}^{k} |\lambda|^{\alpha_i} |t - s|^{\beta_i}.$$

Using the result of A. Yagi [164] some linear degenerate equations are solved by A. Favini and A. Yagi [63],[64]. For degenerate equations see also A. Favini and P. Plazzi [59],[60],[61] as well as papers quoted there.

A. Lunardi, E. Sinestrari and W. von Wahl [105] proved the unique solvability with the optimal Hölder regularity of initial-boundary value problems with inhomogeneous boundary conditions for parabolic equations of first order in the time variable and of arbitrary order in the space variables by a semigroup method. W. von Wahl [157] established an estimate which is applicable to parabolic equations with merely continuous coefficients under the Dirichlet boundary conditions. He also proved an L^p-L^q estimate.

Chapter 7. The first systematic study of evolution equations of hyperbolic type was done by T. Kato [88]. J. R. Dorroh [56] simplified the original proof of [88]. In Kato [89] the stability condition was weakened to quasistability:

$$\left\| \prod_{j=1}^{k} R(\lambda_j, A(t_j)) \right\| \leq M \prod_{j=1}^{k} (\lambda_j - \beta(t_j))^{-1}$$

for $\lambda_1 > \beta(t_1), \cdots, \lambda_k > \beta(\lambda_k)$, where β is an upper-integrable function, and the conditions on $B(\cdot)$ and $S(\cdot)$ in the relation

$$S(t)A(t)S(t)^{-1} = A(t) + B(t)$$

is weakened to the hypotheses that $B(\cdot)$ is strongly measurable with $\|B(\cdot)\|$ upper-integrable and $S(\cdot)$ is the indefinite strong integral of $\dot{S}(\cdot)$ with $\|\dot{S}(\cdot)\|_{\mathcal{L}(Y,X)}$ upper-integrable. Such a generalization is useful in the study of nonlinear equations. The study of nonlinear hyperbolic evolution equations was originated by T. Kato [88] and developed by Kato [89],[90], K. Kobayashi and N. Sanekata [96], Sanekata [131]. For other papers on nonlinear hyperbolic evolution equations see the references of [92]. Assuming that $\{A(t)\}$ is quasistable and only a certain kind of measurability on the mapping $t \mapsto A(t)$ S. Ishii [77] constructed the evolution operator using the Yosida approximation. G. Da Prato and M. Iannelli [41] proved the unique solvabilty under somewhat different assumptions. P. Cannarsa and G. Da Prato [30] established a result which can be applied to Kolmogoroff equations. A. Arosio [18] proved that under the assumption that X is reflexive and $A \in BV(0, +\infty; B(Y, X))$ the evolution operator exists and Duhamel's principle holds if $f \in BV_{loc}([0, +\infty); X)$.

Chapter 8. The well-posedness of retarded functional differential equations with highest order delay was established by G. Di Blasio, K. Kunisch and E. Sinestrari [51]. In [50] they discuss properties of the solution semigroup, and show conditions for its noncompactness, differentiability and nondifferentiability. In [137] Sinestrari solved the problem in the space of Hölder continuous functions with maximal regularity in case $a(\cdot) \in L^1(-h, 0)$ and the coefficient operators are not necessarily densely defined. Nonautonomous equations were solved by Di Blasio [47]. P. Vernole [156] solved second order parabolic differential equations with time dependent coefficients in Hölder spaces. Theorem 8.5 was proved in Y. Fujie and H. Tanabe [66]. In [152] the fundamental solution with strong regularity property is constructed in case where $a(\cdot)$ is Hölder continuous applying the result of J. Prüss [128] in each interval $[nh, (n+1)h], n = 0, 1, 2, \ldots$. But no maximal regularity of solutions is obtained there. Using this fundamental solution the second structural operator is directly defined and its relation with the first structural operator F is established in [82],[114]. S. Nakagiri [113] improved the results of [80],[115] especially in an important case of finite dimensional control spaces. For quasilinear equations we refer to J. Yong and L. Pan [165]. J. K. Hale's book [73] provides excellent information on retarded ordinary functional differential equations.

Bibliography

[1] P. Acquistapace: *Evolution operators and strong solutions of abstract linear parabolic equations*, Diff. Int. Eqns. 1 (1988), 433-457.

[2] P. Acquistapace: *Abstract linear nonautonomous parabolic equations: A survey*, Proc. Bologna Conference on Differential Equations in Banach Spaces, Lecture Notes in Pure and Appl. Math. **148**, ed. G. Dore, A. Favini, E. Obrecht, A. Venni, Marcel Dekker, New York-Basel-Hong Kong, 1993, 1-19.

[3] P. Acquistapace and B. Terreni: *Some existence and regulaity results for abstract non-autonomous parabolic equations*, J. Math. Anal. Appl. **99** (1984), 9-64.

[4] P. Acquistapace and B. Terreni: *Characterization of Hölder and Zygmund classes as interpolation spaces*, preprint, Dipartimento di Math., Univ. Padova, **61** (1984).

[5] P. Acquistapace and B. Terreni: *Maximal space regularity for abstract linear non-autonomous parabolic equations*, J. Func. Anal. **60** (1985), 168-210.

[6] P. Acquistapace and B. Terreni: *On fundamental solutions for abstract parabolic equations*, Differential Equations in Banach spaces, Bologna, 1985, Lecture Notes in Math. **1223**, ed. A. Favini, E. Obrecht, Springer-Verlag, 1986, 1-11.

[7] P. Acquistapace and B. Terreni: *Hölder classes with boundary conditions as interpolation spaces*, Math. Z. **195** (1987), 451-471.

[8] P. Acquistapace and B. Terreni: *A unified approach to abstract linear non-autonomous parabolic equations*, Rend. Sem. Univ. Padova **78** (1987), 47-107.

[9] S. Agmon: *On the eigenfunctions and on the eigenvalues of general elliptic boundary value problems*, Comm. Pure Appl. Math. **15** (1962), 119-147.

[10] S. Agmon: Lectures on Elliptic Boundary Value Problems, D. van Nostrand, Princeton-Toronto-New York-London, 1965.

[11] S. Agmon, A. Douglis and L. Nirenberg: *Estimates near the boundary for solutions of elliptic partial differential equations satisfying general boundary conditions*, I, Comm. Pure Appl. Math. **12** (1959), 623-727.

[12] S. Agmon and L. Nirenberg: *Properties of solutions of ordinary differential equations in Banach space*, Comm. Pure Appl. Math. **16** (1963), 121-239.

[13] H. Amann: *Dual semigroups and second order linear elliptic boundary value problems*, Israel J. Math. **45** (1983), 225-254.

[14] H. Amann: *On abstract parabolic fundamental solutions*, J. Math. Soc. Japan **39** (1987), 93-116.

[15] H. Amann: *Parabolic evolution equations in interpolation and extrapolation spaces*, J. Func. Anal. **78** (1988), 233-270.

[16] H. Amann: Linear and Quasilinear Parabolic Problems, I, Abstract Linear Theory, Monographs in Math. **89**, Birkhäuser Verlag, Basel-Boston-Berlin, 1995.

[17] N. Aronszajn and A. N. Milgram: *Differential operators on Riemannian manifolds*, Rend. Circ. Mat. Palermo, Ser.2 **2** (1953), 266-325.

[18] A. Arosio: *Duhamel's principle for temporally inhomogeneous evolution equations in Banach space*, Nonlinear Anal. T.M.A. **8** (1984), 997-1009.

[19] R. Beals: *Asymptotic behavior of the Green's function and spectral function of an elliptic operator*, J. Func. Anal. **5** (1970), 484-503.

[20] C. Bernier and A. Manitius: *On semigroups in $R^n \times L^p$ corresponding to differential equations with delays*, Canad. J. Math. **30** (1978), 897-914.

[21] F. E. Browder: *On the spectral theory of elliptic differential operators*, I, Math. Anal. **142** (1961), 22-130.

[22] D. L. Burkholder: *A geometrical condition that implies the existence of certain singular integrals of Banach-space-valued functions*, Conferece on Harmonic Analysis in Honour of Antoni Zygmund, Chicago, 1981, ed. W. Beckner, A. P. Calderón, R. Fefferman and P. W. Jones, 270-286, Belmont, Cal., 1983, Wadsworth.

[23] A. Buttu: *On the evolution operator for a class of non-autonomous abstract parabolic equations*, J. Math. Anal. Appl. **170** (1992), 115-137.

[24] A. Buttu: *A construction of the evolution operator for a class of abstract parabolic equations*, Dyn. Syst. Appl. **3** (1994), 221-234.

[25] P. L. Butzer and H. Berens: Semigroups of Operators and Approximation, Springer-Verlag, New York-Berlin-Heidelberg-Tokyo, 1967.

[26] A. P. Calderón: *Intermediate spaces and interpolation, the complex method*, Studia Math. **24** (1964), 113-190.

[27] A. P. Calderón and A. Zygmund: *On the existence of certain singular integrals*, Acta. Math. **88** (1952), 85-139.

[28] S. Campanato: *Generation of analytic semigroups, in the Hölder topology, by elliptic operators of second order with Neumann boundary condition*, Le Matematiche **35** (1980), 61-72.

[29] S. Campanato: *Generation of analytic semigroups by elliptic operators of second order in Hölder spaces*, Ann. Sci. Norm. Sup. Pisa **8** (1981), 495-512.

[30] P. Cannarsa and G. Da Prato: *Some results on abstract evolution equations of hyperbolic type*, Proc. Bologna Conference on Differential Equations in Banach Spaces, Lecture Notes in Pure and Appl. Math. **148**, ed. G. Dore, A. Favini, E. Obrecht, A. Venni, Marcel Dekker, New York-Basel-Hong Kong, 1993, 41-50.

[31] P. Cannarsa, B. Terreni and V. Vespri: *Analytic semigroups generated by non-variational elliptic systems of second order under Dirichlet boundary conditions*, J. Math, Anal. Appl. **112** (1985), 56-103.

[32] P. Cannarsa and V. Vespri: *Analytic semigroups generated on Hölder spaces by second order elliptic systems under Dirichlet boundary conditions*, Ann. Mat. Pura Appl. (IV), **140** (1985), 393-415.

[33] P. Cannarsa and V. Vespri: *Generation of analytic semigroups by elliptic operators with unbounded coefficients*, SIAM J. Math. Anal. **18** (1987), 857-872.

[34] P. Cannarsa and V. Vespri: *Generation of analytic semigroups in the L^p topology by elliptic operators in $\mathbf{R}^n$*, Israel J. Math. **61** (1988), 235-255.

[35] R. W. Carroll: Abstract Methods in Partial Differential Equations, Harper & Row, New York-Evanston-London, 1969.

[36] F. Colombo and V. Vespri: *Generation of analytic semigroups in $W^{k,p}(\Omega)$ and $C^k(\bar\Omega)$*, Diff. Int. Eqns. **9** (1996), 421-436.

[37] G. Da Prato: *Abstract differential equations and extrapolation spaces*, Infinite-Dimensional Systems, Lecture Notes in Math. **1076** ed. A. Dold and B. Eckmann, Springer-Verlag, 1983, 53-61.

[38] G. Da Prato and P. Grisvard: *Sommes d'opérateurs linéaires et équations différentielles opérationelles*, J. Math. Pures Appl. **54** (1975), 305-387.

[39] G. Da Prato and P. Grisvard: *Equations d'évolution abstraites non-linéares de type parabolique*, Ann. Mat. Pura Appl. (IV) **120** (1979), 329-396.

[40] G. Da Prato and P. Grisvard: *Maximal regularity for evolution equations by interpolation and extrapolation*, J. Func. Anal. **58** (1984), 107-124.

[41] G. Da Prato and M. Iannelli: *On a method for studying abstract evolution equation in the hyperbolic case*, Comm. Partial Diff. Eqns. **1** (1976), 585-608.

[42] G. Da Prato and E. Sinestrari: *Hölder regularity for non-autonomous abstract parabolic equations*, Israel J. Math. **42** (1982), 1-19.

[43] G. Da Prato and E. Sinestrari: *Differential operators with nondense domains*, Ann. Scuola Norm. Sup. Pisa, Ser.4 **14** (1987), 285-344.

[44] M. C. Delfour and A. Manitius: *The structural operator F and its role in the theory of retarded systems*, I, J. Math. Anal. Appl. **73** (1980), 466-490.

[45] M. C. Delfour and A. Manitius: *The structural operator F and its role in the theory of retarded systems*, II, J. Math. Anal. Appl. **74** (1980), 359-381.

[46] G. Di Blasio: *Linear parabolic evolution equations in L^p-spaces*, Ann. Mat. Pura Appl. (IV) **138** (1984), 55-104.

[47] G. Di Blasio: *Nonautonomous functional differential equations in Hilbert spaces*, Nonlinear Anal. T.M.A. **9** (1985), 1367-1380.

[48] G. Di Blasio: *Analytic semigroups generated by elliptic operators in L^1 and parabolic equations*, Osaka J. Math. **28** (1991), 367-384.

[49] G. Di Blasio: *Interpolation and extrapolation spaces and parabolic equations*, Proc. Bologna Conference on Differential Equations in Banach Spaces, Lecture Notes in Pure and Appl. Math. **148**, ed. G. Dore, A. Favini, E. Obrecht, A. Venni, Marcel Dekker, New York-Basel-Hong Kong, 1993, 51-58.

[50] G. Di Blasio, K. Kunisch and E. Sinestrari: *Retarded abstract equations in Hilbert spaces*, Infinite-Dimensional Systems, Lecture Notes in Math. **1076** ed. A. Dold and B. Eckmann, Springer-Verlag, 1983, 71-77.

[51] G. Di Blasio, K. Kunisch and E. Sinestrari: *L^2-regularity for parabolic partial integrodifferential equations with delay in the highest-order derivatives*, J. Math. Anal. Appl. **102** (1984), 38-57.

[52] G. Di Blasio, K. Kunisch and E. Sinestrari: *Stability for abstract linear functional differential equations*, Israel J. Math. **50** (1985), 231-263.

[53] G. Dore: *L^p regularity for abstract differential equations*, Lecture Notes in Math. **1540**, ed. H. Komatsu, Springer-Verlag, 1991, 25-38.

[54] G. Dore and A. Venni: *On the closedness of the sum of two closed operators*, Math. Z. **196** (1987), 189-201.

[55] G. Dore and A. Venni: *Maximal regularity for parabolic initial-boundary value problems in Sobolev spaces*, Math. Z. **208** (1991), 297-308.

[56] J. R. Dorroh: *A simplified proof of a theorem of Kato on linear evolution equations*, J. Math. Soc. Japan **27** (1975), 474-478.

[57] S. D. Ĕĭdel'man and S. D. Ivasišen: *Investigation of the Green matrix for a homogeneous parabolic boundary value problem*, Trudy Moscow Mat. Obšč. **23** (1970), 179-234 (in Russian); English translation: Trans. Moscow Math. Soc. **23** (1970), 179-242.

[58] H. O. Fattorini: *On complete controllablity of linear systems*, J. Diff. Eqns. **3** (1967), 391-402.

[59] A. Favini and P. Plazzi: *On some abstract degenerate problems of parabolic type -1:the linear case*, Nonlinear Anal. T.M.A. **12** (1988), 1017-1027.

[60] A. Favini and P. Plazzi: *On some abstract degenerate problems of parabolic type -2:the nonlinear case*, Nonlinear Anal. T.M.A. **13** (1989), 23-31.

[61] A. Favini and P. Plazzi: *On some abstract degenerate problems of parabolic type -3: applications to linear and nonlinear problems*, Osaka J. Math. **27** (1990), 323-359.

[62] A. Favini and H. Tanabe: *Linear parabolic differential equations of higher order in time*, Proc. Bologna Conference on Differential Equations in Banach Spaces, Lecture Notes in Pure and Appl. Math. **148**, ed. G. Dore, A. Favini, E. Obrecht, A. Venni, Marcel Dekker, New York-Basel-Hong Kong, 1993, 85-92.

[63] A. Favini and A. Yagi: *Space and time regularity for degenerate evolution equations*, J. Math. Soc. Japan **44** (1992), 331-350.

[64] A. Favini and A. Yagi: *Multivalued linear operators and degenerate evolution equations*, Annali Mat. Pura Appl. (IV) **163** (1993), 353-384.

[65] A. Friedman: Partial Differential Equations, Krieger, Huntington, New York, 1976.

[66] Y. Fujie and H. Tanabe: *On some parabolic equations of evolution in Hilbert space*, Osaka J. Math. **10** (1973), 115-130.

[67] E. Gagliardo: *Proprietà di alcune classi di funzioni in più variabili*, Ricerche di Mat. **7** (1958), 102-137.

[68] E. Gagliardo: *Ulteriori proprietà di alcune classi di funzioni in più variabili*, Ricerche di Mat. **8** (1959), 24-51.

[69] M. Giga, Y. Giga and H. Sohr: *L^p-estimates for the Stokes system*, Lecture Notes in Math. **1540**, ed. H. Komatsu, Springer-Verlag, 1991, 55-67.

[70] J. A. Goldstein: Semigroups of Linear Operators and Applications, Oxford Univ. Press, New York & Clarendon Press, Oxford, 1985.

[71] P. Grisvard: *Équations différentielles abstraites*, Ann. Sci. Ecol. Norm. Sup., 4e série **2** (1969), 311-395.

[72] D. Guidetti: *On elliptic systems in L^1*, Osaka J. Math. **30** (1993), 397-429.

[73] J. K. Hale: Theory of Functional Differential Equations, Springer-Verlag, New York-Heidelberg-Berlin, 1977.

[74] E. Hille and R. S. Phillips: Functional Analysis and Semi-groups, Amer. Math. Soc., Providence, 1957.

[75] L. Hörmander: *On the Riesz means of spectral functions and eigenfunction expansions for elliptic differential opertors*, Some Recent Advances in the Basic Sciences 2 (Proc. Annual Sci. Conf., Belfer Grad. School Sci., Yeshiva Univ., New York, 1965-1966), 155-202.

[76] N. Ikebe: *On elliptic boundary value problems with discontinuous coefficients*, Mem. Fac. Sci., Kyushu Univ. **21** (1967), 167-184.

[77] S. Ishii: *Linear evolution equations $du/dt + A(t)u = 0$: a case where $A(t)$ is strongly uniform-measurable*, J. Math. Soc. Japan **34** (1982), 413-424.

[78] S. D. Ivasišen: *Green's matrices of boundray value problems for Petrovskiĭ parabolic systems of general form*, I Mat. Sbornik (N.S.) **114(156)** (1981), 110-116 (in Russian); English translation: Math. USSR-Sbornik **42** (1982), 93-144.

[79] S. D. Ivasišen: *Green's matrices of boundary value problems for Petrovskiĭ parabolic systems of general form*, II (Russian), Mat. Sbornik (N.S.) **114(156)** (1981), 523-565, 654-655 (in Russian); English translation: Math. USSR-Sbornik **42** (1982), 461-498.

[80] J.-M. Jeong: *Spectral properties of the operator associated with a retarded functional differential equation in Hilbert space*, Proc. Japan Acad. **65** (1989), 98-101.

[81] J.-M. Jeong: *Retarded functional differential equations with L^1-valued controller*, Funkcial. Ekvac. **36** (1993), 71-93.

[82] J.-M. Jeong, S. Nakagiri and H. Tanabe: *Structural operators and semigroups assiciated with functional differential equations in Hilbert spaces*, Osaka J. Math. **30** (1993), 365-395.

[83] F. John: *General properties of solutions of linear elliptic partial differential equations*, Proc. Symp. Spectral Theory and Differential Problems, Oklahoma Agricultural and Mechanical College, Stillwater, Oklahoma, 1951, 113-175.

[84] F. John: Plane Waves and Sherical Means Applied to Partial Differential Equations, Interscience Publishers, Inc., New York & Interscience Publishers, Ltd., London, 1955.

[85] T. Kato: *Fractional powers of dissipative operators*, J. Math. Soc. Japan **13** (1961), 246-274.

[86] T. Kato: Perturbation Theory for Linear Operators, Springer-Verlag, Belin-Heidelberg-New York, 1966.

[87] T. Kato: *Nonlinear semigroups and evolution equations*, J. Math. Soc. Japan **19** (1967), 508-520.

[88] T. Kato: *Linear evolution equations of "hyperbolic" type*, J. Fac. Sci. Univ. Tokyo, Sec. I, 17 (1970), 241-258.

[89] T. Kato: *Linear evolution equations of "hyperbolic" type, II*, J. Math. Soc. Japan **25** (1973), 648-666.

[90] T. Kato: *The Cauchy probelm for quasi-linear symmetric hyperbolic systems*, Arch. Rat. Mech. Anal. **58** (1975), 181-205.

[91] T. Kato: *Quasi-linear equations of evolution, with applications to partial differential equations*, Lecture Notes in Math. **448**, Springer-Verlag, 1975, 25-70.

[92] T. Kato: *Abstract evolution equations, linear and quasilinear, revisited*, Lecture Notes in Math. **1540**, ed. H. Komatsu, Springer-Verlag, 1991, 103-125.

[93] T. Kato and H. Tanabe: *On the abstract evolution equation*, Osaka Math. J. **14** (1962), 107-133.

[94] N. Kerzman: *Hölder and L^p estimates for solutions of $\bar{\partial}u = f$ in strongly pseudo convex domains*, Comm. Pure Appl. Math. **24** (1971), 301-379.

[95] K. Kobayashi: *On a theorem for linear evolution equations of hyperbolic type*, J. Math. Soc. Japan **31** (1979), 647-654.

[96] K. Kobayashi and N. Sanekata: *A method of iterations for quasilinear evolution equations in nonreflexive Banach sapces*, Hiroshima Math. J. **19** (1989), 521-540.

[97] Y. Kōmura: *Nonlinear semigroups in Hilbert space*, J. Math. Soc. Japan **19** (1967), 493-507.

[98] S. G. Krein: Linear Differential Equations in Banach Space, Nauka, Moscow, 1967 (in Russian); English translation: Transl. Math. Monographs 29, Amer. Math. Soc., Providence, 1972.

[99] J. L. Lions and E. Magenes: Problèmes aux limites non homogènes et applications, Dunod, Paris, vol. 1, 2, 1968, vol. 3, 1970.

[100] J. L. Lions and J. Peetre: *Sur une classe d'espaces d'interpolation*, Inst. Hautes Études Sci. Publ. Math. **19** (1964), 5-68.

[101] A. Lunardi: *Interpolation spaces between domains of elliptic operators and spaces of continuous functions with applications to nonlinear parabolic equations*, Math. Nachr. **121** (1985), 295-318.

[102] A. Lunardi: *On the evolution operator for abstract parabolic equations*, Israel J. Math. **60** (1987), 281-314.

[103] A. Lunardi: *Differentiability with respect to (t, s) of the parabolic evolution operator*, Israel J. Math. **68** (1989), 161-184.

[104] A. Lunardi: Analytic Semigroups and Optimal Regularity in Parabolic Problems, Progress in Nonlinear Diff. Eqns. Appl. **16**, Birkhäuser, Basel-Boston-Berlin, 1995.

[105] A. Lunardi, E. Sinestrari and W. von Wahl: *A semigroup approach to the time dependent parabolic initial-boundary value problem*, Diff. Int. Eqns. **5** (1992), 1275-1306.

[106] A. Lunardi and V. Vespri: *Hölder regularity in variational parabolic non-homogeneous equations*, J. Diff. Eqns. **94** (1991), 1-40.

[107] A. Manitius: *Completeness and F-completeness of eigenfunctions associated with retarded functional differential equations*, J. Diff. Eqns. **35** (1980), 1-29.

[108] A. Manitius: *Necessary and sufficient conditions of approximate controllability for general linear retarded systems*, SIAM J. Control Optim. **19** (1981), 516-532.

[109] A. Manitius: *F-controllability and observability of linear retarded systems*, Appl. Math. Optim. **9** (1982), 73-95.

[110] F. J. Massey III: *Abstract evolution equations and the mixed problem for symmetric hyperbolic systems*, Trans. Amer. Math. Soc. **168** (1972), 165-188.

[111] S. Nakagiri: *Identifyability of linear systems in Hilbert spaces*, SIAM J. Control and Optim. **21** (1983), 501-530.

[112] S. Nakagiri: *Structural properties of functional differential equations in Banach spaces*, Osaka J. Math. **25** (1988), 353-398.

[113] S. Nakagiri: *Controllability and identifiability for linear time-delay systems in Hilbert space*, preprint.

[114] S. Nakagiri and H. Tanabe: *Structural operators and eigenmanifold decomposition for functional differential equations in Hilbert spaces*, to appear in J. Math. Anal. Appl.

[115] S. Nakagiri and M. Yamamoto: *Controllability and observability of linear retarded systems in Banach space*, Int. J. Control. **49** (1989), 1489-1504.

[116] L. Nirenberg: *Remarks on strongly elliptic partial differential equations*, Comm. Pure Appl. Math. **8** (1955), 648-674.

[117] L. Nirenberg: *Estimates and existence of solutions of elliptic equations*, Comm. Pure Appl. Math. **9** (1956), 509-530.

[118] L. Nirenberg: *On elliptic partial differential equations*, Ann. Scuola Norm. Sup. Pisa, Ser. 3 **13** (1959), 113-161.

[119] N. Okazawa: *Remarks on linear evolution equations of hyperbolic type in a Hilbert space*, preprint.

[120] N. Okazawa and A. Unai: *Linear evolution equations of hyperbolic type in Hilbert space*, SUT J. Math. **29** (1993), 51-70.

[121] N. Okazawa and A. Unai: *Abstract quasilinear evolution equations in a Hilbert space, with applications to symmetric hyperbolic systems*, SUT J. Math. **29** (1993), 263-290.

[122] D. G. Park: *Initial-boundary value problem for parabolic equation in L^1*, Proc. Japan Acad. **62** (1986), 178-180.

[123] D. G. Park: *Regularity in time of the solution of parabolic initial-boundary value problem in L^1 space*, Osaka J. Math. **24** (1987), 911-930.

[124] D. G. Park and S. Y. Kim: *Integrodifferential equations with time delay in Hilbert space*, Comm. Korean Math. Soc. **7** (1992), 189-207.

[125] D. G. Park and H. Tanabe: *On the asymptotic behavior of solutions of linear parabolic equations in L^1 space*, Ann. Scuola Norm. Sup. Pisa **14** (1987), 587-611.

[126] J.-Y. Park and J.-M. Jeong: *Supplement to the paper "Retarded functional differential equations with L^1-valued controller"*, Funkcial. Ekvac. **38** (1995), 267-275.

[127] A. Pazy: Semigroups of Linear Operators and Applications to Partial Differential Equations, Springer-Verlag, New York-Berlin-Heidelberg-Tokyo, 1983.

[128] J. Prüss: *On resolvent operators for linear integrodifferential equations of Volterra type*, J. Int. Eqns. **5** (1983), 211-236.

[129] J. Prüss and H. Sohr: *On operators with bounded imaginary powers in Banach spaces*, Math. Z. **203** (1990), 429-452.

[130] J. Prüss and H. Sohr: *Imaginary powers of elliptic second order differential operators in L^p-spaces*, Hiroshima Math. J. **23** (1993), 161-192.

[131] N. Sanekata: *Abstract quasi-linear equations of evolution in nonreflexive Banach spaces*, Hiroshima Math. J. **19** (1989), 109-139.

[132] M. Schechter: *Integral inequalities for partial differential operators and functions satisfying general boundary conditions*, Comm. Pure Appl. Math. **12** (1959), 37-66.

[133] M. Schechter: *General boundary value problems for elliptic partial differential equations*, Comm. Pure Appl. Math. **12** (1959), 457-486.

[134] M. Schechter: *Remarks on elliptic boundary value problems*, Comm. Pure Appl. Math. **12** (1959), 561-578.

[135] R. Seeley: *Norms and domains of the complex power A_B^z*, Amer. J. Math. **93** (1971), 299-309.

[136] R. Seeley: *Interpolation in L^p with boundary conditions*, Studia Math. **44** (1972), 47-60.

[137] E. Sinestrari: *On a class of retarded partial differential equations*, Math. Z. **186** (1984), 223-246.

[138] E. Sinestrari: *On the abstract Cauchy problem of parabolic type in spaces of continuous functions*, J. Math. Anal. Appl. **107** (1985), 16-66.

[139] E. Sinestrari: *On the Hille-Yosida operators*, Evolution Equations, Control Theory, and Biomathematics, Han sur Lesse, 1991, Lecture Notes in Pure Appl. Math. **155**, Marcel Dekker, 1994, 537-543.

[140] S. L. Sobolev: *On a theorem of functional analysis*, Mat. Sbornik (N.S.) **4** (1938), 471-497.

[141] S. L. Sobolev: Application of Functional Analysis in Mathematical Physics, Leningrad National Univ., Leningrad, 1950 (in Russian); English translation: Transl. Math. Monographs **7**, Amer. Math. Soc., Providence, 1963.

[142] P. E. Sobolevskii: *Equations of parabolic type in Banach space*, Trudy Moscow Mat. Obshch. **10** (1961), 297-350 (in Russian); English translation: Amer. Math. Soc. Transl. **49** (1965), 1-62.

[143] V. A. Solonnikov: *On boundary value problems for linear parabolic systems of differential equations of general form*, Trudy Mat. Inst. Steklov **83** (1965), 3-163 (in Russian); English translation: Proc. Steklov Inst. Math. **83** (1965), 1-184.

[144] H. B. Stewart: *Generation of analytic semigroups by strongly elliptic operators*, Trans. Amer. Math. Soc. **199** (1974), 141-162.

[145] H. B. Stewart: *Generation of analytic semigroups by strongly elliptic operators under general boundary conditions*, Trans. Amer. Math. Soc. **259** (1980), 299-310.

[146] T. Suzuki and M. Yamamoto: *Observability, controllability and feedback stabilizability for evolution equation* I, Japan J. Appl. Math. **2** (1985), 211-228.

[147] T. Takagi: Lectures on Algebra, Kyoritsu Shuppan Publ. Company, Tokyo, 1946 (in Japanese).

[148] H. Tanabe: *On Green's functions of elliptic and parabolic boundary value problems*, Proc. Japan Acad. **48** (1972), 709-711.

[149] H. Tanabe: Equations of Evolution, Iwanami Shoten, Tokyo, 1975 (in Japanese); English translation: Pitman, London, 1979.

[150] H. Tanabe: Functional Analysis, II, Jikkyo Shuppan Publ. Company, Tokyo, 1981 (in Japanese).

[151] H. Tanabe: *Structural operators for linear delay-differential equations in Hilbert space*, Proc. Japan Acad. **64** (1988), 265-266.

[152] H. Tanabe: *Fundamental solutions for linear retarded functional differential equations in Banach space*, Funkcial. Ekvac. **35** (1992), 149-177.

[153] E. C. Titchmarsh: Introduction to the Theory of Fourier Integrals, Clarendon Press, Oxford, 1937.

[154] H. Triebel: Interpolation Theory, Function Spaces, Differential Operators, North-Holland, Amsterdam-New York-Oxford, 1978.

[155] A. Venni: *A counterexample concerning imaginary powers of linear operators*, Lecture Notes in Math. **1540**, ed. H. Komatsu, Springer-Verlag, 1991, 381-387.

[156] P. Vernole: *A time dependent parabolic initial boundary value delay problem*, J. Int. Eqns. Appl. **6** (1994), 427-444.

[157] W. von Wahl: *The equation $u' + A(t)u = f$ in a Hilbert space and L^p-estimates for parabolic equations*, J. London Math. Soc., Ser. II **25** (1982), 483-497.

[158] G. N. Watson: A Treatise on the Theory of Bessel Functions, 2nd edition, Cambridge Univ. Press, 1958.

[159] A. Yagi: *On the abstract evolution equation of parabolic type*, Osaka J. Math. **14** (1977), 557-568.

[160] A. Yagi: *Remarks on proof of a theorem of Kato and Kobayashi on linear evolution equations*, Osaka J. Math. **17** (1980), 233-244.

[161] A. Yagi: *Fractional powers of operators and evolution equations of parabolic type*, Proc. Japan Acad. **64** (1988), 227-230.

[162] A. Yagi: *Parabolic evolution equations in which the coefficients are the generators of infinitely differentiable semigroups*, Funkcial. Ekvac. **32** (1989), 107-124.

[163] A. Yagi: *Parabolic evolution equations in which the coefficients are the generators of infinitely differentiable semigroups*, II, Funkcial. Ekvac. **33** (1990), 139-150.

[164] A. Yagi: *Generation theorem of semigroup for multivalued linear operators*, Osaka J. Math. **28** (1991), 385-410.

[165] J. Yong and L. Pan: *Quasi-linear parabolic partial differential equations with delays in the highest order spatial derivatives*, J. Austral. Math. Soc. (Series A) **54** (1993), 174-203.

[166] K. Yosida: Functional Analysis, Springer-Verlag, Berlin-Heidelberg-New York, 1966.

[167] S. D. Zaidman: Abstract Differential Equations, Research Notes in Math. **36**, Pitman, San Francisco-London-Melbourne, 1979.

[168] A. Zygmund: Trigonometrical Series, Cambrigde Univ. Press, 1968.

List of Symbols

411

$(X, Y)_{\theta, p}$, 7
$D_A(\theta, \infty)$, 246
$\Delta(\cdot)$, 350
$\Delta_T(\cdot)$, 350
$\sigma(\Delta), \sigma(\Delta_T)$, 355
$\rho(\Delta), \rho(\Delta_T)$, 355
$\sigma_p(\Delta), \sigma_p(\Delta_T)$, 367
$\sigma_d(A)$, 371
$\sigma_0(A)$, 378
$\sigma_0(\Delta), \sigma_0(\Delta_T)$, 383
E_λ, 352
T_λ, 352
H_λ, 352
K_λ, 352
Cl, 355
Im, 361
Ker, 358
P_λ, 370
P_λ^T, 370
$\mathcal{M}_\lambda$, 370
$\mathcal{M}_\lambda^*$, 370
$\mathcal{M}_\lambda^T$, 370
$\mathcal{M}_\lambda^{T*}$, 370
$\mathcal{M}$, 385
$\mathcal{M}^*$, 385

Index